Fundamental Equations of Dynamics

KINEMATICS

Particle Rectilinear Motion

Variable a

$$a = \frac{dv}{dt}$$

$$v = \frac{ds}{dt}$$

$$v\, dv = a\, ds$$

Constant $a = a_c$

$$v = v_0 + a_c t$$

$$s = s_0 + v_0 t + \tfrac{1}{2} a_c t^2$$

$$v^2 = v_0^2 + 2a_c(s - s_0)$$

Particle Curvilinear Motion

x, y, z Coordinates

$$v_x = \dot{x} \qquad a_x = \ddot{x}$$
$$v_y = \dot{y} \qquad a_y = \ddot{y}$$
$$v_z = \dot{z} \qquad a_z = \ddot{z}$$

r, θ, z Coordinates

$$v_r = \dot{r} \qquad a_r = \ddot{r} - r\dot{\theta}^2$$
$$v_\theta = r\dot{\theta} \qquad a_\theta = r\ddot{\theta} + 2\dot{r}\dot{\theta}$$
$$v_z = \dot{z} \qquad a_z = \ddot{z}$$

n, t, Coordinates

$$v = \dot{s} \quad\Big|\quad a_t = \dot{v} = v\frac{dv}{ds}$$

$$a_n = \frac{v^2}{\rho} \qquad \rho = \left| \frac{[1 + (dy/dx)^2]^{3/2}}{d^2 y/dx^2} \right|$$

Relative Motion

$$\mathbf{v}_B = \mathbf{v}_A + \mathbf{v}_{B/A} \qquad \mathbf{a}_B = \mathbf{a}_A + \mathbf{a}_{B/A}$$

Rigid Body Motion About a Fixed Axis

Variable α

$$\alpha = \frac{d\omega}{dt}$$

$$\omega = \frac{d\theta}{dt}$$

$$\omega\, d\omega = \alpha\, d\theta$$

Constant $\alpha = \alpha_c$

$$\omega = \omega_0 + \alpha_c t$$

$$\theta = \theta_0 + \omega_0 t + \tfrac{1}{2}\alpha_c t^2$$

$$\omega^2 = \omega_0^2 + 2\alpha_c(\theta - \theta_0)$$

For Point P

$$s = \theta r \qquad v = \omega r \qquad a_t = \alpha r \qquad a_n = \omega^2 r$$

Relative General Plane Motion-Translating Axes

$$\mathbf{v}_B = \mathbf{v}_A + \mathbf{v}_{B/A(\text{pin})} \qquad \mathbf{a}_B = \mathbf{a}_A + \mathbf{a}_{B/A(\text{pin})}$$

Relative General Plane Motion-Trans. and Rot. Axis

$$\mathbf{v}_B = \mathbf{v}_A + \mathbf{\Omega} \times \mathbf{r}_{B/A} + (\mathbf{v}_{B/A})_{\text{rel}}$$
$$\mathbf{a}_B = \mathbf{a}_A + \dot{\mathbf{\Omega}} \times \mathbf{r}_{B/A} + \mathbf{\Omega} \times (\mathbf{\Omega} \times \mathbf{r}_{B/A}) + 2\mathbf{\Omega} \times (\mathbf{v}_{B/A})_{\text{rel}} + (\mathbf{a}_{B/A})_{\text{rel}}$$

KINETICS

Mass Moment of Inertia

$$I = \int r^2\, dm$$

Parallel-Axis Theorem

$$I = I_G + md^2$$

Radius of Gyration

$$k = \sqrt{\frac{I}{m}}$$

Equations of Motion

Particle	$\Sigma \mathbf{F} = m\mathbf{a}$
Rigid Body (Plane Motion)	$\Sigma F_x = m(a_G)_x$
	$\Sigma F_y = m(a_G)_y$
	$\Sigma M_G = I_G\alpha \;/\; \Sigma M_P = \Sigma(M_k)_P$

Principle of Work and Energy

$$T_1 + U_{1-2} = T_2$$

Kinetic Energy

Particle	$T = \tfrac{1}{2}mv^2$
Rigid Body (Plane Motion)	$T = \tfrac{1}{2}mv_G^2 + \tfrac{1}{2}I_G\omega^2$

Work

Variable force	$U_F = \int F \cos\theta\, ds$
Constant force	$U_{F_c} = (F_c \cos\theta)\,\Delta s$
Weight	$U_W = -W\,\Delta y$
Spring	$U_s = -(\tfrac{1}{2}ks_2^2 - \tfrac{1}{2}ks_1^2)$
Couple moment	$U_M = M\,\Delta\theta$

Power and Efficiency

$$P = \frac{dU}{dt} = \mathbf{F} \cdot \mathbf{v} \qquad \varepsilon = \frac{P_{\text{out}}}{P_{\text{in}}} = \frac{U_{\text{out}}}{U_{\text{in}}}$$

Conservation of Energy Theorem

$$T_1 + V_1 = T_2 + V_2$$

Potential Energy

$$V = V_g + V_e, \text{ where } V_g = \pm Wy, \; V_e = +\tfrac{1}{2}ks^2$$

Principle of Linear Impulse and Momentum

Particle	$m\mathbf{v}_1 + \Sigma \int \mathbf{F}\, dt = m\mathbf{v}_2$
Rigid Body	$m(\mathbf{v}_G)_1 + \Sigma \int \mathbf{F}\, dt = m(\mathbf{v}_G)_2$

Conservation of Linear Momentum

$$\Sigma(\text{syst. } m\mathbf{v})_1 = \Sigma(\text{syst. } m\mathbf{v})_2$$

Coefficient of Restitution

$$e = \frac{(v_B)_2 - (v_A)_2}{(v_A)_1 - (v_B)_1}$$

Principle of Angular Impulse and Momentum

Particle	$(\mathbf{H}_O)_1 + \Sigma \int \mathbf{M}_O\, dt = (\mathbf{H}_O)_2,$ where $H_O = (d)(mv)$
Rigid Body (Plane Motion)	$(\mathbf{H}_G)_1 + \Sigma \int \mathbf{M}_G\, dt = (\mathbf{H}_G)_2,$ where $H_G = I_G\omega$
	$(\mathbf{H}_O)_1 + \Sigma \int \mathbf{M}_O\, dt = (\mathbf{H}_O)_2,$ where $H_O = I_G\omega + (d)(mv_G)$

Conservation of Angular Momentum

$$\Sigma(\text{syst. } \mathbf{H})_1 = \Sigma(\text{syst. } \mathbf{H})_2$$

Engineering Mechanics
Statics

Engineering Mechanics

Statics

4TH EDITION

R. C. Hibbeler

Macmillan Publishing Company
New York

Collier Macmillan Publishers
London

Macmillan Publishing Company
866 Third Avenue, New York, New York 10022

Collier Macmillan Canada, Inc.

Library of Congress Cataloging in Publication Data

Hibbeler, R. C.
    Engineering mechanics—statics.

    Includes index.
    1. Statics.   I. Title.
TA351.H5   1986   620.1pr03   85-15388
ISBN 0-02-354670-0 (Hardcover Edition)
ISBN 0-02-946260-6 (International Edition)

Printing: 1 2 3 4 5 6 7 8   Year: 6 7 8 9 0 1

ISBN 0-02-354670-0

Preface

The purpose of this book is to provide the student with a clear and thorough presentation of the theory and application of the principles of engineering mechanics. Emphasis is placed on developing the student's ability to analyze problems—a most important skill of any engineer. In this revision, presentation and arrangement of some of the topics have been made more effective, and the examples and problem assignments have been improved. Throughout the book, attention has been given to numerical accuracy and clarity in both defining and developing the concepts.

Organization and Approach. The contents of each chapter are organized into well-defined sections. Selected groups of sections contain an explanation of specific topics, illustrative example problems, and a set of homework problems. The topics within each section are placed into subgroups defined by boldface titles. The purpose of this is to present a structured method for introducing each new definition or concept, and to make the book convenient for later reference and review.

A "procedure for analysis" is used at the end of many sections of the book in order to provide the student with a logical and orderly method to follow when applying the theory. As in the previous editions, the example problems are solved using this outlined method in order to clarify its numerical application. It is to be understood, however, that once the relevant principles have been mastered and enough confidence and judgment have been obtained, the student can then develop his own procedures for solving problems. In most cases it is felt that the first step in any procedure should require drawing a diagram. In doing so, the student forms the habit of tabulating the necessary data while focusing on the physical aspects of the problem and its associated geometry. If this step is correctly performed, applying the relevant equations of mechanics becomes somewhat methodical, since the data can be taken directly from the diagram.

Since mathematics provides a systematic means of applying the principles of mechanics, the student is expected to have prior knowledge of algebra, geometry, trigonometry, and, for complete coverage, some calculus. Vector analysis is introduced at points where it is most applicable. Its use often provides a convenient means for presenting concise derivations of the theory, and it makes possible a simple and systematic solution of many complicated three-dimensional problems. Occasionally, the example problems are solved using more than one method of analysis so that the student develops the ability to use mathematics as a tool whereby the solution of any problem may be carried out in the most direct and effective manner.

Problems.   Numerous problems in the book depict realistic situations encountered in engineering practice. It is hoped that this realism, both in the drawings and photographs, will both stimulate the student's interest in engineering mechanics and provide a means for developing the skill to reduce any such problem from its physical description to a model or symbolic representation to which the principles of mechanics may be applied. Most of the problems in this edition are new and many problems retained from the previous edition have been changed numerically. Also, in this edition an effort has been made to include some problems which may be solved using a numerical procedure executed on either a microcomputer or programmable pocket calculator. Suitable numerical techniques along with associated computer programs are given in Appendix B. The intent here is to broaden the student's capacity for using other forms of mathematical analysis *without* sacrificing the necessary time needed to focus on the application of the principles of mechanics. Problems of this type which either can or must be solved using numerical procedures are identified by a ''square'' symbol (■) preceding the problem number.

Throughout the text there is an approximate balance of problems using either SI or FPS units. Furthermore, in any set, an attempt has been made to arrange the problems in order of increasing difficulty.* The answers to all but every fourth problem are listed in the back of the book. To alert the user to a problem without a reported answer, an asterisk (*) is placed before the problem number.

Contents.   The book is divided into 11 chapters, in which the principles introduced are first applied to simple situations. Most often, each principle is applied first to a particle, then to a rigid body subjected to a coplanar system

---

*Review problems, at the end of each chapter, are presented in random order.

of forces, and finally to the general case of three-dimensional force systems acting on a rigid body.

The text begins in Chapter 1 with an introduction to mechanics and a discussion of units. The notion of a vector and the properties of a concurrent force system are introduced in Chapter 2. This theory is then applied to the equilibrium of particles in Chapter 3. Chapter 4 contains a general discussion of force systems and the methods used to simplify them. The principles of rigid-body equilibrium are developed in Chapter 5 and then applied to specific problems involving the equilibrium of trusses, frames, and machines in Chapter 6. Topics related to the center of gravity, centroid, and distributed loading are treated in Chapter 7. The analysis of internal forces in beams and cables is covered in Chapter 8 and applications to problems involving frictional forces are discussed in Chapter 9. If time permits, sections concerning more advanced topics, indicated by stars ($\star$), may be covered. Most of these topics are included in Chapter 10 (area and mass moments of inertia) and Chapter 11 (virtual work and potential energy). Note that this material also provides a suitable reference for basic principles when it is discussed in more advanced courses.

At the discretion of the instructor, some of the material may be presented in a different sequence with no loss in continuity. For example, it is possible to introduce the concept of a force and all the necessary methods of vector analysis by first covering Chapter 2 and Sec. 4–1. Then, after covering the rest of Chapter 4 (force and moment systems), the equilibrium methods in Chapters 3 and 5 can be discussed.

Acknowledgments. I have endeavored to write this book so that it will appeal to both the student and instructor. Through the years many people have helped in its development and I should like to acknowledge their valued suggestions and comments. Specifically, I wish to personally thank the reviewers who have contributed to this revision, namely, Professors Nicholas J. Altiero, Michigan State University; James G. Andrews, University of Iowa; A. Bedford, University of Texas at Austin; Jack Chen, Northeastern University; K. L. DeVries, University of Utah; Ronald C. Dykhuizen, Arizona State University; C. Stuart Ferrell, Louisiana Tech University; Dale Foremann, Colorado School of Mines; John H. Forrester, University of Tennessee; Ray Hagglund, Worcester Polytechnic Institute; Allen Hoffman, Worcester Polytechnic Institute; Edward E. Hornsey, University of Missouri, Rolla; C. C. Hsiao, University of Minnesota; Terry Ishihara, Saginaw Valley State College; Will Liddell, Auburn University; Benjamin W. Mooring, Texas A & M University; Walter F. Rowland, California State University, Fresno; C. R. Ullrich, University of Louisville; and Warren C. Young, University of Wisconsin, Madison.

Many thanks are also extended to all my students and to members of the teaching profession who have freely taken the time to send me their suggestions and comments. Although the list is too long to mention, it is hoped that those who have given help in this manner will accept this anonymous recognition. Furthermore, I greatly appreciate the personal support given to me by my editors and the staff at Macmillan. Lastly, I should like to acknowledge the assistance of my wife, Conny, who has always been quite helpful in attending to many of the details of preparing the manuscript for publication.

Russell Charles Hibbeler

Contents

General Principles
1

1.1 Mechanics 1
1.2 Fundamental Concepts 2
1.3 Units of Measurement 4
1.4 The International System of Units 5
1.5 Numerical Calculations 7
1.6 General Procedure for Analysis 10

Force Vectors
13

2.1 Scalars and Vectors 13
2.2 Vector Operations 14
2.3 Vector Addition of Forces 16
2.4 Addition of a System of Coplanar Forces 24
2.5 Cartesian Vectors 33
2.6 Addition and Subtraction of Cartesian Vectors 37
2.7 Position Vectors 45
2.8 Force Vector Directed Along a Line 48
2.9 Dot Product 58

3 Equilibrium of a Particle

69

3.1 Condition for the Equilibrium of a Particle   69
3.2 The Free-Body Diagram   70
3.3 Coplanar Force Systems   74
3.4 Three-Dimensional Force Systems   84

4 Force System Resultants

95

4.1 Cross Product   95
4.2 Moment of a Force   98
4.3 Transmissibility of a Force and the Principle of Moments   105
4.4 Moment of a Force About a Specified Axis   113
4.5 Moment of a Couple   119
4.6 Movement of a Force on a Rigid Body   128
4.7 Resultants of a Force and Couple System   129
4.8 Further Reduction of a Force and Couple System   133

5 Equilibrium of a Rigid Body

151

5.1 Conditions for Rigid-Body Equilibrium   151
Equilibrium in Two Dimensions   153
5.2 Free-Body Diagrams   153
5.3 Equations of Equilibrium   167
5.4 Two- and Three-Force Members   176
Equilibrium in Three Dimensions   189
5.5 Free-Body Diagrams   189
5.6 Equations of Equilibrium   194
5.7 Constraints for a Rigid Body   195

6 Structural Analysis

213

6.1 Simple Trusses   213
6.2 The Method of Joints   216
6.3 Zero-Force Members   222
6.4 The Method of Sections   229
*6.5 Space Trusses   239
6.6 Frames and Machines   243

Center of Gravity and Centroid

279

7.1 Center of Gravity and Center of Mass for a System
of Particles   279
7.2 Center of Gravity, Center of Mass,
and Centroid for a Body   280
7.3 Composite Bodies   296
7.4 Theorems of Pappus and Guldinus   306
7.5 Resultant of a Distributed Force System   312
7.6 Fluid Pressure Acting on a Submerged Surface   324

Internal Forces

337

8.1 Internal Forces Developed in Structural Members   337
8.2 Shear and Moment Diagrams for a Beam   356
8.3 Relations Between Distributed Load, Shear, and
Moment   364
8.4 Cables   373

Friction

391

9.1 Characteristics of Dry Friction   391
9.2 Problems Involving Dry Friction   395
9.3 Wedges   415
9.4 Frictional Forces on Screws   417
9.5 Frictional Forces on Flat Belts   424
9.6 Frictional Forces on Collar Bearings, Pivot Bearings,
and Disks   431
9.7 Frictional Forces on Journal Bearings   435
9.8 Rolling Resistance   440

Moments of Inertia

447

10.1 Definition of Moments of Inertia for Areas   447
10.2 Parallel-Axis Theorem for an Area   448
10.3 Radius of Gyration of an Area   449

10.4   Moments of Inertia for an Area by Integration   449
10.5   Moments of Inertia for Composite Areas   458
*10.6   Product of Inertia for an Area   464
*10.7   Moments of Inertia for an Area About Inclined Axes   468
*10.8   Mohr's Circle for Moments of Inertia   471
10.9   Mass Moment of Inertia   478

11

Virtual Work

491

11.1   Definition of Work and Virtual Work   491
11.2   Principle of Virtual Work for a Particle and a Rigid Body   493
11.3   Principle of Virtual Work for a System of Connected Rigid Bodies   494
*11.4   Conservative Forces   504
*11.5   Potential Energy   505
*11.6   Potential-Energy Criterion for Equilibrium   507
*11.7   Stability of Equilibrium   508

APPENDIX

A

Mathematical Expressions

521

B

Numerical and Computer Analysis

524

Answers

533

Index

544

Engineering Mechanics
Statics

General Principles

1

Mechanics 1.1

Mechanics can be defined as that branch of the physical sciences concerned with the state of rest or motion of bodies that are subjected to the action of forces. A thorough understanding of this subject is required for the study of structural engineering, machine design, fluid flow, electrical devices, and even the molecular and atomic behavior of elements.

In general, mechanics is subdivided into three branches: *rigid-body mechanics, deformable-body mechanics,* and *fluid mechanics*. This book treats only rigid-body mechanics, since this subject forms a suitable basis for the design and analysis of many engineering problems, and it provides part of the necessary background for the study of the mechanics of deformable bodies and the mechanics of fluids.

Rigid-body mechanics is divided into two areas: statics and dynamics. *Statics* deals with the equilibrium of bodies, that is, those that are either at rest or move with a constant velocity; whereas *dynamics* is concerned with the accelerated motion of bodies. Although statics can be considered as a special case of dynamics, in which the acceleration is zero, statics deserves separate treatment in engineering education, since most structures are designed with the intention that they remain in equilibrium.

Historical Development.   The subject of statics developed very early in history, because the principles involved could be formulated simply from measurements of geometry and force. For example, the writings of Archimedes (287–212 B.C.) deal with the principle of the lever. Studies of the pulley, inclined plane, and wrench are also recorded in ancient writings—at times when the requirements of engineering were limited primarily to building construction.

Since the principles of dynamics depend upon an accurate measurement of time, this subject developed much later. Galileo Galilei (1564–1642) was one of the first major contributors to this field. His work consisted of experiments using pendulums and falling bodies. The most significant contributions in dynamics, however, were made by Isaac Newton (1642–1727), who is noted for his formulation of the three fundamental laws of motion and the law of universal gravitational attraction. Shortly after these laws were postulated, important techniques for their application were developed by Euler, D'Alembert, Lagrange, and others.

1.2 Fundamental Concepts

Before beginning our study of rigid-body mechanics, it is important to understand the meaning of certain fundamental concepts and principles.

Basic Quantities. The following four quantities are used throughout rigid-body mechanics.

Length. Length is needed to locate the position of a point in space and thereby describe the size of a physical system. Once a standard unit of length is defined, one can then quantitatively define distances and geometrical properties of a body.

Time. Time is conceived by a succession of events. Although the principles of statics are time independent, this quantity does play an important role in the study of dynamics.

Mass. Mass is a property of matter by which we can compare the action of one body with that of another. This property manifests itself as a gravitational attraction between two bodies and provides a quantitative measure of the resistance of matter to a change in velocity.

Force. In general, force is considered as a "push" or "pull" exerted by one body on another. This interaction can occur when there is either direct contact between the bodies, such as a person pushing on a wall, or it can occur through a distance when the bodies are physically separated. Examples of the latter type include gravitational, electrical, and magnetic forces. In any case, a force is completely characterized by its magnitude, direction, and point of application.

Idealizations. In mechanics models or idealizations are used in order to simplify application of the theory. A few of the more important idealizations will now be defined. Others that are of importance will be discussed at points where they are needed.

Particle. A *particle* has a mass but a size that can be neglected in the analysis of the problem. For example, the size of the earth is insignificant compared to the size of its orbit, and therefore the earth can be modeled as a particle when studying its orbital motion. When a body is idealized as a particle, the principles of mechanics reduce to a rather simplified form since the geometry of the body will not be involved in the analysis of the problem.

Rigid Body. A *rigid body* can be considered as a combination of a large number of particles in which all the particles remain at a fixed distance from one another both before and after applying a load. As a result, the material properties of any body that is assumed to be rigid will not have to be considered in the analysis of the forces acting on the body. In most cases the actual deformations occurring in structures, machines, mechanisms, and the like are relatively small, and the rigid-body assumption is suitable for analysis.

Concentrated Force. A *concentrated force* represents the effect of a loading which is assumed to act at a *point* on a body. We can represent the effect of the loading by a concentrated force, provided the area over which the load is applied is *small* compared to the overall size of the body.

Newton's Three Laws of Motion. The entire subject of rigid-body mechanics is formulated on the basis of Newton's three laws of motion. These laws, which apply to the motion of a particle as measured from a nonaccelerating reference frame, may be briefly stated as follows:

First Law. A particle originally at rest, or moving in a straight line with constant velocity, will remain in this state provided the particle is not subjected to an unbalanced force.

Second Law. A particle acted upon by an unbalanced force **F** experiences an acceleration **a** that has the same direction as the force and a magnitude that is directly proportional to the force.* If **F** is applied to a particle of mass m, this law may be expressed mathematically as

$$\mathbf{F} = m\mathbf{a} \tag{1-1}$$

Third Law. For every force acting on a particle, the particle exerts an equal, opposite, and collinear reactive force.

Newton's Law of Gravitational Attraction. Shortly after formulating his three laws of motion, Newton postulated a law governing the gravitational attraction between any two particles. This law can be expressed mathematically as

$$F = G\frac{m_1 m_2}{r^2} \tag{1-2}$$

where
F = force of gravitation between the two particles
G = universal constant of gravitation; according to experimental evidence, $G = 6.673(10^{-11})$ m³/(kg · s²)
m_1, m_2 = mass of each of the two particles
r = distance between the two particles

---

*Stated another way, the unbalanced force acting on the particle is proportional to the time rate of change of the particle's linear momentum.

Weight. Any two particles or bodies have a mutual attractive (gravitational) force acting between them. In the case of a particle located at or near the surface of the earth, however, the only gravitational force having any sizable magnitude is that between the earth and the particle. Consequently, this force, termed the *weight*, will be the only gravitational force considered in our study of mechanics.

From Eq. 1–2, we can develop an approximate expression for finding the weight W of a particle having a mass $m_1 = m$. If we assume the earth to be a nonrotating sphere of constant density and having a mass m_2, then if r is the distance between the earth's center and the particle, we have

$$W = G\frac{mm_2}{r^2}$$

Letting $g = Gm_2/r^2$ yields

$$W = mg \qquad\qquad\qquad (1\text{–}3)$$

By comparison with Eq. 1–1, we term g the acceleration due to gravity. Since it depends upon r, it can be seen that the weight of a body is *not* an absolute quantity. Instead, its magnitude depends upon where the measurement was made. For most engineering calculations, however, g is determined at sea level and at a latitude of 45°, which is considered the "standard location."

1.3 Units of Measurement

The four quantities—length, time, mass, and force—are not all independent from one another; in fact, they are *related* by Newton's second law of motion, **F** = *m***a.** Hence, the units used to define force, mass, length, and time cannot *all* be selected arbitrarily. The equality **F** = *m***a** is maintained only if three of the four units, called *base units*, are *arbitrarily defined* and the fourth unit is *derived* from the equation.

SI Units. The International System of units, abbreviated SI after the French "Système International d'Unités," is a modern version of the metric system which has received worldwide recognition. As shown in Table 1–1, the SI system specifies length in meters (m), time in seconds (s), and mass in kilograms (kg). The unit of force, called a newton (N), is *derived* from **F** = *m***a.** Thus, 1 newton is equal to a force required to give 1 kilogram of mass an acceleration of 1 m/s² (N = kg · m/s²).

If the weight of a body located at the "standard location" is to be determined in newtons, then Eq. 1–3 must be applied. Here $g = 9.806\,65$ m/s²; however, for calculations, the value $g = 9.81$ m/s² will be used. Thus,

$$W = mg \qquad (g = 9.81 \text{ m/s}^2) \qquad\qquad (1\text{–}4)$$

Therefore, a body of mass 1 kg has a weight of 9.81 N, a 2-kg body weighs 19.62 N, and so on.

Table 1–1   Systems of Units

Name	Length	Time	Mass	Force
International System of Units (SI)	meter (m)	second (s)	kilogram (kg)	newton* (N) $\left(\dfrac{kg \cdot m}{s^2}\right)$
U.S. Customary (FPS)	foot (ft)	second (s)	slug* $\left(\dfrac{lb \cdot s^2}{ft}\right)$	pound (lb)

*Derived unit.

U.S. Customary.   In the U.S. Customary system of units (FPS), Table 1–1, length is measured in feet (ft), time in seconds (s), and force in pounds (lb). The unit of mass, called a *slug,* is *derived* from $\mathbf{F} = m\mathbf{a}.$ Hence, 1 slug is equal to the amount of matter accelerated at 1 ft/s² when acted upon by a force of 1 lb (slug = lb · s²/ft).

In order to determine the mass of a body having a weight measured in pounds, we must apply Eq. 1–3. If the measurements are made at the "standard location," then $g = 32.2$ ft/s² will be used for calculations. Therefore,

$$m = \frac{W}{g} \qquad (g = 32.2 \text{ ft/s}^2) \tag{1–5}$$

so that a body weighing 32.2 lb has a mass of 1 slug, a 64.4-lb body has a mass of 2 slugs, and so on.

The International System of Units 1.4

The SI system of units is used extensively in this book since it is intended to become the worldwide standard for measurement. Consequently, the rules for its use and some of its terminology relevant to mechanics will now be presented.

Prefixes.   When a numerical quantity is either very large or very small the units used to define its size may be modified by using a prefix. Some of the prefixes used in the SI system are shown in Table 1–2. Each represents a multiple or submultiple of a unit which, if applied successively, moves the decimal point of a numerical quantity to every third place.* For example, 4 000 000 N = 4 000 kN   (kilo-newton) = 4 MN   (mega-newton),   or 0.005 m = 5 mm (milli-meter). Notice that the SI system does not include the multiple deca (10) or the submultiple centi (0.01), which form part of the old metric system. Except for some volume and area measurements, the use of these prefixes is to be avoided in science and engineering.

*The kilogram is the only base unit that is defined with a prefix.

Table 1–2 Prefixes

	Exponential Form	Prefix	SI Symbol
Multiple			
1 000 000 000	10^9	giga	G
1 000 000	10^6	mega	M
1 000	10^3	kilo	k
Submultiple			
0.001	10^{-3}	milli	m
0.000 001	10^{-6}	micro	μ
0.000 000 001	10^{-9}	nano	n

Attaching a prefix to a unit in effect creates a new unit. Thus, if a multiple or submultiple is raised to a power, the power applies to this new unit, not just to the original unit *without* the multiple or submultiple. For example, $(2 \text{ km})^2 = (2)^2 (\text{km})^2$ and is represented symbolically as 4 km^2. Also, 1 mm^2 represents 1 (mm)^2, not $1 \text{ m(m}^2)$.

Rules for Use. The following rules are given for the proper use of the various SI symbols:

1. A symbol is *never* written with a plural "s," since it may be confused with the unit for second (s).
2. Symbols are always written in lowercase letters, with two exceptions: symbols for the two largest prefixes shown in Table 1–2, giga and mega, are capitalized as G and M, respectively; and symbols named after an individual are capitalized; e.g., N.
3. Quantities defined by several units which are multiples of one another are separated by a *dot* to avoid confusion with prefix notation, as illustrated by $N = kg \cdot m/s^2 = kg \cdot m \cdot s^{-2}$. Also, $m \cdot s$ (meter-second), whereas ms (milli-second).
4. Physical constants or numbers having several digits on either side of the decimal point should be reported with a *space* between every three digits rather than with a comma; e.g., 73 569.213 427. In the case of four digits on either side of the decimal, the spacing is optional; e.g., 8537 or 8 537. Furthermore, always try to use decimals and avoid fractions; that is, write 15.25, *not* $15\frac{1}{4}$.
5. When performing calculations, represent the numbers in terms of their *base or derived units* by converting all prefixes to powers of 10. The final result should then be expressed using a *single prefix*. Also, after calculation, it is best to keep numerical values between 0.1 and 1000; otherwise, a suitable prefix should be chosen. For example,

$$(50 \text{ kN})(60 \text{ nm}) = [50(10^3) \text{ N}][60(10^{-9}) \text{ m}]$$
$$= 3000(10^{-6}) \text{ N} \cdot m = 3 \text{ mN} \cdot m$$

6. Compound prefixes should not be used; e.g., kμs (kilo-micro-second) should be expressed as ms (milli-second) since 1 kμs = $1(10^3)(10^{-6})$ s = $1(10^{-3})$ s = 1 ms.

7. With the exception of the base unit the kilogram, in general avoid the use of a prefix in the denominator of composite units. For example, do not write N/mm, but rather kN/m.

8. Although not expressed in multiples of 10, the minute, hour, etc., are retained for practical purposes as multiples of the second. Furthermore, plane angular measurement is made using radians (rad). In this book, degrees will often be used, where $180° = \pi$ rad. Fractions of a degree, however, should be expressed in decimal form rather than in minutes, as in $10.4°$, not $10°24'$.

Numerical Calculations 1.5

Numerical work in engineering practice is most often performed by using electronic calculators and computers. It is important, however, that the answers be reported with both justifiable accuracy and appropriate significant figures. In this section we will discuss these topics together with some other important aspects involved in all engineering calculations.

Dimensional Homogeneity. The terms of any equation used to describe a physical process must be *dimensionally homogeneous,* that is, each term must be expressed in the same units. Provided this is the case, all the terms of an equation can then be combined if numerical values are substituted for the variables. Consider, for example, the equation $s = vt + \frac{1}{2}at^2$, where, in SI units, s is the position in meters (m), t is time in seconds (s), v is velocity in m/s, and a is acceleration in m/s². Regardless of how this equation is evaluated, it maintains its dimensional homogeneity. In the form stated each of the three terms is expressed in meters [m, (m/$\not{s}$)$\not{s}$, (m/$\not{s}^2$)$\not{s}^2$], or solving for a, $a = 2s/t^2 - 2v/t$, the terms are each expressed in units of m/s² [m/s², m/s², (m/s)/s].

Since mechanics problems involve the solution of dimensionally homogeneous equations, the fact that all terms of an equation are represented by a consistent set of units can be used as a partial check for algebraic manipulations of an equation.

Significant Figures. A *significant figure* is any digit from 1 to 9, or a zero that is *not used* to show the position of the decimal point for a number. For example, the numbers 5605, 34.43, 0.05892, and 500.0 each have four significant figures. The number of significant figures in numbers having zeros can be clarified by using powers of 10. There are two ways of doing this. The format for *scientific notation* specifies one digit to the left of the decimal point, with the remaining digits to the right; for example, 45,000 expressed to three significant figures would be $4.50(10^4)$. Using *engineering notation,* which is preferred here, the exponent is displayed in multiples of three in order to facilitate conversion of SI units to those having an appropriate prefix. Thus, 45,000 expressed to three significant figures would be $45.0(10^3)$.

Rounding-off Numbers. For numerical calculations, the accuracy obtained from the solution of a problem generally can never be better than the accuracy of the problem data. This is what is to be expected, but often computers or hand calculators involve more figures in the answer than the number of significant figures used for the data. For this reason, a calculated result should always be "rounded off" to an appropriate number of significant figures.

To ensure accuracy, the following rules for rounding off a number to n significant figures apply:

1. If the $n + 1$ digit is *less than 5,* the $n + 1$ digit and others following it are dropped. For example, 2.326 and 0.451 rounded off to $n = 2$ significant figures would be 2.3 and 0.45.
2. If the $n + 1$ digit is equal to 5 with zeros following it, then round off the nth digit to an *even number.* For example, 1245 and 0.8655 rounded off to three significant figures become 1240 and 0.866.
3. If the $n + 1$ digit is *greater than 5* or equal to 5 with any nonzero digits following it, then increase the nth digit by 1 and drop the $n + 1$ digit and others following it. For example, 0.007 238 7 and 565.500 3 rounded off to three significant figures become 0.00724 and 566.

Calculations. As a general rule, to ensure accuracy of a final result when performing calculations with numbers of unequal accuracy, always retain one extra significant figure in the more accurate numbers than in the least accurate number *before* beginning the computations. Then round off the final result so that it has the same number of significant figures as the least accurate number. In particular, when adding numbers, retain one more *decimal* digit (digits to the right of the decimal point) in the more accurate numbers than is contained in the least accurate number. Also, try to work out the computations so that numbers which are approximately equal are not subtracted since accuracy is often lost from this calculation.

Most of the example problems in this book are solved with the assumption that any measured *data* are accurate to three significant figures.* Consequently, intermediate calculations are often worked out to four significant figures and the answers are generally reported to three significant figures.

Conversion of Units. In some cases it may be necessary to convert from one system of units to another. In this regard, Table 1–3 provides a set of direct conversion factors between FPS and SI units for the fundamental quantities. Also, in the FPS system, recall that 1 ft = 12 in. (inches), 5280 ft = 1 mi (mile); and 1000 lb = 1 kip (kilo-pound), 2000 lb = 1 ton.

---

*Of course some numbers, such as π, e, or numbers used in derived formulas, are exact and are therefore accurate to an infinite number of significant figures.

Table 1–3 Conversion Factors

Quantity	Unit of Measurement (FPS)	Equals	Unit of Measurement (SI)
Force	lb		4.448 2 N
Mass	slug		14.593 8 kg
Length	ft		0.304 8 m

When derived units are present, a general procedure using a simple cancellation technique should be applied. The following examples illustrate this method.

Example 1–1

Convert 2 km/h to m/s.

Solution

Since 1 km = 1000 m and 1 h = 3600 s, the factors of conversion are arranged as follows:

$$2 \text{ km/h} = \frac{2 \text{ km}}{\text{h}} \left(\frac{1000 \text{ m}}{\text{km}} \right) \left(\frac{1 \text{ h}}{3600 \text{ s}} \right)$$

$$= \frac{2000 \text{ m}}{3600 \text{ s}} = 0.556 \text{ m/s} \qquad \qquad Ans.$$

Example 1–2

Convert the quantity 300 lb · s to appropriate SI units.

Solution

Using Table 1–3, 1 lb = 4.448 2 N, then

$$300 \text{ lb} \cdot \text{s} = 300 \text{ lb} \cdot \text{s} \left(\frac{4.448 \text{ 2 N}}{\text{lb}} \right)$$

$$= 1334.5 \text{ N} \cdot \text{s} = 1.33 \text{ kN} \cdot \text{s} \qquad \qquad Ans.$$

ample 1–3

Evaluate each of the following and express with SI units having an appropriate prefix: (a) (50 mN)(6 GN), and (b) (400 mm)(0.6 MN)^2.

Solution

First convert each number to base units, perform the indicated operations, then choose an appropriate prefix.

Part (a)

$$(50 \text{ mN})(6 \text{ GN}) = [50(10^{-3}) \text{ N}][6(10^9) \text{ N}]$$
$$= 300(10^6) \text{ N}^2$$
$$= 300(10^6) \cancel{N^2} (10^{-3} \text{ mN}/\cancel{N})(10^{-3} \text{ mN}/\cancel{N})$$
$$= 300 \text{ mN}^2 \qquad\qquad Ans.$$

Note carefully the convention mN^2 = (mN)^2 = 10^{-6} N^2.

Part (b)

$$(400 \text{ mm})(0.6 \text{ MN})^2 = [400(10^{-3}) \text{ m}][0.6(10^6) \text{ N}]^2$$
$$= [400(10^{-3}) \text{ m}][0.36(10^{12}) \text{ N}^2]$$
$$= 144(10^9) \text{ m} \cdot \text{N}^2$$
$$= 144 \text{ Gm} \cdot \text{N}^2 \qquad\qquad Ans.$$

Note that we can also write

$$144(10^9)\text{m} \cdot \text{N}^2 = 144(10^9)\text{m} \cdot \cancel{N^2}(1 \text{ MN}/10^6 \cancel{N})(1 \text{ MN}/10^6 \cancel{N})$$
$$= 0.144 \text{ m} \cdot \text{MN}^2$$

General Procedure for Analysis 1.6

The most effective way of learning the principles of engineering mechanics is to *solve problems*. To be successful at this, it is important to present the work in a *logical* and *orderly manner,* as suggested by the following sequence of steps:

1. Read the problem carefully and try to correlate the actual physical situation with the theory studied.
2. Draw any necessary diagrams and tabulate the problem data.
3. Apply the relevant principles, generally in mathematical form.
4. Solve the necessary equations algebraically as far as practical, then making sure they are dimensionally homogeneous, use a consistent set of units and complete the solution numerically. Report the answer with no more significant figures than the accuracy of the given data.
5. Study the answer with technical judgment and common sense to determine whether or not it seems reasonable.
6. Once the solution has been completed, review the problem. Try to think of other ways of obtaining the same solution.

In applying this general procedure, do the work as neatly as possible. Being neat generally stimulates clear and orderly thinking, and vice versa.

Problems

1–1. What is the weight in newtons of an object that has a mass of (a) 8 kg, (b) 0.04 g, and (c) 760 Mg?

1–2. Wood has a density of 4.70 slug/ft^3. What is its density expressed in SI units?

1–3. Using Table 1–3, determine your own mass in kilograms, your weight in newtons, and your height in meters.

*1–4. Represent each of the following as a number between 0.1 and 1000 using an appropriate prefix: (a) 45 320 kN, (b) 568(10^5) mm, and (c) 0.00563 mg.

1–5. Evaluate each of the following and express with an appropriate prefix: (a) (430 kg)^2, (b) (0.002 mg)^2, and (c) (230 m)^3.

1–6. Represent each of the following combinations of units in the correct SI form: (a) GN · μm, (b) kg/μm, (c) N/ks^2, and (d) kN/μs.

1–7. Represent each of the following combinations of units in the correct SI form: (a) kN/μs, (b) Mg/mN, and (c) MN/(kg · ms).

*1–8. The *pascal* (Pa) is actually a very small unit of pressure. To show this, convert 1 Pa = 1 N/m^2 to lb/ft^2. Atmospheric pressure at sea level is 14.7 lb/in^2. How many pascals is this?

1–9. Convert: (a) 20 lb · ft to N · m, (b) 450 lb/ft^3 to kN/m^3, and (c) 15 ft/h to mm/s.

1–10. Determine the number of cubic millimeters contained in one cubic inch.

1–11. A steel disk has a diameter of 500 mm and a thickness of 70 mm. If the density of steel is 7850 kg/m^3, determine the weight of the disk in pounds.

*1–12. If an object has a mass of 40 slugs, determine its mass in kilograms.

1–13. Using the base units of the SI system, show that Eq. 1–2 is a dimensionally homogeneous equation which gives F in newtons. Compute the gravitational force acting between two identical spheres that are touching each other. The mass of each sphere is 150 kg and the radius is 275 mm.

1–14. Two particles have a mass of 8 kg and 12 kg, respectively. If they are 800 mm apart, determine the force of gravity acting between them. Compare this result with the weight of each particle.

1–15. If a man weighs 155 lb on earth, specify (a) his mass in slugs, (b) his mass in kilograms, and (c) his weight in newtons. If the man is on the moon, where the acceleration due to gravity is $g_m = 5.30$ ft/s^2, determine (d) his weight in pounds, and (e) his mass in kilograms.

Force Vectors

In this chapter we will introduce the concept of a concentrated force and give the procedures for adding forces, resolving them into components, and projecting them along an axis. Since force is a vector quantity, we must use the rules of vector algebra whenever forces are considered. We will begin our study by defining scalar and vector quantities and then develop some of the basic rules of vector algebra.

Scalars and Vectors 2.1

Most of the physical quantities in mechanics can be expressed mathematically by means of scalars and vectors.

Scalar. A quantity characterized by a positive or negative number is called a *scalar*. Mass, volume, and length are scalar quantities often used in statics. In this book, scalars are indicated by letters in italic type, such as the scalar A. The mathematical operations involving scalars follow the same rules as those of elementary algebra.

Vector. A *vector* is a quantity that has both a magnitude and direction and obeys the parallelogram law of addition. This law, which will be described later, utilizes a form of construction that accounts for the combined magnitude and direction of the vector. Vector quantities commonly used in statics are position, force, and moment vectors.

A vector is represented graphically by an arrow, which is used to define its magnitude, direction, and sense. The *magnitude* of the vector is indicated by the *length* of the arrow, the *direction* is defined by the *angle* between a

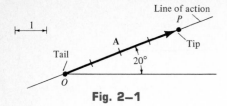

Fig. 2–1

reference axis and the arrow's *line of action,* and the *sense* is indicated by the *arrowhead.* For example, the vector **A** shown in Fig. 2–1 has a *magnitude* of 4 units, a line of action which is *directed* 20° above the horizontal axis, and a *sense* which is upward and to the right. The point O is called the *tail* of the vector, the point P is the *tip.*

For handwritten work, a vector is generally represented by a letter with an arrow written over it, such as $\vec{A}$. The magnitude is designated by $|\vec{A}|$ or simply A. In this book *vectors* will be symbolized in *boldface type;* for example, **A** is used to designate the vector "A". Its *magnitude,* which is *always a positive quantity,* is symbolized in *italic type,* written as $|A|$ or simply A when it is understood that A is a positive scalar.

2.2 Vector Operations

Multiplication and Division of a Vector by a Scalar. The product of vector **A** and scalar a, yielding $a\mathbf{A}$, is defined as a vector having a magnitude $|aA|$. The *sense* of $a\mathbf{A}$ is the *same* as **A** provided a is *positive;* it is *opposite* to **A** if a is *negative.* Consequently, the negative of a vector is formed by multiplying the vector by the scalar (-1), Fig. 2–2. Division of a vector by a scalar can be defined using the laws of multiplication, since $\mathbf{A}/a = (1/a)\mathbf{A}, a \neq 0$. Graphic examples of these operations are shown in Fig. 2–3.

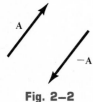

Fig. 2–2

Vector Addition. Two vectors **A** and **B**, Fig. 2–4a, may be added to form a "resultant" vector $\mathbf{R} = \mathbf{A} + \mathbf{B}$ by using the *parallelogram law.* To do this, **A** and **B** are joined at their tails, Fig. 2–4b. Parallel dashed lines drawn from the tip of each vector intersect at a common point, thereby forming the adjacent sides of a parallelogram. As shown, the resultant **R** is the diagonal of this parallelogram. Addition of **A** and **B** is defined in this manner since it gives physically meaningful results.

We can also add **B** to **A** using a *triangle construction,* which is a special case of the parallelogram law, whereby vector **B** is added to vector **A** in a "tip-to-tail" fashion, i.e., by placing the tail of **B** at the tip of **A**, Fig. 2–4c.

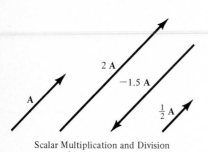

Scalar Multiplication and Division

Fig. 2–3

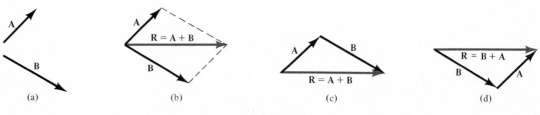

Vector Addition

Fig. 2–4

The resultant **R** extends from the tail of **A** to the tip of **B**. In a similar manner, **R** can also be obtained by adding **A** to **B**, Fig. 2–4*d*. By comparison, it is seen that vector addition is commutative; in other words, the vectors can be added in either order, i.e., **R** = **A** + **B** = **B** + **A**.

If the two vectors **A** and **B** are *collinear*, i.e., both have the same line of action, the parallelogram law reduces to an *algebraic* or *scalar addition* $R = A + B$ as shown in Fig. 2–5.

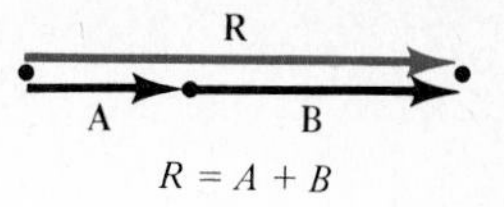

Vector Subtraction. The resultant *difference* between two vectors **A** and **B** may be expressed as

$$\mathbf{R'} = \mathbf{A} - \mathbf{B} = \mathbf{A} + (-\mathbf{B})$$

This vector sum is shown graphically in Fig. 2–6. Subtraction is therefore defined as a special case of addition, so that the rules of vector addition also apply to vector subtraction.

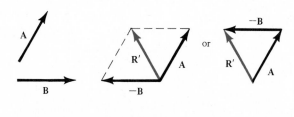

Vector subtraction

Fig. 2–6

Resolution of a Vector. By using the parallelogram law it is possible to resolve a vector into "components" having known lines of action. For example, if **R** in Fig. 2–7 is to be resolved into two components, acting along the lines *a* and *b*, one starts at the *tip* of **R** and extends a dashed line *parallel* to the *a* axis until it intersects the *b* axis. Likewise, a dashed line parallel to the *b* axis is drawn from the tip of **R** to the point of intersection with the *a* axis. The two components **A** and **B** are then drawn such that they extend from the tail of **R** to the points of intersection, as shown in the figure.

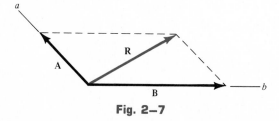

Fig. 2–7

R

A   B

$R = A + B$

Addition of Collinear Vectors

Fig. 2–5

2.3 Vector Addition of Forces

Experimental evidence has shown that a force is a vector quantity since it has a specified magnitude and direction and it adds according to the parallelogram law. Two common problems in statics involve finding either the resultant force, knowing its components, or resolving a known force into components. As described above, both of these problems require application of the parallelogram law.

In particular, if more than two forces are to be added, successive applications of the parallelogram law can be carried out in order to obtain the resultant force. For example, if three forces $\mathbf{F}_1$, $\mathbf{F}_2$, $\mathbf{F}_3$ act at O, Fig. 2–8, the resultant of any two of the forces is found, say $\mathbf{F}_1 + \mathbf{F}_2$, and then this resultant is added to the third force, yielding the resultant of all three forces; i.e., $\mathbf{F}_R = (\mathbf{F}_1 + \mathbf{F}_2) + \mathbf{F}_3$. Using the parallelogram law to add more than two forces, as shown here, often requires extensive geometric and trigonometric calculation for the numerical value of the resultant. Instead, problems of this type are easily solved by using the "rectangular-component method," which is explained in Sec. 2.4.

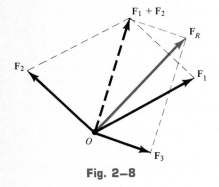

Fig. 2–8

PROCEDURE FOR ANALYSIS

Problems that involve the addition of two forces and contain at most *two unknowns* can be solved by using the following procedure:

Parallelogram Law. Make a sketch showing the vector addition using the parallelogram law. If possible, determine the interior angles of the parallelogram from the geometry of the problem. Recall that the sum total of the angles is to be 360°. Unknown angles, along with known and unknown force magnitudes, should clearly be labeled on this sketch. Redraw a portion of the constructed parallelogram to illustrate the triangular tip-to-tail addition of the components.

Trigonometry. By using trigonometry, the two unknowns can be determined from the data listed on the triangle. If the triangle does *not* contain a 90° angle, the law of sines and/or the law of cosines may be used for the solution. These formulas are given in Fig. 2–9 for the triangle shown.

The following examples numerically illustrate this method for solution.

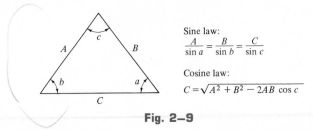

Sine law:
$$\frac{A}{\sin a} = \frac{B}{\sin b} = \frac{C}{\sin c}$$

Cosine law:
$$C = \sqrt{A^2 + B^2 - 2AB\,\cos c}$$

Fig. 2–9

Example 2–1

The screw eye in Fig. 2–10a is subjected to two forces, $\mathbf{F}_1$ and $\mathbf{F}_2$. Determine the magnitude and direction of the resultant force.

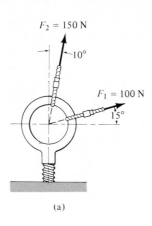

(a)

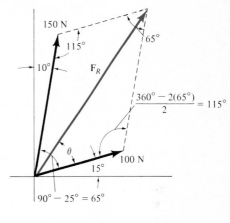

(b)

Fig. 2–10

Solution

Parallelogram Law. The parallelogram law of addition is shown in Fig. 2–10b. The two unknowns are the magnitude of $\mathbf{F}_R$ and the angle θ (theta). From Fig. 2–10b, the force triangle, Fig. 2–10c, is constructed.

Trigonometry. F_R is determined by using the law of cosines:

$$F_R = \sqrt{(100)^2 + (150)^2 - 2(100)(150)\cos 115^\circ}$$
$$= \sqrt{10\,000 + 22\,500 - 30\,000(-0.4226)} = 212.6 \text{ N}$$
$$= 213 \text{ N} \qquad \qquad \textit{Ans.}$$

The angle θ is determined by applying the law of sines, using the computed value of F_R.

$$\frac{150}{\sin \theta} = \frac{212.6}{\sin 115^\circ}$$

$$\sin \theta = \frac{150}{212.6}(0.9063)$$

$$\theta = 39.8^\circ$$

Thus, the direction ϕ (phi) of $\mathbf{F}_R$, measured from the horizontal, is

$$\phi = 39.8^\circ + 15.0^\circ = 54.8^\circ \quad \angle\phi \qquad \textit{Ans.}$$

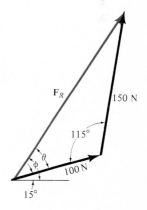

(c)

Example 2–2

Resolve the 200-lb force shown acting on the pin, Fig. 2–11a, into components in the (a) x and y directions, and (b) x' and y directions.

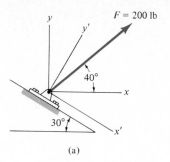

(a)

Solution

In each case the parallelogram law is used to resolve **F** into its two components, and then the triangle law is applied to determine the numerical results by trigonometry.

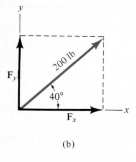

(b)

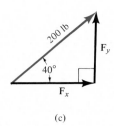

(c)

Fig. 2–11

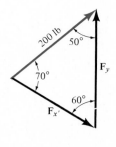

(d)

Part (a). The vector addition $\mathbf{F} = \mathbf{F}_x + \mathbf{F}_y$ is shown in Fig. 2–11b. In particular, note that the length of the components is scaled along the x and y axes by first constructing dashed lines parallel to the axes in accordance with the parallelogram law. From the vector triangle, Fig. 2–11c,

$$F_x = 200 \cos 40° = 153 \text{ lb} \qquad Ans.$$
$$F_y = 200 \sin 40° = 129 \text{ lb} \qquad Ans.$$

Part (b). The vector addition $\mathbf{F} = \mathbf{F}_{x'} + \mathbf{F}_y$ is shown in Fig. 2–11d. Note carefully how the parallelogram is constructed. Applying the law of sines and using the data listed on the vector triangle, Fig. 2–11e, yields

$$\frac{F_{x'}}{\sin 50°} = \frac{200}{\sin 60°}$$
$$F_{x'} = 200\left(\frac{\sin 50°}{\sin 60°}\right) = 177 \text{ lb} \qquad Ans.$$
$$\frac{F_y}{\sin 70°} = \frac{200}{\sin 60°}$$
$$F_y = 200\left(\frac{\sin 70°}{\sin 60°}\right) = 217 \text{ lb} \qquad Ans.$$

(e)

Example 2–3

The force **F** acting on the frame shown in Fig. 2–12a has a magnitude of 500 N and is to be resolved into two components acting along struts AB and AC. Determine the angle θ so that the component $\mathbf{F}_{AC}$ is directed from A toward C and has a magnitude of 400 N.

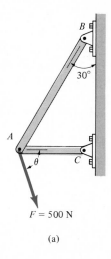

(a)

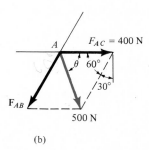

(b)

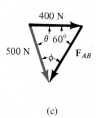

(c)

Solution

By using the parallelogram law, the vector addition of the two components yielding the resultant is shown in Fig. 2–12b. Note carefully how the resultant force is resolved into the two components $\mathbf{F}_{AB}$ and $\mathbf{F}_{AC}$, which have specified lines of action. The corresponding vector triangle is shown in Fig. 2–12c. The angle ϕ can be determined by using the law of sines:

Fig. 2–12

$$\frac{400}{\sin \phi} = \frac{500}{\sin 60°}$$

$$\sin \phi = \left(\frac{400}{500}\right) \sin 60°$$

$$\phi = 43.9°$$

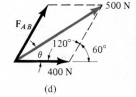

(d)

Hence,

$$\theta = 180° - 60° - 43.9° = 76.1° \quad \text{\reflectbox{θ}} \qquad \textit{Ans.}$$

Show that $\mathbf{F}_{AB}$ has a magnitude of 560 N.

Notice that $\mathbf{F}_{AB}$ can also be directed from A towards B. A sketch of the resulting vector addition is shown in Fig. 2–12d. Show that $\theta = 16.1°$ and $F_{AB} = 161$ N.

Example 2–4

The ring shown in Fig. 2–13a is subjected to two forces, F_1 and F_2. If it is required that the resultant force have a magnitude of 1 kN and be directed vertically downward, determine (a) the magnitudes of F_1 and F_2 provided $\theta = 30°$, and (b) the magnitudes of F_1 and F_2 if F_2 is to be a minimum.

Solution

Part (a). A sketch of the vector addition according to the parallelogram law is shown in Fig. 2–13b. From the vector triangle constructed in Fig. 2–13c the unknown magnitudes F_1 and F_2 can be determined by using the law of sines.

$$\frac{F_1}{\sin 30°} = \frac{1000}{\sin 130°}$$

$$F_1 = 653 \text{ N} \qquad\qquad Ans.$$

$$\frac{F_2}{\sin 20°} = \frac{1000}{\sin 130°}$$

$$F_2 = 446 \text{ N} \qquad\qquad Ans.$$

Part (b). By the vector triangle, F_2 may be added to F_1 in various ways to yield F_R, Fig. 2–13d. In particular, the *minimum* length or magnitude of F_2 will occur when its line of action is *perpendicular* to F_1. Any other direction, such as OA or OB, yields a larger value for F_2. Hence, when $\theta = 90° - 20° = 70°$, F_2 is minimum. From the triangle shown in Fig. 2–13e, it is seen that

$$F_1 = 1000 \sin 70° = 940 \text{ N} \qquad\qquad Ans.$$

$$F_2 = 1000 \sin 20° = 342 \text{ N} \qquad\qquad Ans.$$

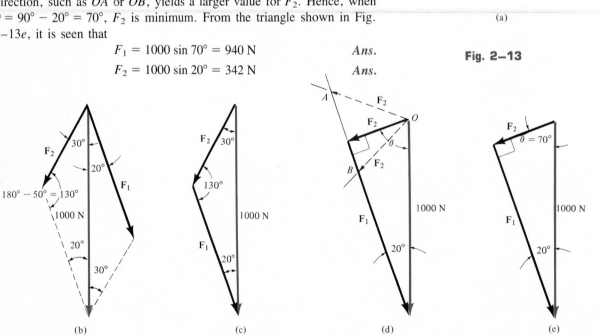

(a)

Fig. 2–13

(b)   (c)   (d)   (e)

Problems

2–1. Determine the magnitude of the resultant force and its direction measured from the positive x axis.

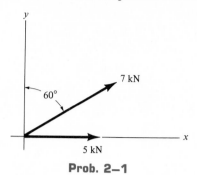

Prob. 2–1

2–2. Determine the magnitude of the resultant force and its direction measured from the positive x axis.

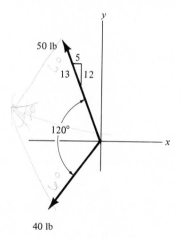

Prob. 2–2

2–3. Determine the x and y components of the 4-kN force.

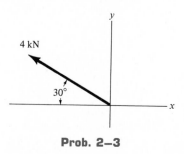

Prob. 2–3

*2–4. Determine the x and y components of the 500-N force.

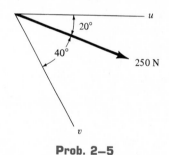

Prob. 2–4

2–5. Resolve the 250-N force into components acting along the u and v axes and determine the magnitudes of the components.

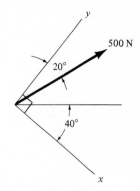

Prob. 2–5

2–6. Resolve the 80-N force into components acting along the u and v axes and determine the magnitudes of the components.

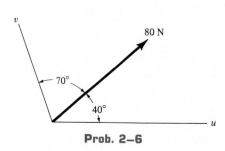

Prob. 2–6

2–7. Determine the components of the 250-N force acting along the u and v axes.

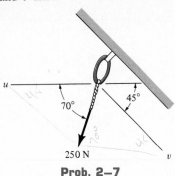

Prob. 2–7

*2–8. Resolve the 60-lb force into components acting along the u and v axes and determine the magnitudes of the components.

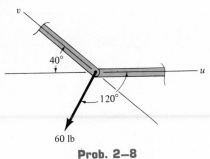

Prob. 2–8

2–9. The wind is deflected by the sail of a boat such that it exerts a resultant force of $F = 110$ lb perpendicular to the sail. Resolve this force into two components, one parallel and one perpendicular to the keel aa of the boat. *Note:* The ability to sail into the wind is known as tacking, made possible by the force parallel to the boat's keel. The perpendicular component tends to tip the boat or push it over.

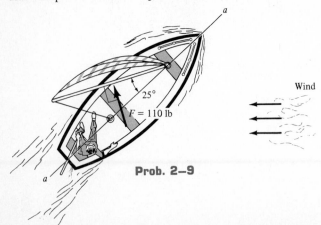

Prob. 2–9

2–10. The plate is subjected to the two forces at A and B as shown. If $\theta = 60°$, determine the magnitude of the resultant of these two forces and its direction measured from the horizontal.

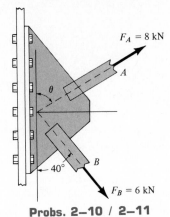

Probs. 2–10 / 2–11

2–11. Determine the angle θ for connecting member A to the plate so that the resultant force of $\mathbf{F}_A$ and $\mathbf{F}_B$ is directed horizontally to the right. Also, what is the magnitude of the resultant force?

*2–12. The post is to be pulled out of the ground using two ropes A and B. Rope A is subjected to a force of 600 lb and is directed at 60° from the horizontal. Determine the force $\mathbf{T}$ in rope B if the post starts to lift out when $\theta = 20°$. For this to occur, the resultant force on the post is to be directed vertically upward. Also calculate the magnitude of the resultant force.

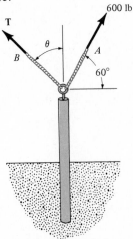

Probs. 2–12 / 2–13

2–13. The post is to be pulled out of the ground using two ropes A and B. Rope A is subjected to a force of 600 lb and is directed at 60° from the horizontal. If the resultant force acting on the post is to be 1200 lb, vertically upward, determine the force T in rope B and the corresponding angle θ.

2–14. The boat is to be pulled onto the shore using two ropes. Determine the magnitudes of forces **T** and **P** acting in each rope in order to develop a resultant force of 80 lb, directed along the keel aa as shown. Take $\theta = 40°$.

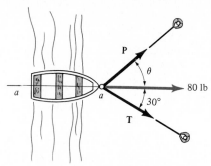

Probs. 2–14 / 2–15

2–15. The boat is to be pulled onto the shore using two ropes. If the resultant force is to be 80 lb, directed along the keel aa, as shown, determine the magnitudes of forces **T** and **P** acting in each rope and the angle θ of **P** so that the magnitude of **P** is a *minimum*. **T** acts at 30° from the keel aa, as shown.

*2–16. Determine the design angle θ (0° ≤ θ ≤ 90°) for strut AB so that the 400-lb horizontal force has a component of 500 lb directed from A towards C. What is the component of force acting along member AB? Take $\phi = 40°$.

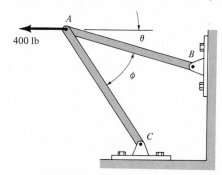

Probs. 2–16 / 2–17

2–17. Determine the design angle ϕ (0° ≤ ϕ ≤ 90°) between struts AB and AC so that the 400-lb horizontal force has a component of 600 lb which acts up to the left, in the same direction as from B towards A. Take $\theta = 30°$.

2–18. Two forces having a magnitude of 10 lb and 6 lb act on the ring. If the resultant force has a magnitude of 14 lb, determine the angle θ between the forces.

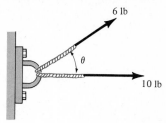

Prob. 2–18

2–19. Two forces $\mathbf{F}_1$ and $\mathbf{F}_2$ act on the hook. If their lines of action are at an angle θ apart and the magnitude of each force is $F_1 = F_2 = F$, determine the magnitude of the resultant force $\mathbf{F}_R$ and the angle between $\mathbf{F}_R$ and $\mathbf{F}_1$.

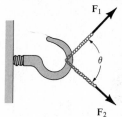

Prob. 2–19

*2–20. Determine the magnitude and direction of the resultant $\mathbf{F}_R = \mathbf{F}_1 + \mathbf{F}_2 + \mathbf{F}_3$ of the three forces by first finding the resultant $\mathbf{F}' = \mathbf{F}_1 + \mathbf{F}_2$ and then forming $\mathbf{F}_R = \mathbf{F}' + \mathbf{F}_3$.

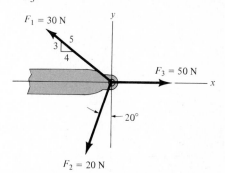

Probs. 2–20 / 2–21

23

2–21. Determine the magnitude and direction of the resultant $\mathbf{F}_R = \mathbf{F}_1 + \mathbf{F}_2 + \mathbf{F}_3$ of the three forces by first finding the resultant $\mathbf{F}' = \mathbf{F}_2 + \mathbf{F}_3$ and then forming $\mathbf{F}_R = \mathbf{F}' + \mathbf{F}_1$.

2–22. The log is being towed by two tractors A and B. Determine the magnitudes of the two towing forces $\mathbf{F}_A$ and $\mathbf{F}_B$ if it is required that the resultant force have a magnitude $F_R = 10$ kN and be directed along the x axis. Set $\theta = 15°$.

*2–24. The power line BC can be hoisted over the pole provided the resultant force acting on it is 150 lb. In order to develop this loading, three cables are attached to the line at A. If two of the cables are subjected to known forces, as shown, determine the direction θ ($0° \leq \theta \leq 90°$) of the third cable so that the magnitude of force $\mathbf{F}_3$ is a minimum. What is the magnitude of $\mathbf{F}_3$? All forces lie in the same plane. *Hint:* First determine the resultant of $\mathbf{F}_1$ and $\mathbf{F}_2$. $\mathbf{F}_3$ acts in the *same direction* θ as this resultant.

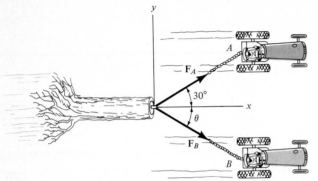

Probs. 2–22 / 2–23

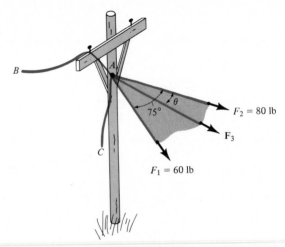

Prob. 2–24

2–23. If the resultant $\mathbf{F}_R$ of the two forces acting on the log is to be directed along the positive x axis and have a magnitude of 10 kN, determine the angle θ of the cable attached to B such that the force $\mathbf{F}_B$ in this cable is a *minimum*. What is the magnitude of force in each cable for this situation?

2.4 Addition of a System of Coplanar Forces

If a force $\mathbf{F}$ lies in the x-y plane, Fig. 2–14, the force may be resolved into two *rectangular components* $\mathbf{F}_x$ and $\mathbf{F}_y$, which lie along the x and y axes, respectively. By the parallelogram law, we require

$$\mathbf{F} = \mathbf{F}_x + \mathbf{F}_y$$

This method of resolving a force into its x and y components can now be used to determine the resultant sum of several *coplanar forces,* i.e., forces that all lie in the same plane. To do this, each force is first resolved into its x and y components and then the respective components are added using *scalar algebra* since they are collinear. The resultant force is formed by adding the

Fig. 2–14

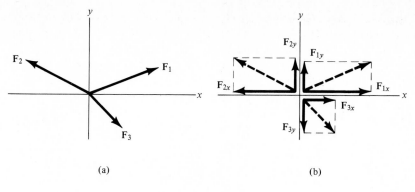

(a)

(b)

Fig. 2–15

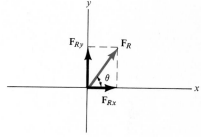

(c)

resultants of the x and y components using the parallelogram law. For example, consider the three forces in Fig. 2–15a, which have x and y components as shown in Fig. 2–15b. The sense (arrowhead) of each component will be designated analytically by a *plus sign* if it acts in the *positive x* or y direction, and by a *negative sign* if it acts in the *negative x* or y direction. Thus, adding the respective x and y components *algebraically,* since they are *collinear,* we have

$$F_{Rx} = F_{1x} - F_{2x} + F_{3x}$$
$$F_{Ry} = F_{1y} + F_{2y} - F_{3y}$$

The resultant force is then determined graphically from the parallelogram law as shown in Fig. 2–15c.

In the general case, the x and y components of the resultant of any number of coplanar forces can be represented symbolically by the scalar sums

$$F_{Rx} = \Sigma F_x$$
$$F_{Ry} = \Sigma F_y$$

$$(2-1)$$

The *magnitude* of $\mathbf{F}_R$ can be determined from the Pythagorean theorem, Fig. 2–15c, that is,

$$F_R = \sqrt{F_{Rx}^2 + F_{Ry}^2}$$

If the angle θ is used to specify the direction of $\mathbf{F}_R$, then by trigonometry

$$\theta = \tan^{-1}\left(\frac{F_{Ry}}{F_{Rx}}\right)$$

The above concepts are illustrated numerically in the following examples.

Example 2–5

Determine the x and y components of $\mathbf{F}_1$ and $\mathbf{F}_2$ shown in Fig. 2–16a.

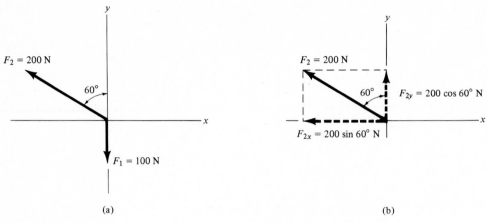

(a)   (b)

Fig. 2–16

Solution

Since $\mathbf{F}_1$ acts along the negative y axis, and the magnitude of $\mathbf{F}_1$ is 100 N, then

$$F_{1x} = 0, \qquad F_{1y} = -100 \text{ N} \qquad \qquad Ans.$$

These results may *also* be expressed using positive scalars to indicate the magnitude and arrows to indicate the direction and sense, so that

$$F_{1x} = 0, \qquad F_{1y} = 100 \text{ N} \downarrow$$
$$Ans.$$

By the parallelogram law, $\mathbf{F}_2$ is resolved into x and y components, Fig. 2–16b. The magnitude of each component is determined by trigonometry. Since $\mathbf{F}_{2x}$ acts in the $-x$ direction, and $\mathbf{F}_{2y}$ acts in the $+y$ direction, we have

$$F_{2x} = -200 \sin 60° = -173 \text{ N} = 173 \text{ N} \leftarrow \qquad Ans.$$
$$F_{2y} = 200 \cos 60° = 100 \text{ N} = 100 \text{ N} \uparrow \qquad Ans.$$

Example 2–6

Determine the x and y components of the force $\mathbf{F}$ shown in Fig. 2–17a.

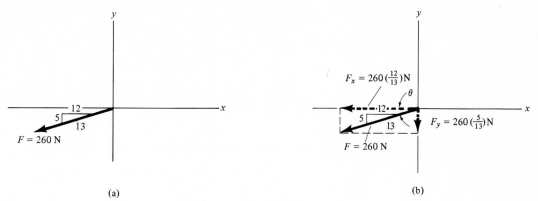

(a)   (b)

Fig. 2–17

Solution

The force is resolved into its x and y components as shown in Fig. 2–17b. Here the *slope* of the line of action for the force is indicated. From this "slope triangle" we could obtain the direction angle θ, e.g., $\theta = \tan^{-1}\left(\frac{5}{12}\right)$, and then proceed to determine the magnitudes of the components in the same manner as for $\mathbf{F}_2$ in Example 2–5. An easier method, however, consists of using proportional parts of similar triangles, i.e.,

$$\frac{F_x}{260} = \frac{12}{13} \qquad F_x = 260\left(\frac{12}{13}\right) = 240 \text{ N}$$

Similarly,

$$F_y = 260\left(\frac{5}{13}\right) = 100 \text{ N}$$

Notice that the magnitude of the *horizontal component*, F_x, was obtained by multiplying the force magnitude by the ratio of the *horizontal leg* of the slope triangle divided by the hypotenuse; whereas the magnitude of the *vertical component*, F_y, was obtained by multiplying the force magnitude by the ratio of the *vertical leg* divided by the hypotenuse. Hence,

$$F_x = -240 \text{ N} = 240 \text{ N} \leftarrow \qquad\qquad Ans.$$
$$F_y = -100 \text{ N} = 100 \text{ N} \downarrow \qquad\qquad Ans.$$

Example 2–7

   The pin in Fig. 2–18*a* is subjected to two forces **F**$_1$ and **F**$_2$. Determine the magnitude and direction of the resultant force.

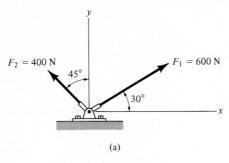

(a)

Fig. 2–18

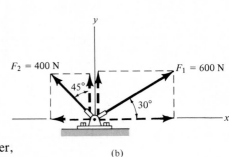

(b)

Solution
   This problem can be solved by using the parallelogram law; however, here we will resolve each force into its *x* and *y* components, Fig. 2–18*b*, and sum these components algebraically. Indicating the "positive" sense of the *x* and *y* force components alongside the equations, we have

$\xrightarrow{+} F_{Rx} = \Sigma F_x;$ $F_{Rx} = 600 \cos 30° - 400 \sin 45°$
$= 236.8 \text{ N} \rightarrow$

$+ \uparrow F_{Ry} = \Sigma F_y;$ $F_{Ry} = 600 \sin 30° + 400 \cos 45°$
$= 582.8 \text{ N} \uparrow$

The resultant force, shown in Fig. 2–18*c*, has a *magnitude* of

$$F_R = \sqrt{(236.8)^2 + (582.8)^2}$$
$$= 629 \text{ N} \qquad\qquad Ans.$$

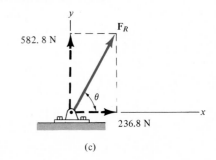

(c)

The *direction*, defined by the angle θ in Fig. 2–18*c*, is

$$\theta = \tan^{-1}\left(\frac{582.8}{236.8}\right) = 67.9° \qquad Ans.$$

Example 2–8

The end of the boom O in Fig. 2–19a is subjected to three concurrent and coplanar forces. Determine the magnitude and direction of the resultant force.

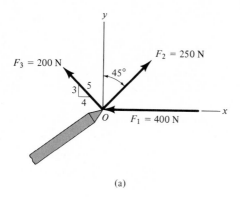

(a)

Fig. 2–19

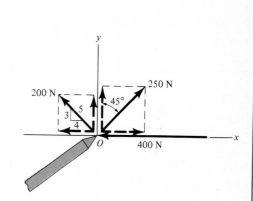

(b)

Solution

Each force is resolved into its x and y components as shown in Fig. 2–19b. Summing the x components, we have

$$\xrightarrow{+} F_{Rx} = \Sigma F_x; \qquad F_{Rx} = -400 + 250 \sin 45° - 200(\tfrac{4}{5})$$
$$= -383.2 \text{ N} = 383.2 \text{ N} \leftarrow$$

The negative sign indicates that F_{Rx} acts to the left, i.e., in the negative x direction as noted by the small arrow.

$$+\uparrow F_{Ry} = \Sigma F_y; \qquad F_{Ry} = 250 \cos 45° + 200(\tfrac{3}{5})$$
$$= 296.8 \text{ N} \uparrow$$

The resultant force, shown in Fig. 2–19c, has a *magnitude* of

$$F_R = \sqrt{(-383.2)^2 + (296.8)^2}$$
$$= 485 \text{ N} \qquad\qquad \textit{Ans.}$$

The *direction*, defined by the angle θ in Fig. 2–19c, is

$$\theta = \tan^{-1}\left(\frac{296.8}{383.2}\right) = 37.8° \qquad \textit{Ans.}$$

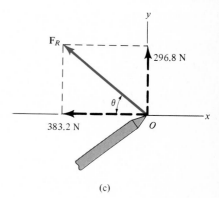

(c)

Note that $\mathbf{F}_R$ creates the *same effect* on the boom as the three forces in Fig. 2–19a.

Problems

2–25. Determine the x and y components of the 30-kN force.

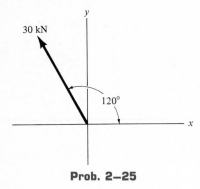

Prob. 2–25

2-26. Determine the x and y components of the 150-lb force.

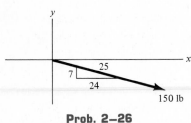

Prob. 2–26

2–27. Determine the magnitude of the resultant force and its direction measured from the positive x axis.

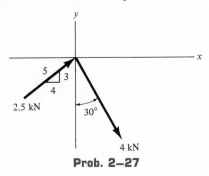

Prob. 2–27

*2–28. Determine the magnitude of the resultant force and its direction measured from the positive x axis.

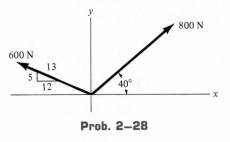

Prob. 2–28

2–29. Determine the magnitude of the resultant force and its direction measured from the positive x axis.

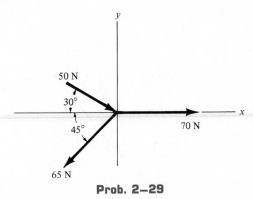

Prob. 2–29

2–30. Determine the magnitude of the resultant force and its direction measured from the positive x axis.

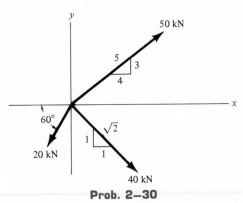

Prob. 2–30

2–31. A force of 23 kN is developed by the main rotor of a helicopter while it is flying forward. Resolve this force into its x and y components and explain what physical effects on the helicopter are caused by each of these components.

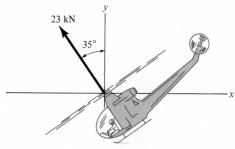

Prob. 2–31

2–33. Three concurrent forces act on the ring as shown. If each has a magnitude of 60 N, find the resultant force.

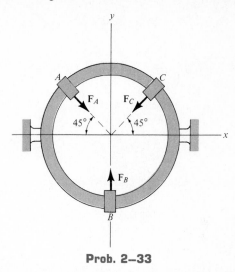

Prob. 2–33

*2–32. Determine the x and y components of the 800-lb force.

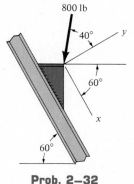

Prob. 2–32

2–34. Determine the x and y components of each force acting on the *gusset plate* of the bridge truss. Show that the resultant force is zero.

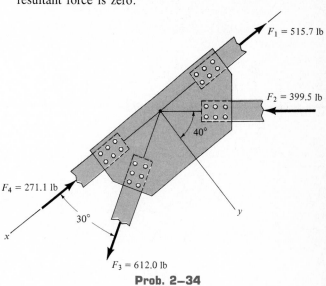

Prob. 2–34

2–35. Determine the magnitude and direction θ of $\mathbf{F}_1$ so that the resultant force is directed vertically upward and has a magnitude of 800 N.

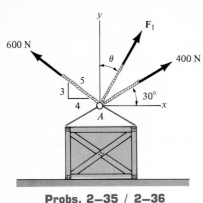

Probs. 2–35 / 2–36

2–39. The three concurrent forces acting on the post produce a resultant force $\mathbf{F}_R = \mathbf{0}$. If $F_2 = \frac{1}{2}F_1$, and F_1 is to be $90°$ from F_2 as shown, determine the required magnitude of F_3 expressed in terms of F_1 and the angle θ.

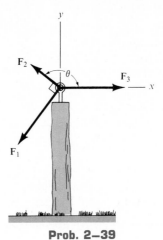

Prob. 2–39

*2–36. Determine the magnitude and direction of the resultant force of the three forces acting on the ring A. Take $F_1 = 500$ N and $\theta = 20°$.

2–37. Three forces act on the bracket. Determine the magnitude and direction θ of $\mathbf{F}_1$ so that the resultant force is directed along the positive x' axis and has a magnitude of 1000 N.

*2–40. Determine the magnitude of force $\mathbf{F}$ so that the resultant $\mathbf{F}_R$ of the three forces is as small as possible. What is the minimum magnitude of $\mathbf{F}_R$?

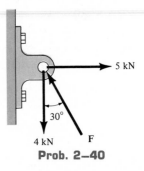

Prob. 2–40

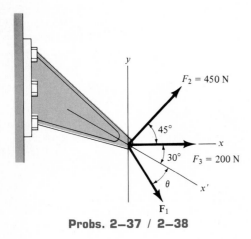

Probs. 2–37 / 2–38

2–38. If $F_1 = 300$ N and $\theta = 20°$, determine the magnitude and direction, measured from the x' axis, of the resultant force of the three forces acting on the bracket.

32

Cartesian Vectors 2.5

The operations of vector algebra, when applied to solving problems in three dimensions, are greatly simplified if the vectors are first represented in Cartesian vector form. In this section we will present a general method for doing this, then in Sec. 2.6 we will apply this method to solving problems involving the addition of forces. Similar applications will be illustrated for the position and moment vectors given in later sections of the text.

Right-Handed Coordinate System. A right-handed coordinate system will be used in developing the theory of vector algebra that follows. A rectangular or Cartesian coordinate system is said to be right-handed provided the thumb of the right hand points in the direction of the positive z axis when the right-hand fingers are curled about the axis and directed from the positive x to the positive y axis, Fig. 2–20. Furthermore, according to this rule, the z axis in Fig. 2–19 is directed outward, perpendicular to the page.

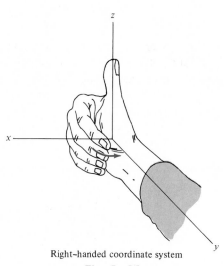

Right-handed coordinate system

Fig. 2–20

Rectangular Components of a Vector. A vector $\mathbf{A}$ may have one, two, or three rectangular components along the x, y, z coordinate axes, depending upon how the vector is oriented relative to the axes. In general, though, when $\mathbf{A}$ is directed within an octant of the x, y, z frame, Fig. 2–21, then by two successive applications of the parallelogram law, we may resolve the components as $\mathbf{A} = \mathbf{A}' + \mathbf{A}_z$ and $\mathbf{A}' = \mathbf{A}_x + \mathbf{A}_y$. Combining these equations, $\mathbf{A}$ is then represented by the vector sum of its *three* rectangular components,

$$\mathbf{A} = \mathbf{A}_x + \mathbf{A}_y + \mathbf{A}_z \qquad (2-2)$$

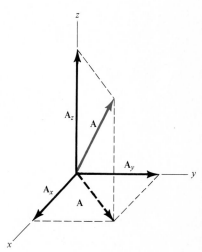

Fig. 2–21

33

Unit Vector. A *unit vector* is a vector having a magnitude of 1. If **A** is a vector having a magnitude $A \neq 0$, then a unit vector having the *same direction* as **A** is represented by*

$$\mathbf{u}_A = \frac{\mathbf{A}}{A} \qquad (2\text{--}3)$$

Rewriting this equation gives

$$\mathbf{A} = A\mathbf{u}_A \qquad (2\text{--}4)$$

Since vector **A** is of a certain kind, e.g., a force vector, it is customary to use the proper set of units for its description. The magnitude A also has this same set of units; hence, from Eq. 2–3, the *unit vector will be dimensionless* since the units will cancel out. Equation 2–4 indicates that vector **A** may therefore be expressed in terms of both its magnitude and direction *separately;* i.e., A (a scalar) defines the *magnitude* of **A**, and $\mathbf{u}_A$ (a dimensionless vector) defines the *direction* of **A**, Fig. 2–22.

Fig. 2–22

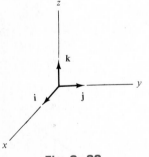

Fig. 2–23

Cartesian Unit Vectors. In order to simplify vector-algebra operations, a set of Cartesian unit vectors, **i, j, k,** will be used to designate the directions of the x, y, z axes respectively. The *sense* (or arrowhead) of these vectors will be described analytically by a plus or minus sign, depending upon whether they are pointing along the positive or negative x, y, or z axis. Thus, the positive unit vectors are shown in Fig. 2–23.

*For handwritten work unit vectors are usually indicated using a circumflex, e.g., $\hat{u}_A$.

Cartesian Vector Representation. Using the Cartesian unit vectors, the three vector components of Eq. 2–2 may be written in "Cartesian vector form." Since the components act in the positive **i, j,** and **k** directions, Fig. 2–24, we have

$$\mathbf{A} = A_x\mathbf{i} + A_y\mathbf{j} + A_z\mathbf{k} \tag{2–5}$$

There is a distinct advantage to writing vectors in terms of their Cartesian components. In doing so, the *magnitude* and *direction* of each *component vector* are *separated,* and as stated previously, this will simplify the operations of vector algebra, particularly in three dimensions.

Magnitude of a Cartesian Vector. It is always possible to obtain the magnitude of **A** if the vector is expressed in Cartesian vector form. As shown in Fig. 2–25 from the shaded right triangle, the magnitude $A' = \sqrt{A_x^2 + A_y^2}$. Similarly, from the colored right triangle, $A = \sqrt{A'^2 + A_z^2}$. Combining these equations yields

$$A = \sqrt{A_x^2 + A_y^2 + A_z^2} \tag{2–6}$$

*Hence, the magnitude of **A** is equal to the positive square root of the sum of the squares of its components.*

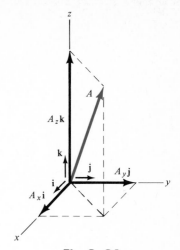

Fig. 2–24

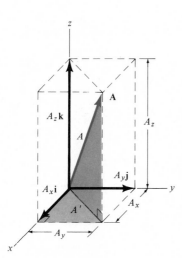

Fig. 2–25

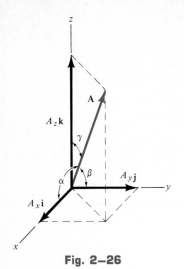

Fig. 2–26

Direction of a Cartesian Vector. The *direction* of vector **A** is defined by the *coordinate direction angles* α (alpha), β (beta), and γ (gamma), measured between the *tail* of **A** and the *positive x, y, z* axes, Fig. 2–26. Note that regardless of where **A** is directed, each of these angles will be between 0° and 180°. To determine α, β, and γ, consider the projection of **A** onto the *x, y, z* axes, Fig. 2–27. Referring to the colored right triangles shown in each figure, we have

$$\cos \alpha = \frac{A_x}{A} \qquad \cos \beta = \frac{A_y}{A} \qquad \cos \gamma = \frac{A_z}{A} \qquad (2\text{–}7)$$

These numbers are known as the *direction cosines* of **A.** Once they have been obtained, the coordinate direction angles α, β, γ can then be determined from the inverse cosines.

An easy way of obtaining the direction cosines of **A** *is to form a unit vector in the direction of* **A.** Provided **A** is expressed in Cartesian vector form as $\mathbf{A} = A_x\mathbf{i} + A_y\mathbf{j} + A_z\mathbf{k}$, (Eq. 2–5) we have

$$\mathbf{u}_A = \frac{\mathbf{A}}{A} = \frac{A_x}{A}\mathbf{i} + \frac{A_y}{A}\mathbf{j} + \frac{A_z}{A}\mathbf{k} \qquad (2\text{–}8)$$

where $A = \sqrt{(A_x)^2 + (A_y)^2 + (A_z)^2}$ (Eq. 2–6). By comparison with Eqs. 2–7, it is seen that *the* **i, j,** *and* **k** *components of* $\mathbf{u}_A$ *represent the direction cosines* of **A,** i.e.,

$$\mathbf{u}_A = \cos \alpha\mathbf{i} + \cos \beta\mathbf{j} + \cos \gamma\mathbf{k} \qquad (2\text{–}9)$$

Since the magnitude of a vector is equal to the square root of the sum of the squares of the magnitudes of its components, and $\mathbf{u}_A$ has a magnitude of 1,

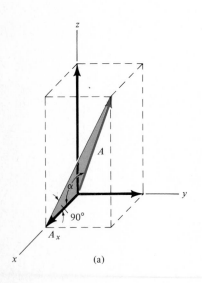

(a)

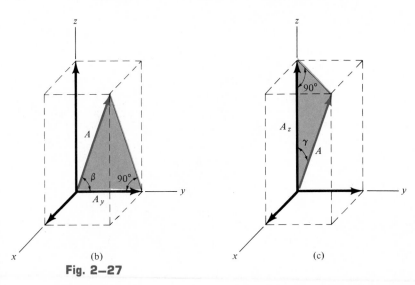

(b) (c)

Fig. 2–27

then from Eq. 2–9 an important relation between the direction cosines can be formulated as

$$\cos^2 \alpha + \cos^2 \beta + \cos^2 \gamma = 1 \tag{2–10}$$

This equation is useful for determining one of the coordinate direction angles if the other two are known.

Finally, if the magnitude and coordinate direction angles of **A** are given, **A** may be expressed in Cartesian vector form as

$$
\begin{aligned}
\mathbf{A} &= A\mathbf{u}_A \\
&= A \cos \alpha \mathbf{i} + A \cos \beta \mathbf{j} + A \cos \gamma \mathbf{k} \\
&= A_x \mathbf{i} + A_y \mathbf{j} + A_z \mathbf{k}
\end{aligned}
\tag{2–11}
$$

Addition and Subtraction of Cartesian Vectors   2.6

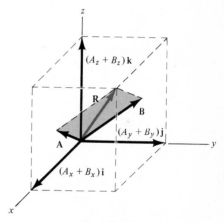

Fig. 2–28

The vector operations of addition and subtraction of two or more vectors are greatly simplified if the vectors are expressed in terms of their Cartesian components. For example, consider the two vectors **A** and **B**, both of which are directed within the positive octant of the x, y, z frame, Fig. 2–28. If $\mathbf{A} = A_x\mathbf{i} + A_y\mathbf{j} + A_z\mathbf{k}$ and $\mathbf{B} = B_x\mathbf{i} + B_y\mathbf{j} + B_z\mathbf{k}$, then the resultant vector, **R**, has components which represent the scalar sums of the **i**, **j** and **k** components of **A** and **B**, i.e.,

$$\mathbf{R} = \mathbf{A} + \mathbf{B} = (A_x + B_x)\mathbf{i} + (A_y + B_y)\mathbf{j} + (A_z + B_z)\mathbf{k}$$

Vector subtraction, being a special case of vector addition, simply requires a scalar subtraction of the respective **i**, **j**, and **k** components of **A** and **B**. For example,

$$\mathbf{R}' = \mathbf{A} - \mathbf{B} = (A_x - B_x)\mathbf{i} + (A_y - B_y)\mathbf{j} + (A_z - B_z)\mathbf{k}$$

Concurrent Force Systems. In particular, the above concept of vector addition may be generalized and applied to a system of several concurrent forces. In this case, the force resultant of the system is the vector sum of all the forces in the system and can be written as

$$\mathbf{F}_R = \Sigma \mathbf{F} = \Sigma F_x \mathbf{i} + \Sigma F_y \mathbf{j} + \Sigma F_z \mathbf{k} \tag{2–12}$$

Here ΣF_x, ΣF_y, and ΣF_z represent the algebraic sums of the respective x, y, z, or **i**, **j**, **k** components of each force in the system.

The following examples numerically illustrate the methods used to apply the above theory to the solution of problems involving force as a vector quantity.

Example 2–9

Express force **F** shown in Fig. 2–29 in Cartesian vector form.

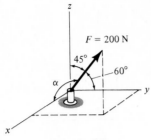

Fig. 2–29

Solution

Since only two coordinate direction angles are specified, Fig. 2–29, the third angle α is determined from Eq. 2–10, i.e.,

$$\cos^2 \alpha + \cos^2 \beta + \cos^2 \gamma = 1$$
$$\cos^2 \alpha + \cos^2 60° + \cos^2 45° = 1$$
$$\cos \alpha = \sqrt{1 - (0.707)^2 - (0.5)^2} = \pm 0.5$$

Hence,

$$\alpha = \cos^{-1}(0.5) = 60° \qquad \text{or} \qquad \alpha = \cos^{-1}(-0.5) = 120°$$

By inspection of Fig. 2–29, however, it is necessary that $\alpha = 60°$.

Using Eq. 2–9, with $F = 200$ N, we have

$$\mathbf{F} = F \cos \alpha \mathbf{i} + F \cos \beta \mathbf{j} + F \cos \gamma \mathbf{k}$$
$$= 200 \cos 60° \mathbf{i} + 200 \cos 60° \mathbf{j} + 200 \cos 45° \mathbf{k}$$
$$= \{100.0\mathbf{i} + 100.0\mathbf{j} + 141.4\mathbf{k}\} \text{ N} \qquad\qquad \textit{Ans.}$$

By applying Eq. 2–6, note that indeed the magnitude of $F = 200$ N.

$$F = \sqrt{(F_x)^2 + (F_y)^2 + (F_z)^2}$$
$$= \sqrt{(100.0)^2 + (100.0)^2 + (141.4)^2} = 200 \text{ N}$$

Example 2–10

Express **F** shown acting on the hook in Fig. 2–30*a* in Cartesian vector form.

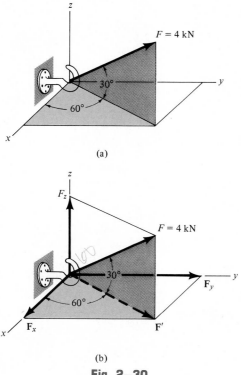

(a)

(b)

Fig. 2–30

Solution

In this case the angles 60° and 30° defining the direction of **F** are *not* coordinate direction angles. Why? By two successive applications of the parallelogram law, however, **F** can be resolved into its *x*, *y*, *z* components as shown in Fig. 2–30*b*. First, from the shaded triangle,

$$F' = 4 \cos 30° = 3.46 \text{ kN}$$
$$F_z = 4 \sin 30° = 2.00 \text{ kN}$$

Next, using **F**′ and the colored triangle,

$$F_x = 3.46 \cos 60° = 1.73 \text{ kN}$$
$$F_y = 3.46 \sin 60° = 3.00 \text{ kN}$$

Thus,

$$\mathbf{F} = \{1.73\mathbf{i} + 3.00\mathbf{j} + 2.00\mathbf{k}\} \text{ kN} \qquad \textit{Ans.}$$

Example 2–11

Express force **F** shown in Fig. 2–31a in Cartesian vector form.

Solution

As in Example 2–10, the angles of 60° and 45° defining the direction of **F** are *not* coordinate direction angles. The two successive applications of the parallelogram law needed to resolve **F** into its x, y, z components are shown in Fig. 2–31b. By trigonometry, the magnitudes of the components are

$$F_z = 100 \sin 60° = 86.6 \text{ lb}$$
$$F' = 100 \cos 60° = 50 \text{ lb}$$
$$F_x = 50 \cos 45° = 35.4 \text{ lb}$$
$$F_y = 50 \sin 45° = 35.4 \text{ lb}$$

Realizing that **F**$_y$ has a direction defined by $-\mathbf{j}$, we have

$$\mathbf{F} = \mathbf{F}_x + \mathbf{F}_y + \mathbf{F}_z$$
$$\mathbf{F} = \{35.4\mathbf{i} - 35.4\mathbf{j} + 86.6\mathbf{k}\} \text{ lb} \qquad \textit{Ans.}$$

To show that the magnitude of this vector is indeed 100 lb, apply Eq. 2–6,

$$F = \sqrt{F_x^2 + F_y^2 + F_z^2}$$
$$= \sqrt{(35.4)^2 + (-35.4)^2 + (86.6)^2} = 100 \text{ lb}$$

If needed, the coordinate direction angles of **F** can be determined from the components of the unit vector acting in the direction of **F**. Hence,

$$\mathbf{u} = \frac{\mathbf{F}}{F} = \frac{F_x}{F}\mathbf{i} + \frac{F_y}{F}\mathbf{j} + \frac{F_z}{F}\mathbf{k}$$

$$= \frac{35.4}{100}\mathbf{i} - \frac{35.4}{100}\mathbf{j} + \frac{86.6}{100}\mathbf{k}$$

$$= 0.354\mathbf{i} - 0.354\mathbf{j} + 0.866\mathbf{k}$$

so that

$$\cos \alpha = 0.354$$
$$\cos \beta = -0.354$$
$$\cos \gamma = 0.866$$

or

$$\alpha = \cos^{-1}(0.354) = 69.3°$$
$$\beta = \cos^{-1}(-0.354) = 110.7°$$
$$\gamma = \cos^{-1}(0.866) = 30.0°$$

These results are shown in Fig. 2–31c.

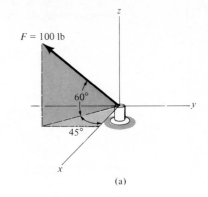

(a)

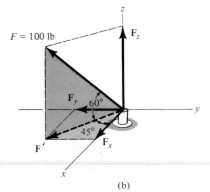

(b)

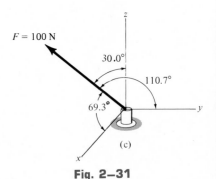

(c)

Fig. 2–31

Example 2–12

Determine the magnitudes and the coordinate direction angles of $\mathbf{F}_1$ and $\mathbf{F}_2$ acting on the ring in Fig. 2–32a. Also, find the resultant force.

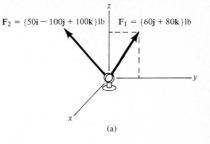

(a)

Solution

Since each force is represented in Cartesian vector form, its magnitude is found from Eq. 2–6. The coordinate direction angles α, β, γ are determined from the components of unit vectors acting in the direction of each force, using Eqs. 2–7 and 2–8. Hence, for $\mathbf{F}_1$,

$$\mathbf{F}_1 = \{60\mathbf{j} + 80\mathbf{k}\} \text{ lb}$$
$$F_1 = \sqrt{(60)^2 + (80)^2} = 100 \text{ lb} \qquad \textit{Ans.}$$

and

$$\mathbf{u}_{F_1} = \frac{\mathbf{F}_1}{F_1} = \frac{60}{100}\mathbf{j} + \frac{80}{100}\mathbf{k}$$
$$= 0.6\mathbf{j} + 0.8\mathbf{k}$$

so that

$$\cos \alpha_1 = 0 \qquad \alpha_1 = 90° \qquad \textit{Ans.}$$
$$\cos \beta_1 = 0.6 \qquad \beta_1 = 53.1° \qquad \textit{Ans.}$$
$$\cos \gamma_1 = 0.8 \qquad \gamma_1 = 36.9° \qquad \textit{Ans.}$$

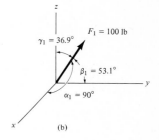

(b)

The results are shown in Fig. 2–32b.

In a similar manner for $\mathbf{F}_2$ we have

$$\mathbf{F}_2 = \{50\mathbf{i} - 100\mathbf{j} + 100\mathbf{k}\} \text{ lb}$$
$$F_2 = \sqrt{(50)^2 + (-100)^2 + (100)^2} = 150 \text{ lb} \qquad \textit{Ans.}$$

and

$$\mathbf{u}_{F_2} = \frac{\mathbf{F}_2}{F_2} = \frac{50}{150}\mathbf{i} - \frac{100}{150}\mathbf{j} + \frac{100}{150}\mathbf{k}$$
$$= 0.333\mathbf{i} - 0.667\mathbf{j} + 0.667\mathbf{k}$$

so that

$$\cos \alpha_2 = 0.333 \qquad \alpha_2 = 70.5° \qquad \textit{Ans.}$$
$$\cos \beta_2 = -0.667 \qquad \beta_2 = 131.8° \qquad \textit{Ans.}$$
$$\cos \gamma_2 = 0.667 \qquad \gamma_2 = 48.2° \qquad \textit{Ans.}$$

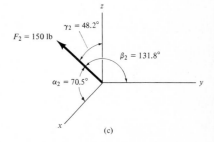

(c)

The results are shown in Fig. 2–32c.

The resultant force, shown in Fig. 2–32d, is

$$\mathbf{F}_R = \Sigma\mathbf{F} = \mathbf{F}_1 + \mathbf{F}_2 = (60\mathbf{j} + 80\mathbf{k}) + (50\mathbf{i} - 100\mathbf{j} + 100\mathbf{k})$$
$$= \{50\mathbf{i} - 40\mathbf{j} + 180\mathbf{k}\} \text{ lb} \qquad \textit{Ans.}$$

How would you find the magnitude and coordinate direction angles of $\mathbf{F}_R$?

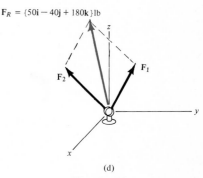

(d)

Fig. 2–32

Example 2–13

Two forces act on the pipe shown in Fig. 2–33*a*. Specify the direction of F_2 so that the resultant force $\mathbf{F}_R$ acts along the positive *y* axis and has a magnitude of 800 N.

Solution

To solve this problem, the resultant force and its two components, $\mathbf{F}_1$ and $\mathbf{F}_2$, will each be expressed in Cartesian vector form. Then, as shown in Fig. 2–33*b*, it is necessary that $\mathbf{F}_R = \mathbf{F}_1 + \mathbf{F}_2$.

Applying Eq. 2–11,

$$\mathbf{F}_1 = F_1 \mathbf{u}_{F_1} = F_1 \cos \alpha_1 \mathbf{i} + F_1 \cos \beta_1 \mathbf{j} + F_1 \cos \gamma_1 \mathbf{k}$$
$$= 300 \cos 45°\mathbf{i} + 300 \cos 60°\mathbf{j} + 300 \cos 120°\mathbf{k}$$
$$= \{212.1\mathbf{i} + 150\mathbf{j} - 150\mathbf{k}\} \text{ N}$$

$$\mathbf{F}_2 = F_2 \mathbf{u}_{F_2} = 700 \cos \alpha_2 \mathbf{i} + 700 \cos \beta_2 \mathbf{j} + 700 \cos \gamma_2 \mathbf{k}$$

According to the problem statement, the resultant force $\mathbf{F}_R$ has a magnitude of 800 N and acts in the $+\mathbf{j}$ direction. Hence,

$$\mathbf{F}_R = (800 \text{ N})(+\mathbf{j}) = \{800\mathbf{j}\} \text{ N}$$

It is required that

$$\mathbf{F}_R = \mathbf{F}_1 + \mathbf{F}_2$$
$$800\mathbf{j} = 212.1\mathbf{i} + 150\mathbf{j} - 150\mathbf{k} + 700 \cos \alpha_2 \mathbf{i} + 700 \cos \beta_2 \mathbf{j}$$
$$+ 700 \cos \gamma_2 \mathbf{k}$$
$$800\mathbf{j} = (212.1 + 700 \cos \alpha_2)\mathbf{i} + (150 + 700 \cos \beta_2)\mathbf{j}$$
$$+ (-150 + 700 \cos \gamma_2)\mathbf{k}$$

To satisfy this equation, the corresponding **i**, **j**, and **k** components on the left and right sides must be equal. This is equivalent to stating that the *x*, *y*, *z* components of $\mathbf{F}_R$ be equal to the corresponding *x*, *y*, *z* components of $(\mathbf{F}_1 + \mathbf{F}_2)$. Hence,

$$0 = 212.1 + 700 \cos \alpha_2 \qquad \alpha_2 = \cos^{-1}\left(\frac{-212.1}{700}\right) = 107.6° \qquad Ans.$$

$$800 = 150 + 700 \cos \beta_2 \qquad \beta_2 = \cos^{-1}\left(\frac{650}{700}\right) = 21.8° \qquad Ans.$$

$$0 = -150 + 700 \cos \gamma_2 \qquad \gamma_2 = \cos^{-1}\left(\frac{150}{700}\right) = 77.6° \qquad Ans.$$

The results are shown in Fig. 2–33*b*.

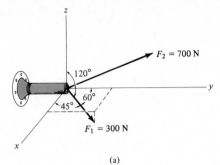

(a)

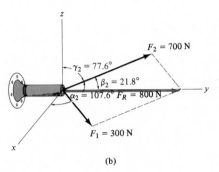

(b)

Fig. 2–33

Problems

2–41. Determine the magnitude and coordinate direction angles of the resultant force.

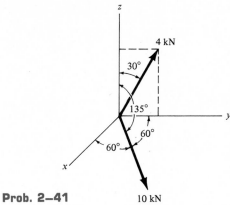

Prob. 2–41

2–42. Determine the magnitude and coordinate direction angles of the resultant force.

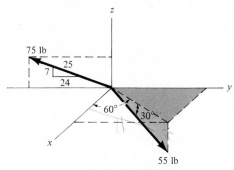

Prob. 2–42

2–43. Express each force in Cartesian vector form.

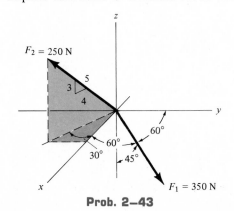

Prob. 2–43

*2–44. The man pulls on the rope with a force of 60 lb. If **F** acts within the octant shown, such that $\alpha = 45°$, $\beta = 60°$, determine the x, y, z components of **F**.

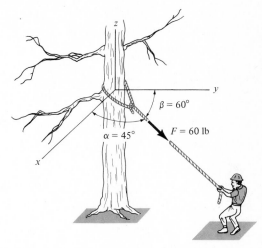

Prob. 2–44

2–45. The beam is subjected to the two forces shown. Express each force in Cartesian vector form and determine the magnitude and coordinate direction angles of the resultant force.

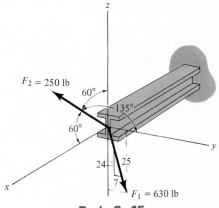

Prob. 2–45

2–46. The cables attached to the screw eye are subjected to the three forces shown. Express each force in Cartesian vector form and determine the magnitude and coordinate direction angles of the resultant force.

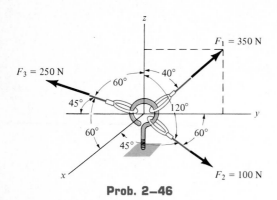

Prob. 2–46

2–47. The mast is subjected to the three forces shown. Determine the coordinate direction angles α_1, β_1, γ_1 of $\mathbf{F}_1$ so that the resultant force acting on the mast is $\mathbf{F}_R = \{350\mathbf{i}\}$ N.

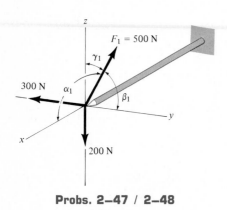

Probs. 2–47 / 2–48

***2–48.** The mast is subjected to the three forces shown. Determine the coordinate direction angles α_1, β_1, γ_1 of $\mathbf{F}_1$ so that the resultant force acting on the mast is zero.

2–49. The hook is subjected to the cable force $\mathbf{F}$ which has a component along the x axis of $F_x = 60$ N, a component along the z axis of $F_z = -80$ N, and a coordinate direction angle of $\beta = 80°$. Determine the magnitude of $\mathbf{F}$.

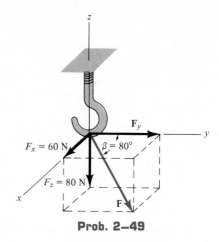

Prob. 2–49

2–50. The bolt is subjected to the force $\mathbf{F}$, which has components acting along the x, y, z axes as shown. If the magnitude of $\mathbf{F}$ is 80 N, and $\alpha = 60°$ and $\gamma = 45°$, determine the magnitudes of its components.

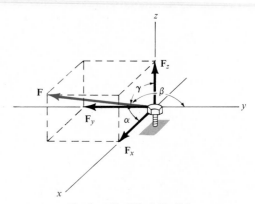

Probs. 2–50 / 2–51

2–51. The bolt is subjected to the force $\mathbf{F}$ which has components $F_x = 20$ N, $F_z = 20$ N. If $\beta = 120°$, determine the magnitudes of $\mathbf{F}$ and $\mathbf{F}_y$.

*2–52. Determine the magnitude and coordinate direction angles of $\mathbf{F}_2$ so that the resultant of the two forces acts along the positive x axis and has a magnitude of 500 N.

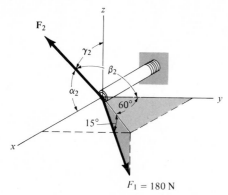

$F_1 = 180$ N

Probs. 2–52 / 2–53

2–53. Determine the magnitude and coordinate direction angles of $\mathbf{F}_2$ so that the resultant of the two forces is zero.

2–54. A force $\mathbf{F}$ is applied at the top of the tower at A. If it acts in the direction shown such that one of its components lying in the shaded y-z plane has a magnitude of 80 lb, determine its magnitude F and coordinate direction angles α, β, γ.

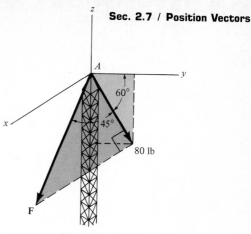

Prob. 2–54

2–55. Two forces $\mathbf{F}_1$ and $\mathbf{F}_2$ act on the bolt. If the resultant force $\mathbf{F}_R$ has a magnitude and coordinate direction angles α and β as shown, determine the magnitude and coordinate direction angles of force $\mathbf{F}_2$.

Prob. 2–55

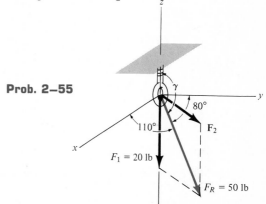

Position Vectors 2.7

In this section we will introduce the concept of a position vector. It will be shown in Sec. 2.8 that this vector is of importance in formulating a Cartesian force vector directed between any two points in space, and in Chapter 4 we will use it for finding the moment of a force.

***x, y, z* Coordinates.** Throughout this text we will use a *right-handed* coordinate system to reference the location of points in space. Furthermore, we will use the convention followed in many technical books, and that is to require the positive z axis to be directed *upward* (the zenith direction) so that it measures the height of an object or the altitude of a point. The x, y axes then lie in the horizontal plane, Fig. 2–34. Points in space are located relative to

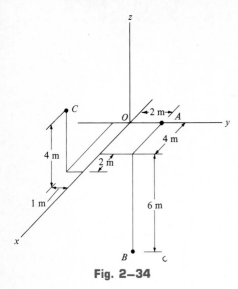

Fig. 2–34

the origin of coordinates, O, by successive measurements along the x, y, z axes. For example, in Fig. 2–34 the coordinates of point A are obtained by starting at O and measuring $x_A = 0$ along the x axis, $y_A = +2$ m along the y axis, and $z_A = 0$ along the z axis. Thus, $A(0, 2, 0)$. In a similar manner, measurements along the x, y, z axes from O to B yield the coordinates of B, i.e., $B(4, 2, -6)$. Also notice that $C(6, -1, 4)$.

Position Vector. The *position vector* $\mathbf{r}$ is defined as a fixed vector which locates a point in space relative to another point. For example, if $\mathbf{r}$ extends from the origin of coordinates, O, to point $P(x, y, z)$, Fig. 2–35a, then $\mathbf{r}$ can be expressed in Cartesian vector form as

$$\mathbf{r} = x\mathbf{i} + y\mathbf{j} + z\mathbf{k}$$

In particular, note how the tip-to-tail vector addition of the three components, going from O to P, yields vector $\mathbf{r}$, as shown in Fig. 2–35b.

In the more general case, the position vector is directed from point A to point B in space, Fig. 2–36a. From the figure, by the tip-to-tail triangle construction, we require

$$\mathbf{r}_A + \mathbf{r} = \mathbf{r}_B$$

Solving for $\mathbf{r}$ and expressing $\mathbf{r}_A$ and $\mathbf{r}_B$ in Cartesian vector form yields

$$\mathbf{r} = \mathbf{r}_B - \mathbf{r}_A = (x_B\mathbf{i} + y_B\mathbf{j} + z_B\mathbf{k}) - (x_A\mathbf{i} + y_A\mathbf{j} + z_A\mathbf{k})$$

or

$$\mathbf{r} = (x_B - x_A)\mathbf{i} + (y_B - y_A)\mathbf{j} + (z_B - z_A)\mathbf{k} \qquad (2\text{–}13)$$

Thus, the $\mathbf{i}$, $\mathbf{j}$, $\mathbf{k}$ components of the position vector $\mathbf{r}$ may be formed by taking the coordinates of the tip of the vector, $B(x_B, y_B, z_B)$, and subtracting from them the corresponding coordinates of the tail, $A(x_A, y_A, z_A)$. Again note how the tip-to-tail addition of these three components, going from A to B, first in the $+x$ direction, then the $+y$ direction, and finally the $+z$ direction, yields $\mathbf{r}$, Fig. 2–36b.

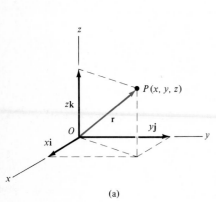

(a)

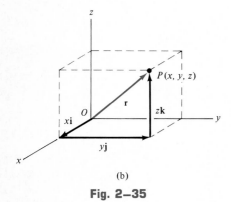

(b)

Fig. 2–35

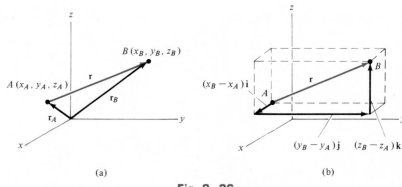

(a)

(b)

Fig. 2–36

Example 2–14

Determine the magnitude and direction of the position vector extending from A to B in Fig. 2–37a.

Solution

The coordinates of A and B are $A(1, 0, -3)$, $B(-2, 2, 3)$. Hence,

$$\mathbf{r} = (-2 - 1)\mathbf{i} + (2 - 0)\mathbf{j} + (3 - (-3))\mathbf{k}$$
$$= \{-3\mathbf{i} + 2\mathbf{j} + 6\mathbf{k}\} \text{ m}$$

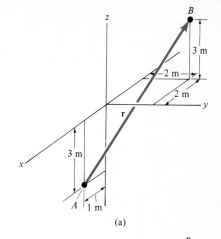

(a)

As shown in Fig. 2–37b, the three components of $\mathbf{r}$ can be obtained in a more *direct manner* by realizing that in going from A to B one must move along the x axis $\{-3\mathbf{i}\}$ m, along the y axis $\{2\mathbf{j}\}$ m, and finally along the z axis $\{6\mathbf{k}\}$ m.

The magnitude of $\mathbf{r}$ is thus

$$r = \sqrt{(-3)^2 + (2)^2 + (6)^2} = 7 \text{ m} \qquad Ans.$$

Formulating a unit vector in the direction of $\mathbf{r}$, we have

$$\mathbf{u} = \frac{\mathbf{r}}{r} = \frac{-3}{7}\mathbf{i} + \frac{2}{7}\mathbf{j} + \frac{6}{7}\mathbf{k}$$

The components of the unit vector yield the coordinate direction angles

$$\alpha = \cos^{-1}\left(\frac{-3}{7}\right) = 115.4° \qquad Ans.$$

$$\beta = \cos^{-1}\left(\frac{2}{7}\right) = 73.4° \qquad Ans.$$

$$\gamma = \cos^{-1}\left(\frac{6}{7}\right) = 31.0° \qquad Ans.$$

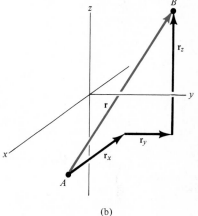

(b)

These angles are measured from a localized coordinate system placed at the tail of $\mathbf{r}$ as shown in Fig. 2–37c.

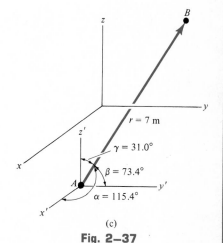

(c)

Fig. 2–37

2.8 Force Vector Directed Along a Line

Quite often in three-dimensional statics problems, the direction of a force is specified by two points through which its line of action passes. Such a situation is shown in Fig. 2–38, where the force **F** is directed along the cord AB. We can formulate **F** as a Cartesian vector by realizing that it acts in the *same direction* as the position vector **r** directed from point A to point B on the cord. This common direction is specified by the *unit vector* **u** = **r**/r. Symbolically,

$$\mathbf{F} = F\mathbf{u} = F\left(\frac{\mathbf{r}}{r}\right)$$

Although the force **F** is shown in Fig. 2–38, note that it does not actually extend from A to B; only the position vector does. This is because **r** has units of length, whereas **F** has units of force.

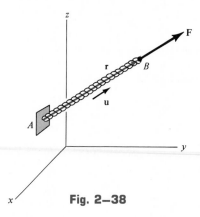

Fig. 2–38

PROCEDURE FOR ANALYSIS

When **F** is directed along a line which extends from point A to point B, then **F** can be expressed in Cartesian vector form as follows:

Position Vector. Determine the position vector **r** directed from A to B, and compute its magnitude r.

Unit Vector. Determine the unit vector **u** = **r**/r which defines the *direction* of *both* **r** and **F**.

Force Vector. Determine **F** by combining its magnitude F and direction **u**, i.e., **F** = F**u.**

This procedure is illustrated numerically in the following example problems.

Example 2–15

The man shown in Fig. 2–39a pulls on the cord with a force of 70 lb. Represent this force, acting on the support A, as a Cartesian vector and determine its direction.

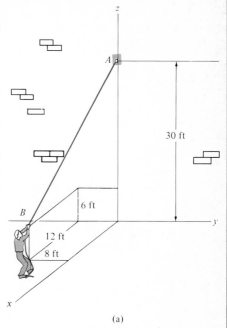

Solution

Force **F** is shown in Fig. 2–39b. The direction of this vector, **u,** is determined from the position vector **r**, which extends from A to B, Fig. 2–39b. To formulate **F** as a Cartesian vector we use the following procedure.

Position Vector. The coordinates of the end points of the cord are A(0, 0, 30) and B(12, −8, 6). Forming the position vector by subtracting the corresponding x, y, and z coordinates of A from those of B, we have

$$\mathbf{r} = (12 - 0)\mathbf{i} + (-8 - 0)\mathbf{j} + (6 - 30)\mathbf{k}$$
$$= \{12\mathbf{i} - 8\mathbf{j} - 24\mathbf{k}\} \text{ ft}$$

Show on Fig. 2–39a how one can write **r** *directly* by going from A $\{12\mathbf{i}\}$ ft, then $\{-8\mathbf{j}\}$ ft, and finally $\{-24\mathbf{k}\}$ ft to get to B.

The magnitude of **r,** which represents the *length* of cord AB, is

$$r = \sqrt{(12)^2 + (-8)^2 + (-24)^2} = 28 \text{ ft}$$

Unit Vector. Forming the unit vector that defines the direction of both **r** and **F** yields

$$\mathbf{u} = \frac{\mathbf{r}}{r} = \frac{12}{28}\mathbf{i} - \frac{8}{28}\mathbf{j} - \frac{24}{28}\mathbf{k}$$

Force Vector. Since **F** has a *magnitude* of 70 lb and a *direction* specified by **u,** then

$$\mathbf{F} = F\mathbf{u} = 70 \text{ lb} \left(\frac{12}{28}\mathbf{i} - \frac{8}{28}\mathbf{j} - \frac{24}{28}\mathbf{k} \right)$$
$$= \{30\mathbf{i} - 20\mathbf{j} - 60\mathbf{k}\} \text{ lb} \qquad\qquad \textit{Ans.}$$

As shown in Fig. 2–39b, the coordinate direction angles are measured between **r** (or **F**) and the *positive axes* of a localized coordinate system with origin placed at A. From the components of the unit vector:

$$\alpha = \cos^{-1}\left(\frac{12}{28}\right) = 64.6° \qquad\qquad \textit{Ans.}$$

$$\beta = \cos^{-1}\left(\frac{-8}{28}\right) = 106.6° \qquad\qquad \textit{Ans.}$$

$$\gamma = \cos^{-1}\left(\frac{-24}{28}\right) = 149.0° \qquad\qquad \textit{Ans.}$$

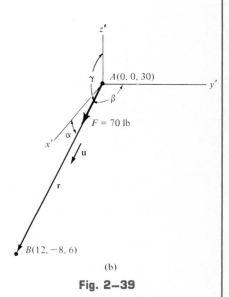

Fig. 2–39

Example 2–16

The circular plate shown in Fig. 2–40*a* is partially supported by the cable *AB*. If the magnitude of force in the cable is 500 N, express **F** as a Cartesian vector.

Solution

As shown in Fig. 2–40*b*, **F** acts in the same direction as the position vector $\mathbf{r}_B$, which extends from *A* to *B*.

Position Vector. The coordinates of the end points of the cable are $A(0, 0, 2)$ and $B(1.707, 0.707, 0)$, as indicated in the figure. Thus,

$$\mathbf{r}_B = (1.707 - 0)\mathbf{i} + (0.707 - 0)\mathbf{j} + (0 - 2)\mathbf{k}$$
$$= \{1.707\mathbf{i} + 0.707\mathbf{j} - 2\mathbf{k}\} \text{ m}$$

Note how one can calculate these components *directly* by going from *A*, $\{-2\mathbf{k}\}$ m along the *z* axis, then $\{1.707\mathbf{i}\}$ m along the *x* axis, and finally $\{0.707\mathbf{j}\}$ m parallel to the *y* axis to get to *B*.

The magnitude of $\mathbf{r}_B$ is

$$r_B = \sqrt{(1.707)^2 + (0.707)^2 + (-2)^2} = 2.72 \text{ m}$$

Unit Vector. Thus,

$$\mathbf{u}_B = \frac{\mathbf{r}_B}{r_B} = \frac{1.707}{2.72}\mathbf{i} + \frac{0.707}{2.72}\mathbf{j} - \frac{2}{2.72}\mathbf{k}$$
$$= 0.627\mathbf{i} + 0.260\mathbf{j} - 0.735\mathbf{k}$$

Force Vector. Since $F = 500$ N and **F** has the direction $\mathbf{u}_B$, we have

$$\mathbf{F} = F\mathbf{u}_B = 500 \text{ N}(0.627\mathbf{i} + 0.260\mathbf{j} - 0.735\mathbf{k})$$
$$= \{313.5\mathbf{i} + 129.8\mathbf{j} - 367.5\mathbf{k}\} \text{ N} \qquad \textit{Ans.}$$

Using these components, notice that indeed the magnitude of **F** is 500 N, i.e.,

$$F = \sqrt{(313.5)^2 + (129.8)^2 + (-367.3)^2} = 500 \text{ N}$$

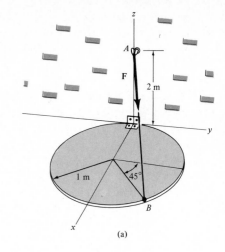

(a)

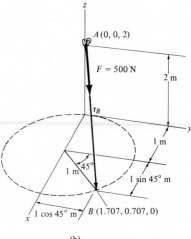

(b)

Fig. 2–40

Example 2-17

The cables exert forces $F_B = 100$ N and $F_C = 120$ N on the ring at A as shown in Fig. 2–41a. Determine the magnitude of the resultant force acting at A.

Solution

The resultant force $\mathbf{F}_R$ is shown graphically in Fig. 2–41b. We can express this force as a Cartesian vector by first formulating $\mathbf{F}_B$ and $\mathbf{F}_C$ as Cartesian vectors and then adding their components. The directions of $\mathbf{F}_B$ and $\mathbf{F}_C$ are specified by forming unit vectors $\mathbf{u}_B$ and $\mathbf{u}_C$ along the cables. These unit vectors are obtained from the associated position vectors $\mathbf{r}_B$ and $\mathbf{r}_C$. With reference to Fig. 2–41b for $\mathbf{F}_B$ we have

$$\mathbf{r}_B = (4 - 0)\mathbf{i} + (0 - 0)\mathbf{j} + (0 - 4)\mathbf{k}$$
$$= \{4\mathbf{i} - 4\mathbf{k}\} \text{ m}$$
$$r_B = \sqrt{(4)^2 + (-4)^2} = 5.66 \text{ m}$$
$$\mathbf{F}_B = 100 \text{ N}\left(\frac{\mathbf{r}_B}{r_B}\right) = 100 \text{ N}\left(\frac{4}{5.66}\mathbf{i} - \frac{4}{5.66}\mathbf{k}\right)$$
$$\mathbf{F}_B = \{70.7\mathbf{i} - 70.7\mathbf{k}\} \text{ N}$$

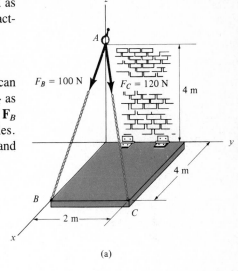

(a)

For $\mathbf{F}_C$ we have

$$\mathbf{r}_C = (4 - 0)\mathbf{i} + (2 - 0)\mathbf{j} + (0 - 4)\mathbf{k}$$
$$= \{4\mathbf{i} + 2\mathbf{j} - 4\mathbf{k}\} \text{ m}$$
$$r_C = \sqrt{(4)^2 + (2)^2 + (-4)^2} = 6 \text{ m}$$
$$\mathbf{F}_C = 120 \text{ N}\left(\frac{\mathbf{r}_C}{r_C}\right) = 120 \text{ N}\left(\frac{4}{6}\mathbf{i} + \frac{2}{6}\mathbf{j} - \frac{4}{6}\mathbf{k}\right)$$
$$= \{80\mathbf{i} + 40\mathbf{j} - 80\mathbf{k}\} \text{ N}$$

The resultant force is therefore

$$\mathbf{F}_R = \mathbf{F}_B + \mathbf{F}_C = (70.7\mathbf{i} - 70.7\mathbf{k}) + (80\mathbf{i} + 40\mathbf{j} - 80\mathbf{k})$$
$$= \{150.7\mathbf{i} + 40\mathbf{j} - 150.7\mathbf{k}\} \text{ N}$$

The magnitude of $\mathbf{F}_R$ is thus

$$F_R = \sqrt{(150.7)^2 + (40)^2 + (-150.7)^2}$$
$$= 217 \text{ N} \qquad\qquad Ans.$$

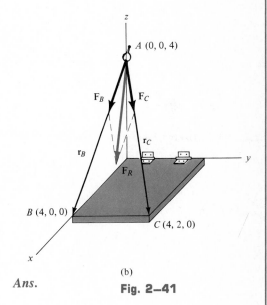

(b)

Fig. 2–41

Problems

*2–56. Express the position vector **r** in Cartesian vector form; then determine its magnitude and coordinate direction angles.

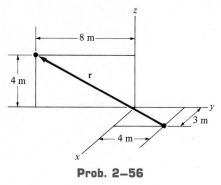

Prob. 2–56

2–57. Express the position vector **r** in Cartesian vector form; then determine its magnitude and coordinate direction angles.

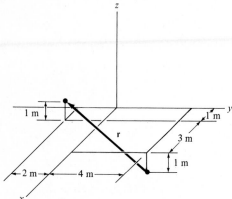

Prob. 2–57

2–58. Express the position vector **r** in Cartesian vector form; then determine its magnitude and coordinate direction angles.

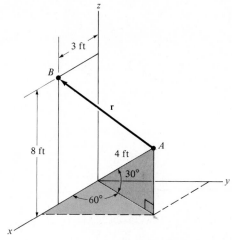

Prob. 2–58

2–59. At the instant shown, position vectors along the robotic arm from O to B and B to A are $\mathbf{r}_{OB} = \{100\mathbf{i} + 300\mathbf{j} + 400\mathbf{k}\}$ mm and $\mathbf{r}_{BA} = \{350\mathbf{i} + 225\mathbf{j} - 640\mathbf{k}\}$ mm, respectively. Determine the distance from O to the grip at A.

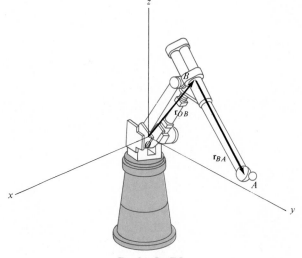

Prob. 2–59

*2–60. Determine the length of member AB of the truss by first establishing a Cartesian position vector from A to B and then determining its magnitude.

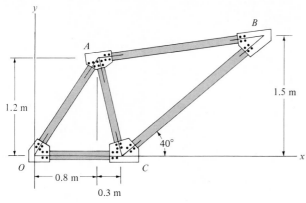

Prob. 2–60

2–61. The 8-m-long cable is anchored to the ground at A. If $x = 4$ m and $y = 2$ m, determine the coordinate z to the highest point of attachment along the column.

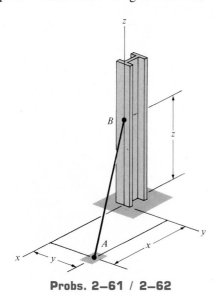

Probs. 2–61 / 2–62

2–62. The 8-m-long cable is anchored to the ground at A. If $z = 5$ m, determine the location $+x$, $+y$ where point A should be anchored to the ground. Choose a value such that $x = y$.

2–63. At a given instant, the positions of a plane at A and a train at B are measured relative to a radar antenna at O. Determine the distance d between A and B at this instant. To solve the problem, formulate a position vector, directed from A to B, and then determine its magnitude.

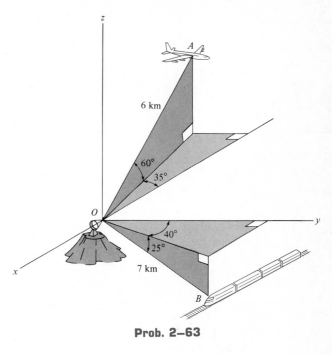

Prob. 2–63

*2–64. Express force $\mathbf{F}$ as a Cartesian vector; then determine its coordinate direction angles.

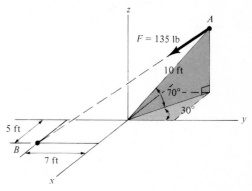

Prob. 2–64

53

2–65. Express force **F** as a Cartesian vector; then determine its coordinate direction angles.

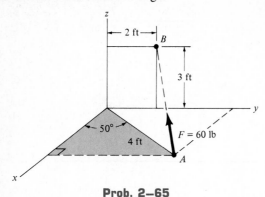

Prob. 2–65

2–66. Express force **F** as a Cartesian vector; then determine its coordinate direction angles.

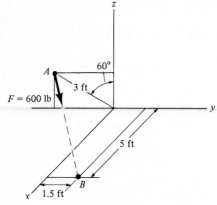

Prob. 2–66

2–67. Determine the magnitude and coordinate direction angles of the resultant force.

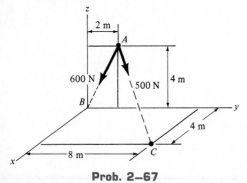

Prob. 2–67

*2–68. Determine the magnitude and coordinate direction angles of the resultant force.

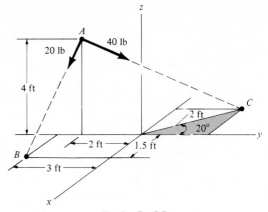

Prob. 2–68

2–69. Express each of the forces in Cartesian vector form and determine the magnitude and coordinate direction angles of the resultant force.

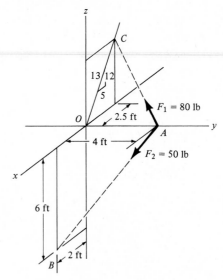

Prob. 2–69

54

2–70. The cable attached to the bulldozer at *B* exerts a force of 350 lb on the framework. Express this force as a Cartesian vector.

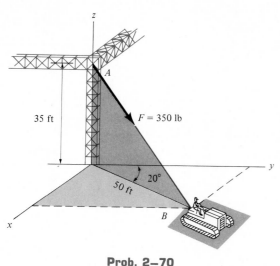

Prob. 2–70

***2–72.** The guy wires are used to support the telephone pole. Represent the force in each wire in Cartesian vector form.

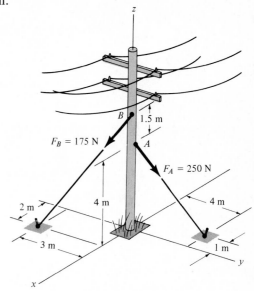

Prob. 2–72

2–71. The two mooring cables exert forces on the stern of a ship as shown. Represent each force as a Cartesian vector and determine the magnitude and direction of the resultant.

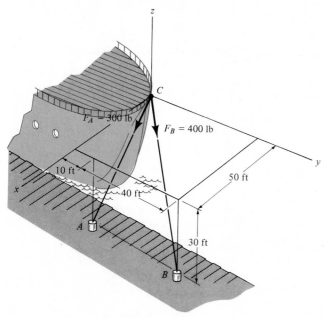

Prob. 2–71

2–73. The window is held open by cable *AB*. Determine the length of the cable and express the 30-N force acting at *A* along the cable as a Cartesian vector.

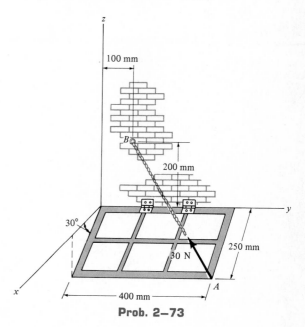

Prob. 2–73

55

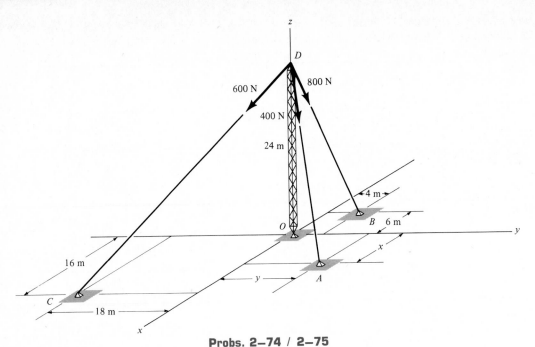

Probs. 2–74 / 2–75

2–74. The tower is held in place by three cables. If the force of each cable acting on the tower is shown, determine the position (x, y) for fixing cable DA so that the resultant force exerted on the tower is directed along its axis, from D toward O.

2–75. The tower is held in place by three cables. If the force of each cable acting on the tower is shown, determine the magnitude and coordinate direction angles α, β, γ of the resultant force. Take $x = 20$ m, $y = 15$ m.

***2–76.** The force acting on the man, caused by his pulling on the anchor cord, is $\mathbf{F} = \{40\mathbf{i} + 20\mathbf{j} - 50\mathbf{k}\}$ N. If the length of the cord is 30 m, determine the coordinates $A(x, y, -z)$ of the anchor. Assume cord OA is along a straight line.

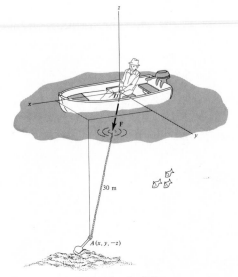

Prob. 2–76

2–77. The cord exerts a force of $\mathbf{F} = \{12\mathbf{i} + 9\mathbf{j} - 8\mathbf{k}\}$ lb on the hook. If the cord is 8 ft long, determine the location x, y of the point of attachment B, and the height z of the hook.

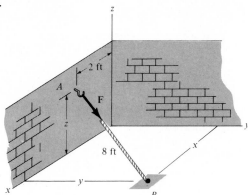

Probs. 2–77 / 2–78

2–78. The cord exerts a force of $F = 30$ lb on the hook. If the cord is 8 ft long, $z = 4$ ft, and the x component of the force is $F_x = 25$ lb, determine the location x, y of the point of attachment B of the cord to the ground.

2–79. The force $\mathbf{F}$ has a magnitude of 80 lb and acts at the midpoint C of the thin rod. Express the force as a Cartesian vector.

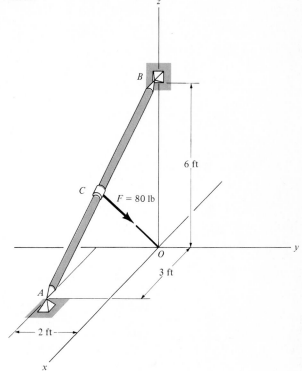

Prob. 2–79

57

2.9 Dot Product

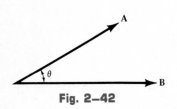

Fig. 2–42

Occasionally in statics there is a need to be able to find the component of a force along a line or to find the angle between two lines. In two dimensions, these problems can readily be solved by trigonometry since the geometry is easy to visualize. In three dimensions, however, this is often difficult so that, instead, vector methods should be employed for the solution. The dot product defines a particular method for "multiplying" two vectors and is used to solve the above-mentioned problems.

The *dot product* of two vectors **A** and **B,** written **A · B,** read "**A** dot **B,**" is defined as the product of the magnitudes of **A** and **B** and the cosine of the angle θ between their tails, Fig. 2–42. Expressed in equation form,

$$\mathbf{A} \cdot \mathbf{B} = AB \cos \theta \qquad (2\text{--}14)$$

where $0° \leq \theta \leq 180°$. The dot product is often referred to as the *scalar product* of vectors, since the result is a *scalar* and not a vector.

Laws of Operation

1. Commutative law:

$$\mathbf{A} \cdot \mathbf{B} = \mathbf{B} \cdot \mathbf{A}$$

2. Multiplication by a scalar:

$$a(\mathbf{A} \cdot \mathbf{B}) = (a\mathbf{A}) \cdot \mathbf{B} = \mathbf{A} \cdot (a\mathbf{B}) = (\mathbf{A} \cdot \mathbf{B})a$$

3. Distributive law:

$$\mathbf{A} \cdot (\mathbf{B} + \mathbf{D}) = (\mathbf{A} \cdot \mathbf{B}) + (\mathbf{A} \cdot \mathbf{D})$$

It is easy to prove the first and second laws by using Eq. 2–14. The proof of the distributive law is left as an exercise (see Prob. 2–80).

Cartesian Unit Vector Formulation. Equation 2–14 may be used to find the dot product of each of the Cartesian unit vectors. For example, $\mathbf{i} \cdot \mathbf{i} = (1)(1) \cos 0° = 1$. In a similar manner,

$$\mathbf{i} \cdot \mathbf{i} = 1 \qquad \mathbf{k} \cdot \mathbf{k} = 1 \qquad \mathbf{j} \cdot \mathbf{k} = 0 \qquad \mathbf{i} \cdot \mathbf{k} = 0 \qquad \mathbf{k} \cdot \mathbf{j} = 0$$
$$\mathbf{j} \cdot \mathbf{j} = 1 \qquad \mathbf{i} \cdot \mathbf{j} = 0 \qquad \mathbf{k} \cdot \mathbf{i} = 0 \qquad \mathbf{j} \cdot \mathbf{i} = 0$$

These results should not be memorized; rather, it should be clearly understood how each is obtained.

Cartesian Vector Formulation. Consider now the dot product of two general vectors **A** and **B** which are expressed in Cartesian vector form. Then

$$\mathbf{A} \cdot \mathbf{B} = (A_x\mathbf{i} + A_y\mathbf{j} + A_z\mathbf{k}) \cdot (B_x\mathbf{i} + B_y\mathbf{j} + B_z\mathbf{k})$$
$$= A_xB_x(\mathbf{i} \cdot \mathbf{i}) + A_xB_y(\mathbf{i} \cdot \mathbf{j}) + A_xB_z(\mathbf{i} \cdot \mathbf{k})$$
$$+ A_yB_x(\mathbf{j} \cdot \mathbf{i}) + A_yB_y(\mathbf{j} \cdot \mathbf{j}) + A_yB_z(\mathbf{j} \cdot \mathbf{k})$$
$$+ A_zB_x(\mathbf{k} \cdot \mathbf{i}) + A_zB_y(\mathbf{k} \cdot \mathbf{j}) + A_zB_z(\mathbf{k} \cdot \mathbf{k})$$

Carrying out the dot-product operations, the final result becomes

$$\mathbf{A} \cdot \mathbf{B} = A_xB_x + A_yB_y + A_zB_z \qquad (2\text{--}15)$$

Thus, to determine the dot product of two Cartesian vectors, multiply their corresponding x, y, z components and sum their products algebraically. Since the result is a scalar, be careful *not* to include any unit vectors in the final result.

Applications.   The dot product has two important applications in mechanics.

1. *The angle formed between two vectors or intersecting lines.* The angle θ between vectors **A** and **B** in Fig. 2–42 is determined from Eq. 2–14 and written as

$$\theta = \cos^{-1}\left(\frac{\mathbf{A} \cdot \mathbf{B}}{AB}\right) \qquad 0° \leqslant \theta \leqslant 180°$$

   Here $\mathbf{A} \cdot \mathbf{B}$ is computed from Eq. 2–15. In particular, notice that if $\mathbf{A} \cdot \mathbf{B} = 0$, $\theta = \cos^{-1} 0 = 90°$, so that **A** will be *perpendicular* to **B**.

2. *The component of a vector parallel to a line.* The component of vector **A** parallel to or collinear with the line *ll* in Fig. 2–43 is defined by $\mathbf{A}_\parallel$, where $A_\parallel = A \cos \theta$. This component is sometimes referred to as the *projection* of **A** onto the line, since a right angle is formed in the construction. If the *direction* of the line is specified by the unit vector **u**, then, since $u = 1$, we can determine $A_\parallel$ directly from the dot product (Eq. 2–14), i.e.,

$$A_\parallel = A \cos \theta = \mathbf{A} \cdot \mathbf{u}$$

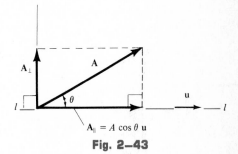

$A_\parallel = A \cos \theta \, \mathbf{u}$

Fig. 2–43

This equation states that *the magnitude of the component $\mathbf{A}_\parallel$ is determined from the dot product of **A** and the unit vector **u** which defines the direction of the line*. The component $\mathbf{A}_\parallel$ represented as a *vector* is therefore

$$\mathbf{A}_\parallel = A \cos \theta \, \mathbf{u} = (\mathbf{A} \cdot \mathbf{u})\mathbf{u}$$

   Note that the component of **A** which is *perpendicular* to line *ll* can also be obtained, Fig. 2–43. Since $\mathbf{A} = \mathbf{A}_\parallel + \mathbf{A}_\perp$, then $\mathbf{A}_\perp = \mathbf{A} - \mathbf{A}_\parallel$. If θ is first determined from the dot product, $\theta = \cos^{-1}(\mathbf{A} \cdot \mathbf{u}/A)$, then $A_\perp = A \sin \theta$. Alternatively, if $A_\parallel$ is known, then by the Pythagorean theorem we can also write $A_\perp = \sqrt{A^2 - A_\parallel^2}$.

   The above two applications are illustrated numerically in the following example problems.

Example 2–18

The frame shown in Fig. 2–44*a* is subjected to a horizontal force $\mathbf{F} = \{300\mathbf{j}\}$ N acting at its corner. Determine the component of this force along member *AB*.

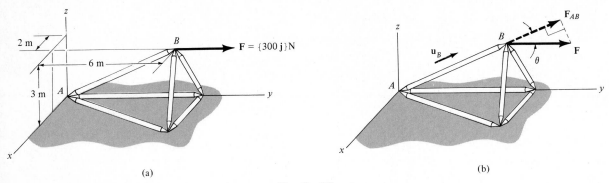

(a)   (b)

Fig. 2–44

Solution

The magnitude of the component of $\mathbf{F}$ along AB is equal to the dot product of $\mathbf{F}$ and the unit vector $\mathbf{u}_B$ which defines the direction of AB, Fig. 2–44*b*. Since

$$\mathbf{u}_B = \frac{\mathbf{r}_B}{r_B} = \frac{2\mathbf{i} + 6\mathbf{j} + 3\mathbf{k}}{\sqrt{(2)^2 + (6)^2 + (3)^2}} = 0.286\mathbf{i} + 0.857\mathbf{j} + 0.429\mathbf{k}$$

Then

$$F_{AB} = F \cos \theta = \mathbf{F} \cdot \mathbf{u}_B = (300\mathbf{j}) \cdot (0.286\mathbf{i} + 0.857\mathbf{j} + 0.429\mathbf{k})$$

$$= (0)(0.286) + (300)(0.857) + (0)(0.429)$$

$$= 257.1 \text{ N}$$

Note the difficulty that would be encountered if one had to determine θ from the geometry of the problem and *then* calculate $F_{AB} = F \cos \theta$. Expressing $\mathbf{F}_{AB}$ in Cartesian vector form, we have

$$\mathbf{F}_{AB} = F_{AB}\mathbf{u}_B = 257.1 \text{ N}(0.286\mathbf{i} + 0.857\mathbf{j} + 0.429\mathbf{k})$$

$$= \{73.5\mathbf{i} + 220\mathbf{j} + 110\mathbf{k}\} \text{ N} \qquad Ans.$$

Example 2–19

The pipe in Fig. 2–45a is subjected to the force $F = 80$ lb at its end B. Determine the angle θ between $\mathbf{F}$ and the pipe segment BA, and the magnitudes of the components of $\mathbf{F}$, which are parallel and perpendicular to BA.

Solution

Angle θ. Unit vectors along BA and BC are first determined.

$$\mathbf{u}_{BA} = \frac{\mathbf{r}_{BA}}{r_{BA}} = -\frac{2}{3}\mathbf{i} - \frac{2}{3}\mathbf{j} + \frac{1}{3}\mathbf{k}$$

$$= -0.667\mathbf{i} - 0.667\mathbf{j} + 0.333\mathbf{k}$$

$$\mathbf{u}_{BC} = \frac{\mathbf{r}_{BC}}{r_{BC}} = -\frac{3}{3.16}\mathbf{j} + \frac{1}{3.16}\mathbf{k}$$

$$= -0.949\mathbf{j} + 0.316\mathbf{k}$$

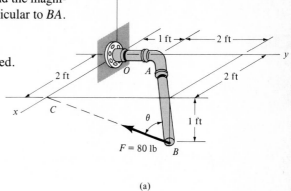

(a)

Since $u_{BA} = u_{BC} = 1$, the dot product $\mathbf{u}_{BA} \cdot \mathbf{u}_{BC} = u_{BA}u_{BC} \cos \theta$ reduces to

$$\cos \theta = \mathbf{u}_{BA} \cdot \mathbf{u}_{BC}$$

$$= (-0.667)(0) + (-0.667)(-0.949) + (0.333)(0.316)$$

$$= 0.738$$

$$\theta = 42.5° \hspace{3cm} Ans.$$

This same result can, of course, be obtained using $\mathbf{r}_{BA} \cdot \mathbf{r}_{BC} = r_{BA}r_{BC} \cos \theta$.

Components of F. Since $\mathbf{F} = (80 \text{ lb})\mathbf{u}_{BC}$, then from Fig. 2–45b, the component of $\mathbf{F}$ in the direction of BA is

$$F_{BA} = \mathbf{F} \cdot \mathbf{u}_{BA} = (80 \text{ lb})\mathbf{u}_{BC} \cdot \mathbf{u}_{BA}$$

$$= (80 \text{ lb})(0.738)$$

$$= 59.0 \text{ lb} \hspace{3cm} Ans.$$

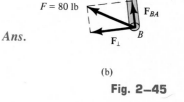

The magnitude of the perpendicular component is

$$F_{\perp} = \sqrt{F^2 - F_{BA}^2}$$

$$= \sqrt{(80)^2 - (59.0)^2}$$

$$= 54.0 \text{ lb} \hspace{3cm} Ans.$$

(b)

Fig. 2–45

Problems

*2–80. Given the three vectors **A, B,** and **D,** show that **A · (B + D) = (A · B) + (A · D).**

2–81. Determine the angle θ between the tails of the two vectors.

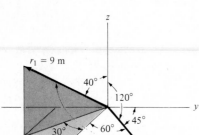

Prob. 2–81

2–82. Determine the angle θ between the tails of the two vectors.

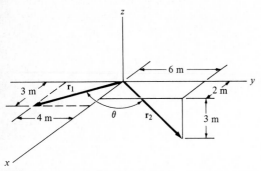

Prob. 2–82

2–83. Determine the magnitude of the component of the position vector **r** along the *Oa* axis.

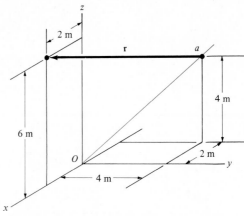

Prob. 2–83

*2–84. Determine the magnitude of the component of the force **F** along the *Aa* axis.

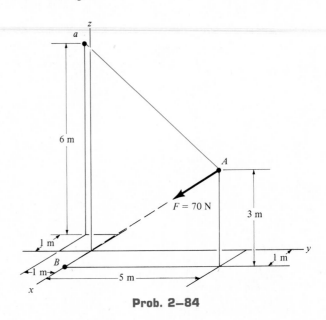

Prob. 2–84

2–85. Determine the two components of the force **F** along the lines *Oa* and *Ob* such that $\mathbf{F} = \mathbf{F}_A + \mathbf{F}_B$. Also find the projections of **F** along *Oa* and *Ob*. Show graphically how the components and projections are constructed.

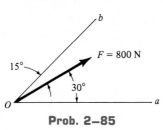

Prob. 2–85

2–86. Determine the magnitude of the component of $\mathbf{F}_1$ along the line of action of $\mathbf{F}_2$.

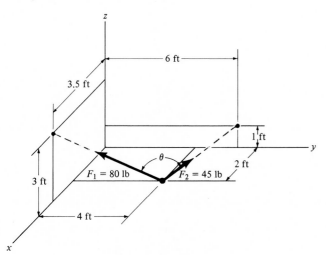

Prob. 2–86

2–87. Cable *AB* exerts a force of 90 N on the end of the 3-m-long boom. Determine the magnitude of the component of this force along *AO*.

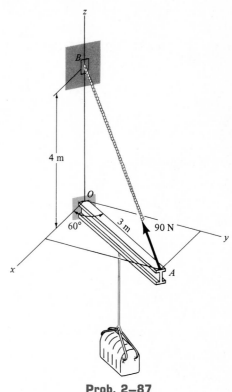

Prob. 2–87

***2–88.** Determine the angle θ between the edges of the sheet-metal bracket.

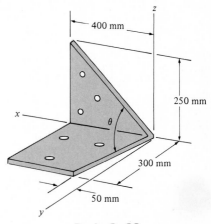

Prob. 2–88

63

2–89. Cable OA is used to support the column OB. Determine the angle θ it makes with beam OC.

2–91. Determine the component of the 80-N force acting along the axis AB of the pipe.

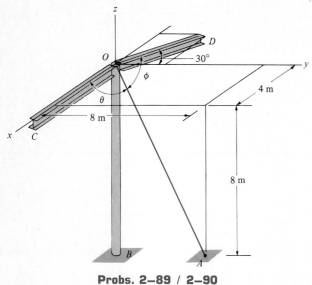

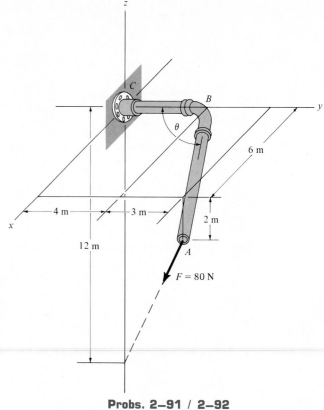

Probs. 2–89 / 2–90

2–90. Cable OA is used to support the column OB. Determine the angle ϕ it makes with beam OD.

Probs. 2–91 / 2–92

*2–92. Determine the angle θ made between pipe segments BA and BC.

2–93. Determine the angles θ and ϕ made between the axes OA of the flag pole and each cable, AB and AC.

2–95. The clamp is used on a jig. If the vertical force acting on the bolt is $\mathbf{F} = \{-500\mathbf{k}\}$ N, determine the magnitudes of the components $\mathbf{F}_1$ and $\mathbf{F}_2$ which act along the OA axis and perpendicular to it.

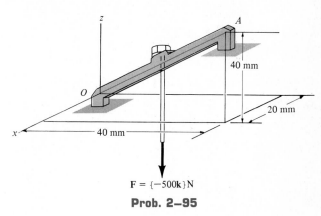

Prob. 2–95

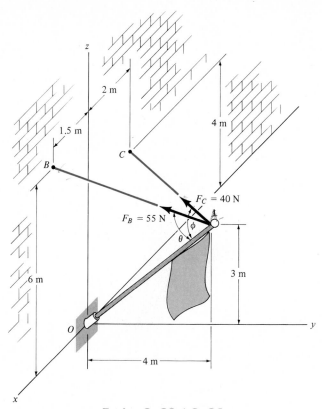

Probs. 2–93 / 2–94

2–94. The two supporting cables exert the forces shown on the flag pole. Determine the component of each force acting along the axis OA of the pole.

*2–96. The force $\mathbf{F}$ acts at the end A of the pipe assembly. Determine the magnitudes of the components $\mathbf{F}_1$ and $\mathbf{F}_2$ which act along the axis of AB and perpendicular to it.

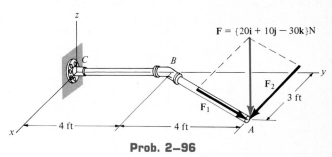

Prob. 2–96

65

Review Problems

2–97. Express each of the three forces acting on the column in Cartesian vector form and compute the magnitude of the resultant force.

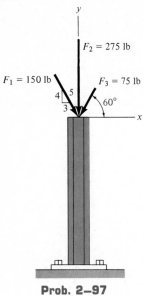

Prob. 2–97

2–99. Resolve the 50-lb force into components along (a) the x and y axes, and (b) the x and y' axes.

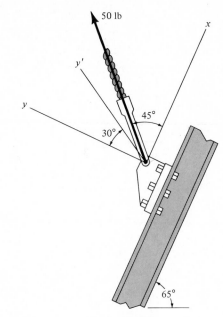

Prob. 2–99

2–98. Determine the magnitudes of the two components of the 600-N force, one directed along the cable AC and the other along the axis of strut AB.

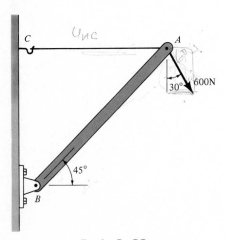

Prob. 2–98

*2–100. The girl exerts a force of 75 N along the handle AB of the lawn mower. Resolve this force into x and y components which act parallel and perpendicular to the ground.

Prob. 2–100

2–101. Using chains, two tractors exert forces of $\mathbf{F}_A$ and $\mathbf{F}_B$ on the trunk of a tree as shown. Determine the magnitude of the resultant force. If the same magnitudes of forces $\mathbf{F}_A$ and $\mathbf{F}_B$ are maintained, in what direction should the tractors pull on the trunk so as to cause the *maximum* resultant force? What is the magnitude of this maximum force?

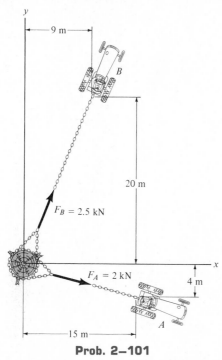

Prob. 2–101

2–102. Express each of the three forces acting on the bracket in Cartesian vector form with respect to the x and y axes. Determine the magnitude and direction θ of $\mathbf{F}_1$ so that the resultant force is directed along the positive x' axis and has a magnitude of $F_R = 600$ N.

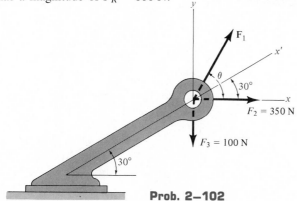

Prob. 2–102

2–103. Each of the four forces acting at E has a magnitude of 28 kN. Express each force as a Cartesian vector and determine the resultant force.

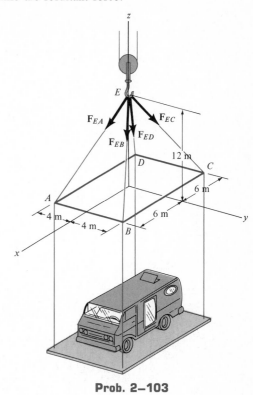

Prob. 2–103

*2–104. Determine the angles θ and ϕ between the wire segments.

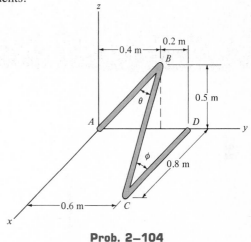

Prob. 2–104

2–105. The vertical force of 60 lb acts downward at A on the two-member frame. Determine the magnitudes of the two components of **F** directed along the axes of members AB and AC. Set $\theta = 45°$.

2–107. The *gusset plate* is subjected to four forces which are concurrent at point O. Determine the magnitude and direction of the resultant force.

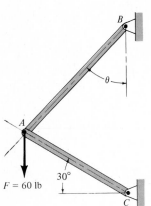

Probs. 2–105 / 2–106

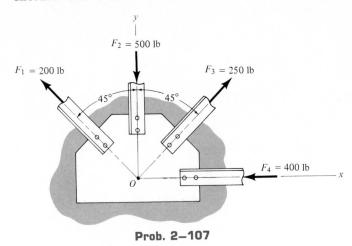

Prob. 2–107

2–106. Determine the angle θ ($0° \leq \theta \leq 90°$) of member AB so that the force acting along the axis of AB is 80 lb. What is the magnitude of force acting along the axis of member AC?

Equilibrium of a Particle

In this chapter the methods of resolving a force into components and expressing a force as a Cartesian vector will be used to solve problems involving the equilibrium of a particle. Recall that the dimensions or size of a particle are assumed to be neglected, and therefore a particle can be subjected *only* to a system of *concurrent forces*. To simplify the discussion, the equilibrium of a particle subjected to a coplanar force system will be considered first. Then, in the last part of the chapter, equilibrium problems involving three-dimensional force systems will be considered.

Condition for the Equilibrium of a Particle   3.1

By definition, *equilibrium* of a particle requires that the particle be either at rest if originally at rest, or have a constant velocity if originally in motion. Most often, however, the term ''equilibrium'' or more specifically ''static equilibrium'' is used to describe an object at rest. To maintain a state of equilibrium, it is *necessary* to satisfy Newton's first law of motion, which states that if the *resultant force* acting on a particle is *zero,* then the particle is in equilibrium. This condition may be stated mathematically as

$$\Sigma \mathbf{F} = \mathbf{0} \tag{3-1}$$

where $\Sigma \mathbf{F}$ is the vector *sum of all the forces* acting on the particle.

Not only is Eq. 3–1 a necessary condition for equilibrium, but it is also a *sufficient* condition. This follows from Newton's second law of motion, which can be written as $\Sigma \mathbf{F} = m\mathbf{a}$. Since the force system satisfies Eq. 3–1, then $m\mathbf{a} = \mathbf{0}$, and therefore the particle's acceleration $\mathbf{a} = \mathbf{0}$, and consequently the particle indeed moves with constant velocity or remains at rest.

3.2 The Free-Body Diagram

To correctly apply the equation of equilibrium, it is necessary to account for *all* the known and unknown forces ($\Sigma \mathbf{F}$) which act *on* the particle. The best way to do this is to draw the particle's free-body diagram. This diagram is a sketch of the particle which represents it as being isolated or "free" from its surroundings. On this sketch one then shows *all* the forces which the surroundings exert *on* the particle. Provided this diagram is correctly drawn, it will then be easy to apply Eq. 3–1.

Before presenting a formal procedure as to how to draw a free-body diagram, we will first discuss two types of supports often encountered in particle equilibrium problems.

Cables and Pulleys.   Throughout this book, except in Sec. 8.4, all cables are assumed to have negligible weight and they cannot be stretched. A cable can support *only* a tension or "pulling" force, and this force always acts in the direction of the cable. Later it will be shown that the tension force developed in a *continuous cable* which passes over a frictionless pulley must have a *constant* magnitude to keep the cable in equilibrium. (see Example 5–6). Hence, for any angle θ, shown in Fig. 3–1, the cable is subjected to a constant tension $\mathbf{T}$ throughout its length.

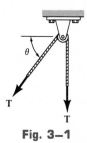

Fig. 3–1

Springs.   If a *linear elastic spring* is used for a support, the length of the spring will change in direct proportion to the force acting on it. A characteristic that defines the "elasticity" of a spring is the *spring constant* or *stiffness k*. Specifically, the magnitude of force developed by a linear elastic spring which has a stiffness k, and is deformed (compressed or elongated) a distance s measured from its unloaded position, is

$$F = ks \tag{3–2}$$

Note that s is determined from the difference in the spring's deformed length l

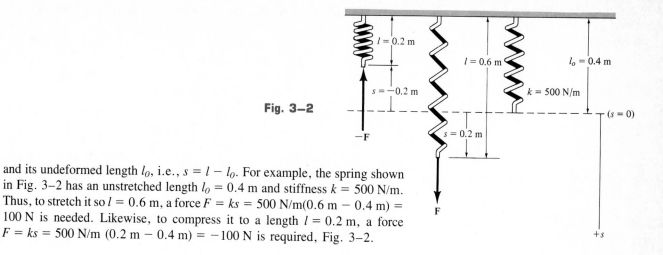

Fig. 3–2

and its undeformed length l_0, i.e., $s = l - l_0$. For example, the spring shown in Fig. 3–2 has an unstretched length $l_0 = 0.4$ m and stiffness $k = 500$ N/m. Thus, to stretch it so $l = 0.6$ m, a force $F = ks = 500$ N/m$(0.6$ m $- 0.4$ m$) = 100$ N is needed. Likewise, to compress it to a length $l = 0.2$ m, a force $F = ks = 500$ N/m $(0.2$ m $- 0.4$ m$) = -100$ N is required, Fig. 3–2.

PROCEDURE FOR DRAWING A FREE-BODY DIAGRAM

The importance of drawing a free-body diagram before applying the equation of equilibrium to the solution of a problem cannot be overemphasized. To construct a free-body diagram, the following three steps are necessary.

Step 1. Imagine the particle to be *isolated* or cut "free" from its surroundings. Draw or sketch its outlined shape.

Step 2. Indicate on this sketch *all* the forces that act *on the particle*. These forces can be *active forces,* which tend to set the particle in motion, such as those caused by attached cables, weight, or magnetic and electrostatic interaction. Also, *reactive forces* will occur, such as those caused by the constraints or supports that tend to prevent motion. To account for all these forces, it may help to trace around the particle's boundary, carefully noting each force acting on it.

Step 3. The forces that are *known* should be labeled with their proper magnitudes and directions. Letters are used to represent the magnitudes and directions of forces that are unknown. In particular, if a force has a known line of action but unknown magnitude, the "arrowhead," which defines the sense of the force, can be *assumed.* The correct sense will become apparent after solving for the unknown magnitude. By definition, the *magnitude* of a force is *always positive* so that, if the solution yields a "negative" scalar, the *minus sign* indicates that the arrowhead or sense of the force is opposite to that which was originally assumed.

Application of the above steps is illustrated in the following two examples.

71

Example 3-1

The crate in Fig. 3-3a has a weight of 20 lb. Draw a free-body diagram of the cord BD and the ring at B.

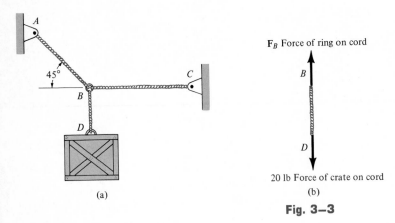

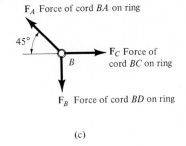

$\mathbf{F}_B$ Force of ring on cord

B

D

20 lb Force of crate on cord

(b)

$\mathbf{F}_A$ Force of cord BA on ring

45°

$\mathbf{F}_C$ Force of cord BC on ring

B

$\mathbf{F}_B$ Force of cord BD on ring

(c)

(a)

Fig. 3-3

Solution

If we imagine the cord BD to be isolated from its surroundings, then by inspection there are only two forces that act on it; namely, the 20-lb force at D caused by the weight of the crate, Fig. 3-3b, and the force $\mathbf{F}_B$ at B which is caused by the ring. Since these forces tend to *stretch* the cord, we can state that the cord is in *tension*. (This must be the case, since otherwise the cord would collapse.)

When the ring at B is isolated from its surroundings, it is realized that three forces act on it. All these forces are caused by the attached cords, Fig. 3-3c. Notice that $\mathbf{F}_B$ shown here is equal but opposite to that shown in Fig. 3-3b, a consequence of Newton's third law.

Example 3-2

The sphere in Fig. 3–4a has a mass of 6 kg and is supported as shown. Draw a free-body diagram of point C.

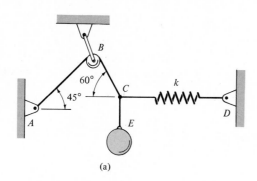

(a)

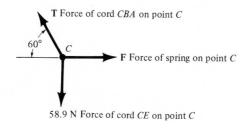

T Force of cord CBA on point C

F Force of spring on point C

58.9 N Force of cord CE on point C

(b)

Fig. 3–4

Solution

By inspection three forces act *on* point C; they are caused by cords CBA and CE, and the spring CD. The weight of the sphere, $W = (6 \text{ kg})(9.81 \text{ m/s}^2) = 58.9$ N, is "transmitted" to point C by the supporting cord CE. Thus, the free-body diagram of point C is shown in Fig. 3–4b.

Notice that since the magnitude of the force **T** in cord CBA is constant, a force of **T** pulls on the support at A. Also, once the force **F** in the spring CD is determined, the stretch s of the spring can be computed using $F = ks$.

3.3 Coplanar Force Systems

Many particle equilibrium problems involve a coplanar force system. If the forces lie in the x-y plane, they can each be resolved into their respective **i** and **j** components and Eq. 3–1 can be written as

$$\Sigma \mathbf{F} = \mathbf{0}$$

$$\Sigma F_x \mathbf{i} + \Sigma F_y \mathbf{j} = \mathbf{0}$$

For this vector equation to be satisfied, both the x and y components must equal zero, otherwise $\Sigma \mathbf{F} \neq \mathbf{0}$. Hence, we require

$$\Sigma F_x = 0$$
$$\Sigma F_y = 0$$

(3–3)

These scalar equilibrium equations state that the algebraic sum of the x and y components of all the forces acting on the particle be equal to zero. As a result, Eqs. 3–3 can be solved for at most two unknowns, generally represented as angles and magnitudes of forces shown on the particle's free-body diagram.

PROCEDURE FOR ANALYSIS

The following procedure provides a method for solving coplanar force equilibrium problems:

Free-Body Diagram. Draw a free-body diagram of the particle. As outlined in Sec. 3.2, this requires that all the known and unknown force magnitudes and angles be labeled on the diagram. The sense of a force having an unknown magnitude can be assumed.

Equations of Equilibrium. Establish the x, y axes and apply the two equations of equilibrium, $\Sigma F_x = 0$ and $\Sigma F_y = 0$. If more than two unknowns exist and the problem involves a spring, apply $F = ks$ (Eq. 3–2) to relate the spring force to the deformation s of the spring.

Since the magnitude of a force is always positive, then if the solution to the equations yields a negative value, it indicates that the sense of the force shown on the free-body diagram is *opposite* to that which was *assumed*.

The following example problems numerically illustrate this solution procedure.

Example 3–3

Determine the tensions in cords AB and AD for equilibrium of the 10-kg block shown in Fig. 3–5a.

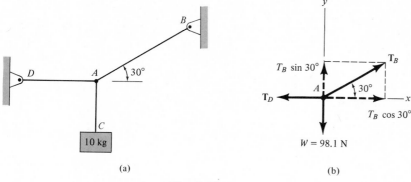

(a)

(b)

Fig. 3–5

Solution

The cord tensions can be obtained by investigating the equilibrium of point A.

Free-Body Diagram. As shown in Fig. 3–5b, there are three concurrent forces *acting on point A*. The cord tension forces $\mathbf{T}_B$ and $\mathbf{T}_D$ have unknown magnitudes but known directions. Cord AC exerts a downward force on A equal to the weight of the block, $W = (10 \text{ kg})(9.81 \text{ m/s}^2) = 98.1 \text{ N}$.

Equations of Equilibrium. Since the equilibrium equations require a summation of the x and y components of each force, $\mathbf{T}_B$ must be resolved into x and y components. These components, shown dashed on the free-body diagram, have magnitudes of $T_B \cos 30°$ and $T_B \sin 30°$, respectively. Equations 3–3 will be applied by assuming that "positive" force components act along the positive x and y axes. Indicating these directions alongside the equations, we have

$$\xrightarrow{+} \Sigma F_x = 0; \qquad\qquad T_B \cos 30° - T_D = 0 \qquad\qquad (1)$$
$$+ \uparrow \Sigma F_y = 0; \qquad\qquad T_B \sin 30° - 98.1 = 0 \qquad\qquad (2)$$

Solving Eq. (2) for T_B and substituting into Eq. (1) to obtain T_D yields

$$T_B = 196 \text{ N} \qquad\qquad\qquad Ans.$$
$$T_D = 170 \text{ N} \qquad\qquad\qquad Ans.$$

Example 3-4

If the cylinder at A in Fig. 3-6a has a weight of 20 lb, determine the weight of B and the force in each cord needed to hold the system in the equilibrium position shown.

Solution

Since the weight of A is known, the unknown tensions in cables EG and EC can be determined by investigating the equilibrium of point E.

Free-Body Diagram. There are three forces acting on point E, as shown in Fig. 3-6b.

Equations of Equilibrium. Resolving each force into its x and y components using trigonometry, and applying the equations of equilibrium, we have

$$\xrightarrow{+}\Sigma F_x = 0; \qquad T_{EG}\sin 30° - T_{EC}\cos 45° = 0 \qquad (1)$$
$$+\uparrow\Sigma F_y = 0; \qquad T_{EG}\cos 30° - T_{EC}\sin 45° - 20 = 0 \qquad (2)$$

Solving Eq. (1) for T_{EG} in terms of T_{EC} and substituting the result into Eq. (2) allows a solution for T_{EC}. One then obtains T_{EG} from Eq. (1). The results are

$$T_{EC} = 38.6 \text{ lb} \qquad\qquad Ans.$$
$$T_{EG} = 54.6 \text{ lb} \qquad\qquad Ans.$$

Using the calculated result for T_{EC}, the equilibrium of point C can now be investigated to determine the tension in CD and the weight of B.

Free-Body Diagram. As shown in Fig. 3-6c, $T_{EC} = 38.6$ lb "pulls" on C. The reason for this becomes clear when one draws the free-body diagram of cord CE and applies the principle of action, equal but opposite force reaction (Newton's third law), Fig. 3-6d.

Equations of Equilibrium. Noting the components of $\mathbf{T}_{CD}$ are proportional to the slope of the cord as defined by the 3-4-5 triangle, we have

$$\xrightarrow{+}\Sigma F_x = 0; \qquad 38.6\cos 45° - (\tfrac{4}{5})T_{CD} = 0 \qquad (3)$$
$$+\uparrow\Sigma F_y = 0; \qquad (\tfrac{3}{5})T_{CD} + 38.6\sin 45° - W_B = 0 \qquad (4)$$

Solving Eq. (3) and substituting the result into Eq. (4) yields

$$T_{CD} = 34.1 \text{ lb} \qquad\qquad Ans.$$
$$W_B = 47.8 \text{ lb} \qquad\qquad Ans.$$

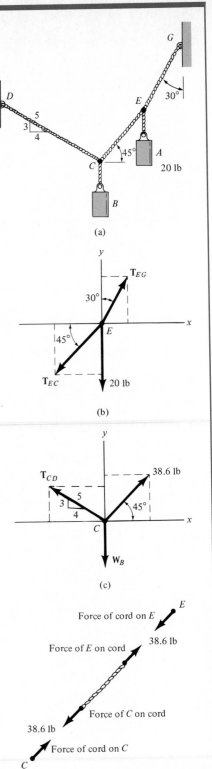

(a)

(b)

(c)

(d)

Fig. 3-6

Example 3–5

Determine the required length of cord AC in Fig. 3–7a so that the 8-kg lamp is suspended in the position shown. The *unstretched* length of the spring AB is $l'_{AB} = 0.4$ m, and the spring has a stiffness of $k_{AB} = 300$ N/m.

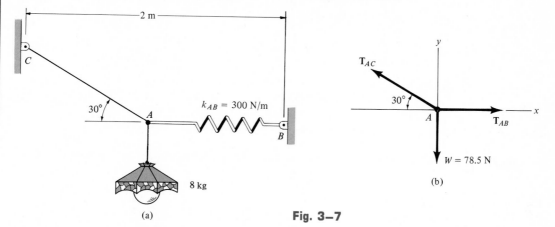

Fig. 3–7

Solution

If the force in spring AB is known, the stretch in the spring can be found $(F = ks)$. Using the problem geometry, it is then possible to calculate the required length of AC.

Free-Body Diagram. The lamp has a weight $W = 8(9.81) = 78.5$ N. The free-body diagram of point A is shown in Fig. 3–7b.

Equations of Equilibrium

$$\xrightarrow{+} \Sigma F_x = 0; \qquad T_{AB} - T_{AC} \cos 30° = 0$$
$$+ \uparrow \Sigma F_y = 0; \qquad T_{AC} \sin 30° - 78.5 = 0$$

Solving, we obtain

$$T_{AC} = 157.0 \text{ N}$$
$$T_{AB} = 136.0 \text{ N}$$

The stretch in spring AB is therefore

$$T_{AB} = k_{AB}s_{AB}; \qquad 136.0 \text{ N} = 300 \text{ N/m}(s_{AB})$$
$$s_{AB} = 0.453 \text{ m}$$

so that the stretched length of spring AB is

$$l_{AB} = l'_{AB} + s_{AB}$$
$$l_{AB} = 0.4 \text{ m} + 0.453 \text{ m} = 0.853 \text{ m}$$

The horizontal distance from C to B, Fig. 3–7a, requires

$$2 \text{ m} = l_{AC} \cos 30° + 0.853 \text{ m}$$
$$l_{AC} = 1.32 \text{ m} \qquad\qquad\qquad \textit{Ans.}$$

Problems

3–1. Determine the magnitudes of $\mathbf{F}_1$ and $\mathbf{F}_2$ so that particle P is in equilibrium.

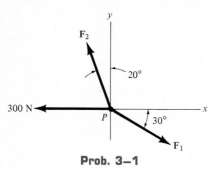

Prob. 3–1

3–2. Determine the magnitude and direction θ of $\mathbf{F}_1$ so that particle P is in equilibrium.

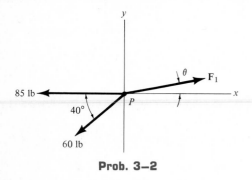

Prob. 3–2

3–3. Determine the magnitude and direction θ of $\mathbf{F}$ so that particle P is in equilibrium.

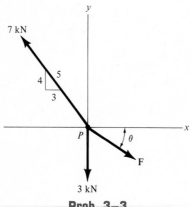

Prob. 3–3

*3–4. The *gusset plate P* is subjected to the forces of three members as shown. Determine the force in member C and its proper orientation θ for equilibrium. The forces are concurrent at point O.

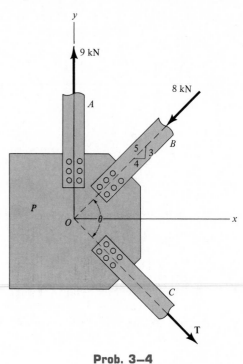

Prob. 3–4

78

3–5. The motor at B winds up the cord attached to the 80-lb crate with a constant speed. Determine the force in cord CD supporting the pulley and the angle θ for equilibrium. Neglect the size of the pulley at C.

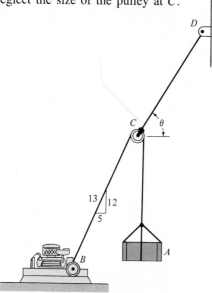

Prob. 3–5

3–6. The device shown is used to straighten the frames of wrecked autos. Determine the tension in each segment of the chain, i.e., AB and BC, if the force which the hydraulic cylinder DB exerts on point B is 3500 N, as shown.

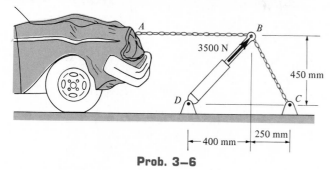

Prob. 3–6

3–7. The members of a truss are pin-connected at joint O as shown. Determine the magnitudes of **F** and **T** for equilibrium.

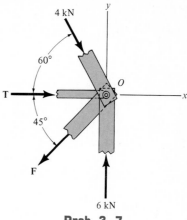

Prob. 3–7

*3–8. Determine the force in cable AB and AC necessary to support the 12-kg traffic light.

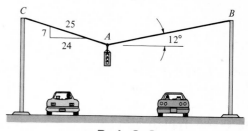

Prob. 3–8

3–9. The members of a truss are connected to the gusset plate. If the forces are concurrent at point O, determine the magnitudes of **F** and **T** for equilibrium.

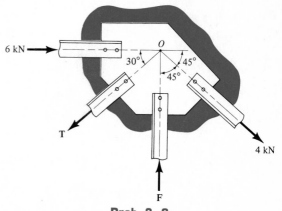

Prob. 3–9

79

3–10. Determine the mass that must be supported at A and the angle θ of the connecting cord in order to hold the system in equilibrium.

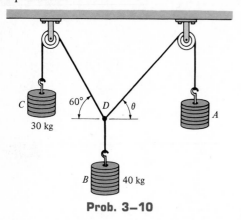

Prob. 3–10

3–11. The 500-lb crate is hoisted using the ropes AB and AC. Each rope can withstand a maximum force of 2500 lb before it breaks. If AB always remains horizontal, determine the smallest angle θ to which the crate can be hoisted before one of the ropes breaks.

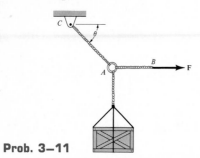

Prob. 3–11

*3–12. Determine the stiffness k_T of the single spring such that the force $\mathbf{F}$ will stretch it by the same amount s as the force $\mathbf{F}$ stretches the two springs. Express the result in terms of the stiffness k_1 and k_2 of the two springs.

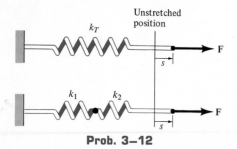

Prob. 3–12

3–13. Two different springs AB and BC are connected together as shown. If forces of 60 lb and 80 lb are applied at their ends, determine the vertical deflection of point A.

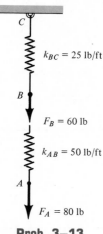

Prob. 3–13

3–14. Determine the stretch in each spring for equilibrium of the 2-kg block. The springs are shown in their equilibrium position.

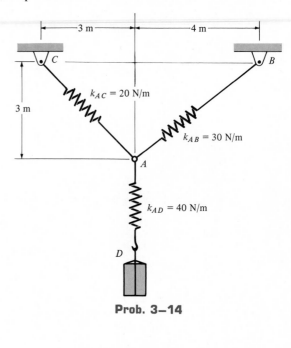

Prob. 3–14

3–15. The elastic cord (or spring) *ABC* has a stiffness of 500 N/m and an unstretched length of 6 m. Determine the horizontal force **F** applied to the cord which is attached to the *small* pulley *B* so that the displacement of the pulley from the wall is *d* = 1.5 m.

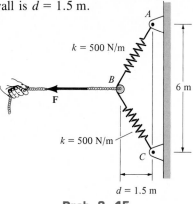

Prob. 3–15

*3–16. Romeo tries to reach Juliet by climbing with constant velocity up a rope which is knotted at point *A*. Any of the three segments of the rope can sustain a maximum force of 2 kN before it breaks. Determine if Romeo, who has a mass of 65 kg, can climb the rope, and if so, can he along with his Juliet, who has a mass of 60 kg, climb down with constant velocity?

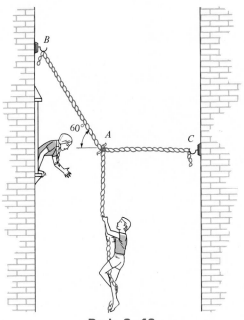

Prob. 3–16

3–17. The bulldozer attempts to pull down the chimney using the cable and *small* pulley arrangement shown. If the tension in *AB* is 600 lb, determine the tension in cable *CAD* and the angle θ which the cable makes at the pulley.

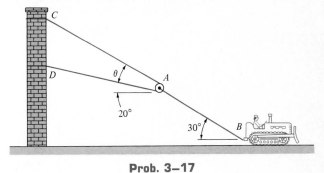

Prob. 3–17

3–18. Two electrically charged pith balls, each having a mass of 0.2 g, are suspended from light threads of equal length. Determine the resultant horizontal force of repulsion, **F**, acting on each ball if the measured distance between them is *r* = 200 mm.

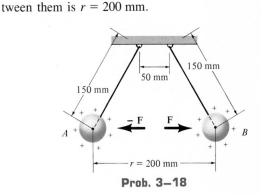

Prob. 3–18

■3–19. A car is to be towed using the rope arrangement shown. The towing force required is 600 lb. Determine the maximum length *l* of rope *AB* so that the tension in either rope *AB* or *AC* does not exceed 750 lb. *Hint:* Use the equilibrium condition at point *A* to determine the required angle *θ* for attachment, then determine *l* using trigonometry applied to triangle *ABC*.

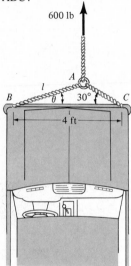

Prob. 3–19

*3–20. The 30-kg pipe is supported at *A* by a system of five cords. Determine the force in each cord for equilibrium.

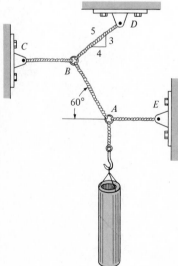

Prob. 3–20

3–21. If the cords suspend the two buckets in the equilibrium position shown, determine the weight of bucket *B*. Bucket *A* has a weight of 60 lb.

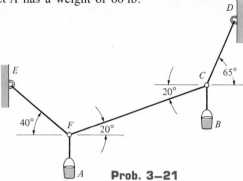

Prob. 3–21

■3–22. A vertical force of 10 lb is applied to the ends of the cord *AB* and spring *AC*. If the spring has an unstretched length of 2 ft, determine the angle *θ* for equilibrium.

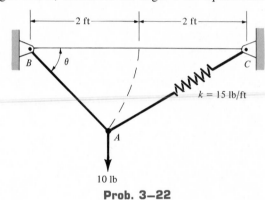

Prob. 3–22

■3–23. The 10-lb lamp fixture is suspended from two springs, each having an unstretched length of 4 ft and stiffness of $k = 5$ lb/ft. Determine the angle *θ* for equilibrium.

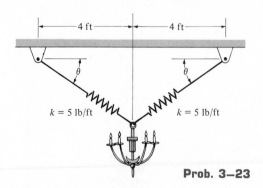

Prob. 3–23

*3–24. The 2-m-long cord AB is attached to a spring BC having an unstretched length of 2 m. If the cord sags downward an amount $\theta = 30°$ as shown, determine the vertical force $\mathbf{F}$ applied. The spring has a stiffness of $k = 50$ N/m.

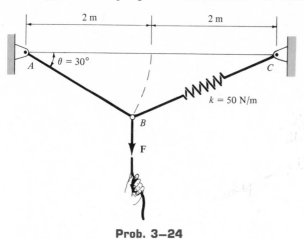

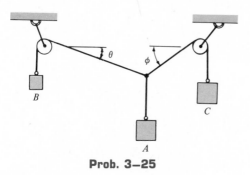

Prob. 3–24

3–25. Three blocks are supported using the cords and two pulleys as shown. If they have weights of $W_A = W$, $W_B = 0.25W$, and $W_C = W$, determine the angle θ for equilibrium.

Prob. 3–25

3–26. When y is zero, the springs sustain a force of 60 lb. Determine the magnitude of the applied vertical forces $\mathbf{F}$ required to pull point A away from point B a distance of $y = 2$ ft. The ends of cords CAD and CBD are attached to rings at C and D.

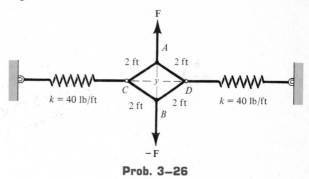

Prob. 3–26

3–27. The pail and its contents have a mass of 60 kg. If the cable is 15 m long, determine the elevation y of the pulley for equilibrium. Neglect the size of the pulley at A.

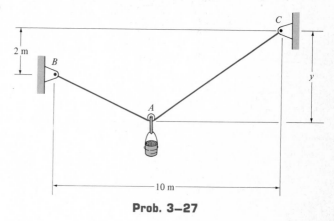

Prob. 3–27

83

3.4 Three-Dimensional Force Systems

It was shown in Sec. 3.1 that particle equilibrium requires

$$\Sigma \mathbf{F} = \mathbf{0} \qquad\qquad (3\text{--}4)$$

If the forces acting on the particle are resolved into their respective **i, j, k** components, we can then write

$$\Sigma F_x \mathbf{i} + \Sigma F_y \mathbf{j} + \Sigma F_z \mathbf{k} = \mathbf{0}$$

To ensure that Eq. 3–4 is satisfied, we must therefore require that the following three scalar component equations be satisfied:

$$\begin{aligned} \Sigma F_x &= 0 \\ \Sigma F_y &= 0 \\ \Sigma F_z &= 0 \end{aligned} \qquad\qquad (3\text{--}5)$$

These equations represent the *algebraic sums* of the x, y, z force components acting on the particle. Using them we can solve for at most three unknowns, generally represented as angles or magnitudes of forces shown on the particle's free-body diagram.

PROCEDURE FOR ANALYSIS

The following procedure provides a method for solving three-dimensional force equilibrium problems.

Free-Body Diagram. Draw a free-body diagram of the particle and label all the known and unknown forces on this diagram.

Equations of Equilibrium. Establish the x, y, z coordinate axes with origin located at the particle and apply the equations of equilibrium. Use the three scalar equations 3–5 in cases where it is easy to resolve each force acting on the particle into its x, y, z components. If this appears difficult, first express each force acting on the particle in Cartesian vector form, and then substitute these vectors into Eq. 3–4. By setting the respective **i, j, k** components equal to zero, the three scalar equations 3–5 can be generated. If more than three unknowns exist and the problem involves a spring, consider using $F = ks$ to relate the spring force to the deformation s of the spring.

The following example problems numerically illustrate this solution procedure.

Example 3–6

The 90-lb cylinder shown in Fig. 3–8a is supported by two cables and a spring having a stiffness $k = 500$ lb/ft. Determine the force in the cables and the stretch of the spring for equilibrium. Cord AD lies in the x-y plane and cord AC lies in the x-z plane.

Solution

The stretch of the spring can be determined once the force in AB is determined.

Free-Body Diagram. Point A is chosen for the analysis since the cable forces are concurrent at this point, Fig. 3–8b.

Equations of Equilibrium. By inspection, each force can easily be resolved into its x, y, z components, and therefore the three scalar equations of equilibrium can be directly applied. Considering components directed along the positive axes as "positive," we have

$$\Sigma F_x = 0; \qquad\qquad F_D \sin 30° - \tfrac{4}{5}F_C = 0 \qquad\qquad (1)$$

$$\Sigma F_y = 0; \qquad\qquad -F_D \cos\ 30° + F_B = 0 \qquad\qquad (2)$$

$$\Sigma F_z = 0; \qquad\qquad \tfrac{3}{5}F_C - 90 \text{ lb} = 0 \qquad\qquad (3)$$

Solving Eq. (3) for F_C, then Eq. (1) for F_D, and finally Eq. (2) for F_B, we get

$$F_C = 150 \text{ lb} \qquad\qquad \textit{Ans.}$$
$$F_D = 240 \text{ lb} \qquad\qquad \textit{Ans.}$$
$$F_B = 208 \text{ lb} \qquad\qquad \textit{Ans.}$$

The stretch of the spring is therefore

$$F_B = ks_{AB}$$
$$208 \text{ lb} = 500 \text{ lb/ft} \ (s_{AB})$$
$$s_{AB} = 0.416 \text{ ft} \qquad\qquad \textit{Ans.}$$

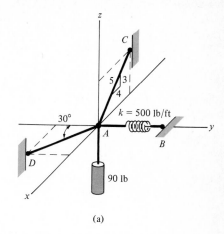

(a)

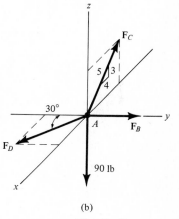

(b)

Fig. 3–8

Example 3-7

Determine the magnitude and coordinate direction angles of **F** in Fig. 3–9a that are required for equilibrium of point O.

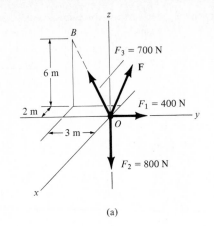

(a)

Solution

Free-Body Diagram. The four forces acting on point O are shown in Fig. 3–9b.

Equations of Equilibrium. A Cartesian vector analysis will be used for the solution in order to formulate the components of **F**$_3$. Hence,

$$\Sigma \mathbf{F} = 0; \qquad \mathbf{F}_1 + \mathbf{F}_2 + \mathbf{F}_3 + \mathbf{F} = 0 \qquad (1)$$

Expressing each of these forces in Cartesian vector form, noting that the coordinates of B are $B(-2, -3, 6)$, we have

$$\mathbf{F}_1 = \{400\mathbf{j}\} \text{ N}$$
$$\mathbf{F}_2 = \{-800\mathbf{k}\} \text{ N}$$
$$\mathbf{F}_3 = F_3\mathbf{u}_B = F_3\left(\frac{\mathbf{r}_B}{r_B}\right) = 700 \text{ N}\left[\frac{-2\mathbf{i} - 3\mathbf{j} + 6\mathbf{k}}{\sqrt{(-2)^2 + (-3)^2 + (6)^2}}\right]$$
$$= \{-200\mathbf{i} - 300\mathbf{j} + 600\mathbf{k}\} \text{ N}$$
$$\mathbf{F} = F_x\mathbf{i} + F_y\mathbf{j} + F_z\mathbf{k}$$

Substituting into Eq. (1) yields

$$400\mathbf{j} - 800\mathbf{k} - 200\mathbf{i} - 300\mathbf{j} + 600\mathbf{k} + F_x\mathbf{i} + F_y\mathbf{j} + F_z\mathbf{k} = 0$$

Equating the respective **i**, **j**, **k** components to zero, we have

$\Sigma F_x = 0;$	$-200 + F_x = 0$	$F_x = 200 \text{ N}$
$\Sigma F_y = 0;$	$400 - 300 + F_y = 0$	$F_y = -100 \text{ N}$
$\Sigma F_z = 0;$	$-800 + 600 + F_z = 0$	$F_z = 200 \text{ N}$

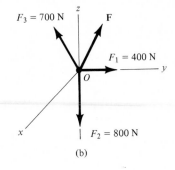

(b)

Thus,

$$\mathbf{F} = \{200\mathbf{i} - 100\mathbf{j} + 200\mathbf{k}\} \text{ N}$$

so that

$$F = \sqrt{(200)^2 + (-100)^2 + (200)^2} = 300 \text{ N} \qquad \textit{Ans.}$$

$$\mathbf{u}_F = \frac{\mathbf{F}}{F} = \frac{200}{300}\mathbf{i} - \frac{100}{300}\mathbf{j} + \frac{200}{300}\mathbf{k}$$

$$\alpha = \cos^{-1}\left(\frac{200}{300}\right) = 48.2° \qquad \textit{Ans.}$$

$$\beta = \cos^{-1}\left(\frac{-100}{300}\right) = 109.5° \qquad \textit{Ans.}$$

$$\gamma = \cos^{-1}\left(\frac{200}{300}\right) = 48.2° \qquad \textit{Ans.}$$

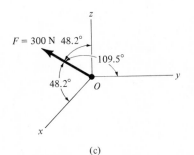

(c)

Fig. 3–9

The magnitude and correct direction of **F** are shown in Fig. 3–9c.

Example 3–8

Determine the force developed in each cable used to support the 40-lb crate shown in Fig. 3–10a.

Solution

Free-Body Diagram.   As shown in Fig. 3–10b, the free-body diagram of point A is considered in order to "expose" the three unknown forces in the cables.

Equations of Equilibrium

$$\Sigma \mathbf{F} = \mathbf{0}; \qquad \mathbf{F}_B + \mathbf{F}_C + \mathbf{F}_D + \mathbf{W} = \mathbf{0} \qquad (1)$$

Since the coordinates of points B and C are $B(-3, -4, 8)$ and $C(-3, 4, 8)$, we have

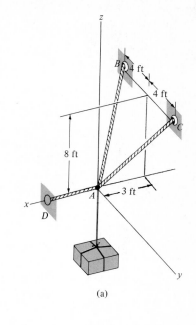

(a)

$$\mathbf{F}_B = F_B \left[\frac{-3\mathbf{i} - 4\mathbf{j} + 8\mathbf{k}}{\sqrt{(-3)^2 + (-4)^2 + (8)^2}} \right]$$

$$= -0.318F_B\mathbf{i} - 0.424F_B\mathbf{j} + 0.848F_B\mathbf{k}$$

$$\mathbf{F}_C = F_C \left[\frac{-3\mathbf{i} + 4\mathbf{j} + 8\mathbf{k}}{\sqrt{(-3)^2 + (4)^2 + (8)^2}} \right]$$

$$= -0.318F_C\mathbf{i} + 0.424F_C\mathbf{j} + 0.848F_C\mathbf{k}$$

$$\mathbf{F}_D = F_D\mathbf{i}$$

$$\mathbf{W} = -40\mathbf{k}$$

Substituting these forces into Eq. (1) gives

$$-0.318F_B\mathbf{i} - 0.424F_B\mathbf{j} + 0.848F_B\mathbf{k} - 0.318F_C\mathbf{i} + 0.424F_C\mathbf{j}$$
$$+ 0.848F_C\mathbf{k} + F_D\mathbf{i} - 40\mathbf{k} = \mathbf{0}$$

Equating the respective $\mathbf{i}$, $\mathbf{j}$, $\mathbf{k}$ components to zero yields

$$\Sigma F_x = 0; \qquad -0.318F_B - 0.318F_C + F_D = 0 \qquad (2)$$
$$\Sigma F_y = 0; \qquad -0.424F_B + 0.424F_C = 0 \qquad (3)$$
$$\Sigma F_z = 0; \qquad 0.848F_B + 0.848F_C - 40 = 0 \qquad (4)$$

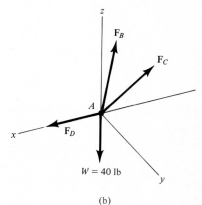

(b)

Fig. 3–10

Equation (3) states $F_B = F_C$. Thus solving Eq. (4) for F_B and F_C and substituting the result into Eq. (2) to obtain F_D, we have

$$F_B = F_C = 23.6 \text{ lb} \qquad \text{Ans.}$$
$$F_D = 15.0 \text{ lb} \qquad \text{Ans.}$$

Example 3–9

The 100-kg cylinder shown in Fig. 3–11a is supported by three cords, one of which is connected to a spring. Determine the tension in each cord for equilibrium and the stretch in the spring.

Solution

Free-Body Diagram. The force in each cord can be determined by investigating the free-body diagram of point A, Fig. 3–11b. The weight of the cylinder is $W = 100(9.81) = 981$ N.

Equations of Equilibrium

$$\Sigma F = 0; \qquad\qquad F_B + F_C + F_D + W = 0 \qquad\qquad (1)$$

Expressing each of these vectors in Cartesian vector form, using Eq. 2–11 for F_C, and noting that the coordinates of point D are $D(-1, 2, 2)$, we have

$$F_B = F_B i$$

$$F_C = F_C \cos 120°i + F_C \cos 135°j + F_C \cos 60°k$$

$$= -0.5F_C i - 0.707F_C j + 0.5F_C k$$

$$F_D = F_D \left[\frac{-1i + 2j + 2k}{\sqrt{(-1)^2 + (2)^2 + (2)^2}} \right]$$

$$= -0.333F_D i + 0.667F_D j + 0.667F_D k$$

$$W = -981k$$

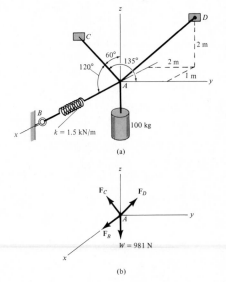

(a)

(b)

Fig. 3–11

Substituting these forces into Eq. (1) yields

$$F_B i - 0.5F_C i - 0.707F_C j + 0.5F_C k - 0.333F_D i + 0.667F_D j$$
$$+ 0.667F_D k - 981k = 0$$

Equating the respective components in the **i, j, k** directions to zero, we have

$$\Sigma F_x = 0; \qquad\qquad F_B - 0.5F_C - 0.333F_D = 0 \qquad\qquad (2)$$
$$\Sigma F_y = 0; \qquad\qquad -0.707F_C + 0.667F_D = 0 \qquad\qquad (3)$$
$$\Sigma F_z = 0; \qquad\qquad 0.5F_C + 0.667F_D - 981 = 0 \qquad\qquad (4)$$

Solving Eq. (3) for F_D in terms of F_C and substituting into Eq. (4) yields F_C. F_D is determined from Eq. (3). Substituting the results into Eq. (2) yields F_B. Hence,

$$F_C = 813 \text{ N} \qquad\qquad Ans.$$
$$F_D = 862 \text{ N} \qquad\qquad Ans.$$
$$F_B = 693.7 \text{ N} = 694 \text{ N} \qquad\qquad Ans.$$

The stretch in the spring is therefore

$$F = ks; \qquad\qquad 693.7 = 1500s$$
$$s = 0.462 \text{ m} \qquad\qquad Ans.$$

Problems

*3–28. Determine the magnitudes of F_1, F_2, and F_3 for equilibrium of particle P.

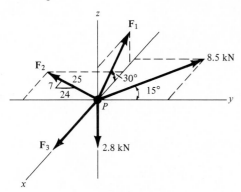

Prob. 3–28

3–29. Determine the magnitudes of F_1, F_2, and F_3 for equilibrium of particle P.

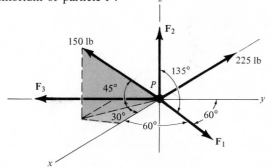

Prob. 3–29

3–30. Determine the magnitudes of F_1, F_2, and F_3 for equilibrium of particle P.

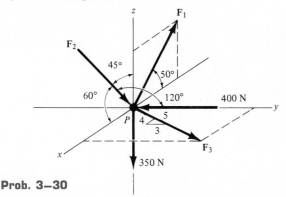

Prob. 3–30

3–31. The three cables are used to support the 800-N force. Determine the force developed in each cable for equilibrium.

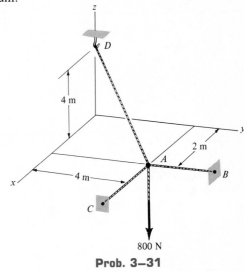

Prob. 3–31

*3–32. If the bucket and its contents have a total weight of 20 lb, determine the force in the supporting cables DA, DB, and DC.

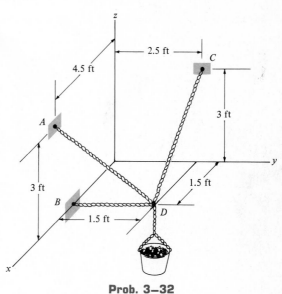

Prob. 3–32

■3–33. The crate is to be hoisted from the hold of a ship using the cable arrangement shown. If a vertical force of 2500 N is applied to the hook at A, determine the tension in each of the three cables for equilibrium.

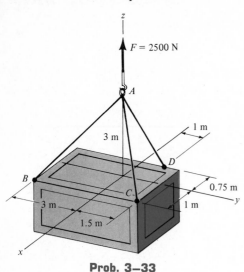

Prob. 3–33

3–34. Determine the tension developed in the three cables required to support the traffic light, which has a mass of 15 kg.

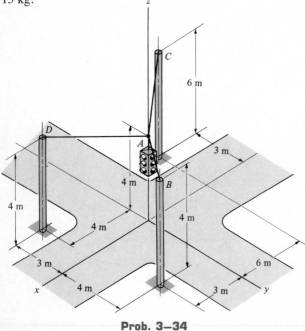

Prob. 3–34

3–35. The 80-lb lamp is supported by three wires as shown. Determine the force in each wire for equilibrium of point A.

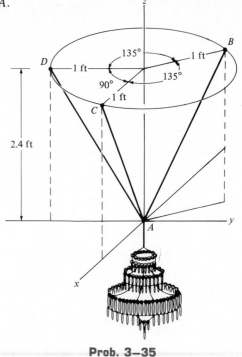

Prob. 3–35

*3–36. The mast OA is supported by three cables. If cable AB is subjected to a tension of 500 N, determine the tension in cables AC and AD and the vertical force F_{AO} which the mast exerts along its axis on the collar at A.

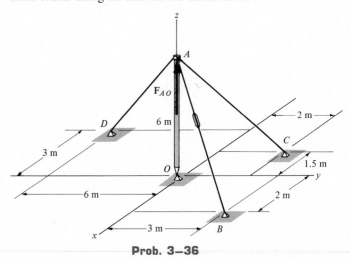

Prob. 3–36

3–37. The 500-lb crate is suspended from the cable system shown. Determine the force in each segment of the cable, i.e., AB, AC, and CD, and the force in cables CE and CF. *Hint:* First analyse the equilibrium of point A, then using the result for AC, analyze the equilibrium of point C.

3–39. The 80-lb ball is suspended from the horizontal ring using three springs each having an unstretched length of 1.5 ft and stiffness of 50 lb/ft. Determine the vertical distance h from the ring to point A for equilibrium.

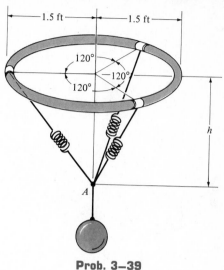

Prob. 3–39

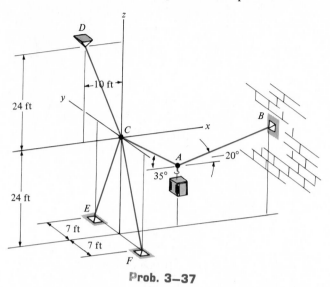

Prob. 3–37

3–38. The 25-kg pot is supported at point A by the three cables. Determine the force acting in each cable for equilibrium.

*3–40. Determine the magnitudes of forces $\mathbf{F}_1$, $\mathbf{F}_2$, and $\mathbf{F}_3$ necessary to hold the force $\mathbf{F} = \{-9\mathbf{i} - 8\mathbf{j} - 5\mathbf{k}\}$ kN in equilibrium.

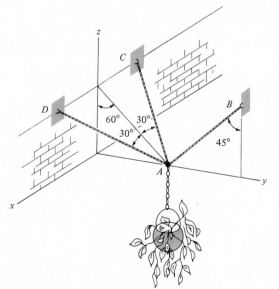

Prob. 3–38

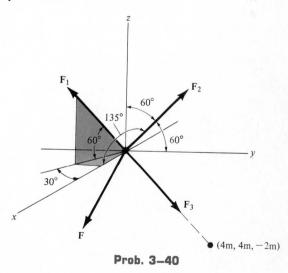

Prob. 3–40

Review Problems

3–41. The uniform 200-lb crate is suspended by using a 6-ft-long cord that is attached to the sides of the crate and passes over the small pulley located at O. If the cord can be attached at either points A and B, or C and D, determine which attachment produces the least amount of tension in the cord and specify the cord tension in this case.

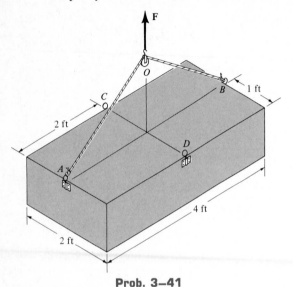

Prob. 3–41

3–42. Blocks D and F weigh 5 lb each and block E weighs 8 lb. If the cord connecting the blocks passes over small frictionless pulleys, determine the sag s for equilibrium.

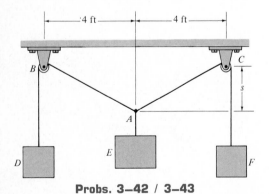

Probs. 3–42 / 3–43

3–43. If blocks D and F weigh 5 lb each, determine the weight of block E if $s = 3$ ft.

*3–44. The 30-kg block is supported by two springs having the stiffness shown. Determine the unstretched length of each spring after the block is removed.

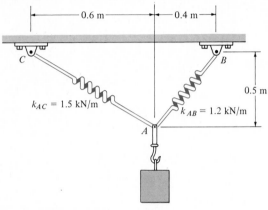

Prob. 3–44

3–45. The spider has a mass of 0.7 g and is suspended from a portion of its web as shown. Determine the force which each of the three web "strings" exerts on the twigs at C, D, and E. String CB is horizontal. *Hint:* First analyze the equilibrium at point A; then, using the result for the force in AB, analyze the equilibrium at B.

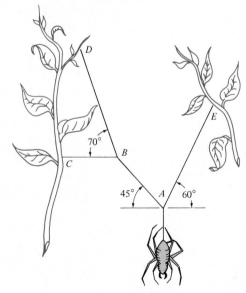

Prob. 3–45

3–46. The ring of negligible size is subjected to a vertical force of 200 lb. Determine the required length l of cord AC such that the tension acting in AC is 160 lb. Also, what is the force acting in cord AB? *Hint:* Use the equilibrium condition to determine the required angle θ for attachment, then determine l using trigonometry applied to $\triangle ABC$.

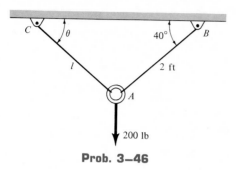

Prob. 3–46

3–47. Determine the tension in cables AB, AC, and AD, required to hold the 60-lb crate in equilibrium.

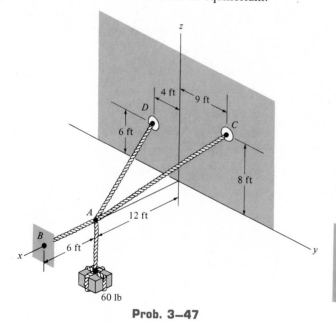

Prob. 3–47

*3–48.** The members of a truss are connected to the gusset plate. If the forces are concurrent at point O, determine the magnitudes of $\mathbf{F}$ and $\mathbf{T}$ for equilibrium.

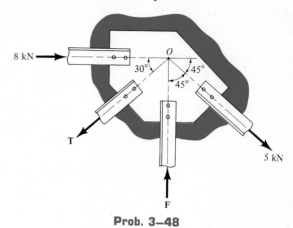

Prob. 3–48

3–49. The shear leg derrick is used to haul the 200-kg net of fish onto the dock. Determine the compressive force in each of the legs AB and CB and the tension in cable DB. Assume the force in each leg acts along its axis.

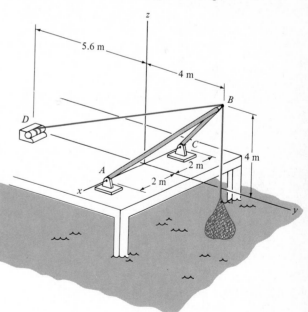

Prob. 3–49

3–50.   The engine E of an automobile has a mass of 60 kg. Determine the force which the man at B must exert on the horizontal rope to position the engine over the truck bed as shown. What force does the man at C exert on the cord? Neglect the size of the pulley at A.

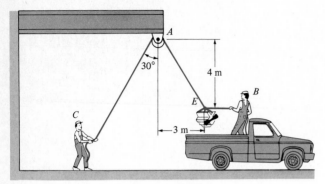

Prob. 3–50

Force System Resultants

4

In Chapter 3 it was shown that the condition for the equilibrium of a particle or a concurrent force system simply requires that the resultant of the force system be equal to zero. In Chapter 5 it will be shown that such a restriction is necessary but not sufficient for the equilibrium of a rigid body. A further restriction must be made with regard to the nonconcurrency of the applied force system, giving rise to the concept of *moment*. In this chapter, a formal definition of a moment will be developed and ways of finding the moment of a force about a point or axis will be discussed. We will also develop methods for determining the resultants of nonconcurrent force systems. This development is important since the methods of force-system simplification are similar to those used for solving rigid-body equilibrium problems, which is discussed in Chapter 5. Furthermore, the resultants will influence the state of equilibrium or motion of a rigid body in the same way as the force system, and therefore we can study rigid-body behavior in a simpler manner by using the resultants.

Cross Product **4.1**

The moment of a force will be formulated using Cartesian vectors in the next section. Before doing this, however, it is first necessary to expand our knowledge of vector algebra and introduce the cross-product method of vector multiplication.

The cross product of two vectors **A** and **B** yields the resultant vector **C**, which is written

$$\mathbf{C} = \mathbf{A} \times \mathbf{B}$$

and is read "**C** equals **A** cross **B**."

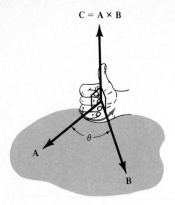

$C = A \times B$

Fig. 4—1

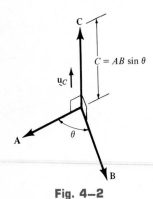

C

$C = AB \sin \theta$

u_C

A

θ

B

Fig. 4—2

Magnitude. The *magnitude* of **C** is defined as the product of the magnitudes of **A** and **B** and the sine of the angle θ made between their tails $(0° \leq \theta \leq 180°)$. Thus, $C = AB \sin \theta$.

Direction. Vector **C** has a *direction* that is perpendicular to the plane containing **A** and **B** such that the direction of **C** is specified by the right-hand rule, i.e., curling the fingers of the right hand from vector **A** (cross) to vector **B**, the thumb then points in the direction of **C**, as shown in Fig. 4–1.

Knowing both the magnitude and direction of **C**, we can write

$$\mathbf{C} = \mathbf{A} \times \mathbf{B} = (AB \sin \theta)\mathbf{u}_C \qquad (4\text{--}1)$$

where the scalar $AB \sin \theta$ defines the *magnitude* of **C** and the unit vector $\mathbf{u}_C$ defines the *direction* of **C**. The terms of Eq. 4–1 are illustrated graphically in Fig. 4–2.

Laws of Operation

1. The commutative law is *not* valid, i.e.,

$$\mathbf{A} \times \mathbf{B} \neq \mathbf{B} \times \mathbf{A}$$

Rather,

$$\mathbf{A} \times \mathbf{B} = -\mathbf{B} \times \mathbf{A}$$

This is shown in Fig. 4–3 by using the right-hand rule. The cross product **B** × **A** yields a vector that acts in the opposite direction to **C**, i.e., **B** × **A** = −**C**.

2. Multiplication by a scalar:

$$a(\mathbf{A} \times \mathbf{B}) = (a\mathbf{A}) \times \mathbf{B} = \mathbf{A} \times (a\mathbf{B}) = (\mathbf{A} \times \mathbf{B})a$$

This property is easily shown, since the magnitude of the resultant vector $(|a|AB \sin \theta)$ and its direction are the same in each case.

3. The distributive law:

$$\mathbf{A} \times (\mathbf{B} + \mathbf{D}) = (\mathbf{A} \times \mathbf{B}) + (\mathbf{A} \times \mathbf{D})$$

The proof of this identity is left as an exercise (see Prob. 4–1). It is important to note that *proper order* of the cross products must be maintained, since they are not commutative.

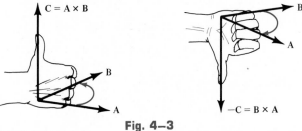

Fig. 4—3

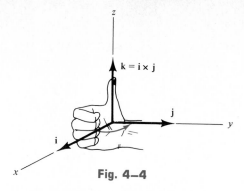

Fig. 4–4

Cartesian Unit Vector Formulation. Equation 4–1 may be used to find the cross product of a pair of Cartesian unit vectors. For example, to find $\mathbf{i} \times \mathbf{j}$, the *magnitude* of the resultant vector is $(i)(j)(\sin 90°) = (1)(1)(1) = 1$, and its *direction* is determined using the right-hand rule. As shown in Fig. 4–4, the resultant vector points in the $+\mathbf{k}$ direction. Thus, $\mathbf{i} \times \mathbf{j} = (1)\mathbf{k}$. In a similar manner,

$$\mathbf{i} \times \mathbf{j} = \mathbf{k} \qquad \mathbf{i} \times \mathbf{k} = -\mathbf{j} \qquad \mathbf{i} \times \mathbf{i} = \mathbf{0}$$
$$\mathbf{j} \times \mathbf{k} = \mathbf{i} \qquad \mathbf{j} \times \mathbf{i} = -\mathbf{k} \qquad \mathbf{j} \times \mathbf{j} = \mathbf{0}$$
$$\mathbf{k} \times \mathbf{i} = \mathbf{j} \qquad \mathbf{k} \times \mathbf{j} = -\mathbf{i} \qquad \mathbf{k} \times \mathbf{k} = \mathbf{0}$$

These results should *not* be memorized; rather, it should be clearly understood how each is obtained by using the right-hand rule and the definition of the cross product. A simple scheme shown in Fig. 4–5 is helpful for obtaining the same results when the need arises. If the circle is constructed as shown, then "crossing" two unit vectors in a *counterclockwise* fashion around the circle yields the *positive* third unit vector, e.g., $\mathbf{k} \times \mathbf{i} = \mathbf{j}$. Moving *clockwise*, a *negative* unit vector is obtained, e.g., $\mathbf{i} \times \mathbf{k} = -\mathbf{j}$.

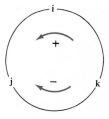

Fig. 4–5

Cartesian Vector Formulation. Consider now the cross product of two general vectors **A** and **B** which are expressed in Cartesian vector form. We have

$$\mathbf{A} \times \mathbf{B} = (A_x\mathbf{i} + A_y\mathbf{j} + A_z\mathbf{k}) \times (B_x\mathbf{i} + B_y\mathbf{j} + B_z\mathbf{k})$$
$$= A_xB_x(\mathbf{i} \times \mathbf{i}) + A_xB_y(\mathbf{i} \times \mathbf{j}) + A_xB_z(\mathbf{i} \times \mathbf{k})$$
$$+ A_yB_x(\mathbf{j} \times \mathbf{i}) + A_yB_y(\mathbf{j} \times \mathbf{j}) + A_yB_z(\mathbf{j} \times \mathbf{k})$$
$$+ A_zB_x(\mathbf{k} \times \mathbf{i}) + A_zB_y(\mathbf{k} \times \mathbf{j}) + A_zB_z(\mathbf{k} \times \mathbf{k})$$

Carrying out the cross-product operations and combining terms yields

$$\mathbf{A} \times \mathbf{B} = (A_yB_z - A_zB_y)\mathbf{i} - (A_xB_z - A_zB_x)\mathbf{j} + (A_xB_y - A_yB_x)\mathbf{k} \qquad (4\text{–}2)$$

This equation may also be written in a more compact determinant form as

$$\mathbf{A} \times \mathbf{B} = \begin{vmatrix} \mathbf{i} & \mathbf{j} & \mathbf{k} \\ A_x & A_y & A_z \\ B_x & B_y & B_z \end{vmatrix} \qquad (4\text{–}3)$$

97

Thus, to find the cross product of any two Cartesian vectors **A** and **B**, it is necessary to expand a determinant whose first row of elements consists of the unit vectors **i**, **j**, and **k** and whose second and third rows represent the x, y, z components of the two vectors **A** and **B**, respectively.*

4.2 Moment of a Force

The *moment* of a force about a point or axis provides a measure of the tendency of the force to cause an object to rotate about the point or axis. For example, consider the force **F** in Fig. 4–6, which acts on the handle of the wrench and is located a distance d from the vertical axis of the pipe. If the force is resolved into three rectangular components, it is seen that *only* $\mathbf{F}_1$ tends to rotate the wrench and pipe about the axis. Obviously, $\mathbf{F}_2$ and $\mathbf{F}_3$ *do not* cause the wrench to turn about this axis since $\mathbf{F}_2$ intersects the axis and $\mathbf{F}_3$ is parallel to it. This rotational effect of $\mathbf{F}_1$ is sometimes called a *torque,* but most often it is called the *moment of a force* or simply the *moment* $\mathbf{M}_O$. In particular, note that the pipe axis is perpendicular to the shaded plane containing both $\mathbf{F}_1$ and d, and that the axis intersects this plane at point O.

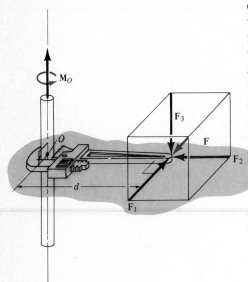

Fig. 4–6

*A determinant having three rows and three columns can be expanded using three minors, each of which is multiplied by one of the three terms in the first row. There are four elements in each minor, e.g.,

$$\begin{vmatrix} A_{11} & A_{12} \\ A_{21} & A_{22} \end{vmatrix}$$

By *definition*, this notation represents the terms $(A_{11}A_{22} - A_{12}A_{21})$, which is simply the product of the two elements of the arrow slanting downward to the right $(A_{11}A_{22})$ *minus* the product of the two elements intersected by the arrow slanting downward to the left $(A_{12}A_{21})$. For a 3×3 determinant, such as Eq. 4–3, the three minors can be generated in accordance with the following scheme:

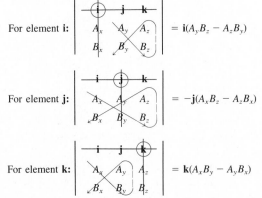

Adding the results and noting that the **j** element *must include the minus sign* yields the expanded form of **A** × **B** given by Eq. 4–2.

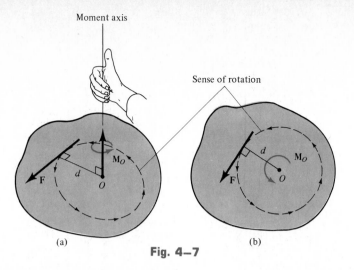

Moment axis

Sense of rotation

M_O

d

F

O

(a)

d

M_O

F

O

(b)

Fig. 4–7

Scalar Formulation. Consider now the general case of the force **F** and point O which lie in a shaded plane as shown in Fig. 4–7a. The moment $\mathbf{M}_O$ about point O, or about an axis passing through O and perpendicular to the plane, is a *vector quantity* since it has a specified magnitude and direction and adds according to the parallelogram law.

Magnitude. The magnitude of $\mathbf{M}_O$ is

$$M_O = Fd \qquad (4-4)$$

where d is referred to as the *moment arm* or perpendicular distance from the axis at point O to the line of action of the force. Units of moment magnitude consist of force times distance, e.g., N · m or lb · ft.

Direction. The direction of $\mathbf{M}_O$ will be specified by using the "right-hand rule." To do this, the fingers of the right hand are curled such that they follow the sense of rotation, which would occur if the force could rotate about point O, Fig. 4–7a. The *thumb* then *points* along the *moment axis* so that it gives the direction and sense of the moment vector, which is *upward* and *perpendicular* to the shaded plane containing **F** and d. By this definition, the moment $\mathbf{M}_O$ can be considered as a *sliding vector* and therefore acts at any point along the moment axis.

In three dimensions, $\mathbf{M}_O$ is illustrated by a vector arrow with a curl on it to *distinguish* it from a force vector, Fig. 4–7a. Many problems in mechanics, however, involve coplanar force systems that may be conveniently viewed in two dimensions. For example, a two-dimensional view of Fig. 4–7a is given in Fig. 4–7b. Here $\mathbf{M}_O$ is simply represented by the (counterclockwise) curl, which indicates the action of **F.** The arrowhead on this curl is used to show the *sense of rotation* caused by **F.** Using the right-hand rule, however, realize that the direction and sense of the moment vector in Fig. 4–7b are indicated by the thumb, which points *out* of the page, since the fingers follow the curl. In two dimensions we will often refer to finding the moment of a force "about a point" (O). Keep in mind that the moment *actually* acts about an axis which is perpendicular to the plane of **F** and d, and intersects the plane at the point (O), Fig. 4–7a.

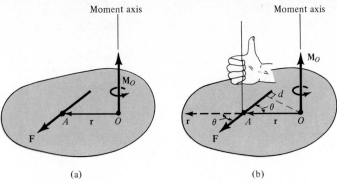

Fig. 4—8

Vector Formulation. The moment of a force **F** about point O, or actually about an axis passing through O and perpendicular to the plane containing O and **F,** Fig. 4–8a, can also be expressed using the vector cross product, namely

$$\mathbf{M}_O = \mathbf{r} \times \mathbf{F} \qquad\qquad (4\text{--}5)$$

Here **r** represents a position vector drawn from O to *any point A* lying on the line of action of **F.** It will now be shown that the moment $\mathbf{M}_O$, when determined by this cross product, has the proper magnitude and direction.

Magnitude. By definition the angle θ for the cross product $\mathbf{r} \times \mathbf{F}$ is measured between the *tails* of **r** and **F.** Hence, **r** must be treated as a sliding vector so that θ can be properly constructed, Fig. 4–8b. As shown, the moment arm $d = r \sin \theta$ so that from Eq. 4–1 we have

$$M_O = rF \sin \theta = F (r \sin \theta) = Fd$$

which agrees with Eq. 4–4.

Direction. The direction and sense of $\mathbf{M}_O$ are determined by the right-hand rule of the cross product. Thus, extending **r** to the dashed position and curling the right-hand fingers from **r** toward **F,** "**r** cross **F,**" the thumb is directed upward or perpendicular to the plane in the *same direction* as $\mathbf{M}_O$, the moment of the force about point O, Fig. 4–8b. Note that the "curl" of the fingers like the curl around the moment vector indicates the sense of rotation caused by the force. Since the cross product is not commutative, it is important that the *proper order* of **r** and **F** be maintained in Eq. 4–5.

Cartesian Vector Formulation. If the position vector **r** and force **F** are expressed in Cartesian vector form, Fig. 4–9, then from Eq. 4–5 we have

$$\mathbf{M}_O = \mathbf{r} \times \mathbf{F} = \begin{vmatrix} \mathbf{i} & \mathbf{j} & \mathbf{k} \\ r_x & r_y & r_z \\ F_x & F_y & F_z \end{vmatrix} \qquad (4\text{--}6)$$

where r_x, r_y, r_z represent the Cartesian components of the position vector drawn from point O to *any point* on the line of action of the force

F_x, F_y, F_z represent the Cartesian components of the force vector

If the determinant is expanded, then like Eq. 4–2 we have

$$\mathbf{M}_O = (r_y F_z - r_z F_y)\mathbf{i} - (r_x F_z - r_z F_x)\mathbf{j} + (r_x F_y - r_y F_x)\mathbf{k} \qquad (4\text{--}7)$$

The physical meaning of these three moment components becomes evident by studying Fig. 4–9a. For example, the **i** component is computed from the moments of forces $\mathbf{F}_z$ and $\mathbf{F}_y$ about the x axis. Since the lines of action of these forces pass through points D and E, respectively, then the magnitude of the moment of $\mathbf{F}_z$ about point A on the x axis is $r_y F_z$. By the right-hand rule this component acts in the positive **i** direction. Likewise, $\mathbf{F}_y$ contributes a moment component of $r_z F_y(-\mathbf{i})$. Force component $\mathbf{F}_x$ does *not* create a moment about the x axis since this force is *parallel* to the x axis. As an exercise, establish the **j** and **k** components of $\mathbf{M}_O$ in this manner, and show that indeed the expanded form of the determinant, Eq. 4–7, represents the moment of $\mathbf{F}$ about point O. Note that once computed, $\mathbf{M}_O$ will be *perpendicular* to the shaded plane containing vectors **r** and **F**, Fig. 4–9b.

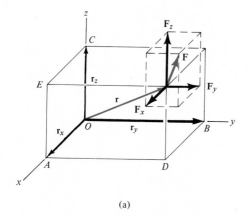

(a)

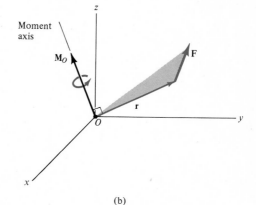

(b)

Fig. 4–9

Example 4–1

For each case illustrated in Fig. 4–10, determine the moment of the force about point O.

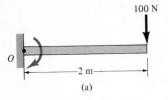
(a)

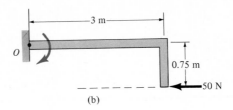
(b)

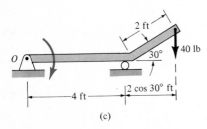
(c)

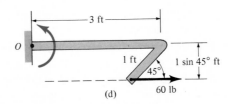
(d)

Fig. 4–10

Solution (Scalar Analysis)

The line of action of each force is extended as a dashed line in order to establish the moment arm d. Also, the tendency of rotation caused by the force is shown as a colored curl. Thus,

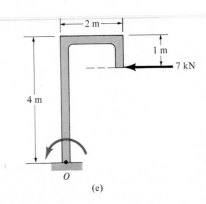
(e)

Fig. 4–10a, $M_O = (100 \text{ N})(2 \text{ m}) = 200 \text{ N} \cdot \text{m} \downarrow$ *Ans.*

Fig. 4–10b, $M_O = (50 \text{ N})(0.75 \text{ m}) = 37.5 \text{ N} \cdot \text{m} \downarrow$ *Ans.*

Fig. 4–10c, $M_O = (40 \text{ lb})(4 \text{ ft} + 2 \cos 30° \text{ ft}) = 229.3 \text{ lb} \cdot \text{ft} \downarrow$ *Ans.*

Fig. 4–10d, $M_O = (60 \text{ lb})(1 \sin 45° \text{ ft}) = 42.4 \text{ lb} \cdot \text{ft} \uparrow$ *Ans.*

Fig. 4–10e, $M_O = (7 \text{ kN})(4 \text{ m} - 1 \text{ m}) = 21 \text{ kN} \cdot \text{m} \uparrow$ *Ans.*

Example 4–2

Determine the moment of the 800-N force acting on the frame in Fig. 4–11 about points A, B, C, and D.

Solution (Scalar Analysis)

In general, $M = Fd$, where d is the moment arm or *perpendicular distance* from the *point* on the moment axis to the *line of action* of the force. Hence,

$M_A = 800 \text{ N}(2.5 \text{ m}) = 2000 \text{ N} \cdot \text{m} \downarrow$ *Ans.*

$M_B = 800 \text{ N}(1.5 \text{ m}) = 1200 \text{ N} \cdot \text{m} \downarrow$ *Ans.*

$M_C = 800 \text{ N}(0) = 0$ (line of action of **F** passes through C)   *Ans.*

$M_D = 800 \text{ N}(0.5 \text{ m}) = 400 \text{ N} \cdot \text{m} \uparrow$ *Ans.*

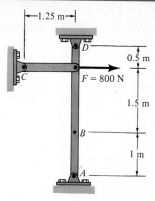

Fig. 4–11

Example 4–3

Determine the point of application P and the direction of a 20-lb force that lies in the plane of the square plate shown in Fig. 4–12a, so that this force creates the greatest counterclockwise moment about point O. What is this moment?

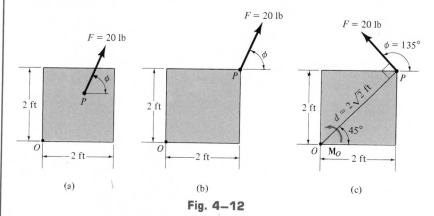

(a)   (b)   (c)

Fig. 4–12

Solution (Scalar Analysis)

Since the maximum moment created by the force is required, the force must act on the plate at a distance *farthest* from point O. As shown in Fig. 4–12b, the point of application must therefore be at the diagonal corner. In order to produce *counterclockwise* rotation of the plate about O, **F** must act at an angle $45° < \phi < 225°$. The greatest moment is produced when the line of action of **F** is *perpendicular* to d, i.e., $\phi = 135°$, Fig. 4–12c. The maximum moment is therefore

$$M_O = Fd = (20 \text{ lb})(2\sqrt{2} \text{ ft}) = 56.6 \text{ lb} \cdot \text{ft} \uparrow \quad Ans.$$

By the right-hand rule, $\mathbf{M}_O$ is directed out of the page.

Example 4–4

The pole in Fig. 4–13a is subjected to a 60-N force that is directed from C to B. Determine the magnitude of the moment created by this force about point A.

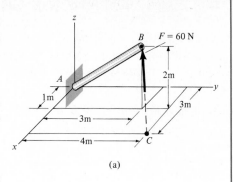

(a)

Solution (Vector Analysis)

As shown in Fig. 4–13b, either one of two position vectors can be used for the solution, since $\mathbf{M}_A = \mathbf{r}_B \times \mathbf{F}$ or $\mathbf{M}_A = \mathbf{r}_C \times \mathbf{F}$. The position vectors are represented as

$$\mathbf{r}_B = \{1\mathbf{i} + 3\mathbf{j} + 2\mathbf{k}\}\,\text{m} \quad \text{and} \quad \mathbf{r}_C = \{3\mathbf{i} + 4\mathbf{j}\}\,\text{m}$$

The force has a magnitude of 60 N and a direction specified by the unit vector $\mathbf{u}_F$, directed from C to B. Thus,

$$\mathbf{F} = (60\ \text{N})\mathbf{u}_F = (60)\left[\frac{(1-3)\mathbf{i} + (3-4)\mathbf{j} + (2-0)\mathbf{k}}{\sqrt{(-2)^2 + (-1)^2 + (2)^2}}\right]$$

$$= \{-40\mathbf{i} - 20\mathbf{j} + 40\mathbf{k}\}\,\text{N}$$

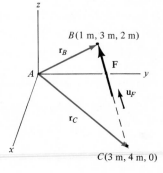

(b)

Substituting into the determinant formulation, Eq. 4–6, and following the scheme for determinant expansion as stated in the footnote on page 98, we have

$$\mathbf{M}_A = \mathbf{r}_B \times \mathbf{F} = \begin{vmatrix} \mathbf{i} & \mathbf{j} & \mathbf{k} \\ 1 & 3 & 2 \\ -40 & -20 & 40 \end{vmatrix}$$

$$\mathbf{M}_A = [3(40) - 2(-20)]\mathbf{i} - [1(40) - 2(-40)]\mathbf{j} + [1(-20) - 3(-40)]\mathbf{k}$$

or

$$\mathbf{M}_A = \mathbf{r}_C \times \mathbf{F} = \begin{vmatrix} \mathbf{i} & \mathbf{j} & \mathbf{k} \\ 3 & 4 & 0 \\ -40 & -20 & 40 \end{vmatrix}$$

$$\mathbf{M}_A = [4(40) - 0(-20)]\mathbf{i} - [3(40) - 0(-40)]\mathbf{j} + [3(-20) - 4(-40)]\mathbf{k}$$

In both cases,

$$\mathbf{M}_A = \{160\mathbf{i} - 120\mathbf{j} + 100\mathbf{k}\}\,\text{N} \cdot \text{m}$$

The *magnitude* of $\mathbf{M}_A$ is therefore

$$M_A = \sqrt{(160)^2 + (-120)^2 + (100)^2} = 223.6\ \text{N} \cdot \text{m} \qquad \textit{Ans.}$$

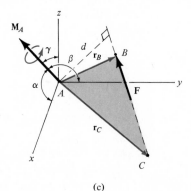

(c)

Fig. 4–13

Note that $\mathbf{M}_A$ acts perpendicular to the shaded plane containing vectors $\mathbf{F}$, $\mathbf{r}_B$, and $\mathbf{r}_C$, Fig. 4–13c. (How would you find its coordinate direction angles $\alpha = 44.3°$, $\beta = 122.5°$, $\gamma = 63.4°$?) Had this problem been worked using a scalar approach, where $M_A = Fd$, notice the difficulty that might arise in obtaining the moment arm d.

Transmissibility of a Force and the Principle of Moments 4.3

Using the definition of a moment formalized by the cross product, we will now present two important concepts often used throughout our study of mechanics.

Transmissibility of a Force. Consider the force $\mathbf{F}$ applied at point A in Fig. 4–14. The moment created by $\mathbf{F}$ about point O is $\mathbf{M}_O = \mathbf{r}_A \times \mathbf{F}$; however, it was shown that the position vector "$\mathbf{r}$" can extend from O to *any point* on the line of action of $\mathbf{F}$. Consequently, $\mathbf{F}$ may be applied at point B and the same moment $\mathbf{M}_O = \mathbf{r}_B \times \mathbf{F}$ will be computed. Hence, $\mathbf{F}$ has the properties of a *sliding vector* and can therefore act at *any point along its line of action and still create the same moment about point O.* We refer to $\mathbf{F}$ in this regard as being "transmissible" and we will discuss this property further in Sec. 4–6.

Principle of Moments. A concept often used in mechanics is the *principle of moments,* which is sometimes referred to as *Varignon's theorem* since it was originally developed by the French mathematician Varignon (1654–1722). It states that *the moment of a force about a point is equal to the sum of the moments of the force's components about the point.* The proof follows directly from the distributive law of the vector cross product. To show this, consider the force $\mathbf{F}$ and two of its components, where $\mathbf{F} = \mathbf{F}_1 + \mathbf{F}_2$, Fig. 4–15. We have

$$\mathbf{M}_O = \mathbf{r} \times \mathbf{F}_1 + \mathbf{r} \times \mathbf{F}_2 = \mathbf{r} \times (\mathbf{F}_1 + \mathbf{F}_2) = \mathbf{r} \times \mathbf{F}$$

This concept has important applications to the solution of problems and proofs of theorems that follow, since it allows us to consider the moments of a force's components rather than the force itself.

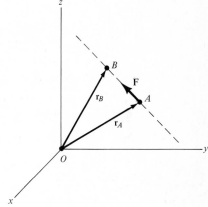

Fig. 4–14

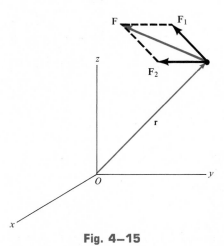

Fig. 4–15

105

Example 4–5

A 200-N force acts on the bracket shown in Fig. 4–16a. Determine the moment of the force about point A.

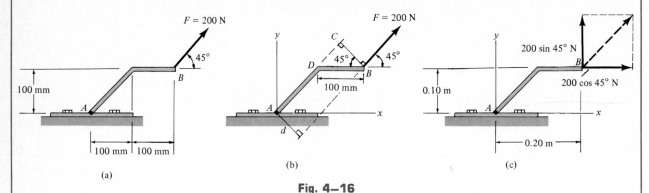

Fig. 4–16

Solution I

The moment arm d can be computed by trigonometry, using the construction shown in Fig. 4–16b. From triangle BCD,

$$CB = d = 100 \cos 45° = 70.71 \text{ mm} = 0.070\,71 \text{ m}$$

Thus,

$$M_A = Fd = 200 \text{ N}(0.070\,71 \text{ m}) = 14.14 \text{ N} \cdot \text{m} \curvearrowleft$$

According to the right-hand rule, M_A is directed in the $+\mathbf{k}$ direction since the force tends to rotate *counterclockwise* about point A. Hence, reporting the moment as a Cartesian vector, we have

$$\mathbf{M}_A = \{14.14\mathbf{k}\} \text{ N} \cdot \text{m} \qquad\qquad Ans.$$

Solution II

The 200-N force may be resolved into x and y components, as shown in Fig. 4–16c. In accordance with the principle of moments, the moment computed about point A is equivalent to the sum of the moments produced by the two force components. Assuming counterclockwise rotation as positive, i.e., in the $+\mathbf{k}$ direction, we can add the moment of each force algebraically, since they both act along the same moment axis (perpendicular to the x-y plane, passing through A).

$$\zeta + M_A = (200 \sin 45°)(0.20) - (200 \cos 45°)(0.10)$$
$$= 14.14 \text{ N} \cdot \text{m} \curvearrowleft$$

Thus

$$\mathbf{M}_A = \{14.14\mathbf{k}\} \text{ N} \cdot \text{m} \qquad\qquad Ans.$$

By comparison, it is seen that Solution II provides a more *convenient method* for analysis than Solution I since the moment arm for each component force is easier to establish.

Example 4–6

The force **F** acts at the end of the beam shown in Fig. 4–17*a*. Determine the moment of the force about point *O*.

Solution I (*Scalar Analysis*)

The force is resolved into its *x* and *y* components as shown in Fig. 4–17*b* and the moments of the components are computed about point *O*. Taking positive moments as counterclockwise, i.e., in the $+\mathbf{k}$ direction, we have

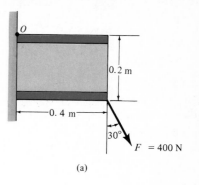

(a)

$$\curvearrowleft+ M_O = 400 \sin 30°(0.2) - 400 \cos 30°(0.4) \qquad (1)$$
$$= -98.6 \text{ N} \cdot \text{m} = 98.6 \text{ N} \cdot \text{m} \downarrow$$

or

$$\mathbf{M}_O = \{-98.6\mathbf{k}\} \text{ N} \cdot \text{m} \qquad \textit{Ans.}$$

Solution II (*Vector Analysis*)

Using a Cartesian vector approach, the force and position vector shown in Fig. 4–17*c* can be represented as

$$\mathbf{r} = \{0.4\mathbf{i} - 0.2\mathbf{j}\} \text{ m}$$
$$\mathbf{F} = \{400 \sin 30°\mathbf{i} - 400 \cos 30°\mathbf{j}\} \text{ N}$$
$$= \{200.0\mathbf{i} - 346.4\mathbf{j}\} \text{ N}$$

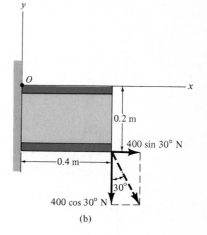

(b)

Hence the required moment becomes

$$\mathbf{M}_O = \mathbf{r} \times \mathbf{F} = \begin{vmatrix} \mathbf{i} & \mathbf{j} & \mathbf{k} \\ 0.4 & -0.2 & 0 \\ 200.0 & -346.4 & 0 \end{vmatrix}$$

$$\mathbf{M}_O = 0\mathbf{i} - 0\mathbf{j} + [0.4(-346.4) - (-0.2)(200.0)]\mathbf{k} \qquad (2)$$
$$\mathbf{M}_O = \{-98.6\mathbf{k}\} \text{ N} \cdot \text{m} \qquad \textit{Ans.}$$

By comparison, it is seen that the scalar analysis (Solution I) provides a more *convenient method* for analysis than Solution II since the moment arm for each component force is easy to establish. Hence, *this method is generally recommended for solving problems displayed in two dimensions*. Although the Cartesian vector analysis (Solution II) *automatically* accounts for the sign of the moment of each component force, Eq. (2), these directions can easily be established *directly* from the diagram, Fig. 4–17*b*, Eq. (1). Hence, there really is no need to set up and evaluate a determinant. As a result, *Cartesian vector analysis is generally recommended only for solving three-dimensional problems,* where the moment arms and force components are often more difficult to determine.

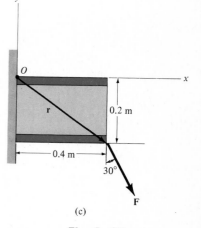

(c)

Fig. 4–17

Problems

4–1. If **A, B,** and **D** are given vectors, prove the distributive law for the vector cross product, i.e., $\mathbf{A} \times (\mathbf{B} + \mathbf{D}) = (\mathbf{A} \times \mathbf{B}) + (\mathbf{A} \times \mathbf{D})$.

4–2. Prove the triple scalar product identity $\mathbf{A} \cdot \mathbf{B} \times \mathbf{C} = \mathbf{A} \times \mathbf{B} \cdot \mathbf{C}$.

4–3. Given the three nonzero vectors **A, B,** and **C,** show that if $\mathbf{A} \cdot (\mathbf{B} \times \mathbf{C}) = 0$, the three vectors *must* lie in the same plane.

*4–4. Determine the magnitude and direction of the moment of the force at A about point P.

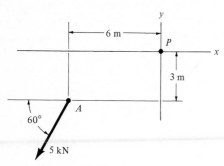

Prob. 4–4

4–5. Determine the magnitude and direction of the moment of the force at A about point P.

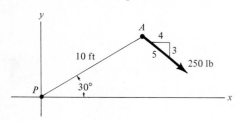

Prob. 4–5

4–6. Determine the moment of the force at A about point P. Express the result as a Cartesian vector.

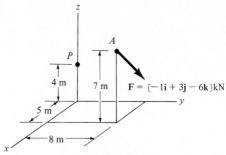

Prob. 4–6

4–7. Determine the moment of the force at A about point P. Express the result as a Cartesian vector.

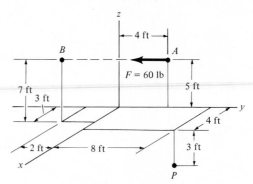

Prob. 4–7

*4–8. Determine the moment of each force about the bolt passing through point A.

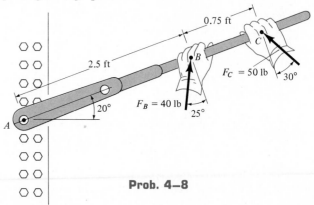

Prob. 4–8

4–9. The 70-N force acts on the end of the pipe at B. Determine (a) the moment of this force about point A, and (b) the magnitude and direction of a horizontal force, applied at C, which produces the same moment.

*4–12. A man holds the end of the stick which supports a 2.2-kg cylinder. If the maximum moment he can develop at his wrist is 13.5 N · m, determine the maximum angle of tilt, θ, which will not allow the cylinder to fall.

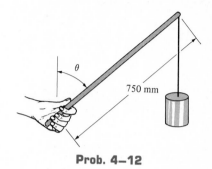

Prob. 4–12

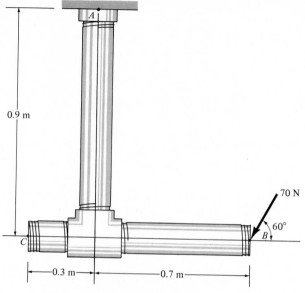

Prob. 4–9

4–10. Determine the moment about point A of each of the three forces acting on the beam.

4–13. Two men exert forces of $F = 80$ lb and $P = 50$ lb on the ropes as shown. Determine the moment of each force about point A. Which way will the pole rotate, clockwise or counterclockwise?

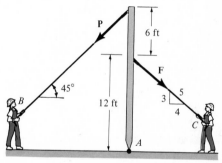

Probs. 4–13 / 4–14

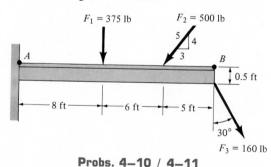

Probs. 4–10 / 4–11

4–14. Two men pull on the ropes shown. If the man at B exerts a force of $P = 30$ lb on his rope, determine the magnitude of the force **F** the man at C must exert to prevent the pole from tipping, i.e., so the resultant moment about A of both forces is zero.

4–11. Determine the moment about point B of each of the three forces acting on the beam.

4–15. Determine the direction θ $(0° \leqslant \theta \leqslant 180°)$ of the 40-lb force **F** so that **F** produces (a) the maximum moment about point A and (b) no moment about point A. Compute the moment in each case.

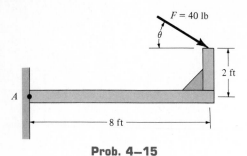

Prob. 4–15

*4–16. The crane can be adjusted for any angle $0° \leqslant \theta \leqslant 90°$ and any extension $0 \leqslant x \leqslant 5$ m. For a suspended mass of 120 kg, determine the moment developed at A as a function of x and θ. What values of both x and θ develop the maximum possible moment at A? Compute this moment. Neglect the size of the pulley at B.

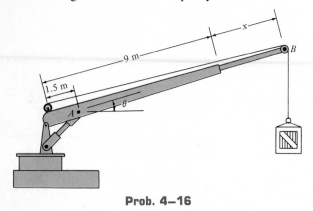

Prob. 4–16

4–17. The hammer is subjected to a horizontal force of $P = 15$ lb at the grip, whereas it takes a force of $F = 110$ lb at the claw to pull the nail out. Find the moment of each force about point A and determine if **P** is sufficient to pull out the nail. The hammer head contacts the board at point A.

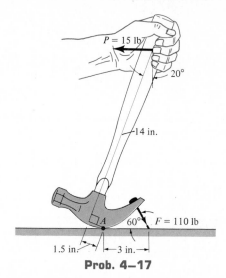

Prob. 4–17

4–18. The man exerts the two forces shown on the handle of the shovel. Determine the moment of each force about the blade at A.

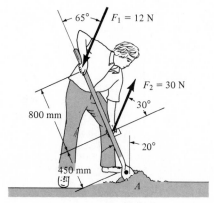

Prob. 4–18

110

4–19. Segments of drill pipe D for an oil well are tightened a prescribed amount by using a set of tongs T, which grip the pipe, and a hydraulic cylinder (not shown) to regulate the force $\mathbf{F}$ applied to the tongs. This force acts along the cable which passes around the small pulley P. If the cable is originally perpendicular to the tongs as shown, determine the magnitude of force $\mathbf{F}$ which must be applied so that the moment about the pipe is $M = 2000$ lb $\cdot$ ft. In order to maintain this same moment what magnitude of $\mathbf{F}$ is required when the tongs rotate 30° to the dashed position? *Note:* The angle DAP is not 90° in this position.

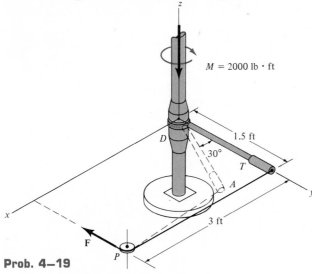

Prob. 4–19

*4–20. The towline exerts a force of $P = 4$ kN at the end of the 20-m long crane boom. If $\theta = 30°$, determine the placement x of the hook at A so that this force creates a maximum moment about point O.

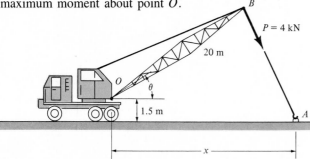

Probs. 4–20 / 4–21

4–21. The towline exerts a force of $P = 4$ kN at the end of the 20-m long crane boom. If $x = 25$ m, determine the position θ of the boom so that this force creates a maximum moment about point O.

4–22. A force of $\mathbf{F} = \{6\mathbf{i} - 2\mathbf{j} + 1\mathbf{k}\}$ kN produces a moment of $\mathbf{M}_O = \{4\mathbf{i} + 5\mathbf{j} - 14\mathbf{k}\}$ kN $\cdot$ m about the origin of coordinates, point O. If the force acts at a point having an x coordinate of $x = 1$ m, determine the y and z coordinates.

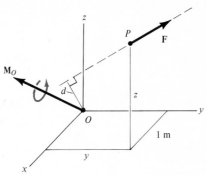

Probs. 4–22 / 4–23

4–23. The force $\mathbf{F} = \{6\mathbf{i} + 8\mathbf{j} + 10\mathbf{k}\}$ N creates a moment about point O of $\mathbf{M}_O = \{-14\mathbf{i} + 8\mathbf{j} + 2\mathbf{k}\}$ N $\cdot$ m. If the force passes through a point having an x coordinate of 1 m, determine the y and z coordinates of the point. Also, realizing that $M_O = Fd$, determine the perpendicular distance d from point O to the line of action of $\mathbf{F}$.

*4–24. The man pulls on the rope with a force of 20 N. Determine the moment that this force exerts about the base of the pole at O. Solve the problem two ways, i.e., by using a position vector from O to A, then O to B.

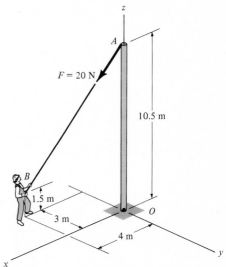

Prob. 4–24

111

4–25. Using Cartesian vector analysis, compute the moment of each of the three forces acting on the column about point A.

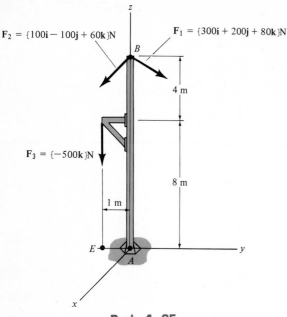

Prob. 4–25

4–27. The pipe assembly is subjected to the 80-N force. Determine the moment of this force about the point of attachment A.

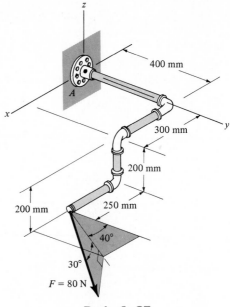

Prob. 4–27

4–26. The curved rod has a radius of 5 ft. If a force of 60 lb acts at its end as shown, determine the moment of this force about point C.

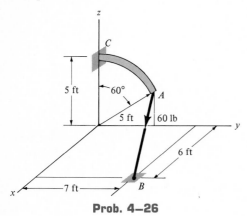

Prob. 4–26

*4–28. The 6-m boom lies in the y-z plane. If a tension force of $F = 4$ kN acts in the cable AB, determine the moment of this force about point O. Realizing that $M = Fd$, find the shortest distance d from point O to the cable.

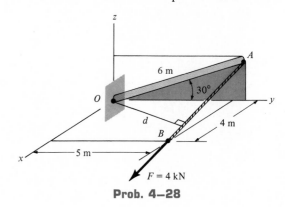

Prob. 4–28

4–29. The rope BC exerts a force of 350 N on the lid at B. Determine the moment of this force about the hinge at A.

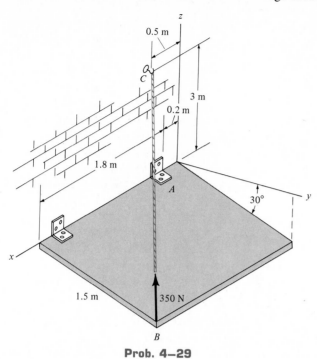

Prob. 4–29

4.4 Moment of a Force About a Specified Axis

Recall that when the moment of a force is computed about a point, the moment and its axis are *always* perpendicular to the plane containing the force and the moment arm. In some problems it is important to find the component of this moment along a specified axis that passes through the point. If the force components or the appropriate moment arms are difficult to determine, vector analysis should be used for the computation.

To see how this is done, consider a rigid body of general shape that is subjected to a force acting at point A, Fig. 4–18. Here we wish to determine the effect of $\mathbf{F}$ in tending to rotate the body about the aa' axis. This tendency for rotation is measured by the moment component $\mathbf{M}_a$. To determine $\mathbf{M}_a$ we first compute the moment of $\mathbf{F}$ about an *arbitrary point* O that lies on the axis. In this case, $\mathbf{M}_O$ is expressed by the cross product $\mathbf{M}_O = \mathbf{r} \times \mathbf{F}$, where $\mathbf{r}$ is directed from O to A. Since $\mathbf{M}_O$ acts along the moment axis bb', which is perpendicular to the plane containing $\mathbf{r}$ and $\mathbf{F}$, the component or projection of $\mathbf{M}_O$ onto the aa' axis is then represented by $\mathbf{M}_a$. The *magnitude* of $\mathbf{M}_a$ can be determined by the dot product, which was discussed in Sec. 2.9. In this case we have $M_a = M_O \cos \theta = \mathbf{u}_a \cdot \mathbf{M}_O$, where $\mathbf{u}_a$ is a unit vector used to define the direction of the aa' axis. As a general expression,

$$M_a = \mathbf{u}_a \cdot (\mathbf{r} \times \mathbf{F}) \tag{4–8}$$

In vector algebra, this combination of dot and cross product yielding the scalar M_a is termed the *mixed triple product*. Provided that the Cartesian components of each of the vectors in Eq. 4–8 are known, the mixed triple product may be written in determinant form as

$$M_a = (u_{a_x}\mathbf{i} + u_{a_y}\mathbf{j} + u_{a_z}\mathbf{k}) \cdot \begin{vmatrix} \mathbf{i} & \mathbf{j} & \mathbf{k} \\ r_x & r_y & r_z \\ F_x & F_y & F_z \end{vmatrix}$$

or simply

$$M_a = \mathbf{u}_a \cdot (\mathbf{r} \times \mathbf{F}) = \begin{vmatrix} u_{a_x} & u_{a_y} & u_{a_z} \\ r_x & r_y & r_z \\ F_x & F_y & F_z \end{vmatrix} \tag{4–9}$$

where u_{a_x}, u_{a_y}, u_{a_z} represent the Cartesian components of the unit vector defining the direction of the aa' axis

r_x, r_y, r_z represent the Cartesian components of the position vector drawn from any point O on the aa' axis to any point A on the line of action of the force

F_x, F_y, F_z represent the Cartesian components of the force vector.

The following examples numerically illustrate application of Eq. 4–9.

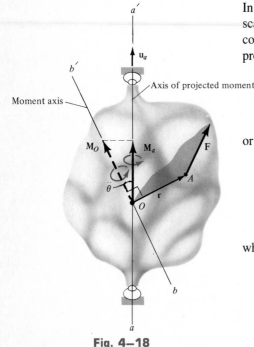

Fig. 4–18

Example 4–7

The force $\mathbf{F} = \{-40\mathbf{i} + 20\mathbf{j} + 10\mathbf{k}\}$ N acts at point A shown in Fig. 4–19a. Compute the moment of this force about the x axis and about the Oa axis.

Solution I *(Vector Analysis)*

We can solve this problem by using the position vector $\mathbf{r}_A$. Why? Since $\mathbf{r}_A = \{-3\mathbf{i} + 4\mathbf{j} + 6\mathbf{k}\}$ m, and $\mathbf{u}_x = \mathbf{i}$, then applying Eq. 4–9,

$$M_x = \mathbf{i} \cdot (\mathbf{r}_A \times \mathbf{F}) = \begin{vmatrix} 1 & 0 & 0 \\ -3 & 4 & 6 \\ -40 & 20 & 10 \end{vmatrix}$$

$$M_x = 1[4(10) - 6(20)] - 0[(-3)(10) - 6(-40)] + 0[(-3)(20) - 4(-40)]$$
$$= -80 \text{ N} \cdot \text{m} \qquad\qquad Ans.$$

The negative sign indicates that the sense of $\mathbf{M}_x$ is opposite to $\mathbf{i}$. Also, since $\mathbf{u}_{Oa} = -\frac{3}{5}(\mathbf{i}) + \frac{4}{5}(\mathbf{j})$, then

$$M_{Oa} = \mathbf{u}_{Oa} \cdot (\mathbf{r}_A \times \mathbf{F}) = \begin{vmatrix} -\frac{3}{5} & \frac{4}{5} & 0 \\ -3 & 4 & 6 \\ -40 & 20 & 10 \end{vmatrix}$$

$$= -\tfrac{3}{5}[4(10) - 6(20)] - \tfrac{4}{5}[(-3)(10) - 6(40)] + 0[(-3)(20) - 4(-40)]$$
$$= -120 \text{ N} \cdot \text{m} \qquad\qquad Ans.$$

The components are shown in Fig. 4–19b.

Solution II *(Scalar Analysis)*

Since the force components and moment arms are easy to determine for the moments about the x axis, a scalar analysis can be used for finding M_x. Referring to Fig. 4–19c, only the 10-N and 20-N forces contribute moments about the x axis. (The line of action of the 40-N force is *parallel* to this axis and hence its moment about x is zero.) Using the right-hand rule, we have

$$M_x = (10 \text{ N})(4 \text{ m}) - (20 \text{ N})(6 \text{ m}) = -80 \text{ N} \cdot \text{m} \qquad Ans.$$

Although not required here, note also that

$$M_y = (10 \text{ N})(3 \text{ m}) - (40 \text{ N})(6 \text{ m}) = -210 \text{ N} \cdot \text{m}$$
$$M_z = (40 \text{ N})(4 \text{ m}) - (20 \text{ N})(3 \text{ m}) = 100 \text{ N} \cdot \text{m}$$

Notice that to determine M_{Oa} by this scalar method requires much more effort, since the force components of 40 N and 20 N are not perpendicular to the direction of Oa. The vector analysis yields a more direct solution.

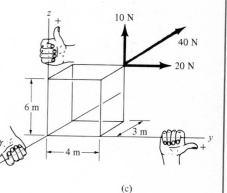

Fig. 4–19

Example 4–8

The rod shown in Fig. 4–20a is supported by two brackets at A and B. Determine the magnitude of the moment $\mathbf{M}_{AB}$ produced by force $\mathbf{F} = \{-600\mathbf{i} + 200\mathbf{j} - 300\mathbf{k}\}$ N, which tends to rotate the rod about the AB axis.

Solution

A vector analysis using Eq. 4–9 will be considered for the solution since the components of the force and the moment arm are difficult to determine. In general, the magnitude of $\mathbf{M}_{AB}$ is

$$M_{AB} = \mathbf{u}_B \cdot (\mathbf{r} \times \mathbf{F}) \tag{1}$$

Each of the terms in this equation will now be identified.

Unit vector $\mathbf{u}_B$ defines the direction of the AB axis of the rod, Fig. 4–20b, where

$$\mathbf{u}_B = \frac{\mathbf{r}_B}{r_B} = \frac{0.4\mathbf{i} + 0.2\mathbf{j}}{\sqrt{(0.4)^2 + (0.2)^2}} = 0.894\mathbf{i} + 0.447\mathbf{j}$$

Vector $\mathbf{r}$ is directed from *any point* on the AB axis to *any* point on the line of action of the force. For example, position vectors $\mathbf{r}_C$ and $\mathbf{r}_D$ are suitable, Fig. 4–20b. For simplicity, choose $\mathbf{r}_D$, where

$$\mathbf{r}_D = \{0.2\mathbf{j}\} \text{ m}$$

The force is

$$\mathbf{F} = \{-600\mathbf{i} + 200\mathbf{j} - 300\mathbf{k}\} \text{ N}$$

Substituting these vectors into the determinant form of Eq. (1) and expanding, we have

$$M_{AB} = \mathbf{u}_B \cdot (\mathbf{r}_D \times \mathbf{F}) = \begin{vmatrix} 0.894 & 0.447 & 0 \\ 0 & 0.2 & 0 \\ -600 & 200 & -300 \end{vmatrix}$$

$$= 0.894[0.2(-300) - 0(200)] - 0.447[0(-300) - 0(-600)]$$
$$+ 0[0(200) - 0.2(-600)]$$

$$= -53.64 \text{ N} \cdot \text{m}$$

The negative sign indicates that the sense of $\mathbf{M}_{AB}$ is opposite to that of $\mathbf{u}_B$. Expressing $\mathbf{M}_{AB}$ as a vector,

$$\mathbf{M}_{AB} = M_{AB}\mathbf{u}_B = (-53.64 \text{ N} \cdot \text{m})(0.894\mathbf{i} + 0.447\mathbf{j})$$
$$= \{-48.0\mathbf{i} - 24.0\mathbf{j}\} \text{ N} \cdot \text{m} \qquad \textit{Ans.}$$

The result is shown in Fig. 4–20b.

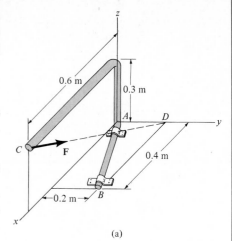

(a)

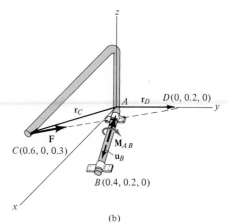

(b)

Fig. 4–20

Problems

4–30. Determine the moment of the force **F** about the Oa axis. Express the result as a Cartesian vector.

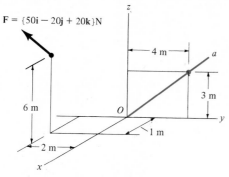

F = {50**i** − 20**j** + 20**k**}N

Prob. 4–30

4–31. Determine the moment of the force **F** about the Aa axis. Express the result as a Cartesian vector.

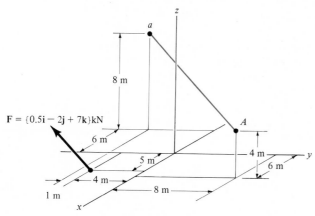

F = {0.5**i** − 2**j** + 7**k**}kN

Prob. 4–31

*4–32. Determine the moment of the 80-lb force about the Oa axis. Express the result as a Cartesian vector.

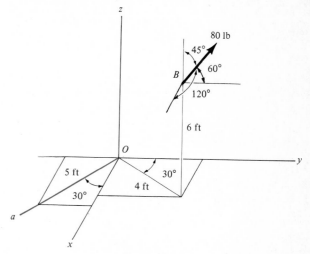

Prob. 4–32

4–33. Determine the moment of the 350-N force acting on the lid at B about the hinged x axis of the lid in Prob. 4–29.

4–34. The metal shaft is held in a lath as shown. The cutting tool exerts a force **F** on the shaft in the direction shown. Determine the moment of this force about the y axis of the shaft.

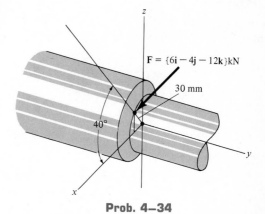

F = {6**i** − 4**j** − 12**k**}kN

Prob. 4–34

117

4–35. The 50-lb force acts on the gear in the direction shown. Determine the moment of this force about the y axis.

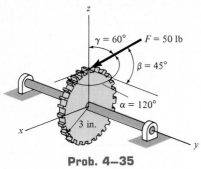

Prob. 4–35

*4–36. The hood of the automobile is supported by the strut AB, which exerts a force of 24 lb on the hood. Determine the moment of this force about the hinged axis y.

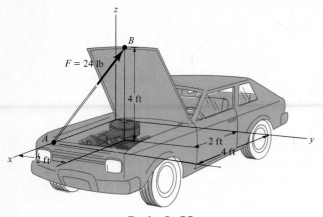

Prob. 4–36

4–37. A vertical force of $F = 60$ N is applied to the handle of the pipe wrench. Determine the moment that this force exerts along the axis AB (x axis) of the pipe assembly. Both the wrench and pipe assembly ABC lie in the x-y plane. *Suggestion:* Use a scalar analysis.

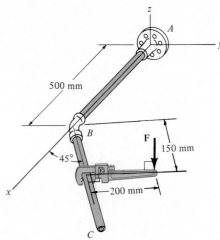

Probs. 4–37 / 4–38

4–38. Determine the magnitude of the vertical force $\mathbf{F}$ acting on the handle of the wrench so that this force produces a component of moment along the AB axis (x axis) of the pipe assembly of $(\mathbf{M}_A)_x = \{-5\mathbf{i}\}$ N $\cdot$ m. Both the pipe assembly ABC and the wrench lie in the x-y plane. *Suggestion:* Use a scalar analysis.

Moment of a Couple 4.5

A *couple* is defined as two parallel forces that have the same magnitude, opposite directions, and are separated by a perpendicular distance d, Fig. 4–21. Since the resultant force of the two forces composing the couple is zero, the only effect of a couple is to produce a rotation or tendency of rotation in a specified direction.

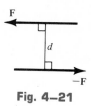

Fig. 4–21

The moment produced by a couple, called a couple moment, is equivalent to the sum of the moments of both couple forces, computed about any arbitrary point O in space. To show this, consider position vectors $\mathbf{r}_A$ and $\mathbf{r}_B$, directed from O to points A and B lying on the line of action of $-\mathbf{F}$ and $\mathbf{F}$, Fig. 4–22. The couple moment computed about O is therefore

$$\mathbf{M} = \mathbf{r}_A \times (-\mathbf{F}) + \mathbf{r}_B \times (\mathbf{F})$$
$$= (\mathbf{r}_B - \mathbf{r}_A) \times \mathbf{F}$$

By the triangle law of vector addition, $\mathbf{r}_A + \mathbf{r} = \mathbf{r}_B$ or $\mathbf{r} = \mathbf{r}_B - \mathbf{r}_A$, so that

$$\mathbf{M} = \mathbf{r} \times \mathbf{F} \qquad (4\text{--}10)$$

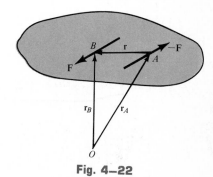

Fig. 4–22

This result indicates that a couple moment is a *free vector*, i.e., it can act at *any point*, since $\mathbf{M}$ depends *only* upon the position vector directed *between* the forces and *not* the position vectors $\mathbf{r}_A$ or $\mathbf{r}_B$, directed from point O to the forces. Consequently, this concept is unlike the moment of a force, which requires a definite point (or axis) about which moments are computed.

Scalar Formulation. The moment of a couple, $\mathbf{M}$, Fig. 4–23, is defined as having a *magnitude* of

$$M = Fd \qquad (4\text{--}11)$$

where F is the magnitude of one of the forces and d is the perpendicular distance or moment arm between the forces. The *direction* and sense of the couple moment are determined by the right-hand rule, where the thumb indicates the direction when the fingers are curled with the sense of rotation caused by the forces. In all cases, $\mathbf{M}$ acts perpendicular to the plane containing the two forces.

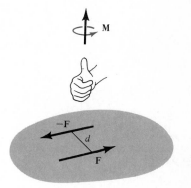

Fig. 4–23

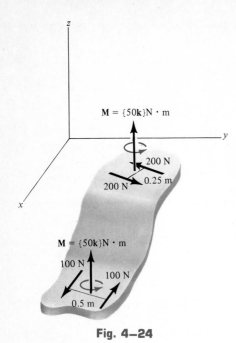

M = {50k}N · m

200 N

200 N 0.25 m

M = {50k}N · m

100 N

100 N

0.5 m

Fig. 4–24

Vector Formulation.   The moment of a couple can also be expressed by the vector cross product using Eq. 4–10,

$$M = r \times F \qquad\qquad (4\text{--}12)$$

Application of this equation is easily remembered if one thinks of taking the moments of both forces about a point lying on the line of action of one of the forces. For example, if moments are taken about point A in Fig. 4–22 the moment of $-F$ is zero and the moment of F is defined from Eq. 4–12.

Equivalent Couples.   Two couples are said to be equivalent if they produce the same moment. Since the moment produced by a couple is always perpendicular to the plane containing the couple forces, it is therefore necessary that the forces of equal couples lie either in the same plane or in planes that are *parallel* to one another. In this way, the direction of each couple moment will be the same, that is, perpendicular to the parallel planes. For example, the two couples shown in Fig. 4–24 are equivalent. One couple is produced by a pair of 100-N forces separated by a distance of $d = 0.5$ m, and the other is produced by a pair of 200-N forces separated by a distance of 0.25 m. Since the planes in which the forces act are parallel to the x-y plane, the moment produced by each of the couples may be expressed as $M = \{50k\}$ N · m.

Resultant Couple.   Since couple moments are free vectors, they may be applied at any point P on a body and added vectorially. For example, the two couples acting on different planes of the rigid body in Fig. 4–25a may be replaced by their corresponding moments M_1 and M_2, shown in Fig. 4–25b. These free vectors may then be moved to the *arbitrary point P* and added to obtain their resultant vector sum $M_R = M_1 + M_2$, shown in Fig. 4–25c.

The following examples numerically illustrate these concepts. In general, problems displayed in two dimensions should be solved using a scalar analysis since the geometry is easy to visualize.

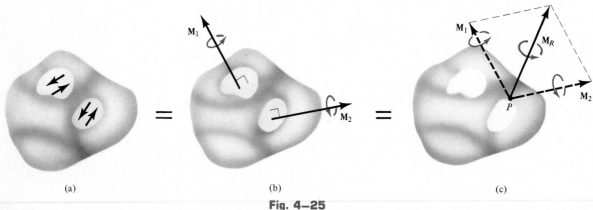

(a)   (b)   (c)

Fig. 4–25

Example 4–9

A couple acts at the end of the beam shown in Fig. 4–26a. Replace it by an equivalent one having a pair of forces that act through (a) points A and B, and (b) points D and E.

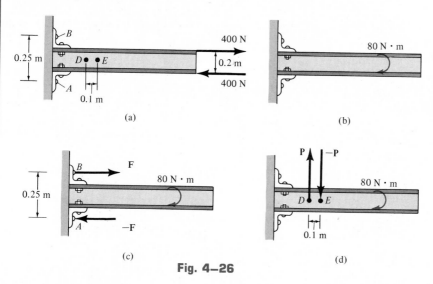

Fig. 4–26

Solution *(Scalar Analysis)*

The couple has a magnitude of $M = Fd = 400(0.2) = 80$ N · m and a direction that is into the page since the forces tend to rotate clockwise. **M** is a free vector so that it can be placed at any point on the beam, Fig. 4–26b.

Part (a). To preserve the direction of **M,** *horizontal* forces acting through points A and B must be directed as shown in Fig. 4–26c. The magnitude of each force is

$$M = Fd$$
$$80 = F(0.25)$$
$$F = 320 \text{ N} \qquad\qquad Ans.$$

Part (b). To generate the required clockwise rotation, forces acting through points D and E must be *vertical* and directed as shown in Fig. 4–26d. The magnitude of each force is

$$M = Pd$$
$$80 = P(0.1)$$
$$P = 800 \text{ N} \qquad\qquad Ans.$$

121

Example 4–10

Determine the moment of the couple acting on the beam shown in Fig. 4–27a.

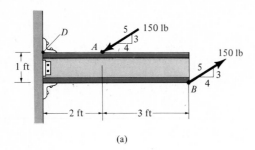

(a)

Fig. 4–27

Solution (Scalar Analysis)

Here it is somewhat difficult to determine the perpendicular distance between the forces and compute the couple moment as $M = Fd$. Instead, we can resolve each force into its horizontal and vertical components, $F_x = \frac{4}{5}(150) = 120$ lb and $F_y = \frac{3}{5}(150) = 90$ lb, Fig. 4–27b, and then use the principle of moments. The couple moment can be computed about any point. For example, if point D is chosen, we have

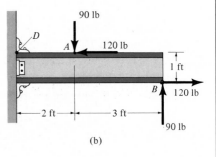

(b)

$$\downarrow+M = 120\ \text{lb}(0\ \text{ft}) - 90\ \text{lb}(2\ \text{ft}) + 90\ \text{lb}(5\ \text{ft}) + 120\ \text{lb}(1\ \text{ft})$$
$$= 390\ \text{lb} \cdot \text{ft} \uparrow \qquad\qquad\qquad Ans.$$

It is easier to compute the moments about point A or B in order to *eliminate* the moment of the force acting at the moment point. For point A, Fig. 4–27b, we have

$$\downarrow+M = 90\ \text{lb}(3\ \text{ft}) + 120\ \text{lb}(1\ \text{ft})$$
$$= 390\ \text{lb} \cdot \text{ft} \uparrow \qquad\qquad\qquad Ans.$$

Show that one obtains this same result if moments are summed about point B. Notice also that the couple in Fig. 4–27a has been replaced by *two* couples in Fig. 4–27b. Using $M = Fd$, one couple has a moment of $M_1 = 90\ \text{lb}(3\ \text{ft}) = 270\ \text{lb} \cdot \text{ft}$ and the other has a moment of $M_2 = 120\ \text{lb}(1\ \text{ft}) = 120\ \text{lb} \cdot \text{ft}$. By the right-hand rule both couple moments are counterclockwise and are therefore directed out of the page. Since these couples are free vectors they can be moved to any point and added, which yields $M = 270\ \text{lb} \cdot \text{ft} + 120\ \text{lb} \cdot \text{ft} = 390\ \text{lb} \cdot \text{ft} \uparrow$, the same result computed above. **M** is a free vector and can therefore act at any point on the beam, Fig. 4–27c.

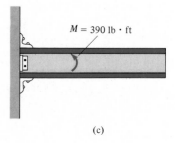

(c)

Example 4–11

Determine the couple moment acting on the shaft shown in Fig. 4–28a. Segment AB is directed 30° below the x-y plane.

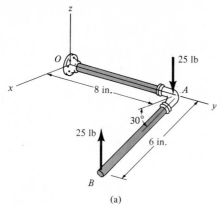

(a)

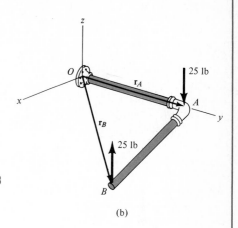

(b)

Fig. 4–28

Solution I (*Vector Analysis*)

The moment of the two couple forces can be computed about *any point*. If point O is used, Fig. 4–28b, we have

$$\mathbf{M} = \mathbf{r}_A \times (-25\mathbf{k}) + \mathbf{r}_B \times (25\mathbf{k})$$
$$= (8\mathbf{j}) \times (-25\mathbf{k}) + (6\cos 30°\mathbf{i} + 8\mathbf{j} - 6\sin 30°\mathbf{k}) \times (25\mathbf{k})$$
$$= -200\mathbf{i} - 129.9\mathbf{j} + 200\mathbf{i}$$
$$= \{-129.9\mathbf{j}\} \text{ lb} \cdot \text{in.} \qquad\qquad Ans.$$

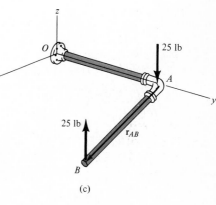

(c)

It is *easier* to take moments of the couple forces about a point lying on the line of action of one of the forces, e.g., point A, Fig. 4–28c. In this case the moment of the force at A is zero, so that

$$\mathbf{M} = \mathbf{r}_{AB} \times (25\mathbf{k})$$
$$= (6\cos 30°\mathbf{i} - 6\sin 30°\mathbf{k}) \times (25\mathbf{k})$$
$$= \{-129.9\mathbf{j}\} \text{ lb} \cdot \text{in.} \qquad\qquad Ans.$$

Solution II (*Scalar Analysis*)

Although this problem is shown in three dimensions, the geometry is simple enough to use the scalar equation $M = Fd$. The perpendicular distance between the lines of action of the forces is $d = 6\cos 30° = 5.20$ in., Fig. 4–28d. Hence, taking moments of the forces about either point A or B yields

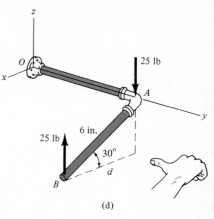

$$M = Fd = 25(5.20) = 129.9 \text{ lb} \cdot \text{in.}$$

Applying the right-hand rule, $\mathbf{M}$ acts in the $-\mathbf{j}$ direction. Thus,

$$\mathbf{M} = \{-129.9\mathbf{j}\} \text{ lb} \cdot \text{in.} \qquad\qquad Ans.$$

(d)

123

Example 4–12

Replace the two couples acting on the triangular block in Fig. 4–29*a* by a single resultant couple.

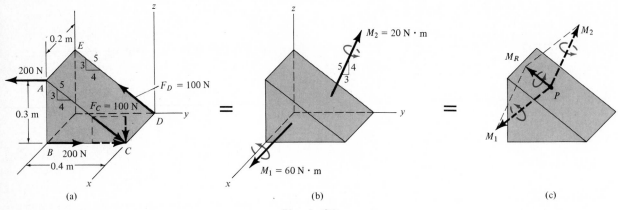

Fig. 4–29

Solution *(Vector Analysis)*

The couple $\mathbf{M}_1$, caused by the forces at A and B, can easily be determined from a scalar formulation.

$$M_1 = Fd = 200(0.3) = 60 \text{ N} \cdot \text{m}$$

By the right-hand rule, $\mathbf{M}_1$ acts in the $+\mathbf{i}$ direction, Fig. 4–29*b*. Hence,

$$\mathbf{M}_1 = \{60\mathbf{i}\} \text{ N} \cdot \text{m}$$

Vector analysis will be used to determine $\mathbf{M}_2$, caused by forces at C and D. If moments are computed about point D, Fig. 4–29*a*, $\mathbf{M}_2 = \mathbf{r}_C \times \mathbf{F}_C$, where $\mathbf{r}_C$ extends from the line of action of $\mathbf{F}_D$ to the line of action of $\mathbf{F}_C$. We have

$$\mathbf{M}_2 = \mathbf{r}_C \times \mathbf{F}_C = (0.2\mathbf{i}) \times [100(\tfrac{4}{5})\mathbf{j} - 100(\tfrac{3}{5})\mathbf{k}]$$
$$= (0.2\mathbf{i}) \times [80\mathbf{j} - 60\mathbf{k}] = 16(\mathbf{i} \times \mathbf{j}) - 12(\mathbf{i} \times \mathbf{k})$$
$$= \{12\mathbf{j} + 16\mathbf{k}\} \text{ N} \cdot \text{m}$$

Try to establish $\mathbf{M}_2$ by using a scalar formulation, Fig. 4–29*b*.

Since $\mathbf{M}_1$ and $\mathbf{M}_2$ are free vectors, they may be moved to some arbitrary point P on the block and added vectorially, Fig. 4–29*c*. The resultant couple moment becomes

$$\mathbf{M}_R = \mathbf{M}_1 + \mathbf{M}_2 = \{60\mathbf{i} + 12\mathbf{j} + 16\mathbf{k}\} \text{ N} \cdot \text{m} \qquad \textit{Ans.}$$

Problems

4–39. Determine the magnitude and sense of the couple moment.

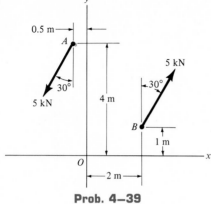

Prob. 4–39

4–42. Determine the couple moment. Express the result as a Cartesian vector.

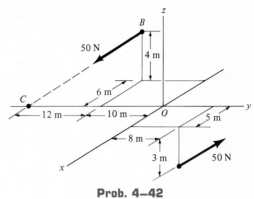

Prob. 4–42

*4–40. Determine the magnitude and sense of the couple moment.

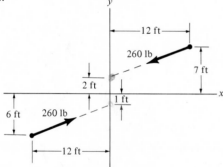

Prob. 4–40

4–41. Determine the couple moment. Express the result as a Cartesian vector.

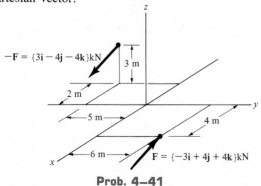

Prob. 4–41

4–43. The crossbar wrench is used to remove a lug nut from the automobile wheel. The mechanic applies a couple to the wrench such that his hands are a constant distance apart. Is it necessary that $a = b$ in order to produce the most effective turning of the nut? Explain. Also, what is the effect of changing the shaft dimension c in this regard? The forces act in the vertical plane.

Prob. 4–43

*4–44. Two couples act on the beam as shown. Determine the magnitude of **F** so that the resultant couple moment is 300 lb · ft counterclockwise. Where on the beam does the resultant couple act?

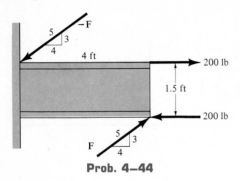

Prob. 4–44

4–45. The ends of the triangular plate are subjected to three couples. Determine the plate dimension d so that the resultant couple is 350 N · m clockwise.

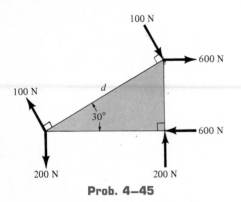

Prob. 4–45

4–46. Three couple moments act on the pipe assembly shown. Determine the magnitude of M_3 and the bend angle θ so that the resultant couple moment is zero.

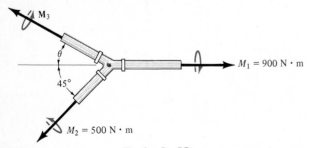

Prob. 4–46

4–47. Two couples act on the beam. One couple is formed by the forces at A and B, and the other by the forces at C and D. If the resultant couple is to be zero, determine the magnitudes of **P** and **F**, and the distance d between A and B.

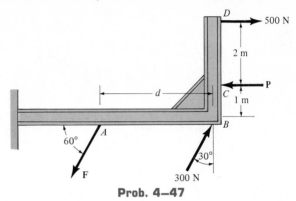

Prob. 4–47

*4–48. The gear reducer is subjected to the couple moments shown. Determine the resultant couple moment and specify its magnitude and coordinate direction angles.

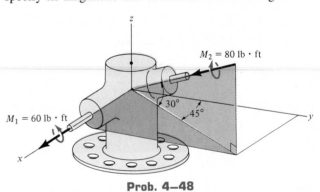

Prob. 4–48

4–49. The meshed gears are subjected to the couple moments shown. Determine the magnitude of the resultant couple moment and specify its coordinate direction angles.

4–51. Express the moment of the couple acting on the pipe assembly in Cartesian vector form. What is the magnitude of the couple moment?

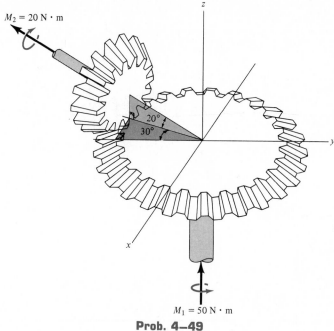

Prob. 4–49

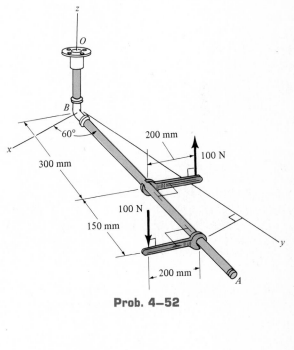

Prob. 4–51

4–50. Determine the resultant couple of the two couples that act on the assembly. Member *OB* lies in the *x-z* plane.

***4–52.** Determine the couple that acts on the assembly. Express the result as a Cartesian vector. Member *BA* lies in the *x-y* plane.

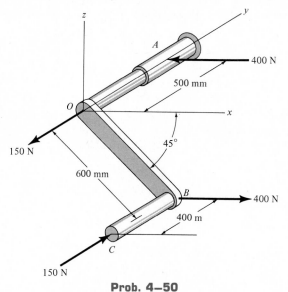

Prob. 4–50

Prob. 4–52

127

4.6 Movement of a Force on a Rigid Body

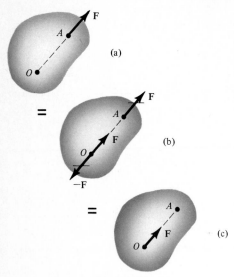

(a)

=

(b)

=

(c)

Fig. 4–30

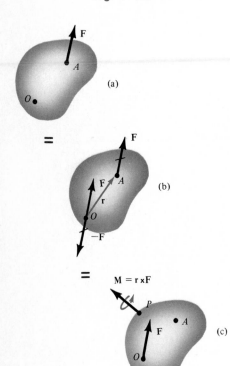

(a)

=

(b)

=

$M = r \times F$

(c)

Fig. 4–31

Many problems in statics, including the reduction of a force system to its simplest possible form, require moving a force from one point to another on a rigid body. Since a force tends to *translate* a body in the direction of the force and *rotate* it about an axis not located on the line of action of the force, it is important that these two "external" effects remain the same if the force is moved from one point to another on the body. Two cases for the location of the point O to which the force is moved will now be considered.

1. *Point O is on the line of action of the force*. Consider the rigid body shown in Fig. 4–30a, which is subjected to the force **F** applied to point A. In order to move the force to point O we will first apply equal but opposite forces **F** and −**F** at O, as shown in Fig. 4–30b. This step in no way alters the external effects on the body; however, the two forces indicated by the slash across them can be canceled, leaving the force at point O as required, Fig. 4–30c. By using this construction procedure, an *equivalent system* has been maintained between each of the diagrams, as indicated by the equal sign. Note that the force has simply been "transmitted" along its line of action, from point A, Fig. 4–30a, to point O, Fig. 4–30c. In other words, the force can be considered as a *sliding vector* since it can act at any point O along its line of action. This concept is referred to as the *principle of transmissibility*, which can be formally stated as follows: *The external effects on a rigid body remain unchanged when a force, acting at a given point on the body, is applied to another point lying on the line of action of the force*. It is important to realize that only the *external effects* caused by the force remain *unchanged* after moving it. Certainly the *internal effects* depend upon where **F** is located. For example, when **F** acts at A, the internal forces in the body have a high intensity around A; whereas if **F** acts at O, the effect of **F** on generating internal forces at A will be less.

2. *Point O is not on the line of action of the force*. This case is shown in Fig. 4–31a. Following the same procedure as above, we first apply equal but opposite forces **F** and −**F** at O, Fig. 4–31b. Here the two forces indicated by a slash across them form a couple which has a moment defined by the cross product $M = r \times F$. Since the couple moment is a *free vector*, it may be applied at any point P on the body as shown in Fig. 4–31c. In addition to this couple, **F** now acts at point O as required. From the construction, note that the couple moment is determined by taking the moment of **F** about O when the force was located at its original point A. Furthermore, the line of action of **M** is *always* perpendicular to the plane containing **F** and **r**.

The foregoing concepts regarding the movement of a force to any point on a body may be summarized by the following two statements:

1. If the force is to be moved to a *point O located on its line of action*, by the principle of transmissibility, simply move the force to the point.
2. If the force is to be moved to a *point O that is not located on its line of action*, an equivalent system is maintained when the force is moved to point O and a couple moment is placed on the body. The magnitude and

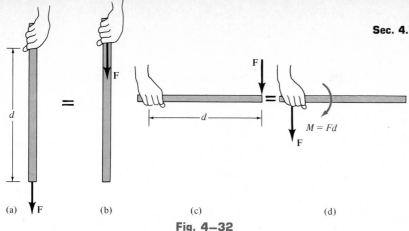

(a)   (b)   (c)   (d)

Fig. 4–32

direction of the couple moment are determined by finding the moment about point O of the force when it is at its original location. In all cases the couple moment will be *perpendicular* to the force.

As a physical illustration of these two statements, consider holding the end of a stick of negligible weight. If a vertical force $\mathbf{F}$ is applied at the other end, and the stick is held in the vertical position, Fig. 4–32a, then, by the principle of transmissibility, the same force is felt at the grip, Fig. 4–32b. When the stick is held in the horizontal position, Fig. 4–32c, the force has the effect of producing *both* a downward force at the grip and a clockwise twist, Fig. 4–32d. The twist can be thought of as being caused by a *couple* that is produced when $\mathbf{F}$ is moved to the grip. The couple moment has the *same* magnitude and direction as the moment of $\mathbf{F}$ about the grip, Fig. 4–32c.

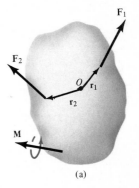

(a)

Resultants of a Force and Couple System 4.7

When a rigid body is subjected to a system of forces and couples, it is often simpler to study their effects by using their resultants rather than the force–couple system itself. To show how to simplify a system of forces and couples to their resultants, consider the general force–couple system acting on the rigid body in Fig. 4–33a. This system can be simplified by moving the forces and couples to the arbitrary point O. In this regard, the couple moment $\mathbf{M}$ is simply moved to O since it is a free vector. Forces $\mathbf{F}_1$ and $\mathbf{F}_2$ are sliding vectors and since O does not lie on the line of action of these forces, each must be moved to O in accordance with the procedure stated in Sec. 4.6. For example, when $\mathbf{F}_1$ is applied at O, a corresponding couple moment $\mathbf{M}_1 = \mathbf{r}_1 \times \mathbf{F}_1$ must also be applied to the body, Fig. 4–33b. By vector addition, the force and couple system shown in Fig. 4–33b can now be reduced to an *equivalent* resultant force $\mathbf{F}_R = \mathbf{F}_1 + \mathbf{F}_2$ and resultant couple moment $\mathbf{M}_{R_O} = \mathbf{M} + \mathbf{M}_1 + \mathbf{M}_2$ as shown in Fig. 4–33c. Note that both the magnitude and direction of $\mathbf{F}_R$ are independent of the location of point O; however, $\mathbf{M}_{R_O}$ depends upon this location, since the moments $\mathbf{M}_1$ and $\mathbf{M}_2$ are computed using the position vectors $\mathbf{r}_1$ and $\mathbf{r}_2$. Realize also that $\mathbf{M}_{R_O}$ is a free vector and can act at any point on the body, although point O is generally chosen as its point of application.

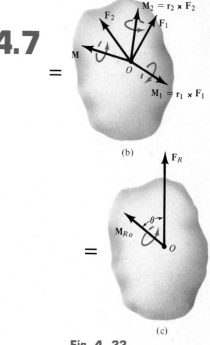

(b)

(c)

Fig. 4–33

PROCEDURE FOR ANALYSIS

The above method for simplifying any force and couple system to a single resultant force acting at point O and a resultant couple moment will now be stated in general terms.

Three-Dimensional Systems. A Cartesian vector analysis is generally used to solve problems involving three-dimensional force–couple systems for which the force components and moment arms are difficult to determine.

Force Summation. The resultant force is equivalent to the vector sum of all the forces in the system, i.e.,

$$\mathbf{F}_R = \Sigma \mathbf{F} \qquad (4\text{--}13)$$

Moment Summation. The resultant couple moment is equivalent to the vector sum of all the couples in the system plus the moments about point O of all the forces in the system, i.e.,

$$\mathbf{M}_{R_O} = \Sigma \mathbf{M}_O \qquad (4\text{--}14)$$

Coplanar Force Systems. Since force components and moment arms are easy to determine in two dimensions, a scalar analysis provides the most convenient solution to problems involving coplanar force systems.

Force Summation. The *resultant force* $\mathbf{F}_R$ is equivalent to the vector sum of its two components $\mathbf{F}_{R_x}$ and $\mathbf{F}_{R_y}$. Each component is found from the scalar (algebraic) sum of the components of all the forces in the system that act in the same direction, i.e.,

$$\begin{aligned} F_{R_x} &= \Sigma F_x \\ F_{R_y} &= \Sigma F_y \end{aligned} \qquad (4\text{--}15)$$

Moment Summation. The *resultant couple moment* $\mathbf{M}_{R_O}$ is perpendicular to the plane containing the forces and is equivalent to the scalar (algebraic) sum of all the couples in the system *plus* the moments about point O of all the forces in the system, i.e.,

$$M_{R_O} = \Sigma M_O \qquad (4\text{--}16)$$

When computing the moments of the forces about O, it is generally advantageous to use the *principle of moments*, i.e., compute the moments of the *components* of each force rather than the moment of the force itself.

The following two examples numerically illustrate these procedures.

Example 4–13

Replace the forces acting on the pipe shown in Fig. 4–34a by an equivalent single resultant force and couple system acting at point A.

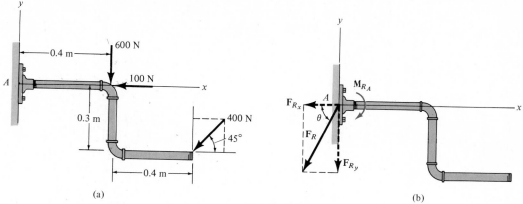

(a)

(b)

Fig. 4–34

Solution *(Scalar Analysis)*

The principle of moments will be applied to the 400-N force, whereby the moments of its two rectangular components will be considered.

Force Summation. The resultant force has x and y components of

$$\xrightarrow{+} F_{R_x} = \Sigma F_x; \quad F_{R_x} = -100 - 400\cos 45° = -382.8 \text{ N} = 382.8 \text{ N} \leftarrow$$
$$+ \uparrow F_{R_y} = \Sigma F_y; \quad F_{R_y} = -600 - 400\sin 45° = -882.8 \text{ N} = 882.8 \text{ N} \downarrow$$

As shown in Fig. 4–34b, F_R has a magnitude of

$$F_R = \sqrt{(F_{R_x})^2 + (F_{R_y})^2} = \sqrt{(382.8)^2 + (882.8)^2} = 962 \text{ N} \quad \textit{Ans.}$$

and a direction defined by

$$\theta = \tan^{-1}\left(\frac{F_{R_y}}{F_{R_x}}\right) = \tan^{-1}\left(\frac{882.8}{382.8}\right) = 66.6° \quad \textit{Ans.}$$

Moment Summation. The resultant couple moment M_{R_A} is determined by summing moments about point A. Noting that the moment of each force in the system is perpendicular to the plane of the forces and assuming that positive moments act counterclockwise, i.e., in the $+\mathbf{k}$ direction, we have

$$\downarrow + M_{R_A} = \Sigma M_A;$$
$$M_{R_A} = 100(0) - 600(0.4) - (400\sin 45°)(0.8) - (400\cos 45°)(0.3)$$
$$= -551 \text{ N} \cdot \text{m} = 551 \text{ N} \cdot \text{m} \downarrow \quad \textit{Ans.}$$

In conclusion, when M_{R_A} and F_R act on the pipe, Fig. 4–34b, they will produce the *same* reaction at the support A (an external effect) as that produced by the force system in Fig. 4–34a.

131

Example 4–14

A beam is subjected to a couple moment $\mathbf{M}$ and forces $\mathbf{F}_1$ and $\mathbf{F}_2$ as shown in Fig. 4–35a. Replace this system by an equivalent single resultant force and couple moment acting at point O.

Solution (Vector Analysis)

Expressing the forces and couple as Cartesian vectors, we have

$$\mathbf{F}_1 = \{-800\mathbf{k}\} \text{ N}$$

$$\mathbf{F}_2 = (300 \text{ N})\mathbf{u}_{CB} = (300 \text{ N})\left(\frac{\mathbf{r}_{CB}}{r_{CB}}\right)$$

$$= 300\left[\frac{-0.15\mathbf{i} + 0.1\mathbf{j}}{\sqrt{(-0.15)^2 + (0.1)^2}}\right] = \{-249.6\mathbf{i} + 166.4\mathbf{j}\} \text{ N}$$

$$\mathbf{M} = -500(\tfrac{4}{5})\mathbf{j} + 500(\tfrac{3}{5})\mathbf{k} = \{-400\mathbf{j} + 300\mathbf{k}\} \text{ N} \cdot \text{m}$$

Force Summation

$$\mathbf{F}_R = \Sigma\mathbf{F}; \qquad \mathbf{F}_R = \mathbf{F}_1 + \mathbf{F}_2 = -800\mathbf{k} - 249.6\mathbf{i} + 166.4\mathbf{j}$$

$$= \{-249.6\mathbf{i} + 166.4\mathbf{j} - 800\mathbf{k}\} \text{ N} \qquad \textit{Ans.}$$

Moment Summation

$$\mathbf{M}_{R_O} = \Sigma\mathbf{M}_O;$$

$$\mathbf{M}_{R_O} = \mathbf{M} + \mathbf{r}_C \times \mathbf{F}_1 + \mathbf{r}_B \times \mathbf{F}_2$$

$$= (-400\mathbf{j} + 300\mathbf{k}) + (1\mathbf{k}) \times (-800\mathbf{k}) + \begin{vmatrix} \mathbf{i} & \mathbf{j} & \mathbf{k} \\ -0.15 & 0.1 & 1 \\ -249.6 & 166.4 & 0 \end{vmatrix}$$

$$= (-400\mathbf{j} + 300\mathbf{k}) + (\mathbf{0}) + (-166.4\mathbf{i} - 249.6\mathbf{j})$$

$$= \{-166.4\mathbf{i} - 649.6\mathbf{j} + 300\mathbf{k}\} \text{ N} \cdot \text{m} \qquad \textit{Ans.}$$

The results are shown in Fig. 4–35b.

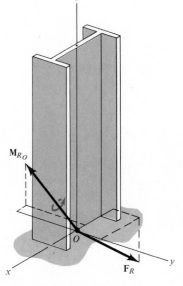

(a)

(b)

Fig. 4–35

Further Reduction of a Force and Couple System 4.8

Simplification to a Single Resultant Force. Consider now a special case for which the system of forces and couples acting on a rigid body, Fig. 4–36a, reduces at point O to a resultant force $\mathbf{F}_R = \Sigma\mathbf{F}$ and resultant couple $\mathbf{M}_{R_O} = \Sigma\mathbf{M}_O$, which are *perpendicular* to one another, Fig. 4–36b. Using the method of Sec. 4–6, $\mathbf{M}_{R_O}$ can always be *eliminated* by moving $\mathbf{F}_R$ to a point P located on axis bb, which is *perpendicular* to both $\mathbf{F}_R$ and $\mathbf{M}_{R_O}$, Fig. 4–36c. Point P is chosen such that the moment arm d satisfies the scalar equation $M_{R_O} = F_R d$ or $d = M_{R_O}/F_R$. Furthermore, P is located, in this case, to the *left* of O to preserve the correct direction of $\mathbf{M}_{R_O}$ when $\mathbf{F}_R$ is moved *back* to O, Fig. 4–36b. (By the right-hand rule, $\mathbf{F}_R$, acting through P in Fig. 4–36c, tends to rotate the body about the aa axis, which is in the same direction as $\mathbf{M}_{R_O}$.)

We refer to the force–couple system in Fig. 4–36a as being *equivalent or equipollent* to the single force "system" in Fig. 4–36c, because each system produces the *same* resultant force and resultant moment when replaced at point O.

If a system of forces is either concurrent, coplanar, or parallel, it can always be reduced, as in the above case, to a single resultant force $\mathbf{F}_R$ acting through a unique point P. This is because in each of these cases $\mathbf{F}_R$ and $\mathbf{M}_{R_O}$ will always be perpendicular to each other when the force system is simplified at any point O.

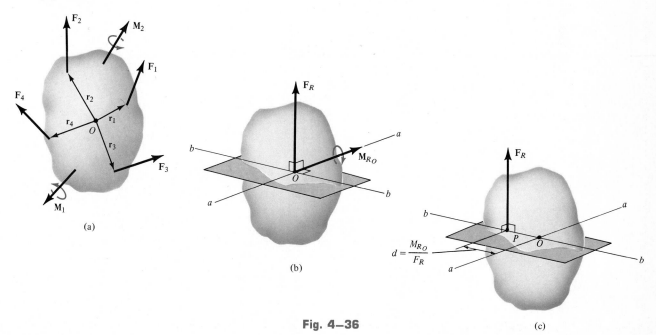

(a)

(b)

(c)

Fig. 4–36

133

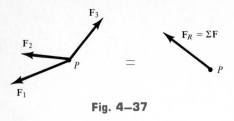

Fig. 4–37

Concurrent force systems have been treated in detail in Chapter 2. Obviously, all the forces act at a point for which there is no resultant couple, so the point P is automatically specified, Fig. 4–37.

Coplanar force systems, such as shown in Fig. 4–38a, can be reduced to a single resultant force, because when each force in the system is moved to any point O in the x-y plane, it produces a couple moment that is *perpendicular* to the plane, i.e., in the $\pm\mathbf{k}$ direction. The resultant moment $\mathbf{M}_{R_O} = \Sigma\mathbf{r} \times \mathbf{F}$ is thus perpendicular to the resultant force $\mathbf{F}_R$, Fig. 4–38b.

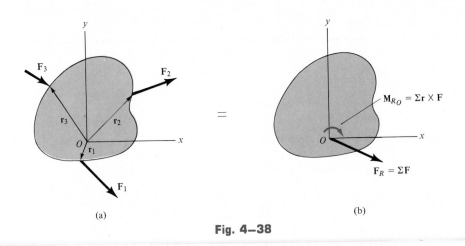

(a)

(b)

Fig. 4–38

Parallel force systems, such as shown in Fig. 4–39a, can be reduced to a single resultant force, because when each (vertical) force is moved to any point O in the x-y plane, it produces a couple that has moment components only about the x and y axes. The resultant moment $\mathbf{M}_{R_O} = \Sigma\mathbf{r} \times \mathbf{F}$ is thus perpendicular to the resultant force $\mathbf{F}_R$, Fig. 4–39b.

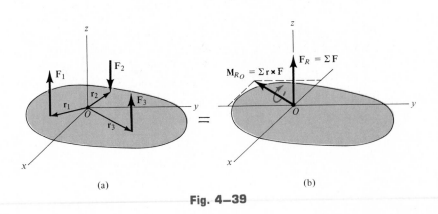

(a)

(b)

Fig. 4–39

PROCEDURE FOR ANALYSIS

The technique used to reduce a coplanar or parallel force system to a single resultant force follows the general procedure outlined in the previous section. The simplification requires the following two steps:

Force Summation. The resultant force $\mathbf{F}_R$ equals the sum of all the forces of the system, i.e.,

$$\mathbf{F}_R = \Sigma \mathbf{F}$$

Moment Summation. The line of action of $\mathbf{F}_R$ is determined by equating the moment of $\mathbf{F}_R$ about an arbitrary point O, $\mathbf{M}_{R_O}$, to the sum of the moments about point O of all the forces in the system, $\Sigma \mathbf{M}_O$, i.e.,

$$\mathbf{M}_{R_O} = \Sigma \mathbf{M}_O$$

Most often a scalar analysis can be used to apply these equations, since the force components and the moment arms are easily determined for either coplanar or parallel force systems.

Reduction to a Wrench. In the general case, the force and couple system acting on a body, Fig. 4–33a, will reduce to a single resultant force $\mathbf{F}_R$ and couple $\mathbf{M}_{R_O}$ at O which are *not* perpendicular. Instead, $\mathbf{F}_R$ will act at an angle θ from $\mathbf{M}_{R_O}$. Fig. 4–33c. As shown in Fig. 4–40a, however, $\mathbf{M}_{R_O}$ may be resolved into two components: one perpendicular, $\mathbf{M}_\perp$, and the other parallel, $\mathbf{M}_\parallel$, to the line of action of $\mathbf{F}_R$. The perpendicular component $\mathbf{M}_\perp$ may be *eliminated* by moving $\mathbf{F}_R$ to point P, as shown in Fig. 4–40b. This point lies on axis bb, which is perpendicular to both $\mathbf{M}_{R_O}$ and $\mathbf{F}_R$. In order to maintain an equivalency of loading, the distance from O to P is $d = M_\perp/F_R$. Furthermore, when $\mathbf{F}_R$ is applied at P, the moment of $\mathbf{F}_R$ tending to cause rotation of the body *about O* is in the *same direction* as $\mathbf{M}_\perp$, Fig. 4–40b. Finally, since $\mathbf{M}_\parallel$ is a free vector, it may be moved to P so that it coincides with $\mathbf{F}_R$, Fig. 4–40c. This combination of a *collinear* force and couple is called a *wrench*. The *axis of the wrench* has the same line of action as the force. Hence, the wrench tends to cause both a translation along and a rotation about this axis. Comparing Fig. 4–40a to Fig. 4–40c, it is seen that a general force–couple system acting on a body can be reduced to a wrench. The axis of the wrench and a point through which this axis passes are unique and can always be determined.

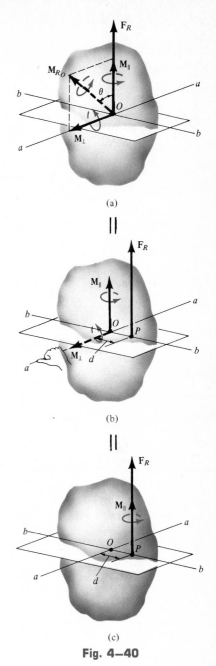

(a)

||

(b)

||

(c)

Fig. 4–40

Example 4–15

Replace the system of forces acting on the beam shown in Fig. 4–41a by an equivalent single resultant force. Specify the distance the force acts from point A.

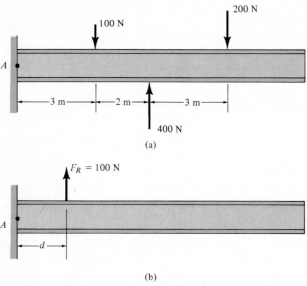

(a)

(b)

Fig. 4–41

Solution

Force Summation. If "positive" forces are assumed to act upward, then from Fig. 4–41a the force resultant $\mathbf{F}_R$ is

$$+\uparrow F_R = \Sigma F; \quad F_R = -100 \text{ N} + 400 \text{ N} - 200 \text{ N} = 100 \text{ N}\uparrow \qquad Ans.$$

Moment Summation. Moments will be summed about point A. Considering counterclockwise rotations as positive, i.e., positive moment vectors are directed out of the page, then from Fig. 4–41a and b we require the moment of $\mathbf{F}_R$ about A to equal the moments of the force system about A, i.e.,

$$\zeta + M_{R_A} = \Sigma M_A;$$
$$100 \text{ N}(d) = -(100 \text{ N})(3 \text{ m}) + (400 \text{ N})(5 \text{ m}) - (200 \text{ N})(8 \text{ m})$$
$$(100)d = 100$$
$$d = 1 \text{ m} \qquad Ans.$$

Using a clockwise sign convention would yield the same result. Since *d* is *positive*, $\mathbf{F}_R$ acts to the right of A as shown in Fig. 4–41b.

Example 4–16

The beam AE in Fig. 4–42a is subjected to a system of coplanar forces. Determine the magnitude, direction, and location on the beam of a single resultant force which is equivalent to the given system of forces.

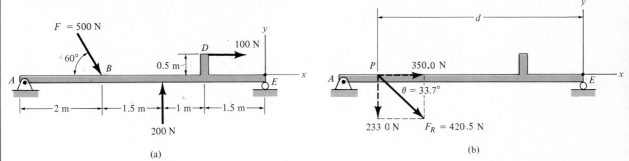

Fig. 4–42

Solution

Force Summation. The origin of coordinates is arbitrarily located at point E as shown in Fig. 4–42a. Resolving the 500-N force into x and y components, and summing the force components, yields

$$\xrightarrow{+} F_{R_x} = \Sigma F_x; \qquad F_{R_x} = 500 \cos 60° + 100 = 350.0 \text{ N} \rightarrow$$

$$+\uparrow F_{R_y} = \Sigma F_y; \qquad F_{R_y} = -500 \sin 60° + 200 = -233.0 \text{ N}$$
$$= 233.0 \text{ N} \downarrow$$

The magnitude and direction of the resultant force shown in Fig. 4–42b are, therefore,

$$F_R = \sqrt{(350.0)^2 + (233.0)^2} = 420.5 \text{ N} \qquad\qquad Ans.$$

$$\theta = \tan^{-1}\left(\frac{233.0}{350.0}\right) = 33.7° \qquad\qquad Ans.$$

Moment Summation. Moments will be summed about point E. Since P lies on the x axis, $\mathbf{F}_{R_x}$ (350.0 N) does not create a moment about E (principle of transmissibility), only $\mathbf{F}_{R_y}$ (233.0 N) does. Hence, from Fig. 4–42a and b, we require that the moment of $\mathbf{F}_R$ about point E equal the moments of the force system about E, i.e.,

$$\zeta + M_{R_E} = \Sigma M_E;$$
$$233.0(d) = (500 \sin 60°)(4) + (500 \cos 60°)(0) - (100)(0.5) - (200)(2.5)$$

$$d = \frac{1182.1}{233.0} = 5.07 \text{ m} \qquad\qquad Ans.$$

Example 4–17

The frame shown in Fig. 4–43a is subjected to three coplanar forces. Replace this loading by an equivalent single resultant force and specify where the resultant's line of action intersects members AB and BC.

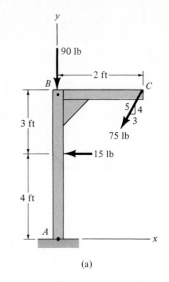

(a)

Solution

Force Summation. Resolving the 75-lb force into x and y components and summing the force components yields

$$\xleftarrow{+}F_{R_x} = \Sigma F_x; \qquad F_{R_x} = 75(\tfrac{3}{5}) + 15 = 60 \text{ lb} \leftarrow$$
$$+\downarrow F_{R_y} = \Sigma F_y; \qquad F_{R_y} = 75(\tfrac{4}{5}) + 90 = 150 \text{ lb} \downarrow$$

As shown in Fig. 4–43b,

$$F_R = \sqrt{(60)^2 + (150)^2} = 161.6 \text{ lb} \qquad \textit{Ans.}$$

$$\theta = \tan^{-1}\left(\frac{150}{60}\right) = 68.2° \quad {}^{\theta}\!\nearrow \qquad \textit{Ans.}$$

Moment Summation. Moments will be summed about point A. If the line of action of $\mathbf{F}_R$ intersects AB, Fig. 4–43b, we require the moment of the components of $\mathbf{F}_R$ in Fig. 4–43b about A to equal the moments of the force system in Fig. 4–43a about A, i.e.,

$$\zeta + M_{R_A} = \Sigma M_A;$$
$$0(150) + y(60) = 4(15) + 0(90) + 7(75)(\tfrac{3}{5}) - 2(75)(\tfrac{4}{5})$$
$$y = 4.25 \text{ ft} \qquad \textit{Ans.}$$

By the principle of transmissibility, $\mathbf{F}_R$ can also intersect BC, Fig. 4–43b, in which case we have

$$\zeta + M_{R_A} = \Sigma M_A;$$
$$7(60) - x(150) = 4(15) + 0(90) + 7(75)(\tfrac{3}{5}) - 2(75)(\tfrac{4}{5})$$
$$x = 1.10 \text{ ft} \qquad \textit{Ans.}$$

We can also solve for these positions by assuming $\mathbf{F}_R$ acts at the arbitrary point (x, y) on its line of action, Fig. 4–43b. Then moments about A yield

$$\zeta + M_{R_A} = \Sigma M_A;$$
$$y(60) - x(150) = 4(15) + 0(90) + 7(75)(\tfrac{3}{5}) - 2(75)(\tfrac{4}{5})$$
$$60y - 150x = 255$$

which is the equation of the dashed line in Fig. 4–43b. To find the points of intersection with the frame, set $x = 0$, then $y = 4.25$ ft, and set $y = 7$ ft, then $x = 1.10$ ft.

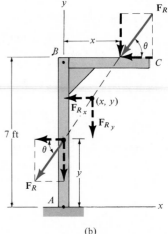

(b)

Fig. 4–43

Example 4–18

The slab in Fig. 4–44a is subjected to four parallel forces. Determine the magnitude and direction of a single resultant force equivalent to the given force system, and locate its point of application on the slab.

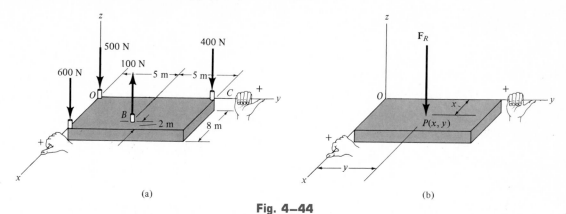

(a)   (b)

Fig. 4–44

Solution *(Scalar Analysis)*
Force Summation. From Fig. 4–44a, the resultant force is

$$+\downarrow F_R = \Sigma F; \qquad F_R = 600 - 100 + 400 + 500$$
$$= 1400 \text{ N} \downarrow \qquad\qquad Ans.$$

Moment Summation. We require the moment about the x axis of the resultant force, Fig. 4–44b, to be equal to the sum of the moments about the x axis of all the forces in the system, Fig. 4–44a. The moment arms are determined from the y coordinates since these coordinates represent the *perpendicular distances* from the x axis to the lines of action of the forces. Using the right-hand rule, where positive moments act in the $+\mathbf{i}$ direction, we have

$$M_{R_x} = \Sigma M_x;$$
$$-1400y = 600(0) + 100(5) - 400(10) + 500(0)$$
$$-1400y = -3500 \qquad y = 2.50 \text{ m} \qquad Ans.$$

In a similar manner, assuming that positive moments act in the $+\mathbf{j}$ direction, a moment equation can be written about the y axis using moment arms defined by the x coordinates of each force.

$$M_{Ry} = \Sigma M_y;$$
$$1400x = 600(8) - 100(6) + 400(0) + 500(0)$$
$$1400x = 4200 \qquad x = 3.00 \text{ m} \qquad Ans.$$

Hence, a force of $F_R = 1400$ N placed at point $P(3.00$ m, 2.50 m$)$ on the slab, Fig. 4–44b, is equivalent to the parallel force system acting on the slab in Fig. 4–44a.

Example 4–19

Three parallel forces act on the rim of the circular plate in Fig. 4–45a. Determine the magnitude and direction of a single resultant force equivalent to the given force system and locate its point of application, P, on the plate.

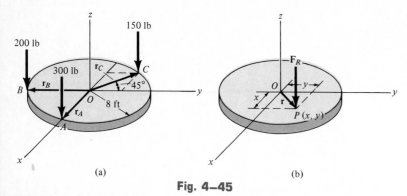

(a)   (b)

Fig. 4–45

Solution *(Vector Analysis)*

Force Summation. From Fig. 4–45a, the force resultant $\mathbf{F}_R$ is

$$\mathbf{F}_R = \Sigma \mathbf{F}; \qquad \mathbf{F}_R = -300\mathbf{k} - 200\mathbf{k} - 150\mathbf{k}$$
$$= \{-650\mathbf{k}\}\,\text{lb} \qquad\qquad Ans.$$

Moment Summation. Choosing point O as a reference for computing moments and assuming that $\mathbf{F}_R$ acts at a point $P(x, y)$, Fig. 4–45b, we require

$$\mathbf{M}_{R_O} = \Sigma \mathbf{M}_O; \quad \mathbf{r} \times \mathbf{F}_R = \mathbf{r}_A \times (-300\mathbf{k}) + \mathbf{r}_B \times (-200\mathbf{k}) + \mathbf{r}_C \times (-150\mathbf{k})$$
$$(x\mathbf{i} + y\mathbf{j}) \times (-650\mathbf{k}) = (8\mathbf{i}) \times (-300\mathbf{k}) + (-8\mathbf{j}) \times (-200\mathbf{k})$$
$$+ (-8 \sin 45°\mathbf{i} + 8 \cos 45°\mathbf{j}) \times (-150\mathbf{k})$$
$$650x\mathbf{j} - 650y\mathbf{i} = 2400\mathbf{j} + 1600\mathbf{i} - 848.5\mathbf{j} - 848.5\mathbf{i}$$

Equating the corresponding $\mathbf{j}$ and $\mathbf{i}$ components yields

$$650x = 2400 - 848.5 \qquad\qquad (1)$$
$$-650y = 1600 - 848.5 \qquad\qquad (2)$$

Solving these equations, we obtain the coordinates of point P,

$$x = 2.39\,\text{ft} \qquad y = -1.16\,\text{ft} \qquad\qquad Ans.$$

The negative sign indicates that it was wrong to have assumed a $+y$ position for $\mathbf{F}_R$ as shown in Fig. 4–45b.

As a review, try to establish Eqs. (1) and (2) by using a scalar analysis, i.e., apply the sum of moments about the x and y axes, respectively.

Problems

4–53. Replace the force at *A* by an equivalent force and couple moment at point *P*.

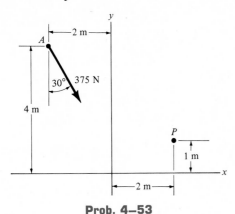

Prob. 4–53

4–54. Replace the force at *A* by an equivalent force and couple moment at point *P*.

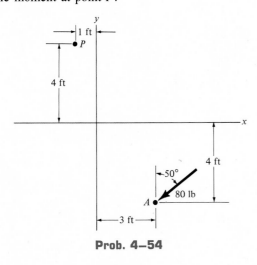

Prob. 4–54

4–55. Replace the force at *A* by an equivalent force and couple moment at point *P*.

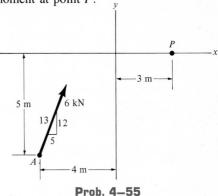

Prob. 4–55

*4–56. Replace the force and couple system by an equivalent force and couple moment acting at point *P*.

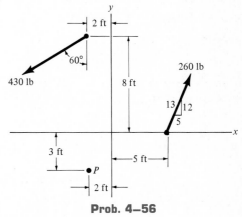

Prob. 4–56

4–57. Replace the force system by a single force resultant and specify its point of application measured along the *x* axis from point *O*.

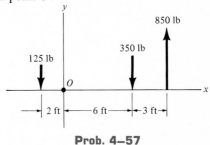

Prob. 4–57

141

4–58. Replace the force system by a single force resultant and specify its coordinate point of application $(x, 0)$ on the x axis.

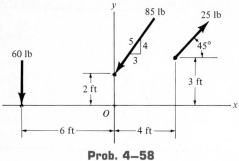

Prob. 4–58

4–59. Replace the force system by a single force resultant and specify its coordinate point of application $(x, 0)$ on the x axis.

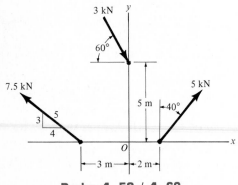

Probs. 4–59 / 4–60

*4–60. Replace the force and couple system by a single force resultant and specify its coordinate point of application $(0, y)$ on the y axis.

4–61. Replace the force at A by an equivalent force and couple moment at point P. Express the results in Cartesian vector form.

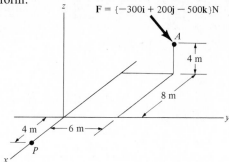

Prob. 4–61

4–62. Replace the force at A by an equivalent force and couple moment at point P. Express the results in Cartesian vector form.

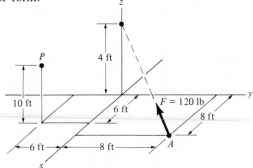

Prob. 4–62

4–63. Replace the force and couple system by an equivalent force and couple moment at point P. Express the results in Cartesian vector form.

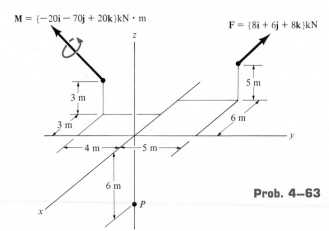

Prob. 4–63

*4–64. Replace the force and couple system by an equivalent force and couple moment at point P. Express the results in Cartesian vector form.

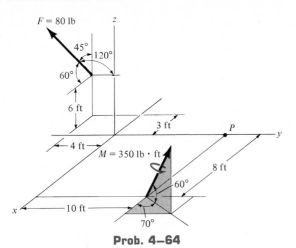

Prob. 4–64

4–65. The force of 40 N is exerted on the handle of the monkey wrench. Replace this force by an equivalent force and couple moment acting through the axis of the bolt at O.

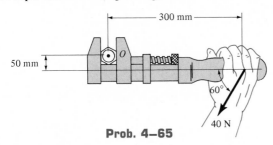

Prob. 4–65

4–66. Replace the loading system acting on the beam by an equivalent force and couple moment at point O.

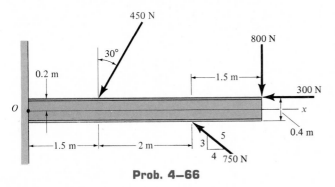

Prob. 4–66

4–67. Replace the loading system acting on the post by a single resultant force and specify its point of application on the post measured from point O.

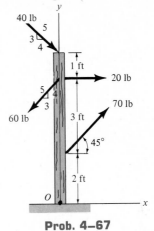

Prob. 4–67

*4–68. The gear is subjected to the two forces shown. Replace these forces by an equivalent force and couple moment acting at point O.

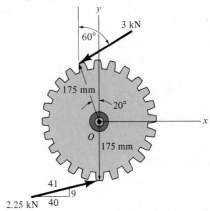

Prob. 4–68

4–69. The tires of a truck exert the forces shown on the road. Replace this system of forces by a single resultant force and specify its location measured from point A.

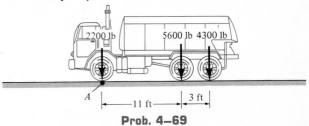

Prob. 4–69

143

4–70. The system of four forces acts on the roof. Determine the resultant force and specify its location along AB, measured from point A.

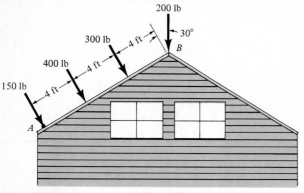

Prob. 4–70

4–71. Determine the magnitude and direction θ of force **F** and its placement d on the beam so that the loading system is equivalent to a resultant force of 12 kN acting vertically downward at point A and a clockwise couple moment of 50 kN · m.

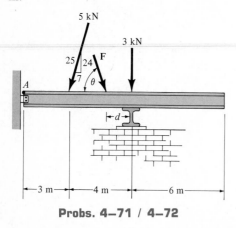

Probs. 4–71 / 4–72

*4–72. Determine the magnitude and direction θ of force **F** and its placement d on the beam so that the loading system creates a zero resultant force and couple moment at point A.

4–73. Replace the force–couple system acting on the frame by a single resultant force and specify where the resultant's line of action intersects member AB measured from A.

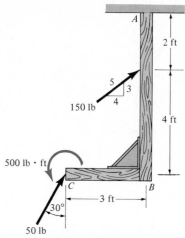

Probs. 4–73 / 4–74

4–74. Replace the force–couple system acting on the frame by a single resultant force and specify where the resultant's line of action intersects member BC measured from B.

4–75. Determine the magnitudes of $\mathbf{F}_1$ and $\mathbf{F}_2$ and the direction of $\mathbf{F}_1$ so that the loading creates a zero resultant force and couple on the wheel.

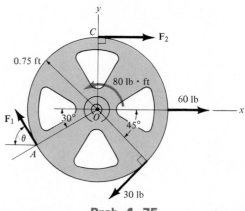

Prob. 4–75

*4–76. Replace the force system acting on the frame by a single resultant force and specify where the resultant's line of action intersects member *AB* measured from point *A*.

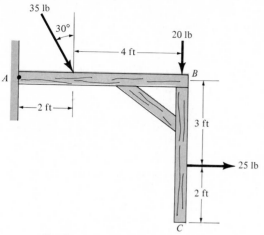

35 lb
30°
20 lb
4 ft
B
A
2 ft
3 ft
25 lb
2 ft
C

Probs. 4–76 / 4–77 / 4–78

4–77. Replace the force system acting on the frame by a single resultant force and specify where the resultant's line of action intersects member *BC* measured from point *B*.

4–78. Replace the force system acting on the frame by an equivalent force and couple moment acting at point *A*.

4–79. A fish exerts a momentary pull on the line of a fishing pole of 50 N. Replace this force by an equivalent force and couple moment at the fisherman's grip *A*. Solve the problem two ways: (a) by placing the force **F** at the fish *C*, so $\mathbf{M}_A = \mathbf{r}_{AC} \times \mathbf{F}$; and (b) by keeping it at *B*, so $\mathbf{M}_A = \mathbf{r}_{AB} \times \mathbf{F}$.

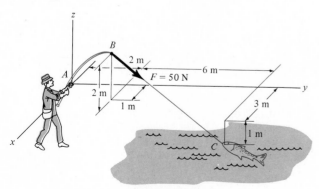

B
2 m
6 m
A
F = 50 N
2 m
1 m
3 m
1 m
C

Prob. 4–79

*4–80. The bracket is subjected to the force **F**. Replace this force by an equivalent force and couple moment acting at point *A*.

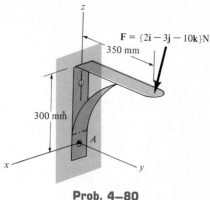

$\mathbf{F} = \{2\mathbf{i} - 3\mathbf{j} - 10\mathbf{k}\}N$
350 mm
300 mm
A

Prob. 4–80

4–81. Replace the two forces acting on the tree by an equivalent force and couple moment acting at point *O*.

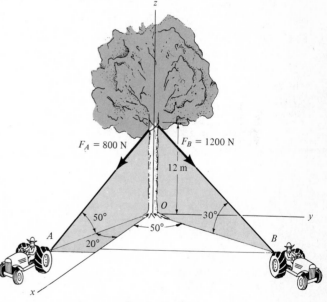

$F_A = 800$ N
$F_B = 1200$ N
12 m
50°
50°
30°
A
20°
O
B

Prob. 4–81

145

4–82. The slab is to be hoisted using the three slings shown. Replace the system of forces acting on slings by an equivalent force and couple moment at point O. The force $\mathbf{F}_1$, is vertical.

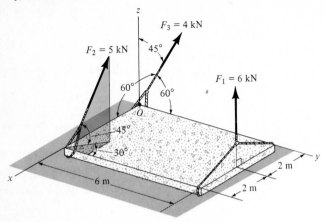

Prob. 4–82

4–83. The building slab is subjected to four parallel column loadings. Determine the resultant force and specify its location (x, y) on the slab.

*4–84. The three parallel bolting forces act on the circular plate. If the force at A is 200 lb, determine the magnitudes of $\mathbf{F}_B$ and $\mathbf{F}_C$ so that the resultant force $\mathbf{F}_R$ of the system has a line of action that coincides with the y axis. *Hint:* This requires $\Sigma M_x = 0$ and $\Sigma M_z = 0$. Why?

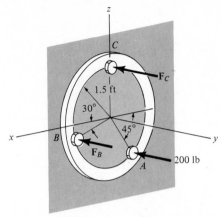

Prob. 4–84

4–85. Three parallel forces act on the circular slab. Determine the resultant force, and specify its location (x, y) on the slab.

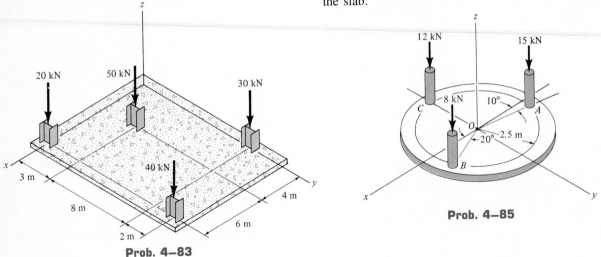

Prob. 4–83

Prob. 4–85

4–86. A force and couple act on the pipe assembly. Replace this system by an equivalent single resultant force. Specify the location of the resultant force along the y axis, measured from A. The pipe lies in the x-y plane.

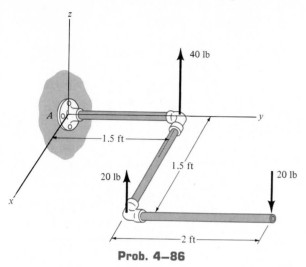

Prob. 4–86

4–87. Replace the two wrenches $F_1 = 60$ lb, $M_1 = 40$ lb · ft and $F_2 = 35$ lb, $M_2 = 40$ lb · ft and the force $F_3 = 30$ lb, acting on the pipe assembly, by an equivalent force and couple moment acting at point O. Note that $\mathbf{F}_2$ and $\mathbf{M}_2$ act in a plane which is parallel to the x-z plane.

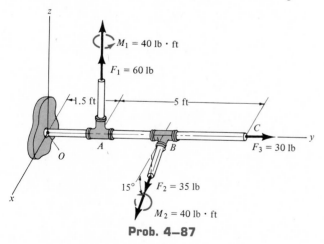

Prob. 4–87

*4–88. The pipe assembly is subjected to the action of a wrench at B and a couple at A. Simplify this system to a single resultant wrench and specify the location of the wrench along the axis of pipe CD, measured from point C.

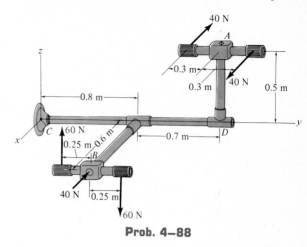

Prob. 4–88

4–89. The three forces acting on the block each have a magnitude of F. Replace this system by a wrench and specify the point where the wrench intersects the z axis.

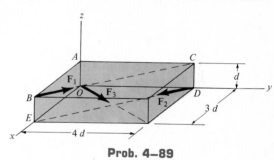

Prob. 4–89

Review Problems

4–90. A force of 90 N acts on the handle of the paper cutter at *A*. Determine the moment created by this force about the hinge at *O* if $\theta = 60°$. At what angle θ should the force be applied so that the clockwise moment it creates about point *O* is a maximum?

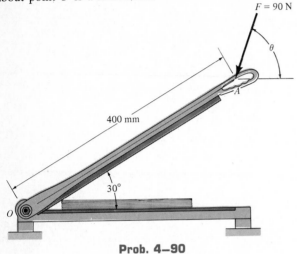

Prob. 4–90

4–91. The A frame is being hoisted into an upright position by the *vertical* towing force of $F = 80$ lb. Determine the moment of this force about the x' axis passing through points *A* and *B* when the frame is in the position shown. *Suggestion:* Use a scalar analysis.

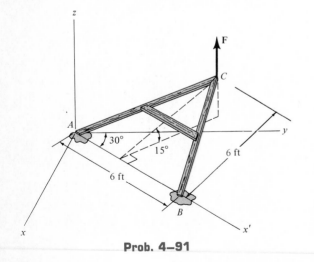

Prob. 4–91

*4–92. Determine the moment of force **F** about point *O*. Set $F = 12$ lb.

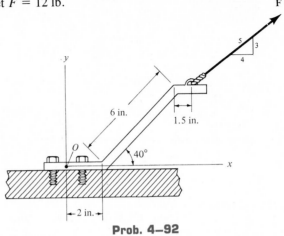

Prob. 4–92

4–93. Determine the magnitude and direction θ of force **F** and the couple moment **M** such that the loading system is equivalent to a resultant force of 600 N, acting vertically downward at *O*, and a clockwise couple moment of 4000 N · m.

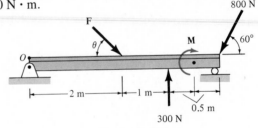

Prob. 4–93

4–94. The system of parallel forces acts on the deck of the bridge. Determine the resultant force of the system and specify its location measured from point *A*.

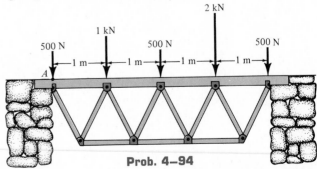

Prob. 4–94

4–95. The x-ray machine is used for medical diagnosis. If the equipment and housing at C have a mass of 150 kg and a mass center at G, determine the moment of its weight about point O when it is in the position shown.

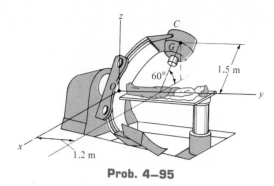

Prob. 4–95

*4–96. A wire cable passes through the bearing support. If the cable ends are subjected to the couple moments shown, determine the resultant couple moment on the bearing. Specify its magnitude and direction.

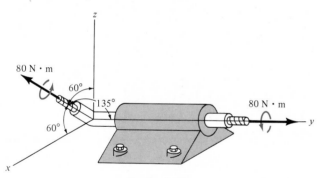

Prob. 4–96

4–97. The main beam along the wing of an airplane is swept back at an angle of 25°. From load calculations it is determined that the beam is subjected to couple moments $M_x = 25,000$ lb · ft and $M_y = 17,000$ lb · ft. Determine the equivalent couple moments created about the x' and y' axes.

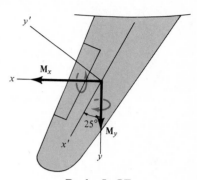

Prob. 4–97

4–98. The building column is subjected to a force of 1500 lb acting at the end of the supporting bracket. Replace this force by an equivalent force and couple moment acting at (a) point O and (b) point A.

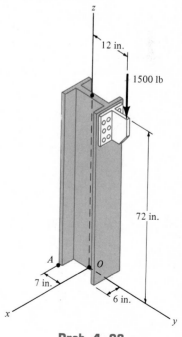

Prob. 4–98

149

4–99. The hollow concrete column supports the four parallel forces shown. Determine the magnitudes of forces $\mathbf{F}_C$ and $\mathbf{F}_D$ acting at C and D so that the resultant force of the system acts through the midpoint O of the column.

*4–100. The force $\mathbf{F} = \{600\mathbf{i} + 300\mathbf{j} - 600\mathbf{k}\}$ N acts at the end B of the beam. Determine the moment of this force about point O.

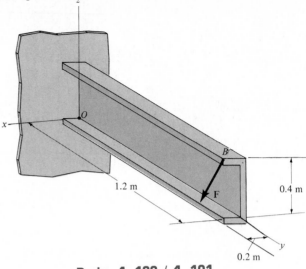

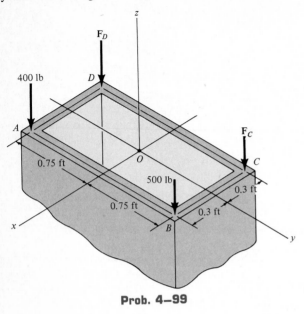

Prob. 4–99

Probs. 4–100 / 4–101

4–101. The force $\mathbf{F} = \{600\mathbf{i} + 300\mathbf{j} - 600\mathbf{k}\}$ N acts on the end of the beam. Determine the moment of the force about the y axis.

4–102. The horizontal 30-N force acts on the handle of the wrench. Determine the moment of this force about point O. Specify the coordinate direction angles α, β, γ of the moment axis.

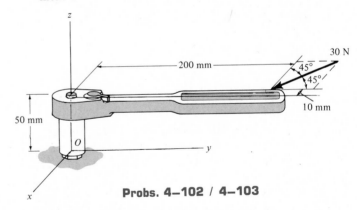

Probs. 4–102 / 4–103

4–103. The horizontal 30-N force acts on the handle of the wrench. What is the magnitude of the moment of this force about the z axis?

Equilibrium of a Rigid Body

In this chapter the fundamental concepts of rigid-body equilibrium will be discussed. It will be shown that equilibrium requires both a *balance of forces,* to prevent the body from translating or moving along a straight or curved path, and a *balance of moments,* to prevent the body from rotating.

Many types of engineering problems involve symmetric loadings and can be solved by projecting all the forces acting on a body onto a single plane. Hence, in the first part of this chapter, the equilibrium of a body subjected to a *coplanar* or *two-dimensional force system* will be considered. Ordinarily the geometry of such problems is not very complex, so a scalar solution is suitable for analysis. The more general discussion of rigid bodies subjected to *three-dimensional force systems* is given in the second part of this chapter. It will be seen that many of these types of problems can best be solved by using vector analysis.

Conditions for Rigid-Body Equilibrium 5.1

In Chapter 3 it was stated that a particle is in equilibrium if it remains at rest or moves with constant velocity. For this to be the case, it is both necessary and sufficient to require the resultant force acting on the particle to be equal to zero. Using this fact, we will now develop the conditions required to maintain equilibrium for a rigid body. To do this, consider the rigid body in Fig. 5–1a, which is fixed in the x, y, z reference and is either at rest or moves with the reference at constant velocity. A free-body diagram of the arbitrary ith particle of the body is shown in Fig. 5–1b. There are two types of forces which act on it. The *internal forces,* represented symbolically as

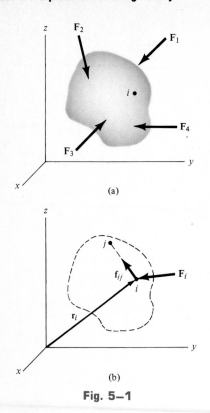

(a)

(b)

Fig. 5–1

$$\sum_{\substack{j=1 \\ (j \neq i)}}^{n} \mathbf{f}_{ij} = \mathbf{f}_i$$

are forces which the other particles exert on the *i*th particle. The summation extends over all *n* particles composing the body; however, it is meaningless for $i = j$ since the *i*th particle cannot exert a force on itself. The resultant *external force* $\mathbf{F}_i$ represents, for example, the effect of gravitational, electrical, magnetic, or contact forces between the *i*th particle and adjacent bodies or particles *not* included within the body. If the particle is in equilibrium, then applying Newton's first law we have

$$\mathbf{F}_i + \mathbf{f}_i = \mathbf{0}$$

When the equation of equilibrium is applied to each of the other particles of the body, similar equations will result. If all these equations are added together *vectorially,* we obtain

$$\Sigma \mathbf{F}_i + \Sigma \mathbf{f}_i = \mathbf{0}$$

The summation of the internal forces if carried out will equal zero since the internal forces between particles will occur in equal but opposite collinear pairs, Newton's third law. Consequently, only the sum of the *external forces* will remain; and therefore, letting $\Sigma \mathbf{F}_i = \Sigma \mathbf{F}$, the above equation can be written as

$$\Sigma \mathbf{F} = \mathbf{0}$$

Let us now consider the moments of the forces acting on the particle about the arbitrary point O, Fig. 5–1*b*. Using the particle equilibrium equation and the distributive law of the vector cross product yields

$$\mathbf{r}_i \times (\mathbf{F}_i + \mathbf{f}_i) = \mathbf{r}_i \times \mathbf{F}_i + \mathbf{r}_i \times \mathbf{f}_i = \mathbf{0}$$

Similar equations can be written for the other particles of the body, and adding them together vectorially, we obtain

$$\Sigma \mathbf{r}_i \times \mathbf{F}_i + \Sigma \mathbf{r}_i \times \mathbf{f}_i = \mathbf{0}$$

The second term is zero since, as stated above, the internal forces occur in equal but opposite collinear pairs, and by the transmissibility of a force, as discussed in Sec. 4.3, the moment of each pair of forces about point O is therefore zero. Hence, using the notation $\Sigma \mathbf{M}_O = \Sigma \mathbf{r}_i \times \mathbf{F}_i$, we can write the above equation as

$$\Sigma \mathbf{M}_O = \mathbf{0}$$

Hence the *equations of equilibrium* for a rigid body can be summarized as follows:

$$\Sigma \mathbf{F} = \mathbf{0}$$
$$\Sigma \mathbf{M}_O = \mathbf{0}$$

(5–1)

These equations require that a rigid body will be in equilibrium provided the sum of all the external forces acting on the body is equal to zero and the sum of the moments of the external forces about a point is equal to zero. The fact that these conditions are *necessary* for equilibrium has now been proven. They are also *sufficient* conditions. To show this, let us assume that the body is *not* in equilibrium, and yet the force system acting on it satisfies Eqs. 5–1. Suppose that an *additional force* $\mathbf{F}'$ is required to hold the body in equilibrium. As a result, the equilibrium equations become

$$\Sigma\mathbf{F} + \mathbf{F}' = \mathbf{0}$$
$$\Sigma\mathbf{M}_O + \mathbf{M}'_O = \mathbf{0}$$

where $\mathbf{M}'_O$ is the moment of $\mathbf{F}'$ about O. Since $\Sigma\mathbf{F} = \mathbf{0}$ and $\Sigma\mathbf{M}_O = \mathbf{0}$, then we require $\mathbf{F}' = \mathbf{0}$ (also $\mathbf{M}'_O = \mathbf{0}$). Consequently, the additional force $\mathbf{F}'$ is not required for holding the body, and indeed Eqs. 5–1 are also sufficient conditions.

Equilibrium in Two Dimensions

Free-Body Diagrams 5.2

Successful application of the equations of equilibrium requires a complete specification of *all* the known and unknown forces that act *on* the body. The best way to account for these forces is to draw the body's free-body diagram. This diagram is a sketch of the outlined shape of the body which represents it as being *isolated* or "free" from its surroundings. On this sketch one then shows *all* the forces (or their resultants) which the surroundings exert *on the body*. Obviously, if the free-body diagram is correctly drawn, the effects of all the applied forces and couples acting on the body can be accounted for when the equations of equilibrium are applied. For this reason, *a thorough understanding of how to draw a free-body diagram is of primary importance for solving problems in mechanics*. It is the preliminary step before applying the equations of equilibrium.

Support Reactions. Before presenting a formal procedure as to how to draw a free-body diagram, we will first consider the various types of reactions that occur at supports and points of support between bodies. For two-dimensional problems, i.e., bodies subjected to coplanar force systems, the supports most commonly encountered are given in Table 5–1. (In all cases the angle θ is assumed to be known.) Carefully study each of the symbols used to represent these supports and the types of reactions they exert on their contacting members. One possible way to determine the type of support reaction is to *imagine* the attached member as being translated or rotated in a particular direction. If the support *prevents translation* in a given direction, then a *force* is developed

Table 5–1 Supports for Rigid Bodies Subjected to Two-Dimensional Force Systems

Types of Connection	Reaction	Number of Unknowns
(1) cable		One unknown. The reaction is a tension force which acts away from the member in the direction of the cable.
(2) weightless link	or	One unknown. The reaction is a force which acts along the axis of the link.
(3) roller		One unknown. The reaction is a force which acts perpendicular to the surface at the point of contact.
(4) roller or pin in confined smooth slot	or	One unknown. The reaction is a force which acts perpendicular to the slot.
(5) rocker		One unknown. The reaction is a force which acts perpendicular to the surface at the point of contact.
(6) smooth contacting surface		One unknown. The reaction is a force which acts perpendicular to the surface at the point of contact.
(7) member pin connected to collar on smooth rod	or	One unknown. The reaction is a force which acts perpendidcular to the rod.

154

Table 5–1 (Contd.)

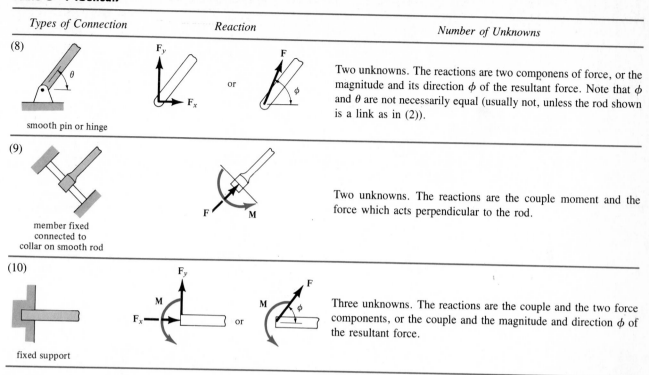

Types of Connection	Reaction	Number of Unknowns
(8) smooth pin or hinge	F_y ... F_x or F ... ϕ	Two unknowns. The reactions are two componens of force, or the magnitude and its direction ϕ of the resultant force. Note that ϕ and θ are not necessarily equal (usually not, unless the rod shown is a link as in (2)).
(9) member fixed connected to collar on smooth rod	F ... M	Two unknowns. The reactions are the couple moment and the force which acts perpendicular to the rod.
(10) fixed support	F_y ... M ... F_x or M ... F ... ϕ	Three unknowns. The reactions are the couple and the two force components, or the couple and the magnitude and direction ϕ of the resultant force.

on the member in that direction. Likewise, if *rotation* is prevented, a *couple moment* is exerted on the member. For example, a member in contact with the smooth surface (6) is prevented from translating *only* in the contact direction, perpendicular or normal to the surface. Hence, the surface exerts only a *normal force* **F** on the member at the point of contact. The magnitude of this force represents *one unknown*. Since the member is free to rotate on the surface, a couple cannot be developed by the surface on the member at the point of contact. The pin or hinge support (8) prevents translation of the connecting member at its point of connection. In this case, unlike that of the smooth surface, translation is prevented in any direction. Hence, a force **F** must be developed at the support such that it has *two unknowns*, its magnitude F and direction ϕ, or, equivalently, the magnitudes of its two components $\mathbf{F}_x$ and $\mathbf{F}_y$. Since the connecting member is allowed to rotate freely in the plane about the pin, a pin support does not resist a couple moment acting perpendicular to this plane. The fixed support (10), however, prevents *both* planar translation and rotation of the connecting member at the point of connection. Therefore, this type of support exerts both a force and a couple moment on the member. Note that the moment acts *perpendicular* to the plane of the page since rotation is prevented in the plane. Hence, there are *three unknowns* at a fixed support.

The concentrated forces and couple moments shown in Table 5–1 actually represent the *resultants* of *distributed surface loads* that exist between each support and its contacting member. Although it is these resultants which are determined in practice, it is generally not important to determine the actual load distribution, since the surface area over which it acts is considerably *smaller* than the *total surface area* of the contacting member.

External and Internal Forces. Since a rigid body is a composition of particles, both *external* and *internal* loadings may act on it. It is important to realize, however, that if the free-body diagram for the body is drawn, the forces that are *internal* to the body are *not represented* on the free-body diagram. As discussed in Sec. 5.1, these forces always occur in equal but opposite collinear pairs, and therefore their *net effect* on the body is zero.

In some problems, a free-body diagram for a ''system'' of connected bodies may be used for an analysis. An example would be the free-body diagram of an entire automobile (system) composed of its many parts. Obviously, the contacting forces between its parts would represent internal forces which would *not* be included on the free-body diagram of the automobile. To summarize then, internal forces act between particles which are located *within* a specified system which is contained within the boundary of the free-body diagram. Particles or bodies outside this boundary exert external forces on the system and these alone must be shown on the free-body diagram.

Weight and the Center of Gravity. When a body is subjected to a gravitational field, each of its particles has a specified weight as defined by Newton's law of gravitation, $F = Gm_1m_2/r^2$, Eq. 1–2. If we assume the size of the body to be ''small'' in relation to the size of the earth, then it is appropriate to consider these gravitational forces to be represented as a *system of parallel forces* acting on the particles contained within the boundary of the body. It was shown in Sec. 4.8 that such a system can be reduced to a single resultant force acting through a specified point. We refer to this force resultant as the *weight* **W** of the body, and to the location of its point of application as the *center of gravity G*. The methods used for its calculation will be developed in Chapter 7.

In the examples and problems that follow, if the weight of the body is important for the analysis, this force will then be reported in the problem statement. Also, when the body is *uniform* or made of homogeneous material, the center of gravity will be located at the body's *geometric center;* however, if the body is nonhomogeneous or has an unusual shape, then its center of gravity will be given.

PROCEDURE FOR DRAWING A FREE-BODY DIAGRAM

To construct a free-body diagram for a rigid body or group of bodies considered as a single system, the following steps should be performed:

Step 1. Imagine the body to be *isolated* or cut "free" from its surroundings, and draw (sketch) its outlined shape.

Step 2. Identify all the external forces and couples that act on the body. Forces and couples generally encountered are those due to (1) applied loadings, (2) reactions occurring at the supports or at points of contact with other bodies, and (3) the weight of the body. To account for all these forces, it may help to trace over the boundary, carefully noting each force or couple acting on it.

Step 3. Indicate the dimensions of the body necessary for computing the moments of forces. The forces and couples that are known should be labeled with their proper magnitudes and directions. Letters are used to represent the magnitudes and direction angles of forces and couples that are *unknown*. In particular, if a force or couple has a known line of action but unknown magnitude, the arrowhead which defines the sense of the vector can be assumed. The correctness of the assumed sense will become apparent after solving the equilibrium equations for the unknown magnitude. By definition, the *magnitude* of a vector is *always positive,* so that if the solution yields a "negative" scalar, the *minus sign* indicates that the vector's sense is *opposite* to that which was originally assumed.

Before proceeding, review this section; then carefully study the following examples. Afterward, attempt to draw the free-body diagrams for the objects in Figs. 5–2 through 5–5 without "looking" at the solutions. Further practice in drawing free-body diagrams should be gained by solving *all* the problems given at the end of this section.

Example 5–1

Draw the free-body diagram of the uniform beam shown in Fig. 5–2a. The beam has a mass of 100 kg.

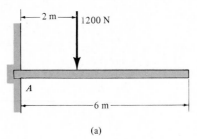

(a)

Fig. 5–2a

Solution

The free-body diagram of the beam is shown in Fig. 5–2b. Since the support at A is a fixed wall, there are three reactions acting *on the beam* at A, denoted as A_x, A_y, and M_A. The magnitudes of these vectors are *unknown,* and their sense has been *assumed.* (How does one obtain the *correct* sense of these vectors?) The weight of the beam, $W = 100(9.81) = 981$ N, acts through the beam's center of gravity G, 3 m from A since the beam is uniform.

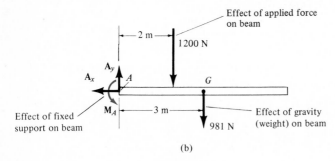

(b)

Fig. 5–2b

Example 5–2 ▬▬▬▬▬

Draw the free-body diagram for the bell crank *ABC* shown in Fig. 5–3*a*.

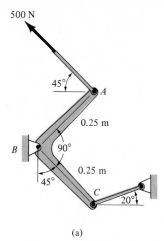

(a)

Fig. 5—3a

Solution

The free-body diagram is shown in Fig. 5–3*b*. The pin support at *B* exerts force components **B**$_x$ and **B**$_y$ *on the crank,* each having a known line of action but unknown magnitude. The link at *C* exerts a force **F**$_C$ acting in the direction of the link and having an unknown magnitude. The dimensions of the crank are also labeled on the free-body diagram, since this information will be useful in computing the moments of the forces. As usual, the sense of the three unknown forces has been assumed. The correct sense will become apparent after solving the equilibrium equations.

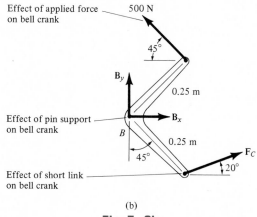

(b)

Fig. 5—3b

Example 5–3

Two smooth balls A and B, each having a mass of 2 kg, rest between the inclined planes shown in Fig. 5–4a. Draw the free-body diagrams for ball A, ball B, and balls A and B together.

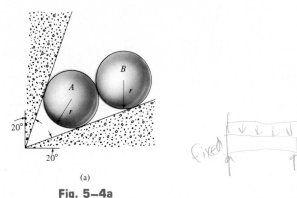

(a)

Fig. 5–4a

Solution

The free-body diagram for ball A is shown in Fig. 5–4b. Note that the weight of the ball is calculated as $W = 2(9.81) = 19.62$ N. Since all contacting surfaces are *smooth,* the reactive forces **T, F, R** act in a direction *normal* to the tangent at their surfaces of contact.

The free-body diagram of ball B is shown in Fig. 5–4c. Can you identify each of the three forces acting *on the ball?* In particular, note that **R,** representing the force of ball A on ball B, Fig. 5–4c, is equal and opposite to **R** representing the force of ball B on ball A, Fig. 5–4b. This is a consequence of Newton's third law of motion.

The free-body diagram of both balls combined ("system") is shown in Fig. 5–4d. Here the contact force **R,** which acts between A and B, is considered as an *internal* force and hence is not shown on the free-body diagram. That is, it represents a pair of equal but opposite collinear forces which cancel each other.

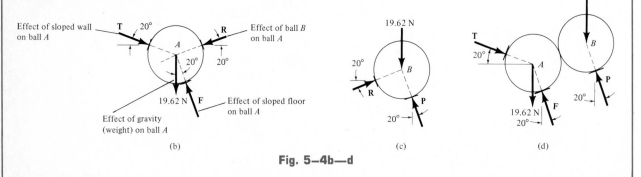

(b)

(c)

(d)

Fig. 5–4b—d

Example 5–4

The free-body diagram of each object in Fig. 5–5 is drawn and the forces acting on the object are identified. The weights of the objects are neglected except where indicated.

Solution

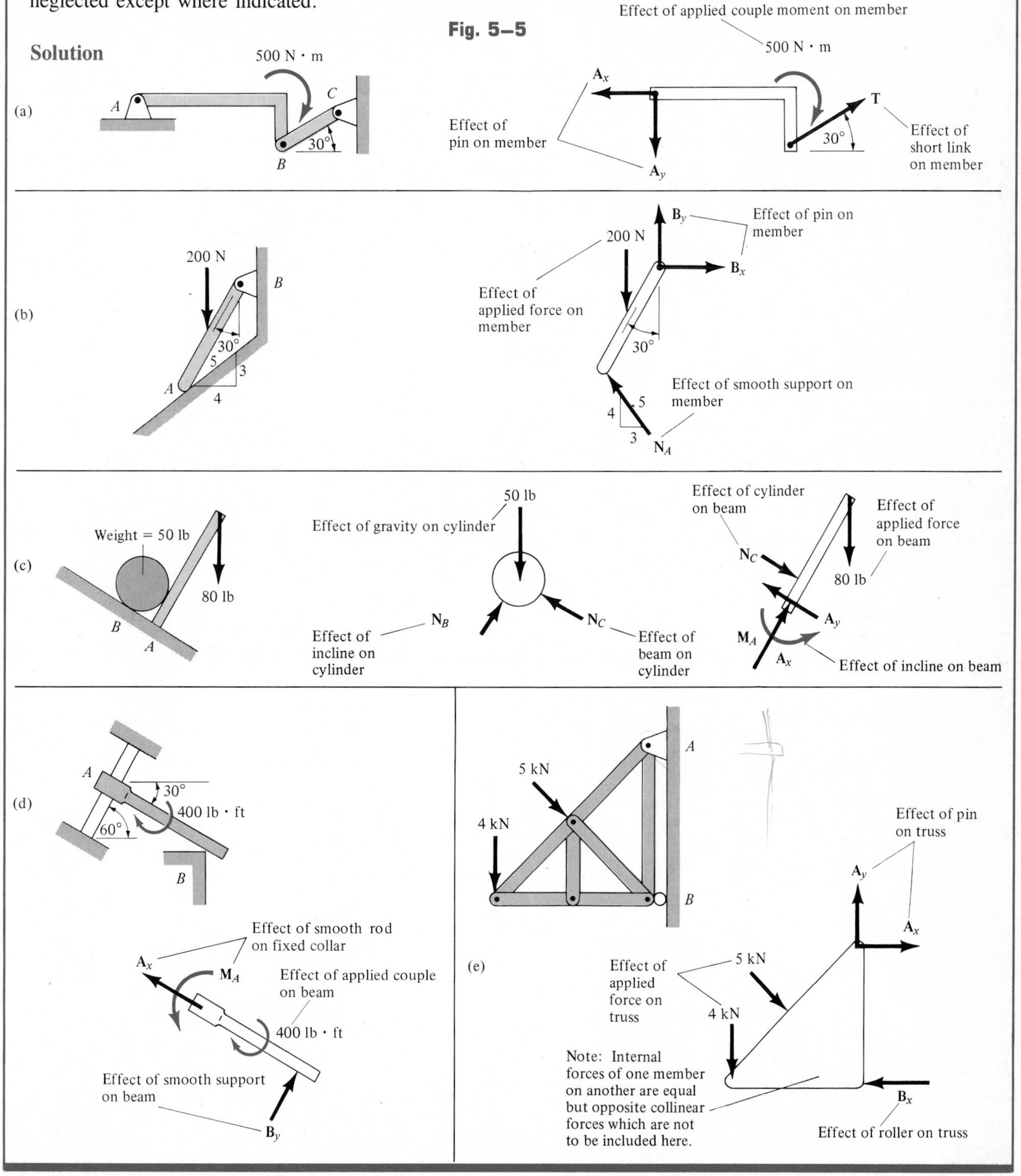

Fig. 5–5

(a)

Effect of applied couple moment on member

500 N · m

Effect of pin on member

A_x

A_y

T

Effect of short link on member

30°

(b)

200 N

B_y — Effect of pin on member

200 N

B_x

Effect of applied force on member

30°

Effect of smooth support on member

5
4
3

N_A

(c)

Weight = 50 lb

80 lb

B

A

50 lb

Effect of gravity on cylinder

Effect of incline on cylinder

N_B

N_C

Effect of beam on cylinder

Effect of cylinder on beam

Effect of applied force on beam

N_C

80 lb

A_y

M_A

A_x

Effect of incline on beam

(d)

A

30°

400 lb · ft

60°

B

Effect of smooth rod on fixed collar

A_x

M_A

Effect of applied couple on beam

400 lb · ft

Effect of smooth support on beam

B_y

(e)

5 kN

4 kN

A

B

Effect of pin on truss

A_y

A_x

Effect of applied force on truss

5 kN

4 kN

B_x

Effect of roller on truss

Note: Internal forces of one member on another are equal but opposite collinear forces which are not to be included here.

Problems

Draw the free-body diagram in each of the following problems and determine the total number of unknown force and couple magnitudes and/or directions. Neglect the weight of the members unless otherwise stated.

5–1. The uniform beam that has a mass of 100 kg, center of gravity at G, and is supported at the smooth surfaces A, B, and C.

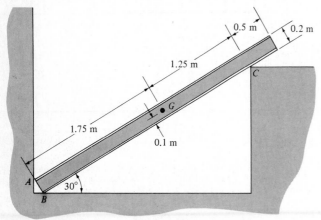

Prob. 5–1

5–2. The 50-kg uniform pipe, which is supported by the smooth contacts at A and B.

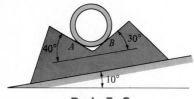

Prob. 5–2

5–3. The hand punch, which is pinned at A and bears down on the smooth surface at B.

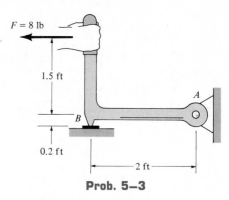

Prob. 5–3

*5–4. The automobile, which is being towed at a constant velocity up the incline using the cable at C. The automobile has a mass of 5 Mg and center of mass at G. The tires are free to roll.

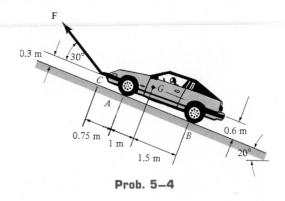

Prob. 5–4

5–5. The jib crane *AB*, which is pin-connected at *A* and supported by member (link) *BC*.

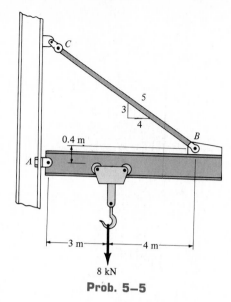

Prob. 5–5

5–6. The pulley, pin-connected at its center and in contact with a cable sustaining a tension of 40 N.

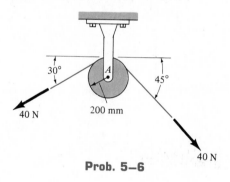

Prob. 5–6

5–7. The link *CAB*, which is pin-connected at *A* and rests on the smooth cam at *B*.

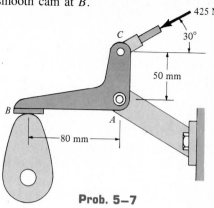

Prob. 5–7

*5–8. The uniform rod *ABC* supported by a pin at *A* and a short link *BD*.

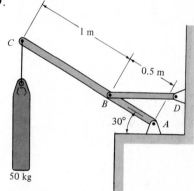

Prob. 5–8

5–9. The dumpster *D* of the truck, which has a weight of 5000 lb and a center of gravity at *G*. It is supported by a pin at *A* and a (short) pin-connected hydraulic cylinder *BC*.

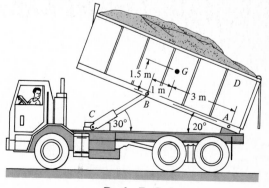

Prob. 5–9

5–10. The beam subjected to the forces and couple moment and supported by a pin at A and short link CB.

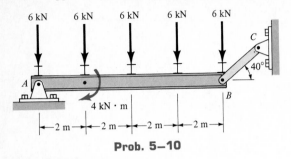

Prob. 5–10

5–11. The small lift bridge ADC, having a deck that weighs 400 lb and center of gravity at G. The bridge is supported by a pin at C and a lift cable AB.

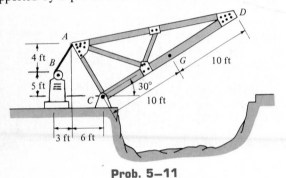

Prob. 5–11

*5–12. The two-hinged gate, which has a weight of 30 lb and center of gravity at G. Due to the contact, the hinge at A is designed to carry the entire vertical load, whereas the hinge at B acts as a collar.

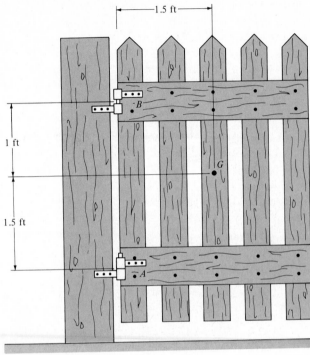

Prob. 5–12

5–13. Draw the specified free-body diagram in each of the following problems and determine the total number of unknown force and couple-moment magnitudes and/or directions. Neglect the weights of the members unless otherwise stated. Although not shown on each photo, assume the geometry (size and angles) is known.

a. Draw the free-body diagram of the beam AB which is used to support the tank. The beam can be assumed pin connected at A and roller supported at B. The tank and its contents exert vertical forces of 750 lb on the beam at points C, D, and E. The pipe FG is not connected to the beam.

c. Draw a free-body diagram of the dragline bucket. The bucket and its contents have a mass of 215 kg and mass center at G. It is supported by chains AB and CD and a cable ED which wraps over the pulley at B.

Prob. 5–13a

Prob. 5–13c

b. Draw the free-body diagram of the structure supporting the railroad signal lights. The trussed portion AB has a mass of 130 kg with mass center at G_1 and the column section BC has a mass of 90 kg with mass center at G_2. The column is fixed connected to a footing at C.

d. The frame F of a railroad car transmits a vertical load of 11 kN to *each* of the three coil springs. Draw a free-body diagram of the wheel housing H. The wheels are pin connected to the housing at the axles A and B. Also draw a free-body diagram of one of the wheels. The track is smooth.

Prob. 5–13b

Prob. 5–13d

5–14. Draw the specified free-body diagram in each of the following problems and determine the total number of unknown force and couple-moment magnitudes and/or directions. Neglect the weights of the members unless otherwise stated. Although not shown on each photo, assume the geometry (size and angles) is known.

a. Draw the free-body diagram for the uniform 250-lb drum which is supported at each of its ends by chains hooked to its rim at A and B.

Prob. 5–14a

c. Draw a free-body diagram of the crane boom ABC which has a mass of 45 kg, center of mass at G, and supports a load of 30 kg. The boom is pin connected to the frame at B and to a vertical link CD. (Consider the link to act as a cable.) The cable supporting the load is attached to the boom at A.

Prob. 5–14c

b. Draw the free-body diagram of the crane boom AB which has a weight of 650 lb and center of gravity at G. The boom is supported by a pin at A and cable BC. The load of 280 lb is suspended from a cable at D which passes over a pulley at E and wraps around a hoist at F. The telescoping member HI is used for safety purposes to prevent the boom from rising to the vertical. It does not contribute a significant force on the boom so its effect can be neglected.

Prob. 5–14b

d. Draw a free-body diagram of the crescent wrench which is subjected to a pulling force of 65 N applied perpendicular to the handle. The *smooth* jaws contact the bolt at points A and B.

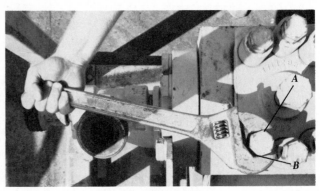

Prob. 5–14d

166

Equations of Equilibrium **5.3**

In Sec. 5.1 we stated the two equations which are both necessary and sufficient for the equilibrium of a rigid body, namely, $\Sigma \mathbf{F} = \mathbf{0}$ and $\Sigma \mathbf{M}_O = \mathbf{0}$. When the body is subjected to a system of forces, which all lie in the x-y plane, then the forces can be resolved into their x and y components. Consequently, the conditions for equilibrium in two dimensions are

$$\Sigma F_x = 0$$
$$\Sigma F_y = 0 \qquad (5\text{--}2)$$
$$\Sigma M_O = 0$$

Here ΣF_x and ΣF_y represent, respectively, the algebraic sums of the x and y components of all the forces acting on the body, and ΣM_O represents the algebraic sum of the moments of all these force components about an axis perpendicular to the x-y plane and passing through the arbitrary point O, which may lie either on or off the body.

Alternative Sets of Equilibrium Equations. Although Eqs. 5–2 are *most often* used for solving equilibrium problems involving coplanar force systems, two *alternative* sets of three independent equilibrium equations may also be used. One such set is

$$\Sigma F_a = 0$$
$$\Sigma M_A = 0 \qquad (5\text{--}3)$$
$$\Sigma M_B = 0$$

When using these equations it is required that the moment points A and B do *not* lie on a line that is *perpendicular* to the a axis. To prove that Eqs. 5–3 provide the *conditions* for equilibrium, consider the free-body diagram of an arbitrarily shaped body shown in Fig. 5–6a. Using the methods of Sec. 4.7, the system of forces may be replaced by a single resultant force $\mathbf{F}_R = \Sigma \mathbf{F}$, acting at point A, and a resultant couple $\mathbf{M}_{R_A} = \Sigma \mathbf{M}_A$, Fig. 5–6b. For equilibrium it is required that $\mathbf{F}_R = \mathbf{0}$ and $\mathbf{M}_{R_A} = \mathbf{0}$. Hence, if $\Sigma M_A = 0$, Eq. 5–3, it is necessary that $\mathbf{M}_{R_A} = \mathbf{0}$. Furthermore, in order that $\mathbf{F}_R$ satisfy $\Sigma F_a = 0$, it must have no component along the a axis, and therefore its line of action must be perpendicular to the a axis, Fig. 5–6c. Finally, if it is required that $\Sigma M_B = 0$, where B does not lie on the line of action of $\mathbf{F}_R$, then $\mathbf{F}_R = \mathbf{0}$, and indeed the body shown in Fig. 5–6a must be in equilibrium.

A second alternative set of equilibrium equations is

$$\Sigma M_A = 0$$
$$\Sigma M_B = 0 \qquad (5\text{--}4)$$
$$\Sigma M_C = 0$$

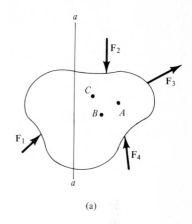

(a)

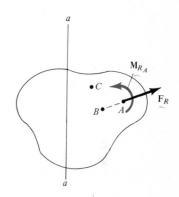

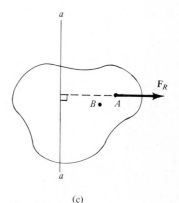

(c)

Fig. 5–6

Here it is necessary that points A, B, and C do not lie on the same line. To prove that these equations when satisfied ensure equilibrium, consider again the free-body diagram in Fig. 5–6b. If it is required that $\Sigma M_A = 0$, then the resultant couple $\mathbf{M}_{R_A} = \mathbf{0}$. $\Sigma M_B = 0$ is satisfied if the line of action of $\mathbf{F}_R$ passes through point B as shown, and finally, if $\Sigma M_C = 0$, where C does not lie on line AB, it is necessary that $\mathbf{F}_R = \mathbf{0}$, and the body in Fig. 5–6a must then be in equilibrium.

PROCEDURE FOR ANALYSIS

The following procedure provides a method for solving coplanar force equilibrium problems:

Free-Body Diagram. Draw a free-body diagram of the body as discussed in Sec. 5.2. Briefly this requires showing all the external forces and couple moments acting *on the body*. The magnitudes of these vectors must be labeled and their directions specified. The sense of a force or couple having an *unknown* magnitude but known line of action can be *assumed*. Dimensions of the body, necessary for computing the moments of forces, are also included on the free-body diagram.

Equations of Equilibrium. Establish the x, y, z axes and apply the equations of equilibrium: $\Sigma F_x = 0$, $\Sigma F_y = 0$, $\Sigma M_O = 0$ (or the alternative sets of Eqs. 5–3 or 5–4). To *avoid* having to solve simultaneous equations, apply the moment equation $\Sigma M_O = 0$ about a point (O) *that lies at the intersection of the lines of action of two of the three unknown forces*. In this way, the moments of these unknowns are *zero* about O, and one can obtain a *direct solution* for the third unknown. When applying the force equations $\Sigma F_x = 0$ and $\Sigma F_y = 0$, orient the x and y axes along lines that will provide the simplest reduction of the forces into their x and y components. If the solution of the equilibrium equations yields a *negative* scalar for an unknown force or couple, it indicates that the sense is *opposite* to that which was assumed on the free-body diagram.

The following example problems numerically illustrate this procedure.

Example 5–5

Determine the horizontal and vertical components of reaction for the beam loaded as shown in Fig. 5–7a. Neglect the weight of the beam in the calculations.

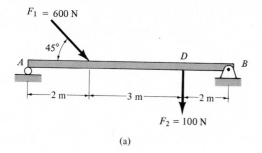

(a)

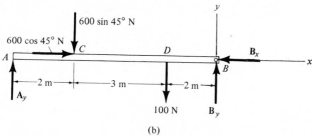

(b)

Fig. 5–7

Solution

Free-Body Diagram. Can you identify each of the forces shown on the free-body diagram of the beam, Fig. 5–7b? For simplicity in applying the equilibrium equations, $\mathbf{F}_1$ is represented by its x and y components as shown.

Equations of Equilibrium. Applying the equilibrium equation $\Sigma F_x = 0$ to the force system on the free-body diagram in Fig. 5–7b yields

$$\overset{+}{\rightarrow}\Sigma F_x = 0; \qquad 600 \cos 45° \text{ N} - B_x = 0$$
$$B_x = 424.3 \text{ N} \qquad \qquad Ans.$$

A direct solution for $\mathbf{A}_y$ can be obtained by applying the moment equation $\Sigma M_B = 0$ about point B. For the calculation, it should be apparent that forces $600 \cos 45°$ N, $\mathbf{B}_x$, and $\mathbf{B}_y$ create zero moment about B. Assuming counterclockwise rotation about B to be positive (in the $+\mathbf{k}$ direction), Fig. 5–7b, we have

$$\underset{\downarrow}{+}\Sigma M_B = 0; \quad 100 \text{ N}(2 \text{ m}) + (600 \sin 45° \text{ N})(5 \text{ m}) - A_y(7 \text{ m}) = 0$$
$$A_y = 331.6 \text{ N} \qquad \qquad Ans.$$

Summing forces in the y direction, using the result $A_y = 331.6$ N, gives

$$+\uparrow\Sigma F_y = 0; \quad 331.6 \text{ N} - 600 \sin 45° \text{ N} - 100 \text{ N} + B_y = 0$$
$$B_y = 192.6 \text{ N} \qquad \qquad Ans.$$

Example 5–6

The cord shown in Fig. 5–8a supports a force of 100 lb and wraps over the frictionless pulley. Determine the tension in the cord and the horizontal and vertical components of reaction at pin A.

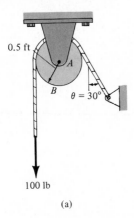

0.5 ft

A

B

$\theta = 30°$

100 lb

(a)

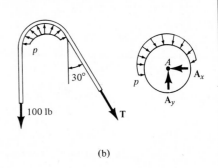

p

30°

100 lb

T

p

A

A_x

A_y

(b)

Fig. 5–8

Solution

Free-Body Diagrams. The free-body diagrams of the cord and pulley are shown in Fig. 5–8b. Note that the principle of action, equal but opposite reaction, must be carefully observed when drawing each of these diagrams: the cord exerts an unknown load distribution p along part of the pulley surface, whereas the pulley exerts an equal but opposite effect on the cord. For the solution, however, it is simpler to *combine* the free-body diagrams of the pulley and a portion of the cord, so that this distribution becomes *internal* to the system and is therefore eliminated from the analysis, Fig. 5–8c.

Equations of Equilibrium. Summing moments about point A to eliminate A_x and A_y, Fig. 5–8c, we have

$$\zeta + \Sigma M_A = 0; \qquad 100 \text{ lb}(0.5 \text{ ft}) - T(0.5 \text{ ft}) = 0$$
$$T = 100 \text{ lb} \qquad\qquad Ans.$$

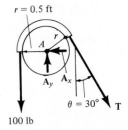

$r = 0.5$ ft

r

A

A_y A_x

$\theta = 30°$ T

100 lb

(c)

It is seen that the tension remains *constant* as the cord passes over the pulley. (This of course is true for *any angle θ* at which the cord is directed and for *any radius r* of the pulley.) Using the result for T, a force summation is applied to determine the reaction at pin A.

$$\xrightarrow{+} \Sigma F_x = 0; \qquad -A_x + 100 \sin 30° \text{ lb} = 0$$
$$A_x = 50.0 \text{ lb} \qquad\qquad Ans.$$

$$+\uparrow \Sigma F_y = 0; \qquad A_y - 100 \text{ lb} - 100 \cos 30° \text{ lb} = 0$$
$$A_y = 186.6 \text{ lb} \qquad\qquad Ans.$$

Example 5–7

Determine the reactions at the fixed support A for the loaded frame shown in Fig. 5–9a.

Solution

Free-Body Diagram. As shown in Fig. 5–9b, there are three unknowns at the fixed support, represented by the magnitudes of $\mathbf{A}_x$, $\mathbf{A}_y$, and $\mathbf{M}_A$.

Equations of Equilibrium

$$\xrightarrow{+}\Sigma F_x = 0; \qquad\qquad A_x = 0 \qquad\qquad Ans.$$

$$+\uparrow \Sigma F_y = 0; \qquad A_y - 400 \text{ N} - 200 \text{ N} = 0$$

$$A_y = 600 \text{ N} \qquad\qquad Ans.$$

$$\downdownarrows +\Sigma M_A = 0; \qquad M_A - 400 \text{ N}(1 \text{ m}) - 200 \text{ N}(2 \text{ m}) = 0$$

$$M_A = 800 \text{ N} \cdot \text{m} \qquad\qquad Ans.$$

Point A was chosen for summing moments since the lines of action of the *unknown* forces $\mathbf{A}_x$ and $\mathbf{A}_y$ pass through this point, and therefore these forces were not included in the moment summation. Note, however, that $\mathbf{M}_A$ must be *included* in the moment summation. This couple moment is a free vector and represents an effect of the fixed support on the frame.

Although only *three* independent equilibrium equations can be written for a rigid body, it is a good practice to *check* all calculations using a fourth equilibrium equation. The latter equation may be obtained from one of the other sets of equilibrium equations, 5–3 or 5–4. For example, the above computations may be verified by summing moments about point C:

$$\downdownarrows +\Sigma M_C = 0; \quad 400 \text{ N}(1 \text{ m}) - 600 \text{ N}(2 \text{ m}) + 800 \text{ N} \cdot \text{m} \equiv 0$$

$$400 \text{ N} \cdot \text{m} - 1200 \text{ N} \cdot \text{m} + 800 \text{ N} \cdot \text{m} \equiv 0$$

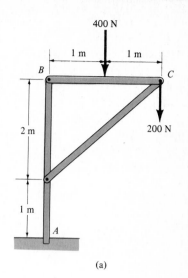

(a)

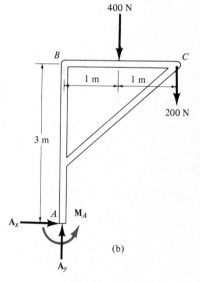

(b)

Fig. 5–9

Example 5-8

The uniform beam shown in Fig. 5–10a is subjected to a force and couple. If the beam is supported at A by a smooth wall and at B and C either at the top or bottom by smooth contacts, determine the reactions at these supports. Neglect the weight of the beam.

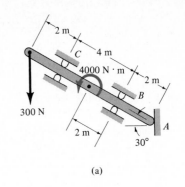

(a)

Solution

Free-Body Diagram. As shown in Fig. 5–10b, all the support reactions act normal to the surface of contact since the contacting surfaces are smooth. The reactions at B and C are shown acting in the positive y' direction. This assumes that only the contacts located on the bottom of the beam are used for support.

Equations of Equilibrium. Using the x, y coordinate system in Fig. 5–10b, we have

$$\overset{+}{\rightarrow}\Sigma F_x = 0; \qquad C_{y'} \sin 30° + B_{y'} \sin 30° - A_x = 0 \qquad (1)$$

$$+\uparrow\Sigma F_y = 0; \qquad -300\ \text{N} + C_{y'} \cos 30° + B_{y'} \cos 30° = 0 \qquad (2)$$

$$\zeta +\Sigma M_A = 0; \qquad -B_{y'}(2\ \text{m}) + 4000\ \text{N} \cdot \text{m} - C_{y'}(6\ \text{m})$$
$$+(300 \cos 30°\ \text{N})(8\ \text{m}) = 0 \qquad (3)$$

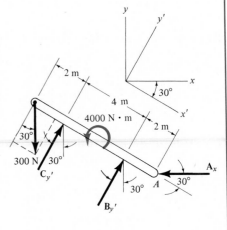

(b)

Fig. 5–10

When writing the moment equation, it should be noticed that the line of action of the force component $300 \sin 30°$ N passes through point A, and therefore this force is not included in the moment equation.

Solving Eqs. (2) and (3) simultaneously, we obtain

$$B_{y'} = -1000.0\ \text{N} \qquad\qquad \textit{Ans.}$$
$$C_{y'} = 1346.4\ \text{N} \qquad\qquad \textit{Ans.}$$

Since $B_{y'}$ is a negative quantity, the sense of $\mathbf{B}_{y'}$ is opposite to that shown on the free-body diagram in Fig. 5–10b. Therefore, the top contact at B serves as the support rather than the bottom one. Retaining the negative sign for $B_{y'}$ and substituting the results into Eq. (1), we obtain

$$1346.4 \sin 30°\ \text{N} - 1000.0 \sin 30°\ \text{N} - A_x = 0$$
$$A_x = 173.2\ \text{N} \qquad\qquad \textit{Ans.}$$

Example 5–9

The beam shown in Fig. 5–11a is pin-connected at A and rests against a roller at B. Compute the horizontal and vertical components of reaction at the pin A.

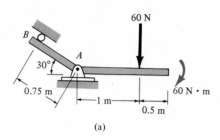

(a)

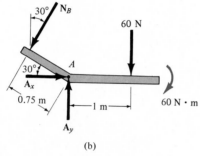

(b)

Fig. 5–11

Solution

Free-Body Diagram. As shown in Fig. 5–11b, the reaction $\mathbf{N}_B$ is perpendicular to the beam at B, since the support is a roller.

Equations of Equilibrium. Summing moments about A, we obtain a direct solution for N_B,

$$\zeta +\Sigma M_A = 0; \quad -60 \text{ N} \cdot \text{m} - 60 \text{ N}(1 \text{ m}) + N_B(0.75 \text{ m}) = 0$$

$$N_B = 160 \text{ N}$$

Using this result,

$$\xrightarrow{+}\Sigma F_x = 0; \qquad A_x - 160 \sin 30° \text{ N} = 0$$

$$A_x = 80.0 \text{ N} \qquad\qquad \textit{Ans.}$$

$$+\uparrow\Sigma F_y = 0; \qquad A_y - 160 \cos 30° \text{ N} - 60 \text{ N} = 0$$

$$A_y = 198.6 \text{ N} \qquad\qquad \textit{Ans.}$$

173

Example 5–10

The man has a weight of 150 lb and stands at the end of the beam shown in Fig. 5–12a. Determine the magnitude and direction of the reaction at the pin A and the tension in the cable.

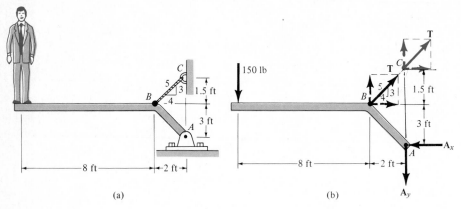

(a)   (b)

Fig. 5–12

Solution

Free-Body Diagram. The forces acting on the beam are shown in Fig. 5–12b.

Equations of Equilibrium. Summing moments about point A to obtain a direct solution for the cable tension yields

$$\zeta +\Sigma M_A = 0; \quad -(\tfrac{3}{5}T)(2 \text{ ft}) - (\tfrac{4}{5}T)(3 \text{ ft}) + 150 \text{ lb}(10 \text{ ft}) = 0$$

$$-3.6T + 150 \text{ lb}(10 \text{ ft}) = 0 \qquad (1)$$

$$T = 416.7 \text{ lb} \qquad Ans.$$

Note that by the principle of transmissibility it is also possible to locate $\mathbf{T}$ at C, even though this point is not on the beam, Fig. 5–12b. In this case, the vertical component of $\mathbf{T}$ creates *zero moment* about A and the moment arm of the horizontal component $(\tfrac{4}{5}T)$ becomes 4.5 ft. Hence, $\Sigma M_A = 0$ yields Eq. (1) directly since $(\tfrac{4}{5}T)(4.5) = 3.6T$.

Summing forces to obtain A_x and A_y, using the result for T, we have

$$\xrightarrow{+} \Sigma F_x = 0; \qquad -A_x + (\tfrac{4}{5})(416.7 \text{ lb}) = 0$$

$$A_x = 333.3 \text{ lb}$$

$$+\uparrow \Sigma F_y = 0; \qquad (\tfrac{3}{5})416.7 \text{ lb} - 150 \text{ lb} - A_y = 0$$

$$A_y = 100 \text{ lb}$$

Thus,

$$F_A = \sqrt{(333.3 \text{ lb})^2 + (100 \text{ lb})^2}$$

$$= 348.0 \text{ lb} \qquad Ans.$$

$$\theta = \tan^{-1} \frac{100 \text{ lb}}{333.3 \text{ lb}} = 16.7° \qquad Ans.$$

Example 5-11

The 100-kg uniform beam AB shown in Fig. 5–13a is supported at A by a pin and at B and C by a continuous cable which wraps around a frictionless pulley located at D. If a maximum tension force of 800 N can be developed in the cable before it breaks, determine the greatest distance d at which the 6-kN force can be placed on the beam. What are the horizontal and vertical components of reaction at A just before the cable breaks?

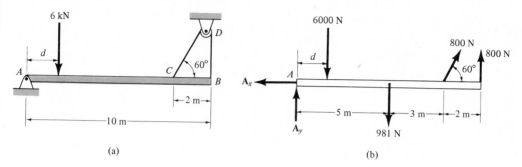

(a)   (b)

Fig. 5–13

Solution

Free-Body Diagram. Since the cable is continuous and passes over a frictionless pulley, the entire cable is subjected to its maximum tension of 800 N when the 6-kN force acts at d. Hence, the cable exerts an 800-N force at points C and B *on the beam* in the direction of the cable, Fig. 5–13b.

Equations of Equilibrium. By summing moments of the force system about point A it is possible to obtain a direct solution for the dimension d. Why?

$$\zeta +\Sigma M_A = 0$$
$$-(6000 \text{ N})(d) - 981 \text{ N}(5 \text{ m}) + (800 \sin 60° \text{ N})(8 \text{ m}) + 800 \text{ N}(10 \text{ m}) = 0$$
$$d = 1.44 \text{ m} \qquad\qquad Ans.$$

Summing forces in the x and y directions, we have

$$\xrightarrow{+}\Sigma F_x = 0; \qquad -A_x + 800 \cos 60° \text{ N} = 0$$
$$A_x = 400 \text{ N} \qquad\qquad Ans.$$

$$+\uparrow \Sigma F_y = 0;$$
$$A_y - 6000 \text{ N} - 981 \text{ N} + 800 \sin 60° \text{ N} + 800 \text{ N} = 0$$
$$A_y = 5.49 \text{ kN} \qquad\qquad Ans.$$

5.4 Two- and Three-Force Members

The solution to some equilibrium problems can be simplified if one is able to recognize members that are subjected to only two or three forces.

Two-Force Members. When a member is subjected to *no* couples and forces are applied at only two points on a member, the member is called a *two-force member*. An example of this situation is shown in Fig. 5–14a. The forces at A and B are first summed to obtain their respective *resultants* $\mathbf{F}_A$ and $\mathbf{F}_B$, Fig. 5–14b. These two forces will maintain *translational or force equilibrium* provided $\mathbf{F}_A$ is of equal magnitude and opposite direction to $\mathbf{F}_B$. Furthermore, *rotational or moment equilibrium* is satisfied if $\mathbf{F}_A$ is *collinear* with $\mathbf{F}_B$. As a result, the line of action of both forces is known, since it always passes through A and B. Hence, only the force magnitude must be determined or stated. Other examples of two-force members held in equilibrium are shown

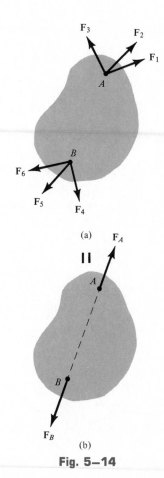

(a)

(b)

Fig. 5–14

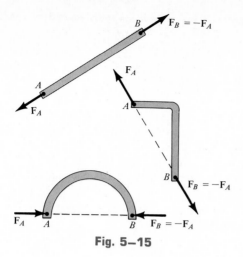

Fig. 5–15

in Fig. 5–15. As will be shown throughout Chapter 6, this concept of first identifying a two-force member and then applying equal but opposite collinear forces at its ends will be of great importance to us in simplifying the force analysis of members which are pin-connected.

Three-Force Members.   *If a member is subjected to three coplanar forces, then it is necessary that the forces be either concurrent or parallel if the member is to be in equilibrium.* To show this, consider the body in Fig. 5–16 and suppose that any two of the three forces acting on the body have lines of action that intersect at point O. To satisfy moment equilibrium about O, i.e., $\Sigma M_O = 0$, the third force must also pass through O, which then makes the force system *concurrent*. If two of the three forces are parallel, the point of concurrency, O, is considered to be at "infinity" and the third force must be parallel to the other two forces to intersect at this "point." Since the three forces are concurrent at a point, only the force equilibrium equations ($\Sigma F_x = 0$, $\Sigma F_y = 0$) must be satisfied.

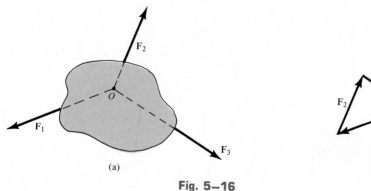

(a)

(b)

Fig. 5–16

Example 5–12

The lever *ABC* is pin-supported at *A* and connected to a short link *BD* as shown in Fig. 5–17a. If the weight of the members is negligible, determine the force developed on the lever at *A*.

Solution

Free-Body Diagrams. As shown by the free-body diagram, Fig. 5–17b, the short link *BD* is a *two-force member,* so the *resultant forces* at pins *D* and *B* must be equal, opposite, and collinear. Although the magnitude of the force is unknown, the line of action is known, since it passes through *B* and *D*.

Lever *ABC* is a *three-force member,* and therefore the three nonparallel forces acting on it must be concurrent at *O*, Fig. 5–17c. In particular, note that the force **F** on the lever is equal but opposite to **F** acting at *B* on the link. Why? The distance *CO* must be 0.5 m, since the lines of action of **F** and the 400-N force are known.

Equations of Equilibrium. Requiring the force system to be concurrent at *O* necessitates $\Sigma M_O = 0$. Hence, the angle θ which defines the line of action of $\mathbf{F}_A$ can be determined from trigonometry,

$$\theta = \tan^{-1}\left(\frac{0.7}{0.4}\right) = 60.3° \qquad \textit{Ans.}$$

Applying the force equilibrium equations yields

$$\xrightarrow{+}\Sigma F_x = 0; \qquad F_A \cos 60.3° - F \cos 45° + 400 \text{ N} = 0$$
$$+\uparrow \Sigma F_y = 0; \qquad F_A \sin 60.3° - F \sin 45° = 0$$

Solving, we get

$$F_A = 1075 \text{ N} \qquad \textit{Ans.}$$
$$F = 1320 \text{ N}$$

We can also solve this problem by representing the force at *A* by its two components $\mathbf{A}_x$ and $\mathbf{A}_y$ and applying $\Sigma M_A = 0$, $\Sigma F_x = 0$, $\Sigma F_y = 0$. Once A_x and A_y are determined, how would you find F_A and θ?

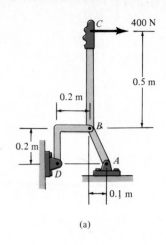

(a)

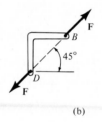

(b)

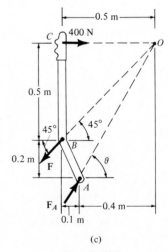

(c)

Fig. 5–17

178

Problems

5–15. Determine the force in the towing cable at *C* and the reactions on *each* of the four tires of the automobile in Prob. 5–4.

*5–16. Determine the reactions at the supports *A* and *B* of the jib crane in Prob. 5–5.

5–17. Determine the resultant force acting at pin *A* of the pulley in Prob. 5–6.

5–18. Determine the reactions at the supports for the rod in Prob. 5–8.

5–19. Determine the reactions at the pin *A* and the force on the hydraulic cylinder of the truck dumpster in Prob. 5–9.

*5–20. Determine the support reactions on the beam in Prob. 5–10.

5–21. Determine the force in the cable *AB* and the reactions at the pin *C* needed to support the bridge in Prob. 5–11.

5–22. Determine the reactions at the supports *A* and *B* of the gate in Prob. 5–12.

5–23. Determine the reactions at *A*, *B*, and *C* needed to hold the beam in Prob. 5–13 in equilibrium.

*5–24. Determine the reactions at the pin and roller supports.

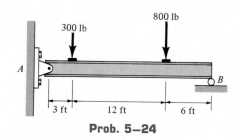

Prob. 5–24

5–25. Determine the reactions at the roller *A* and pin *B*.

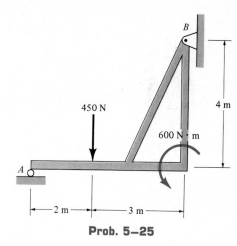

Prob. 5–25

5–26. Determine the reactions at the pin *A* and the tension in cord *BC*.

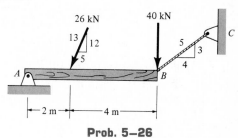

Prob. 5–26

5–27. Determine the reactions at the roller *A* and pin *B*.

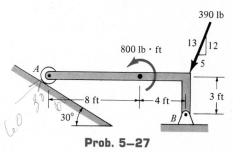

Prob. 5–27

179

*5–28. Determine the reactions where the beam contacts the smooth plane A, the roller B, and the rocker C.

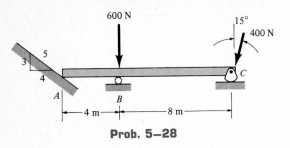

600 N

15°
400 N

5
3
4

C

A
B

|← 4 m →|← 8 m →|

Prob. 5–28

5–29. Determine the reactions at the smooth collar A, roller B, and short link CD.

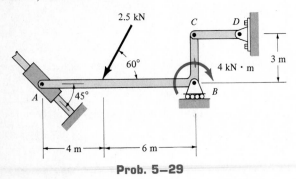

2.5 kN

C   D

60°

4 kN · m

3 m

A
45°
B

|← 4 m →|← 6 m →|

Prob. 5–29

5–30. Determine the reactions at the supports necessary for equilibrium of the beam.

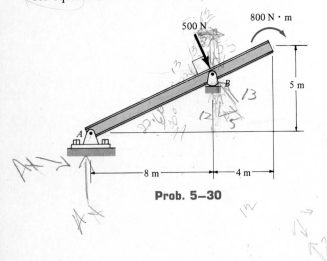

800 N · m

500 N

B

5 m

A

|← 8 m →|← 4 m →|

Prob. 5–30

5–31. Determine the reactions at the supports. The incline at B is smooth.

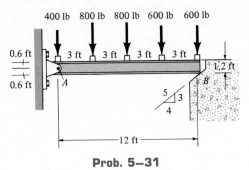

400 lb   800 lb   800 lb   600 lb   600 lb

0.6 ft

|← 3 ft →|← 3 ft →|← 3 ft →|← 3 ft →|

1.2 ft

0.6 ft

A

B

5
3
4

|← 12 ft →|

Prob. 5–31

*5–32. The crane supports a load of 1500 lb which is suspended from cable BD. Determine the horizontal and vertical components of force acting on the pin A of the crane and the tension in the supporting cable BC. Note that BC and BD are separate cables.

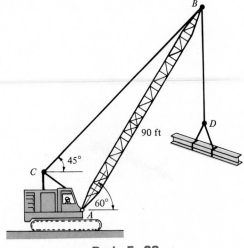

B

D

90 ft

45°

C

60°

A

Prob. 5–32

5–33. The shelf supports the electric motor which has a mass of 15 kg and mass center at G_m. The platform upon which it rests has a mass of 4 kg and mass center at G_p. Assuming that a single bolt B holds the shelf up and the bracket bears against the smooth wall at A, determine this normal force at A and the horizontal and vertical components of reaction of the bolt on the bracket.

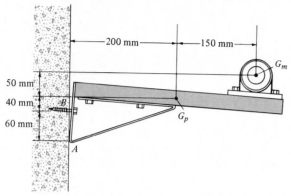

Prob. 5–33

5–34. The oil rig is supported on the trailer by the pin or axle at A and the frame at B. If the rig has a weight of 125,000 lb and center of gravity at G, determine the force $\mathbf{F}$ that must be developed along the hydraulic cylinder CD in order to *start* lifting the rig (slowly) off B toward the vertical. Also compute the horizontal and vertical components of reaction at the pin A.

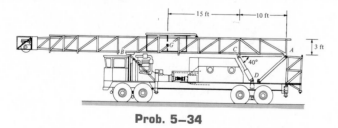

Prob. 5–34

5–35. Davit ABC is subjected to a 400-lb force caused by the boat. If it is supported by a pin at A and a smooth collar at B, determine the components of reaction at these supports.

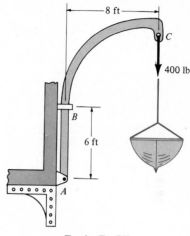

Prob. 5–35

*5–36. The power pole supports the three lines, each line exerting a vertical force on the pole due to its weight as shown. Determine the reactions at the fixed support D. If it is possible for wind or ice to snap the lines, determine which line(s) when removed create(s) a condition for the greatest moment reaction at D.

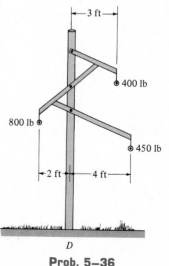

Prob. 5–36

5–37. The uniform cylinder C has a mass of 80 kg and is hoisted using two hooks at A and B which are pin-connected to the ends of a beam. The beam has a mass of 20 kg and a mass center at G. If the mass of the hooks can be neglected, determine the magnitudes of force at A and B and the force $\mathbf{F}$ necessary to hold the system in equilibrium. *Hint:* Draw the free-body diagram of the beam and note that the hooks act as two-force members.

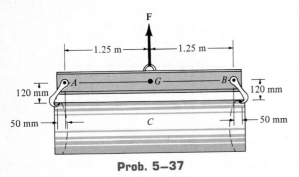

Prob. 5–37

5–38. Outriggers A and B are used to stabilize the crane from overturning when lifting large loads. If the load to be lifted is 3 Mg, determine the *maximum* boom angle θ so that the crane does not overturn. The crane has a mass of 5 Mg and center of mass at G_C, whereas the boom has a mass of 0.6 Mg and center of mass at G_B.

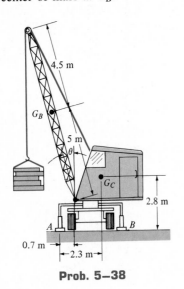

Prob. 5–38

5–39. A skeletal diagram of the lower leg is shown in the lower figure. Here it can be noted that the lower portion of the leg is lifted by the quadriceps muscle attached to the lower hip at A and to the patella bone at B. This bone slides freely over cartilage at the knee joint. The quadriceps is further extended and attached to the tibia at C. Using the mechanical system shown in the upper figure to model the lower leg, determine the tension $\mathbf{T}$ in the quadriceps and the magnitude of the resultant force at the femur (pin), D, in order to hold the lower leg in the position shown. The lower leg has a mass of 3.2 kg and a mass center at G_1, the foot has a mass of 1.6 kg and a mass center at G_2.

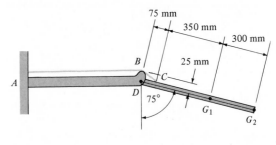

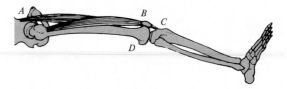

Prob. 5–39

*5–40. Determine the horizontal and vertical reactions at the pin B and the tension in the cable AC necessary to hold the derrick AB in the equilibrium position shown. The 5-kN load is suspended from a *separate* cable which is attached to the boom at A.

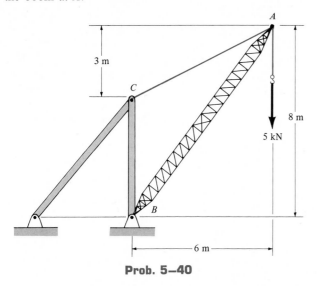

Prob. 5–40

5–42. The winch consists of a drum of radius 4 in., which is pin-connected at its center C. At its outer rim is a ratchet gear having a mean radius of 6 in. The pawl AB serves as a two-force member (short link) and holds the drum from rotating. If the suspended load is 500 lb, determine the horizontal and vertical components of reaction at the pin C.

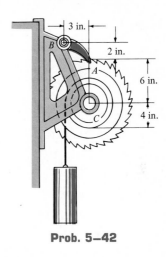

Prob. 5–42

5–41. If the wheelbarrow and its contents have a mass of 60 kg and center of mass at G, determine the magnitude of the resultant force which the man must exert on *each* of the two handles in order to hold the wheelbarrow in equilibrium.

Prob. 5–41

5–43. The jib crane is pin-connected at A and supported by a smooth collar at B. Determine the trolley placement x of the 5000-lb load so that it gives the maximum and minimum reactions at the supports. Calculate these reactions in each case. Neglect the weight of the crane. Require 4 ft $\leqslant x \leqslant$ 10 ft.

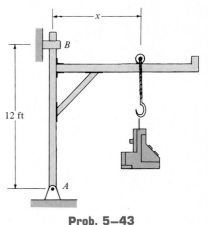

Prob. 5–43

183

*5–44. Determine the horizontal and vertical components of force the man must exert on the hand truck at *A* in order to maintain equilibrium of the load. The hand truck and load have a total mass of 75 kg and a center of mass at *G*. The wheels are free to roll on the ramp. Are these forces any different if the hand truck is moving up or down the ramp at constant velocity?

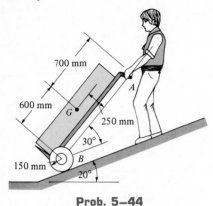

Prob. 5–44

5–45. It is easier to mow a lawn by pulling a lawn mower rather than pushing it. To illustrate this, consider the four-wheel mower shown in the figure having a mass of 6 kg and center of mass at *G*. Determine the magnitude of force **P** required to push it over the smooth obstruction at *A* and compare the result with pulling it over *A*. In both cases the force **P** is to be directed $\theta = 30°$ along the handle as shown. *Hint:* It is required that the vertical reaction of the *ground* at *A* be equal to zero. Only the normal force at the obstruction contact needs to be considered.

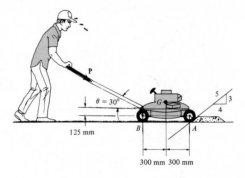

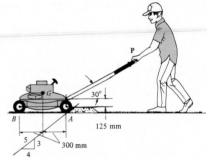

Prob. 5–45

5–46. The cup is filled with 125 g of liquid. The mass center is located at G as shown. If a vertical force $\mathbf{F}$ is applied to the rim of the cup, determine its magnitude so the cup is on the verge of tipping over. *Hint:* When tipping is about to occur, the normal reaction is at A.

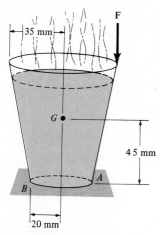

Prob. 5–46

5–47. A sack D having a weight of 8 lb is supported on the scale. Determine the largest number of 0.5-lb weights which should be suspended from the hook at A and the distance x the 0.25-lb bulb B must be placed to keep the scale beam horizontal. The beam remains horizontal when the weights at D, A, and B are removed.

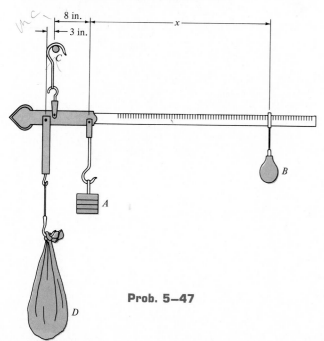

Prob. 5–47

*5–48. A man stands out at the end of the diving board, which is supported by two springs A and B, each having a stiffness of $k = 15$ kN/m. In the position shown the board is horizontal. If the man has a mass of 40 kg, determine the angle of tilt which the board makes with the horizontal after he jumps off. Neglect the weight of the board.

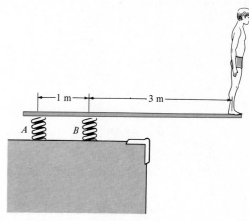

Prob. 5–48

■5–49. The smooth cylinder has a weight of 20 lb and is suspended from a spring which has a stiffness of $k = 5$ lb/ft. If the spring has an unstretched length of $l_o = 1$ ft, determine the angle θ for equilibrium.

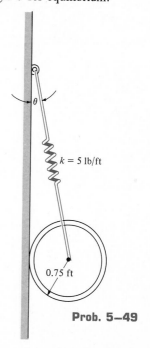

Prob. 5–49

185

5–50. The experiment as shown in the figure is useful for estimating the weight of the man's legs. When the man is lying flat with his legs in the dashed position, the scale reading at B is S_1. By estimation it is determined that the center of gravity of the legs is at G. When his legs are brought to a vertical position, as shown, the center of gravity is displaced horizontally a distance of d, which gives a new scale reading of S_2. Determine the weight of his legs in terms of the scale readings and the dimensions d and l.

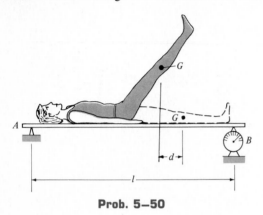

Prob. 5–50

5–51. Three bricks, each having a weight W and length a, are stacked as shown. Determine the maximum distance d the top brick can extend out from the bottom one so the stack does not topple over.

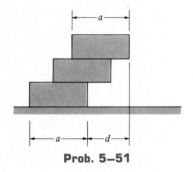

Prob. 5–51

*5–52. The wheel of radius R has a spring attached to its central hub. If the spring has a stiffness k and unstretched length R, show that the *horizontal* force $\mathbf{F}$ needed to pull the wheel forward so that the spring makes an angle θ with the horizontal is $F = kR(\cot \theta - \cos \theta)$.

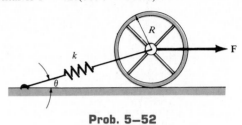

Prob. 5–52

5–53. The disk has a mass of 20 kg and rests on the smooth inclined surface. One end of a spring is attached to the center of the disk and the other end is attached to a roller at A. Consequently, the spring rests in the horizontal position when the disk is in equilibrium. If the unstretched length of the spring is 200 mm, determine its (total) length when the disk is in equilibrium. Neglect the size and weight of the roller.

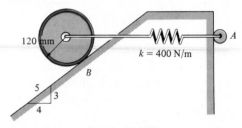

Prob. 5–53

■5–54. A linear *torsional* spring deforms such that an applied couple moment M is related to the spring's rotation θ in radians by the equation $M = (20 \, \theta)$ N · m. If such a spring is attached to the end of a pin-connected uniform 10-kg rod, determine the angle θ for equilibrium. The spring is undeformed when $\theta = 0°$.

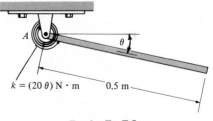

Prob. 5–54

5–55. The solid cone has a weight W and center of gravity G located as shown. If the base of the cone rests against the smooth wall, show that the maximum length l of the string tied to its vertex B and the wall at C will hold the cone in equilibrium provided that $l = \sqrt{\frac{16}{9}r^2 + h^2}$.

5–57. The 400-kg beam is hoisted by two cables at A and B. If the cables are directed as shown when the beam is supported horizontally, determine the location $\bar{x}$ of the center of gravity G of the beam and compute the tension, $\mathbf{T}_1$ and $\mathbf{T}_2$, in each cable.

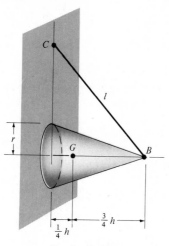

Prob. 5–55

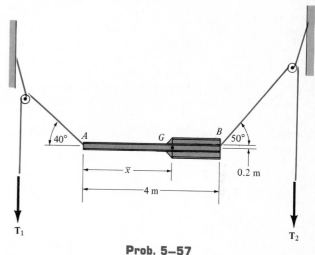

Prob. 5–57

*5–56. The two wheels each have a weight W and are connected to each other by an inextensible cord. If they rest against smooth planes, one of which is inclined at an angle ϕ as shown and the other is vertical, determine the angle θ for equilibrium (a) in terms of the parameters shown, and (b) with $W = 10$ lb, $\phi = 60°$. *Hint:* Draw a separate free-body diagram for each disk.

5–58. The uniform rod has a length l and weight W. It is supported at one end A by a smooth wall and the other end by a cord of length s which is attached to the wall as shown. Show that for equilibrium it is required that $h = [(s^2 - l^2)/3]^{1/2}$.

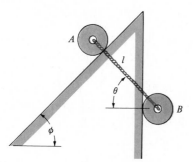

Prob. 5–56

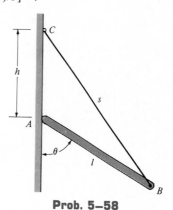

Prob. 5–58

5–59. The uniform ladder of length l rests against the smooth wall at A and smooth inclined roof at B. If the roof is inclined at an angle θ with the horizontal, determine the angle ϕ of the ladder for equilibrium.

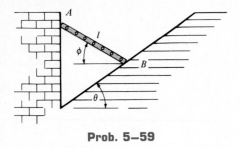

Prob. 5–59

*5–60. The light rod of length $2l$ is suspended from a fixed point O using two cords OA and OB each having a length $1.5l$. If weights of $3W$ and W are suspended from the ends A and B, respectively, determine the angle θ of inclination.

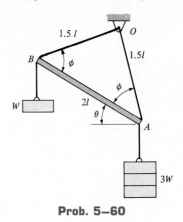

Prob. 5–60

Equilibrium in Three Dimensions

Free-Body Diagrams 5.5

The first step in solving three-dimensional equilibrium problems, as in the case of two dimensions, is to draw a free-body diagram of the body (or group of bodies considered as a system). When drawing this diagram, it is necessary to include *all the reactions* that can occur at each point of support.

Support Reactions. The reactive forces and couples acting at various types of supports and connections, when the members are viewed in three dimensions, are listed in Table 5–2. It is important to recognize the symbols used to represent each of the supports and to clearly understand how the forces and couples are developed by each support. As in the two-dimensional case, *a force is developed by a support that restricts the translation of the attached member, whereas a couple is developed when rotation of the attached member is prevented*. For example, in Table 5–2, the ball-and-socket joint (4) prevents any translation of the connecting member; therefore, a force must act on the member at the point of connection. This force has three components having unknown magnitudes, F_x, F_y, F_z. Provided these components are known, one can obtain the magnitude of force, $F = \sqrt{F_x^2 + F_y^2 + F_z^2}$, and the force's orientation defined by the coordinate direction angles α, β, γ, Eqs. 2–7.* Since the connecting member is allowed to rotate freely about *any* axis, no moment is resisted by a ball-and-socket joint.

It should be noted that the *single* bearing supports (5) and (6), the *single* pin (7), and the *single* hinge (8) can support both force and couple-moment components. Most often, however, these supports are used in conjunction with other bearings, pins, or hinges to hold the body in equilibrium. If this is the case, the *force reactions* at the supports may *alone* be adequate for supporting the body. When this occurs, the couple moments become redundant and may be neglected on the free-body diagram.† The reason for this should become clear after studying the examples which follow, but essentially the couples will not be developed at these supports since rotation of the body will be prevented not by the supporting couples, but rather by the reactions developed at the other supports.

---

*The three unknowns may also be represented as an unknown force magnitude F and two unknown coordinate direction angles. The third direction angle is obtained using the identity $\cos^2 \alpha + \cos^2 \beta + \cos^2 \gamma = 1$, Eq. 2–10.

†It is also necessary that the physical body *maintain* its rigidity when loaded and the supports be *properly aligned* when connected to the body.

upports for Rigid Bodies Subjectd to Three-Dimensional Force Systems.

nection	Reaction	Number of Unknowns
(1) cable		One unknown. The reaction is a force which acts away from the member in the direction of the cable.
(2) smooth surface support		One unknown. The reaction is a force which acts perpendicular to the surfact at the point of contact.
(3) roller on a smooth surface		One unknown. The reaction is a force which acts perpendicular to the surfact at the point of contact.
(4) ball and socket		Three unknowns. The reactions are three rectangular force components.
(5) single journal bearing		Four unknowns. The reactions are two force and two couple components which act perpendicular to the shaft.
(6) single thrust bearing		Five unknowns. The reactions are three force and two couple components.

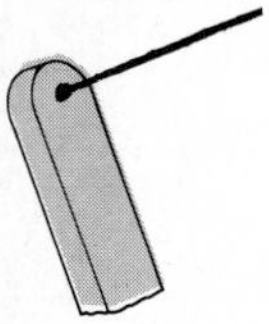

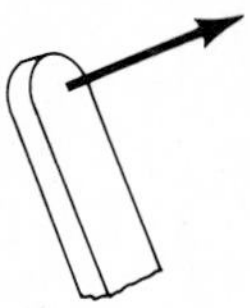

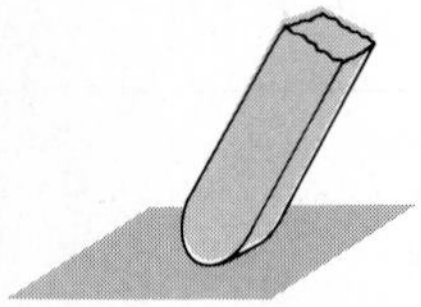

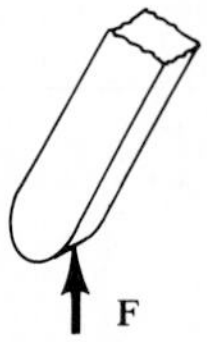

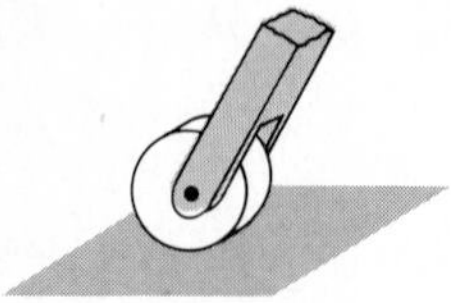

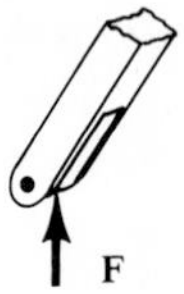

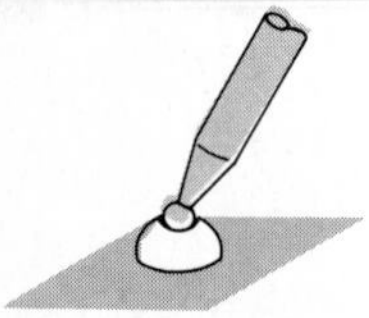

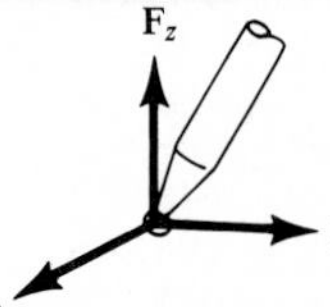

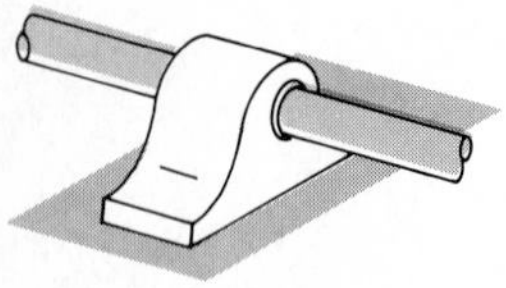

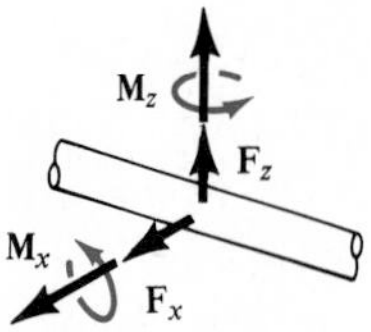

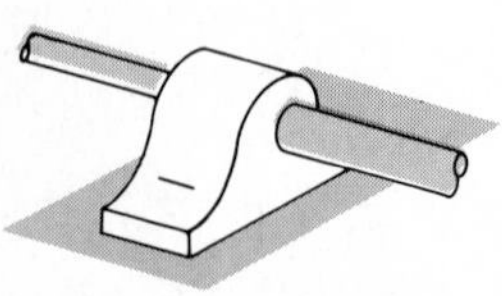

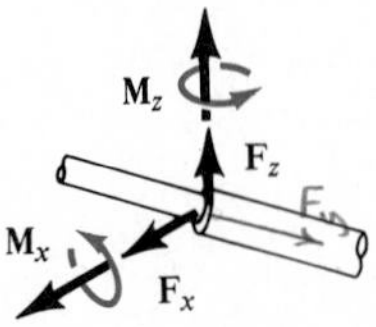

Table 5—2 (Contd.)

Types of Connection	Reaction	Number of Unknowns
(7) single smooth pin		Five unknowns. The reactions are three force and two couple components.
(8) single hinge		Five unknowns. The reactions are three force and two couple components. Delete M_x and M_z if hinge is properly aligned.
(9) fixed support		Six unknowns. The reactions are three force and three couple components.

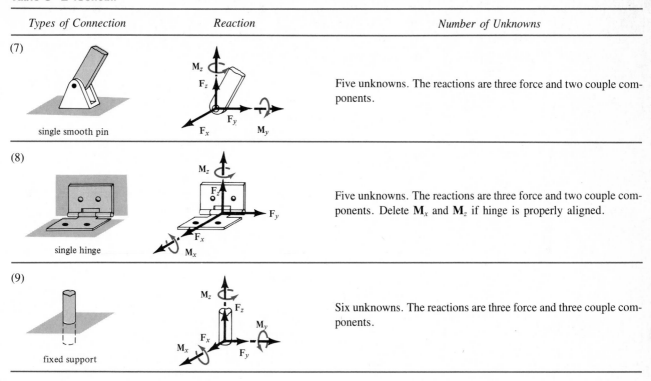

Free-Body Diagrams. The general procedure used to draw the free-body diagram of a rigid body has been outlined in Sec. 5.2. The importance of this step cannot be overemphasized when solving equilibrium problems. Essentially it requires first "isolating" the body by drawing its outlined shape. This is followed by a careful *labeling* of *all* the applied external forces and couples and the support reactions. To do this, x, y, z axes are established with the origin located at a convenient point. As a general rule, *components of reaction* having an *unknown magnitude* are shown acting on the free-body diagram in the *positive sense*. In this way, if any negative values are obtained, they will indicate that the components act in the negative coordinate direction.

191

Example 5–13

Several examples of objects along with their associated free-body diagrams are shown in Fig. 5–18. In all cases, the x, y, z axes are established and the unknown reaction components are indicated in the positive sense. The weights of the members are neglected.

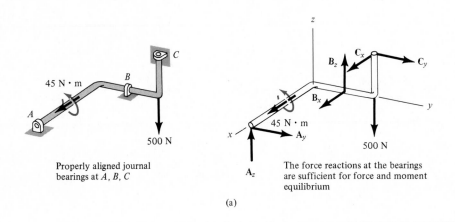

Properly aligned journal bearings at A, B, C

The force reactions at the bearings are sufficient for force and moment equilibrium

(a)

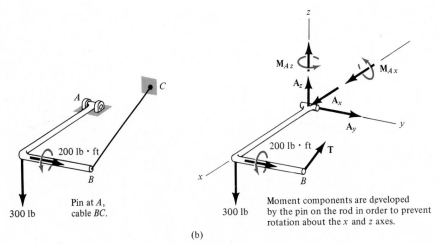

Pin at A, cable BC.

Moment components are developed by the pin on the rod in order to prevent rotation about the x and z axes.

(b)

Fig. 5–18a and b

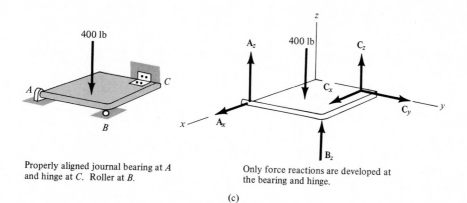

400 lb

A

C

B

Properly aligned journal bearing at *A* and hinge at *C*. Roller at *B*.

z

A_z

400 lb

C_z

C_x

C_y

y

x

A_x

B_z

Only force reactions are developed at the bearing and hinge.

(c)

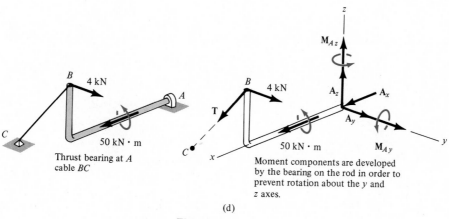

B 4 kN

C

50 kN · m

A

Thrust bearing at *A* cable *BC*

z

M_{Az}

B 4 kN

A_z A_x

T

A_y

C

x

50 kN · m

y

M_{Ay}

Moment components are developed by the bearing on the rod in order to prevent rotation about the *y* and *z* axes.

(d)

Fig. 5—18c and d

5.6 Equations of Equilibrium

As stated in Sec. 5.1, the conditions for equilibrium of a rigid body subjected to a three-dimensional force system require that both the *resultant* force and *resultant* couple moment acting on the body be equal to *zero*.

Vector Equations of Equilibrium.   The two conditions for equilibrium of a rigid body may be expressed mathematically in vector form as

$$\Sigma \mathbf{F} = \mathbf{0}$$
$$\Sigma \mathbf{M}_O = \mathbf{0} \tag{5-5}$$

where $\Sigma \mathbf{F}$ is the vector sum of all the external forces acting on the body and $\Sigma \mathbf{M}_O$ is the sum of the moments of all these forces about any point O located either on or off the body.

Scalar Equations of Equilibrium.   If all the applied external forces and moments are expressed in Cartesian vector form and substituted into Eqs. 5-5, we have

$$\Sigma \mathbf{F} = \Sigma F_x \mathbf{i} + \Sigma F_y \mathbf{j} + \Sigma F_z \mathbf{k} = \mathbf{0}$$
$$\Sigma \mathbf{M}_O = \Sigma M_x \mathbf{i} + \Sigma M_y \mathbf{j} + \Sigma M_z \mathbf{k} = \mathbf{0}$$

Since the $\mathbf{i}, \mathbf{j}$, and $\mathbf{k}$ components are independent from one another, the above equations are satisfied provided

$$\Sigma F_x = 0$$
$$\Sigma F_y = 0$$
$$\Sigma F_z = 0 \tag{5-6a}$$

and

$$\Sigma M_x = 0$$
$$\Sigma M_y = 0$$
$$\Sigma M_z = 0 \tag{5-6b}$$

These *six scalar equilibrium equations* may be used to solve for at most six unknowns shown on the free-body diagram. Equations 5-6a express the fact that the sum of the external force components acting in the x, y, and z directions must be zero, and Eqs. 5-6b require the sum of the moment components about the x, y, and z *axes* to be zero.

PROCEDURE FOR ANALYSIS

The following procedure provides a method for solving three-dimensional equilibrium problems.

Free-Body Diagram. Construct the free-body diagram for the body. Be sure to include *all* the forces and couples that act *on* the body. These interactions are commonly caused by the externally applied loadings, contact forces exerted by adjacent bodies, support reactions, and the weight of the body if it is significant compared to the magnitudes of the other applied forces. Establish the x, y, z axes with origin located at a convenient point. In general, show all the unknown components having a positive sense if the sense cannot be determined. Dimensions of the body, necessary for computing the moments of forces, are also included on the free-body diagram.

Equations of Equilibrium. Apply the equations of equilibrium. In many cases, problems can be solved by *direct application* of the six scalar equations $\Sigma F_x = 0$, $\Sigma F_y = 0$, $\Sigma F_z = 0$, $\Sigma M_x = 0$, $\Sigma M_y = 0$, $\Sigma M_z = 0$, Eqs. 5–6; however, if the force components or moment arms seem difficult to determine, it is recommended that the solution be obtained by using vector equations: $\Sigma \mathbf{F} = \mathbf{0}$, $\Sigma \mathbf{M}_O = \mathbf{0}$, Eqs. 5–5. In any case, it is *not necessary* that the set of axes chosen for force summation *coincide* with the set of axes chosen for moment summation. Instead, it is recommended that one *choose the direction of a moment axis such that it intersects the lines of action of as many unknown forces as possible.* The moments of forces passing through points on this axis or forces which are parallel to the axis will then be zero. Furthermore, *any set of three nonorthogonal axes* may be chosen for either the force or moment summations. By the proper choice of axes, it may be possible to solve directly for an unknown quantity, or at least reduce the need for solving a large number of simultaneous equations for the unknowns.

Constraints for a Rigid Body 5.7

To ensure the equilibrium of a rigid body, it is not only necessary to satisfy the equations of equilibrium, but the body must also be properly held or constrained by its supports. Three situations may occur where the conditions for proper constraint have not been met.

Redundant Constraints. When all the reactive forces on a body can be determined from the equations of equilibrium, the problem is called *statically determinate*. Problems having more unknown forces than equilibrium equations are called *statically indeterminate*. Statical indeterminacy can arise if a body has redundant constraints, i.e., more than are necessary to maintain equilibrium. For example, the two-dimensional problem, Fig. 5–19a, and the three-dimensional problem, Fig. 5–19b, shown together with their free-body diagrams, are both statically indeterminate because of the redundancy in the number of support reactions. In the two-dimensional case, there are five unknowns, that is, M_A, A_x, A_y, B_y, and C_y, for which only three equilibrium equations can be written ($\Sigma F_x = 0$, $\Sigma F_y = 0$, and $\Sigma M_O = 0$, Eqs. 5–2). Two more equations are needed for a complete solution. Consequently, this problem is termed "indeterminate to the second degree" (five unknowns minus three equations). The three-dimensional problem has nine unknowns, for which only six equilibrium equations can be written, Eqs. 5–6. It is "indeterminate to the third degree." The additional equations needed to solve indeterminate problems of the type shown in Fig. 5–19 are generally obtained from the deformation conditions at the points of support. These equations involve the physical properties of the body which are studied in subjects dealing with the mechanics of deformation, such as "strength of materials."

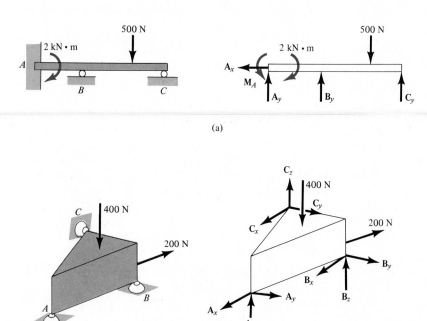

(a)

(b)

Fig. 5–19

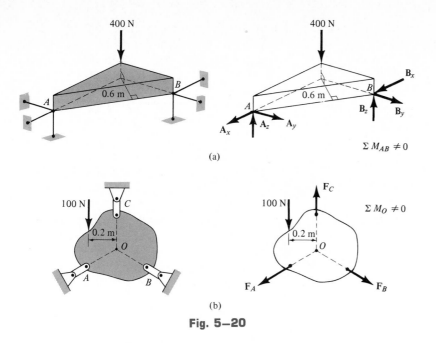

(a)

(b)

Fig. 5–20

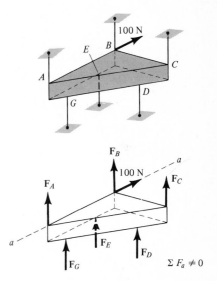

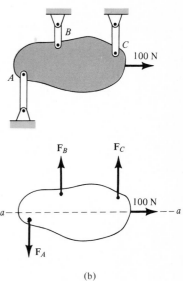

Improper Constraints. In some cases, there may be as many unknown forces as there are equations of equilibrium; however, *instability* of the body can develop because of *improper constraining* action by the supports. In the case of three-dimensional problems, the body is improperly constrained if the support reactions *all intersect a common axis*. For two-dimensional problems, this axis is *perpendicular* to the plane of the forces and therefore appears as a point. Hence, when all the reactive forces are *concurrent* at this point, the body is improperly constrained. Examples of both cases are given in Fig. 5–20. From the free-body diagrams it is seen that the summation of moments about axis AB, Fig. 5–20a, or point O, Fig. 5–20b, will *not* be equal to zero; thus rotation about axis AB or point O will take place.* Furthermore, in both cases it becomes *impossible* to solve *completely* for all the unknowns, since one can write a moment equation that does not involve any of the unknown support reactions. (This reduces by one the number of available equilibrium equations.)

Another way in which improper constraining leads to instability occurs when the *reactive forces* are all *parallel*. Three- and two-dimensional examples of this are shown in Fig. 5–21. In both cases, the summation of forces along the horizontal *aa* axis will not equal zero.

*For the three-dimensional problem, $\Sigma M_{AB} = (400 \text{ N})(0.6 \text{ m}) \neq 0$, and for the two-dimensional problem, $\Sigma M_O = (100 \text{ N})(0.2 \text{ m}) \neq 0$.

(b)

Fig. 5–21

In some cases, a body may have *fewer* reactive forces than equations of equilibrium that must be satisfied. The body then becomes only *partially constrained*. For example, consider the body shown in Fig. 5–22a with its corresponding free-body diagram in Fig. 5–22b. If O is a point not located on the line AB, the equations $\Sigma F_x = 0$ and $\Sigma M_O = 0$ will be satisfied by proper choice of the reactions $\mathbf{F}_A$ and $\mathbf{F}_B$. The equation $\Sigma F_y = 0$, however, will not be satisfied for the loading conditions and therefore equilibrium will not be maintained.

Proper constraining requires that (1) the lines of action of the reactive forces do not intersect points on a common axis, and (2) the reactive forces must not all be parallel to one another. When the number of reactive forces needed to properly constrain the body in question is a *minimum*, the equations of equilibrium can be used to determine *all* the reactive forces.

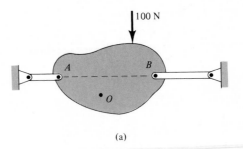

(a)

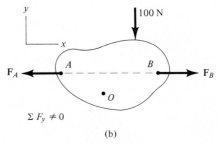

(b)

Fig. 5–22

Example 5-14

The homogeneous plate shown in Fig. 5–23a has a mass of 100 kg and is subjected to a force and couple moment along its edges. If it is supported in the horizontal plane by means of a roller at A, a ball-and-socket joint at B, and a cord at C, determine the components of reaction at the supports.

Solution *(Scalar Analysis)*

Free-Body Diagram. Can you identify each of the forces shown on the free-body diagram of the plate, Fig. 5–23b?

Equations of Equilibrium. Since the three-dimensional geometry is rather simple, a *scalar analysis* provides a *direct solution* to this problem. A force summation yields

$\Sigma F_x = 0;$ $B_x = 0$ *Ans.*

$\Sigma F_y = 0;$ $B_y = 0$ *Ans.*

$\Sigma F_z = 0;$ $A_z + B_z + T_C - 300 \text{ N} - 981 \text{ N} = 0$ (1)

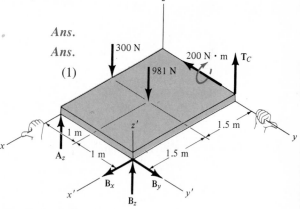

(a)

(b)

Fig. 5–23

Recall that the moment of a force about an axis is equal to the product of the force magnitude and the perpendicular distance (moment arm) from the line of action of the force to the axis. The sense of the moment is determined by the right-hand rule. Hence, summing moments of the forces on the free-body diagram, with positive moments acting along the positive x or y axis, we have

$\Sigma M_x = 0;$ $T_C(2 \text{ m}) - 981 \text{ N}(1 \text{ m}) + B_z(2 \text{ m}) = 0$ (2)

$\Sigma M_y = 0$

$300 \text{ N}(1.5 \text{ m}) + 981 \text{ N}(1.5 \text{ m}) - B_z(3 \text{ m}) - A_z(3 \text{ m}) - 200 \text{ N} \cdot \text{m} = 0$ (3)

The components of force at B can be eliminated if the x', y', z' axes are used. We obtain

$\Sigma M_{x'} = 0;$ $981 \text{ N}(1 \text{ m}) + 300 \text{ N}(2 \text{ m}) - A_z(2 \text{ m}) = 0$ (4)

$\Sigma M_{y'} = 0$

$-300 \text{ N}(1.5 \text{ m}) - 981 \text{ N}(1.5 \text{ m}) - 200 \text{ N} \cdot \text{m} + T_C(3 \text{ m}) = 0$ (5)

Solving Eqs. (1) to (3) or the more convenient Eqs. (1), (4), and (5) yields

$A_z = 790 \text{ N}$ $B_z = -217 \text{ N}$ $T_C = 707 \text{ N}$ *Ans.*

The negative sign indicates that $\mathbf{B}_z$ acts downward.

Example 5–15

The windlass shown in Fig. 5–24a is supported by two smooth journal bearings A and B which are properly aligned on the shaft. Determine the magnitude of the vertical force P that must be applied to the handle to maintain equilibrium of the 100-kg crate. Also calculate the reactions at the bearings.

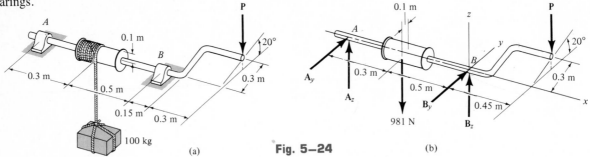

Fig. 5–24

(a)   (b)

Solution (Scalar Analysis)

Free-Body Diagram. Since the bearings at A and B are aligned correctly, *only* force reactions occur at these supports, Fig. 5–24b. Why are there no moment reactions?

Equations of Equilibrium. Summing moments about the x axis yields a direct solution for P. Why? For a scalar moment summation, it is necessary to compute the moment of each force as the product of the force magnitude and the *perpendicular distance* from the x axis to the line of action of the force. Using the right-hand rule and assuming positive moments act in the $+\mathbf{i}$ direction, we have

$$\Sigma M_x = 0; \qquad 981 \text{ N}(0.1 \text{ m}) - P(0.3 \cos 20° \text{ m}) = 0$$

$$P = 348.0 \text{ N} \qquad\qquad Ans.$$

Using this result and summing moments about the y and z axes yields

$$\Sigma M_y = 0; \quad -981 \text{ N}(0.5 \text{ m}) + A_z(0.8 \text{ m}) + (348.0 \text{ N})(0.45 \text{ m}) = 0$$

$$A_z = 417.4 \text{ N} \qquad\qquad Ans.$$

$$\Sigma M_z = 0; \qquad\qquad -A_y(0.8 \text{ m}) = 0 \qquad A_y = 0 \qquad\qquad Ans.$$

The reactions at B are obtained by a force summation, using the results computed above:

$$\Sigma F_y = 0; \qquad\qquad 0 + B_y = 0 \qquad B_y = 0 \qquad\qquad Ans.$$

$$\Sigma F_z = 0; \quad 417.4 - 981 + B_z - 348.0 = 0 \qquad B_z = 911.6 \text{ N} \qquad Ans.$$

As shown on the free-body diagram, the *supports* do not provide resistance against translation in the x direction. Hence, the windlass is only partially constrained.

Example 5–16

Determine the tension in cables BC and BD and the reactions at the ball-and-socket joint A for the mast shown in Fig. 5–25a.

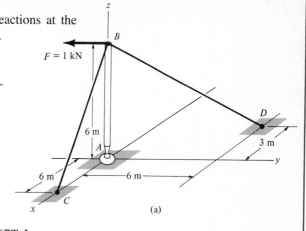

(a)

Solution *(Vector Analysis)*
Free-Body Diagram. There are five unknown force magnitudes shown on the free-body diagram, Fig. 5–25b.

Equations of Equilibrium. Expressing each force in Cartesian vector form, we have

$$\mathbf{F} = \{-1000\mathbf{j}\}\ \text{N}$$
$$\mathbf{F}_A = A_x\mathbf{i} + A_y\mathbf{j} + A_z\mathbf{k}$$
$$\mathbf{T}_C = 0.707T_C\mathbf{i} - 0.707T_C\mathbf{k}$$
$$\mathbf{T}_D = T_D\left(\frac{\mathbf{r}_{BD}}{r_{BD}}\right) = -0.333T_D\mathbf{i} + 0.667T_D\mathbf{j} - 0.667T_D\mathbf{k}$$

Applying the force equation of equilibrium gives

$$\Sigma\mathbf{F} = \mathbf{0}; \qquad \mathbf{F} + \mathbf{F}_A + \mathbf{T}_C + \mathbf{T}_D = \mathbf{0}$$
$$(A_x + 0.707T_C - 0.333T_D)\mathbf{i} + (-1000 + A_y + 0.667T_D)\mathbf{j}$$
$$+ (A_z - 0.707T_C - 0.667T_D)\mathbf{k} = \mathbf{0}$$

$$\Sigma F_x = 0; \qquad A_x + 0.707T_C - 0.333T_D = 0 \tag{1}$$
$$\Sigma F_y = 0; \qquad A_y + 0.667T_D - 1000 = 0 \tag{2}$$
$$\Sigma F_z = 0; \qquad A_z - 0.707T_C - 0.667T_D = 0 \tag{3}$$

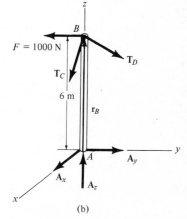

(b)

Fig. 5–25

Summing moments about point A, we have

$$\Sigma\mathbf{M}_A = \mathbf{0}; \qquad \mathbf{r}_B \times (\mathbf{F} + \mathbf{T}_C + \mathbf{T}_D) = \mathbf{0}$$
$$6\mathbf{k} \times (-1000\mathbf{j} + 0.707T_C\mathbf{i} - 0.707T_C\mathbf{k}$$
$$- 0.333T_D\mathbf{i} + 0.667T_D\mathbf{j} - 0.667T_D\mathbf{k}) = \mathbf{0}$$

Evaluating the cross product and combining terms yields

$$(-4T_D + 6000)\mathbf{i} + (4.24T_C - 2T_D)\mathbf{j} = \mathbf{0}$$

$$\Sigma M_x = 0; \qquad -4T_D + 6000 = 0 \tag{4}$$
$$\Sigma M_y = 0; \qquad 4.24T_C - 2T_D = 0 \tag{5}$$

The moment equation about the z axis, $\Sigma M_z = 0$, is automatically satisfied. Why? Solving Eqs. (1) to (5) yields

$$T_C = 707.2\ \text{N} \qquad T_D = 1500.0\ \text{N} \qquad \qquad \textit{Ans.}$$
$$A_x = 0.0\ \text{N} \qquad A_y = 0.0\ \text{N} \qquad A_z = 1500.0\ \text{N} \qquad \textit{Ans.}$$

Since the mast is a two-force member, note that the values of $A_x = A_y = 0.0$ could have been determined *by inspection.*

Example 5–17

Rod AB shown in Fig. 5–26a is used to support the 200-N force. Determine the reactions at the ball-and-socket joint A and at the smooth collar B.

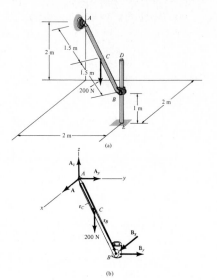

Solution (*Vector Analysis*)
Free-Body Diagram. Here we have considered the rod and attached collar together as shown in Fig. 5–26b.

Equations of Equilibrium. Representing each force on the free-body diagram in Cartesian vector form, we have

$$\mathbf{F}_A = A_x\mathbf{i} + A_y\mathbf{j} + A_z\mathbf{k}$$
$$\mathbf{F}_B = B_x\mathbf{i} + B_y\mathbf{j}$$
$$\mathbf{F} = \{-200\mathbf{k}\}\ \text{N}$$

Applying the force equation of equilibrium,

$$\Sigma\mathbf{F} = 0; \qquad \mathbf{F}_A + \mathbf{F}_B + \mathbf{F} = 0$$
$$(A_x + B_x)\mathbf{i} + (A_y + B_y)\mathbf{j} + (A_z - 200)\mathbf{k} = 0$$

$\Sigma F_x = 0;$ $\qquad\qquad A_x + B_x = 0$ $\qquad$ (1)
$\Sigma F_y = 0;$ $\qquad\qquad A_y + B_y = 0$ $\qquad$ (2)
$\Sigma F_z = 0;$ $\qquad\qquad A_z - 200 = 0$ $\qquad$ (3)

Summing moments about point A yields

$$\Sigma\mathbf{M}_A = 0; \qquad (\mathbf{r}_C \times \mathbf{F}) + (\mathbf{r}_B \times \mathbf{F}_B) = 0$$

Since $\mathbf{r}_C = \frac{1}{2}\mathbf{r}_B$, then

$$(1\mathbf{i} + 1\mathbf{j} - 0.5\mathbf{k}) \times (-200\mathbf{k}) + (2\mathbf{i} + 2\mathbf{j} - 1\mathbf{k}) \times (B_x\mathbf{i} + B_y\mathbf{j}) = 0$$

Expanding and rearranging terms gives

$$(B_y - 200)\mathbf{i} + (-B_x + 200)\mathbf{j} + (2B_y - 2B_x)\mathbf{k} = 0$$

$\Sigma M_x = 0;$ $\qquad\qquad B_y - 200 = 0$ $\qquad$ (4)
$\Sigma M_y = 0;$ $\qquad\qquad -B_x + 200 = 0$ $\qquad$ (5)
$\Sigma M_z = 0;$ $\qquad\qquad 2B_y - 2B_x = 0$ $\qquad$ (6)

Solving Eqs. (1) to (6), we get

$$A_x = A_y = -200\ \text{N} \qquad\qquad Ans.$$
$$A_z = B_x = B_y = 200\ \text{N} \qquad\qquad Ans.$$

The negative sign indicates that $\mathbf{A}_x$ and $\mathbf{A}_y$ have a sense which is opposite to that shown on the free-body diagram, Fig. 5–26b.

Example 5–18

Ch. 5

The bent rod in Fig. 5–27a is supported at *A* by a journal bearing, at *D* by a ball-and-socket joint, and at *B* by means of cable *BC*. Using only *one equilibrium equation,* obtain a direct solution for the tension in cable *BC*. The bearing at *A* is capable of exerting force components only in the *z* and *y* directions, since it is properly aligned on the shaft.

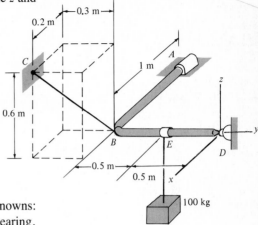

(a)

Solution *(Vector Analysis)*

Free-Body Diagram. As shown in Fig. 5–27b, there are six unknowns: the three force components at the ball-and-socket joint, two at the bearing, and the tension force in the cable.

Equations of Equilibrium. The cable tension $\mathbf{T}_B$ may be obtained *directly* by summing moments about an axis passing through points *D* and *A*. Why? The direction of the axis is defined by the unit vector $\mathbf{u}$, where

$$\mathbf{u} = \frac{\mathbf{r}_{DA}}{r_{DA}} = -\frac{1}{\sqrt{2}}\mathbf{i} - \frac{1}{\sqrt{2}}\mathbf{j}$$

$$= -0.707\mathbf{i} - 0.707\mathbf{j}$$

Hence, the sum of the moments about this axis is zero provided

$$\Sigma M_{DA} = \mathbf{u} \cdot \Sigma(\mathbf{r} \times \mathbf{F}) = 0$$

Here $\mathbf{r}$ represents a position vector drawn from *any point* on the axis *DA* to any point on the line of action of force $\mathbf{F}$ (see Eq. 4–9). With reference to Fig. 5–27b, we can therefore write

$$\mathbf{u} \cdot (\mathbf{r}_B \times \mathbf{T}_B + \mathbf{r}_E \times \mathbf{W}) = 0$$

$$(-0.707\mathbf{i} - 0.707\mathbf{j}) \cdot [(-1\mathbf{j}) \times \left(\tfrac{0.2}{0.7}T_B\mathbf{i} - \tfrac{0.3}{0.7}T_B\mathbf{j} + \tfrac{0.6}{0.7}T_B\mathbf{k}\right)$$

$$+(-0.5\mathbf{j}) \times (-981\mathbf{k})] = 0$$

$$(-0.707\mathbf{i} - 0.707\mathbf{j}) \cdot [(-0.857T_B + 490.5)\mathbf{i} + 0.286T_B\mathbf{k}] = 0$$

$$-0.707(-0.857T_B + 490.5) + 0 + 0 = 0$$

$$T_B = \frac{490.5}{0.857} = 572.3 \text{ N} \qquad\qquad \textit{Ans.}$$

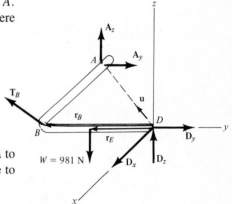

(b)

Fig. 5–27

The advantage of using Cartesian vectors for this solution should be noted. Obviously, it would be very tedious to determine the perpendicular distances from the *DA* axis to the lines of action of $\mathbf{T}_B$ and $\mathbf{W}$ using scalar methods.

Problems

5–61. Draw the specified free-body diagram in each of the following problems and determine the total number of unknown force and couple-moment magnitudes and/or directions. Neglect the weights of the members unless otherwise stated. Although not shown on each photo, assume the geometry (size and angles) is known.

a. Draw a free-body diagram of the telephone pole which has a mass of 200 kg and mass center at G. The weight of the top three wires subjects the pole to vertical loads of 500 N each at the insulators A, B, and C. The wires connected at D and E are subjected to tensions of $T_D = 800$ N and $T_E = 600$ N, respectively. The guy wire FH is used for support. The pole is fixed connected to the ground at I.

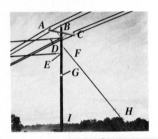

Prob. 5–61a

b. Draw the free-body diagram of the grinding wheel, wirebrush, and axle assembly. The axle is supported by a journal bearing at A and a thrust bearing at C. The belt B wrapped around the pulley is subjected to a tension of 20 N.

Prob. 5–61b

c. Draw a free-body diagram of the gin-pole boom ABC. The boom is pin connected to the truckbed at A and C and supported by chains BD and BE. The load has a weight of 750 lb and is suspended from a cable FGH which wraps over the pulley at G and is attached to a winch at H.

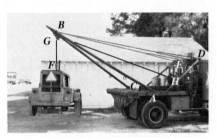

Prob. 5–61c

d. Draw the free-body diagram of the soil auger. The auger has a weight of 135 lb and a center of gravity at G. It is supported by a cable AB and a short link CD which is *pin connected* to the auger at C. Note that the boom EF, link DC, and auger do not lie in the same plane.

Prob. 5–61d

5–62. A vertical force of 50 lb acts on the crankshaft. Determine the horizontal equilibrium force **P** that must be applied to the handle and the x, y, z components of reaction at the journal bearing A and thrust bearing B. The bearings are properly aligned and exert only force reactions on the shaft.

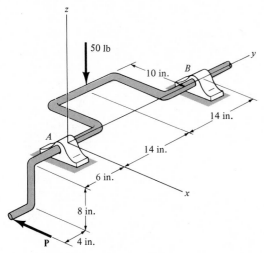

Prob. 5–62

5–63. Determine the force components acting on the ball-and-socket at A, the reaction at the roller B and the tension in the cord CD needed for equilibrium of the quarter circular plate.

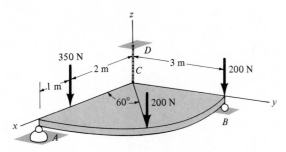

Prob. 5–63

***5–64.** The wrench is used to turn the square bolt at A about the z axis. If the force $F = 6$ lb is applied to the handle as shown, determine the magnitudes of the resultant force and moment that the bolt head exerts on the wrench. The force **F** is in a plane parallel to the x-z plane.

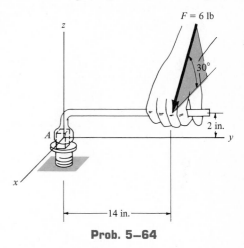

Prob. 5–64

5–65. The air-conditioning unit is hoisted to the roof of a building using the three cables. If the unit has a weight of 900 lb and a center of gravity at G, determine the tension in each of the cables for equilibrium.

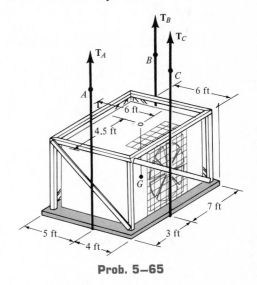

Prob. 5–65

5–66. Due to an unequal distribution of fuel in the wing tanks the centers of gravity for the airplane fuselage A and wings B and C are located as shown. If these components have weights $W_A = 45,000$ lb, $W_B = 8000$ lb, and $W_C = 6000$ lb, determine the normal reactions of the wheels on the ground.

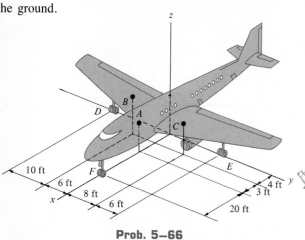

Prob. 5–66

5–67. The cable of the tower crane is subjected to a force of 840 N. Determine the x, y, z components of reaction at the fixed base A.

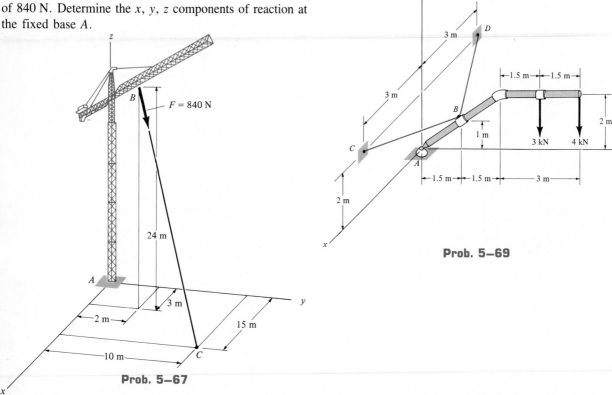

Prob. 5–67

*5–68. The cart supports the uniform crate having a mass of 85 kg. Determine the vertical reactions on the three casters at A, B, and C.

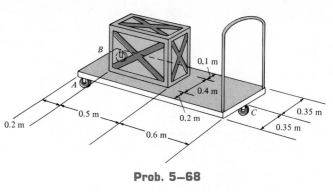

Prob. 5–68

5–69. The pipe assembly supports the vertical loads shown. Determine the components of reaction at the ball-and-socket joint A and the tension in the supporting cables BC and BD.

Prob. 5–69

5–70. The concrete slab is being hoisted with constant velocity using the cable arrangement shown. If the slab has a weight of 32,000 lb and a center of gravity at G, determine the tension in the supporting cables AC, BC, and DE.

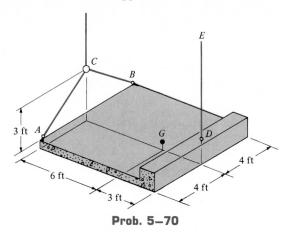

Prob. 5–70

5–71. The shaft assembly is supported by two smooth journal bearings A and B and a short link DC. If a couple moment is applied to it as shown, determine the components of force reaction at the bearings and the force in the link. The link lies in a plane parallel to the y-z plane and the bearings are properly aligned on the shaft.

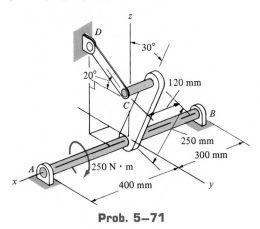

Prob. 5–71

*5–72. Determine the tensions in the cables and the components of reaction acting on the smooth collar at A necessary to hold the 50-lb sign in equilibrium. The center of gravity for the sign is at G.

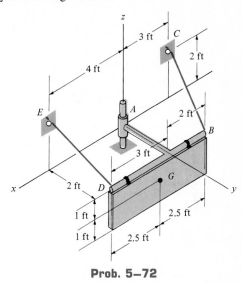

Prob. 5–72

5–73. The member AB is supported by a cable BC and at A by a *square* rod which fits loosely through the square hole at the end joint of the member as shown. Determine the components of reaction at A and the tension in the cable needed to hold the 800-lb cylinder in equilibrium.

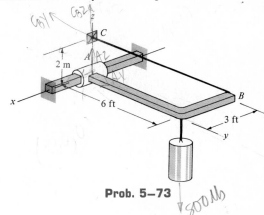

Prob. 5–73

207

5–74. The boom AB is held in equilibrium by a ball-and-socket joint A and a pulley and cord system shown. Determine the reaction components at A and the tension in cable DEC.

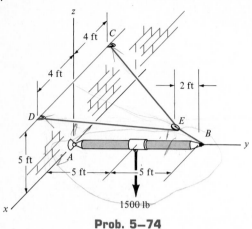

Prob. 5–74

5–75. Both pulleys are fixed to the shaft and as the shaft turns with constant angular velocity, the power of pulley A is transmitted to pulley B. Determine the horizontal tension **T** in the belt at pulley B and the reactive forces at the journal bearing C and thrust bearing D. The bearings are properly aligned on the shaft.

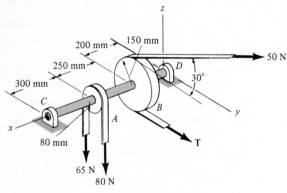

Prob. 5–75

*5–76. The power line is subjected to a tension of 1700 lb. If the insulator AB weighs 50 lb, with center of gravity at G, determine the angle θ it makes with the pole.

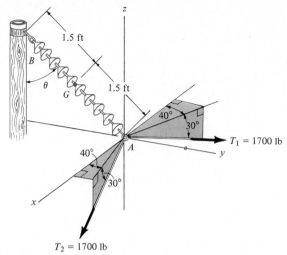

Prob. 5–76

5–77. Determine the components of force acting at the smooth journal bearing A, ball-and-socket joint D, and the tension in the cable BC. The bearing at A only exerts force reactions on the shaft.

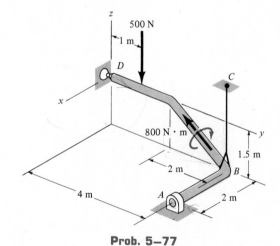

Prob. 5–77

5–78. The boom is supported by a ball-and-socket joint at *A* and a guy wire at *B*. If the loads in the cables are each 5 kN and they lie in a plane which is parallel to the *x-z* plane, determine the components of reaction at *A* for equilibrium.

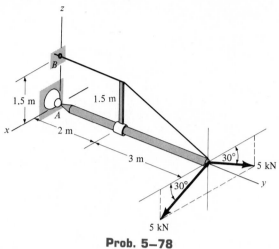

Prob. 5–78

5–79. The pole is subjected to the two forces shown. Determine the components of reaction at *A* assuming it to be a ball-and-socket joint. Also, compute the tension in each of the guy wires, *BC* and *ED*.

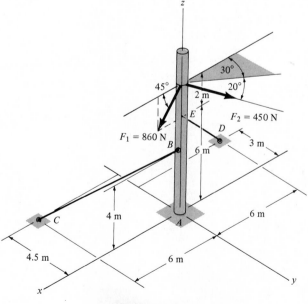

Prob. 5–79

*5–80. The stiff-leg derrick used on ships is supported by a ball-and-socket joint at *D* and two cables *BA* and *BC*. The cables are attached to a smooth collar ring at *B*, which allows rotation of the derrick about the *z* axis. If the derrick supports a crate having a mass of 200 kg, determine the tension in the cables and the *x*, *y*, *z* components of reaction at *D*.

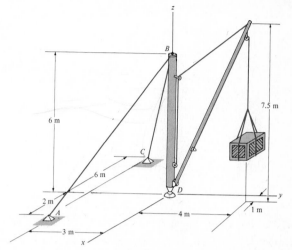

Prob. 5–80

5–81. The boom *AC* is supported at *A* by a ball-and-socket joint and by two cables *BDC* and *CE*. Cable *BDC* is continuous and passes over a frictionless pulley at *D*. Calculate the tension in the cables and the *x*, *y*, *z* components of reaction at *A* if a crate, having a weight of 100 lb, is suspended from the boom.

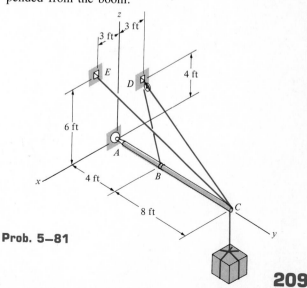

Prob. 5–81

209

Review Problems

5–82. The wing of the jet aircraft is subjected to a thrust force of $T = 8$ kN from its engine and the resultant lift force $L = 45$ kN. If the mass of the wing is 2.1 Mg and the mass center is at G, determine the x, y, z components of reaction where the wing is fixed to the fuselage at A.

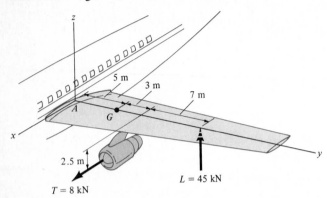

Prob. 5–82

5–83. Determine the reactions at the roller A and pin B for equilibrium of the member.

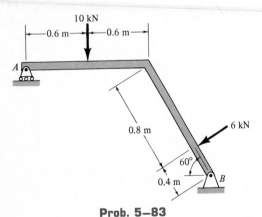

Prob. 5–83

*5–84. Determine the x, y, z components of reaction at the fixed wall A. The 150-N force is parallel to the z axis and the 200-N force is parallel to the y axis.

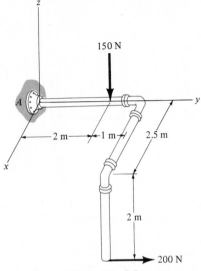

Prob. 5–84

5–85. The support is connected to three springs, which are in the unstretched position when $s = 30$ mm and $P = 0$. The two inclined springs are attached to pivots B,C and D,E at their ends, which allows free rotation of the springs. Determine the load P that must be applied to the support A in order that $s = 0$. Also, compute the force in each spring when $s = 0$.

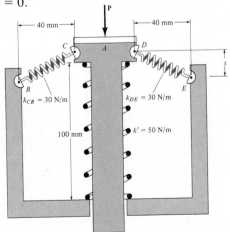

Prob. 5–85

5–86. The windlass is subjected to a load of 150 lb. Determine the horizontal force **P** needed to hold the handle in the position shown, and the components of reaction at the ball-and-socket joint A and the smooth journal bearing B. The bearing at B is in proper alignment and exerts only force reactions on the windlass.

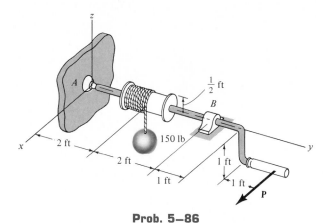

Prob. 5–86

5–87. When holding the 5-lb stone in equilibrium, the humerus H, assumed to be smooth, exerts normal forces $\mathbf{F}_C$ and $\mathbf{F}_A$ on the radius C and ulna A as shown. Determine these forces and the force $\mathbf{F}_B$ that the biceps B exerts on the radius for equilibrium. The stone has a center of mass at G.

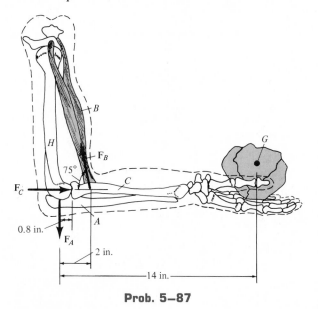

Prob. 5–87

***5–88.** The nonhomogeneous door of a large pressure vessel has a weight of 125 lb and a center of gravity at G. Determine the magnitudes of the resultant force and resultant couple moment developed at the hinge A, needed to support the door in any open position.

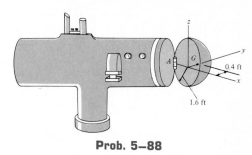

Prob. 5–88

5–89. The forces acting on the plane while it is flying at constant velocity are shown. If the engine thrust is $F_T = 110$ kip and the plane's weight is $W = 170$ kip, determine the atmospheric drag $\mathbf{F}_D$ and the wing lift $\mathbf{F}_L$. Also, determine the distance s to the line of action of the drag force.

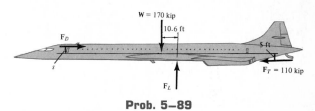

Prob. 5–89

5–90. The semicircular plate supports a load $F = 150$ lb. Determine the tension in cables BD and CD and the components of reaction at the ball-and-socket joint at A for equilibrium.

*5–92. The bent rod is supported at A, B, and C by smooth bearings. Compute the x, y, z components of reaction at the bearings if the rod is subjected to a 200-lb vertical force and a 30-lb $\cdot$ ft couple as shown. The journal bearings are in proper alignment and exert only force reactions on the rod.

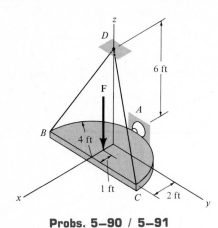

Probs. 5–90 / 5–91

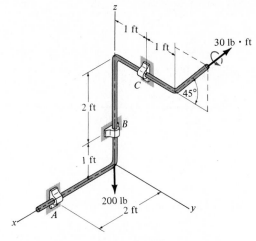

Prob. 5–92

5–91. Determine the magnitude of the force $\mathbf{F}$ acting on the semicircular plate so the force in each of the cables BD and CD is 200 lb. What are the components of reaction at the ball-and-socket joint for this loading?

Structural Analysis

In this chapter we will use the equations of equilibrium to analyze structures composed of pin-connected members. The analysis is based upon the principle that if a structure is in equilibrium, then each of its members is also in equilibrium. By applying the equations of equilibrium to the various parts of a simple truss, frame, or machine, we will be able to determine all the forces acting at the connections. Later, in Chapter 8, we extend this analysis and show how to determine the internal forces in a member.

The topics in this chapter are very important since they provide practice in drawing free-body diagrams, using the principle of action, equal but opposite collinear force reaction, and applying the equations of equilibrium.

Simple Trusses 6.1

A *truss* is a structure composed of slender members joined together at their end points. The members commonly used in construction consist of wooden struts, metal bars, angles, or channels. The joint connections are usually formed by bolting or welding the ends of the members to a common plate, called a *gusset plate*, as shown in Fig. 6–1, or by simply passing a large bolt or pin through each of the members.

Planar Trusses. *Planar* trusses lie in a single plane and are often used to support roofs and bridges. The truss *ABCDE*, shown in Fig. 6–2a, is an example of a typical roof-supporting truss. In this figure, the roof load is transmitted to the truss *at the joints* by means of a series of *purlins*, such as

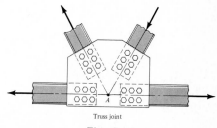

Truss joint

Fig. 6–1

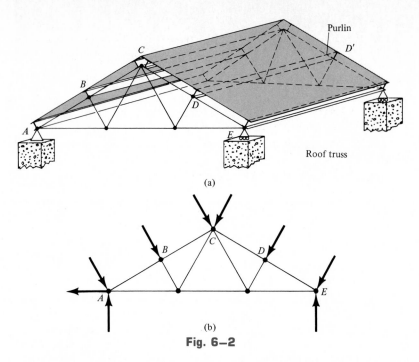

(a)

(b)

Fig. 6–2

DD'. Since the imposed loading acts in the same plane as the truss, Fig. 6–2*b*, the analysis of the forces developed in the truss members is two-dimensional.

In the case of a bridge, such as shown in Fig. 6–3*a*, the load on the deck is first transmitted to *stringers,* then to *floor beams,* and finally to the *joints B, C,* and *D* of the two supporting side trusses. Like the roof truss, the bridge truss loading is also coplanar, Fig. 6–3*b*.

When bridge or roof trusses extend over large distances, a rocker or roller is commonly used for supporting one end, joint *E* in Figs. 6–2*a* and 6–3*a*. This type of support allows freedom for expansion or contraction due to temperature or application of loads.

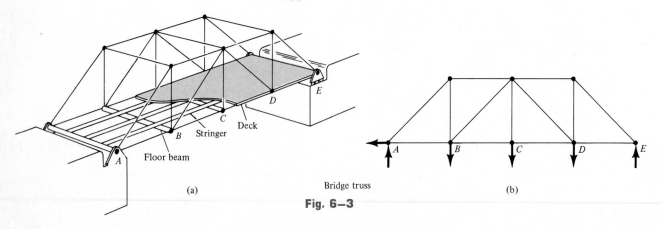

(a)

Bridge truss

(b)

Fig. 6–3

Assumptions for Design. To design both the members and the connections of a truss, it is first necessary to determine the *force* developed in each member when the truss is subjected to a given loading. In this regard, two important assumptions will be made:

1. *All loadings are applied at the joints.* In most situations, such as for bridge and roof trusses, this assumption is true. Frequently in the force analysis the weights of the members are neglected, since the forces supported by the members are usually large in comparison with their weights. If the member's weight is to be included in the analysis, it is generally satisfactory to apply it as a vertical force, half of its magnitude applied at each end of the member.
2. *The members are joined together by smooth pins.* In cases where bolted or welded joint connections are used, this assumption is satisfactory provided the center lines of the joining members are *concurrent*, as in the case of point *A* in Fig. 6–1.

Because of these two assumptions, *each truss member acts as a two-force member,* and therefore the forces acting at the ends of the member must be directed along the axis of the member. If the force tends to *elongate* the member, it is a *tensile force* (**T**), Fig. 6–4a; whereas if the force tends to *shorten* the member, it is a *compressive force* (**C**), Fig. 6–4b. In the actual design of a truss it is important to state whether the nature of the force is tensile or compressive. Most often, compression members must be made *thicker* than tension members, because of the buckling or column effect that occurs when a member is in compression.

Fig. 6–4

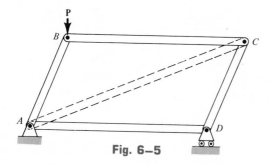

Fig. 6–5

Simple Truss. To prevent collapse, the framework of a truss must be rigid. Obviously, the four-bar frame *ABCD* in Fig. 6–5 will collapse unless a diagonal, such as *AC*, is added for support. The simplest framework which is rigid or stable is a *triangle*. Consequently, a *simple truss* is constructed by *starting* with a basic triangular element, such as *ABC* in Fig. 6–6, and connecting two members (*AD* and *BD*) to form an additional element. Thus it is seen that as each additional element of two members is placed on the truss, the number of joints is increased by one.

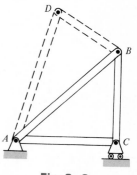

Fig. 6–6

215

6.2 The Method of Joints

If a truss is in equilibrium, then each of its joints must also be in equilibrium. Hence, the method of joints consists of satisfying the equilibrium conditions for the forces exerted *on the pin* at each joint of the truss. Since the truss members are all straight two-force members lying in the same plane, the force system acting at each pin is *coplanar and concurrent*. Consequently, rotational or moment equilibrium is automatically satisfied at the joint (or pin), and it is only necessary to satisfy $\Sigma F_x = 0$ and $\Sigma F_y = 0$ to ensure translational or force equilibrium.

When using the method of joints, it is *first* necessary to draw the joint's free-body diagram before applying the equilibrium equations. In this regard, recall that the *line of action* of each member force acting on the joint is *specified* from the geometry of the truss, since the force in a member passes along the axis of the member. As an example, consider the pin at joint B of the truss in Fig. 6–7a. Three forces act on the pin, namely, the 500-N force and the forces exerted by members BA and BC. The free-body diagram is shown in Fig. 6–7b. As shown, $\mathbf{F}_{BA}$ is "pulling" on the pin, which means that member BA is in *tension*; whereas $\mathbf{F}_{BC}$ is "pushing" on the pin, and consequently member BC is in *compression*. These effects are clearly demonstrated by isolating the joint with small segments of the member connected to the pin, Fig. 6–7c. Notice that pushing or pulling on these small segments indicates the effect of the member being either in compression or tension.

In all cases, the analysis should start at a joint having at least one known force and at most two unknown forces, as in Fig. 6–7b. In this way, applica-

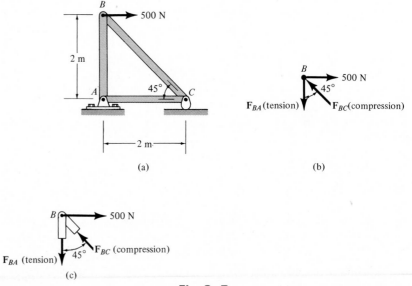

Fig. 6–7

tion of $\Sigma F_x = 0$ and $\Sigma F_y = 0$ yields two algebraic equations which can be solved for the two unknowns. When applying these equations, the correct sense of an unknown member force can be determined using one of two possible methods.

1. *Always assume* the *unknown member forces* acting on the joint's free-body diagram to be in *tension,* i.e., "pulling" on the pin. If this is done, then numerical solution of the equilibrium equations will yield *positive scalars for members in tension and negative scalars for members in compression.* Once an unknown member force is found, use its *correct* magnitude and sense (T or C) on subsequent joint free-body diagrams.

2. The *correct* sense of direction of an unknown member force can, in many cases, be determined "by inspection." For example, F_{BC} in Fig. 6–7*b* must push on the pin (compression) since its horizontal component $F_{BC} \sin 45°$ must balance the 500-N force ($\Sigma F_x = 0$). Likewise, F_{BA} is a tensile force since it balances the vertical component $F_{BC} \cos 45° (\Sigma F_y = 0)$. In more complicated cases, the sense of an unknown member force can be *assumed;* then, after applying the equilibrium equations, the assumed sense can be verified from the numerical results. A *positive* answer indicates that the sense is *correct,* whereas a *negative* answer indicates that the sense shown on the free-body diagram must be *reversed.* This is the method we will use in the example problems which follow.

PROCEDURE FOR ANALYSIS

The following procedure provides a typical means for analyzing a truss using the method of joints.

Draw the free-body diagram of a joint having at least one known force and at most two unknown forces. (If this joint is at one of the supports, it generally will be necessary to know the external reactions at the truss support.) Use one of the two methods described above for establishing the sense of an unknown force. Orient the x and y axes such that the forces on the free-body diagram can be easily resolved into their x and y components and then apply the two force equilibrium equations $\Sigma F_x = 0$ and $\Sigma F_y = 0$. Solve for the two unknown member forces and verify their correct sense.

Continue to analyze each of the other joints, where again it is necessary to choose a joint having at most two unknowns and at least one known force. In this regard, realize that once the force in a member is found from the analysis of a joint at one of its ends, the result can be used to analyze the forces acting on the joint at its other end. Strict adherence to the principle of action, equal but opposite reaction must, of course, be observed. Remember, a member in *compression* "pushes" on the joint and a member in *tension* "pulls" on the joint.

Once the force analysis of the truss has been completed, the size of the members and their connections can be determined using the theory of strength of materials.

Example 6-1

Determine the force in each member of the truss shown in Fig. 6–8a and indicate whether the members are in tension or compression.

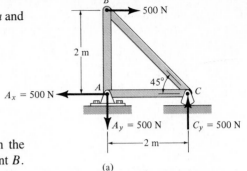

(a)

Solution

If the reactions at A and C had not been computed (although on the figure they are shown), the joint analysis would have to begin at joint B. Why?

Joint B. The free-body diagram of the pin at B is shown in Fig. 6–8b. Three forces act on the pin: the external force of 500 N and the *two* unknown forces developed by members BA and BC. Applying the equations of joint equilibrium, we have

$\xrightarrow{+} \Sigma F_x = 0$; $\quad 500 - F_{BC} \sin 45° = 0 \quad F_{BC} = 707.1 \text{ N} \quad (C) \quad$ *Ans.*

$+ \uparrow \Sigma F_y = 0$; $\quad F_{BC} \cos 45° - F_{BA} = 0 \quad F_{BA} = 500 \text{ N} \quad (T) \quad$ *Ans.*

Since the forces in members BA and BC have been determined, one can proceed to analyze the forces at joints A and C.

Joint A. The free-body diagram of the pin at A is shown in Fig. 6–8c. Here the effect of the *support* on the pin is represented by the reactions $A_x = 500$ N and $A_y = 500$ N. Since BA was found to be in tension, $\mathbf{F}_{BA}$ (=500 N) pulls on the pin (refer to Fig. 6–8e). Applying the equilibrium equations yields

$\xrightarrow{+} \Sigma F_x = 0$; $\quad F_{AC} - 500 = 0 \quad F_{AC} = 500 \text{ N} \quad (T) \quad$ *Ans.*

$+ \uparrow \Sigma F_y = 0$; $\quad 500 - 500 \equiv 0 \quad$ (check)

Joint C. With the forces in all the members known, the results can be checked, in part, from the free-body diagram of the last joint C, Fig. 6–8d.

$\xrightarrow{+} \Sigma F_x = 0$; $\quad -500 + 707.1 \cos 45° \equiv 0 \quad$ (check)

$+ \uparrow \Sigma F_y = 0$; $\quad 500 - 707.1 \sin 45° \equiv 0 \quad$ (check)

The results of the analysis are summarized in Fig. 6–8e. Note that the free-body diagram of each pin shows the effects of all the connected members and external forces applied to the pin, whereas the free-body diagram of each member shows the effects of the end pins on the member.

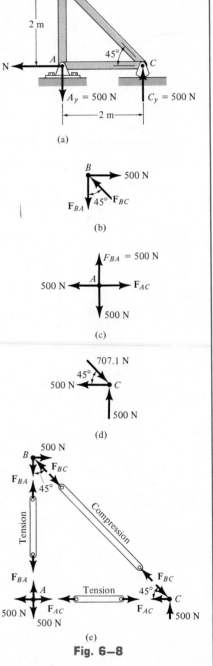

(b)

(c)

(d)

(e)

Fig. 6–8

Example 6–2

Determine the forces acting in all the members of the truss shown in Fig. 6–9a. The reactions at the supports are shown in the figure.

Solution

Since the reactions have been computed, the analysis may begin at either joint C or A. Why?

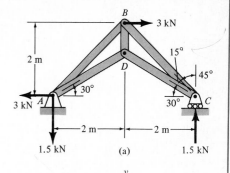

(a)

Joint C (Fig. 6–9b)

$$\overset{+}{\rightarrow}\Sigma F_x = 0; \qquad -F_{CD}\cos 30° + F_{CB}\sin 45° = 0$$

$$+\uparrow\Sigma F_y = 0; \quad 1.5 + F_{CD}\sin 30° - F_{CB}\cos 45° = 0$$

These two equations must be solved *simultaneously* for each of the two unknowns. Note, however, that a *direct solution* for one of the unknown forces may be obtained by applying a force summation along an axis that is *perpendicular* to the direction of the other unknown force. For example, summing forces along the y' axis, which is perpendicular to the direction of $\mathbf{F}_{CD}$, Fig. 6–9c, yields a direct solution for F_{CB}.

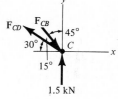

(b)

$$+\nearrow\Sigma F_{y'} = 0;$$

$$1.5\cos 30° - F_{CB}\sin 15° = 0 \qquad F_{CB} = 5.02 \text{ kN} \quad (C) \qquad Ans.$$

In a similar fashion, summing forces along the y'' axis, Fig. 6–9d, yields a direct solution for F_{CD}.

$$+\nearrow\Sigma F_{y''} = 0;$$

$$1.5\cos 45° - F_{CD}\sin 15° = 0 \qquad F_{CD} = 4.10 \text{ kN} \quad (T) \qquad Ans.$$

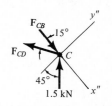

(c)

Joint D (Fig. 6–9e)

$$\overset{+}{\rightarrow}\Sigma F_x = 0; \qquad -F_{DA}\cos 30° + 4.10\cos 30° = 0$$

$$F_{DA} = 4.10 \text{ kN} \quad (T) \qquad\qquad Ans.$$

$$+\uparrow\Sigma F_y = 0; \quad F_{DB} - 2(4.10\sin 30°) = 0 \quad F_{DB} = 4.10 \text{ kN} \quad (T) \quad Ans.$$

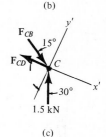

(d)

Joint B (Fig. 6–9f)

$$\overset{+}{\rightarrow}\Sigma F_x = 0; \qquad F_{BA}\cos 45° - 5.02\cos 45° + 3 = 0$$

$$F_{BA} = 0.777 \text{ kN} \quad (C) \qquad\qquad Ans.$$

$$+\uparrow\Sigma F_y = 0; \quad 0.777\sin 45° - 4.10 + 5.02\sin 45° \equiv 0 \quad (check)$$

Check these results in part by analyzing the "last joint" A.

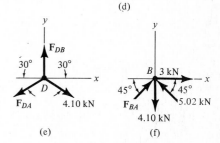

(e)　　　　(f)

Fig. 6–9

Example 6-3

Determine the force in each member of the truss shown in Fig. 6–10a.
Indicate whether the members are in tension or compression.

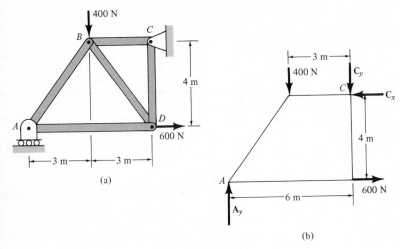

(a)

(b)

Fig. 6—10a and b

Solution

Support Reactions. No joint can be analyzed until the support reactions
are first determined. Why? A free-body diagram of the entire truss is given
in Fig. 6–10b. Applying the equations of equilibrium, we have

$$\xrightarrow{+}\Sigma F_x = 0; \quad 600 - C_x = 0 \quad C_x = 600 \text{ N}$$
$$\zeta+\Sigma M_C = 0; \quad -A_y(6) + 400(3) + 600(4) = 0 \quad A_y = 600 \text{ N}$$
$$+\uparrow\Sigma F_y = 0; \quad 600 - 400 - C_y = 0 \quad C_y = 200 \text{ N}$$

The analysis can now start at either joint A or C. The choice is arbitrary,
since there are one known and two unknown member forces acting on the
pin at each of these joints.

Joint A (Fig. 6–10c). As shown on the free-body diagram, there are three
forces that act on the pin at joint A. The inclination of $\mathbf{F}_{AB}$ is determined
from the geometry of the truss. By inspection, can you see why this force is
assumed to be compressive and $\mathbf{F}_{AD}$ tensile? Applying the equations of
equilibrium, we have

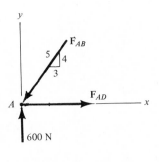

(c)

Fig. 6—10c

$$+\uparrow\Sigma F_y = 0; \quad 600 - \tfrac{4}{5}F_{AB} = 0 \quad F_{AB} = 750 \text{ N} \quad (C) \qquad \textit{Ans.}$$
$$\xrightarrow{+}\Sigma F_x = 0; \quad F_{AD} - \tfrac{3}{5}(750) = 0 \quad F_{AD} = 450 \text{ N} \quad (T) \qquad \textit{Ans.}$$

Joint D (Fig. 6–10*d*). The pin at this joint is chosen next since, by inspection of Fig. 6–10*a*, the force in *AD* is known and two unknown forces in *DB* and *DC* can be determined. Applying the equations of equilibrium, Fig. 6–10*d*, we have

$$\xrightarrow{+}\Sigma F_x = 0; \quad -450 + \tfrac{3}{5}F_{DB} + 600 = 0 \quad F_{DB} = -250 \text{ N}$$

The negative sign indicates $\mathbf{F}_{DB}$ acts in the *opposite sense* to that shown in Fig. 6–10*d*.* Hence,

$$F_{DB} = 250 \text{ N} \quad \text{(T)} \qquad\qquad Ans.$$

To determine $\mathbf{F}_{DC}$, we can either correct the sense of $\mathbf{F}_{DB}$ and then apply $\Sigma F_y = 0$, or apply this equation and retain the negative sign for F_{DB}, i.e.,

$$+\uparrow\Sigma F_y = 0; \quad -F_{DC} - \tfrac{4}{5}(-250) = 0 \quad F_{DC} = 200 \text{ N} \quad \text{(C)} \qquad Ans.$$

Joint C (Fig. 6–10*e*)

$$\xrightarrow{+}\Sigma F_x = 0; \qquad F_{CB} - 600 = 0 \qquad F_{CB} = 600 \text{ N} \quad \text{(C)} \qquad\qquad Ans.$$
$$+\uparrow\Sigma F_y = 0; \qquad\qquad 200 - 200 \equiv 0 \quad \text{(check)}$$

The results may be checked in part by analyzing the pin at the "last joint" *B*. The analysis is summarized in Fig. 6–10*f*, which shows the correct free-body diagram for each pin and member.

*The proper sense could have been determined by inspection, prior to applying $\Sigma F_x = 0$.

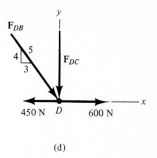

(d)

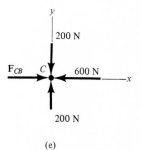

(e)

Fig. 6–10d and e

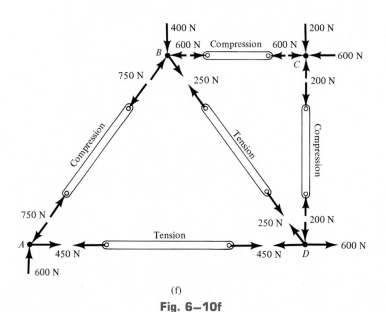

(f)

Fig. 6–10f

6.3 Zero-Force Members

Truss analysis using the method of joints is greatly simplified if one is able to first determine those members which support *no loading*. These *zero-force members* are used to increase the stability of the truss during construction and to provide support if the applied loading is changed.

The zero-force members of a truss can generally be determined *by inspection* of each of its joints. For example, consider the truss shown in Fig. 6–11a. If a free-body diagram of the pin at joint A is drawn, Fig. 6–11b, it is seen that members AB and AF are zero-force members. Note that this is also the case for members DC and DE, Fig. 6–11c. As a general rule then, *if only two members form a truss joint and no external load or support reaction is applied to the joint, then the members must be zero-force members*. The load on the truss in Fig. 6–11a is therefore supported by only five members as shown in Fig. 6–11d. Now consider the truss shown in Fig. 6–12a. The free-body diagram of the pin at joint D is shown in Fig. 6–12b. By orienting

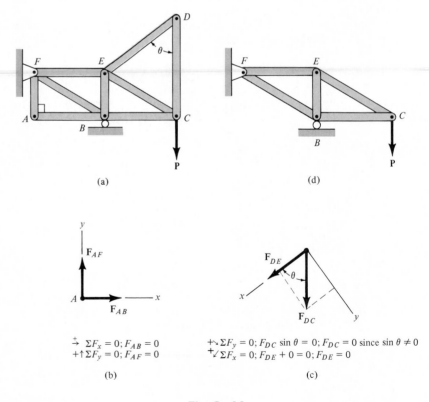

(a)   (d)

$$\xrightarrow{+} \ \Sigma F_x = 0; F_{AB} = 0$$
$$+\uparrow \Sigma F_y = 0; F_{AF} = 0$$

(b)

$$+\nwarrow \Sigma F_y = 0; F_{DC} \sin \theta = 0; F_{DC} = 0 \text{ since } \sin \theta \neq 0$$
$$+\swarrow \Sigma F_x = 0; F_{DE} + 0 = 0; F_{DE} = 0$$

(c)

Fig. 6–11

the y axis along members DC and DE and the x axis along member DA, it is seen that DA is a zero-force member. Note that this is also the case for member CA, Fig. 6–12c. In general, then, *if three members form a truss joint for which two of the members are collinear, the third member is a zero-force member provided no external force or support reaction is applied to the joint.* The truss shown in Fig. 6–12d is therefore suitable for supporting the load **P**.

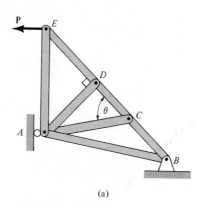

(a)

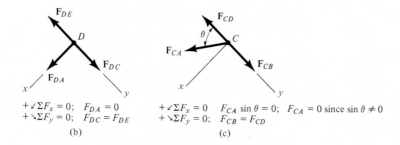

$$+\swarrow\Sigma F_x = 0; \quad F_{DA} = 0$$
$$+\searrow\Sigma F_y = 0; \quad F_{DC} = F_{DE}$$

(b)

$$+\swarrow\Sigma F_x = 0 \quad F_{CA}\sin\theta = 0; \quad F_{CA} = 0 \text{ since } \sin\theta \neq 0$$
$$+\searrow\Sigma F_y = 0; \quad F_{CB} = F_{CD}$$

(c)

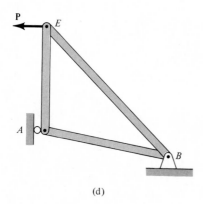

(d)

Fig. 6–12

Example 6–4

Using the method of joints, determine all the members of the *Fink truss* shown in Fig. 6–13a which are subjected to zero force.

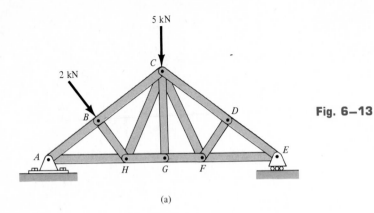

(a)

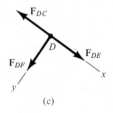

(b)

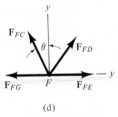

(c)

Fig. 6–13

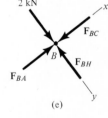

(d)

Solution

Looking for joint situations which are similar to those outlined in Figs. 6–11 and 6–12, we have

Joint G (Fig.6–13b)

$+\uparrow \Sigma F_y = 0;$ $\qquad\qquad F_{GC} = 0$ $\qquad\qquad$ *Ans.*

Joint D (Fig. 6–13c)

$+\nearrow \Sigma F_y = 0;$ $\qquad\qquad F_{DF} = 0$ $\qquad\qquad$ *Ans.*

Joint F (Fig. 6–13d)

$+\uparrow \Sigma F_y = 0;$ $\quad F_{FC} \cos \theta = 0,$ since $\theta \neq 0;$ $\quad F_{FC} = 0$ $\quad$ *Ans.*

Note that if joint B is analyzed, Fig. 6–13e,

$+\searrow \Sigma F_y = 0;$ $\quad 2 - F_{BH} = 0$ $\quad F_{BH} = 2$ kN $\quad$ (C)

Consequently, the numerical value of F_{HC} must satisfy $\Sigma F_y = 0$, Fig. 6–13f, and therefore HC is *not* a zero-force member.

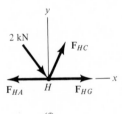

(e)

(f)

Problems

6–1. Determine the force in each member of the truss and state if the members are in tension or compression.

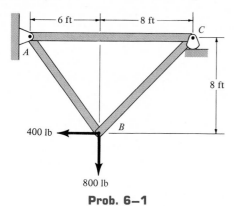

Prob. 6–1

6–2. The truss, used to support a balcony, is subjected to the loading shown. Approximate each joint as a pin and determine the force in each member. State whether the members are in tension or compression.

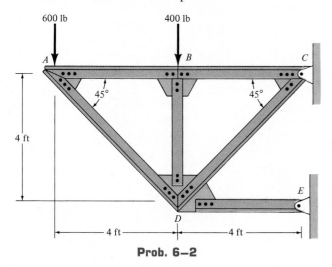

Prob. 6–2

6–3. Determine the force in each member of the truss and state if the members are in tension or compression.

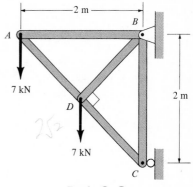

Prob. 6–3

*6–4. Determine the force in each member of the truss and state if the members are in tension or compression.

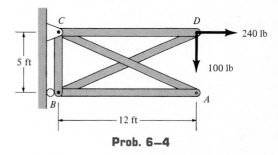

Prob. 6–4

6–5. Determine the force in each member of the truss in terms of the load P, and indicate whether the members are in tension or compression.

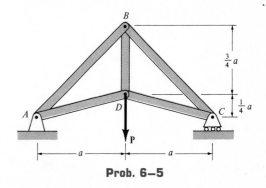

Prob. 6–5

6–6. Determine the force in each member of the platform truss and state if the members are in tension or compression. Approximate each joint as a pin.

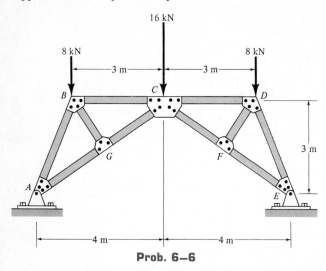

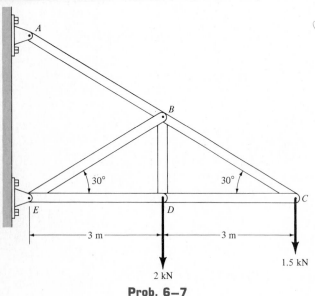

Prob. 6–6

6–7. Determine the force in each member of the truss and indicate whether the members are in tension or compression.

Prob. 6–7

*6–8. Determine the force in each member of the truss and state if the members are in tension or compression. Approximate each joint as a pin. Set $P = 4$ kN.

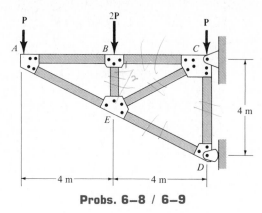

Probs. 6–8 / 6–9

6–9. Assume that each member of the truss is made of steel having a mass per length of 4 kg/m. Set $P = 0$, determine the force in each member, and indicate if the members are in tension or compression. Neglect the weight of the gusset plates and approximate each joint as a pin. Solve the problem by assuming the weight of each member can be represented as a vertical force, half of which is applied at the end of each member.

6–10. The *Warren bridge truss* is subjected to the loadings shown. Determine the force in each member and indicate if the members are in tension or compression. Approximate each joint as a pin.

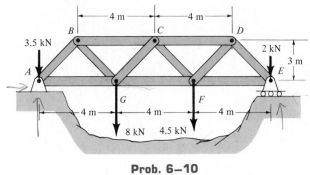

Prob. 6–10

6–11. Each member of the *Warren truss* is uniform and has a weight of 20 lb. Determine the force in each member due to its weight and indicate whether the members are in tension or compression. Neglect the weight of the gusset plates and approximate each joint as a pin. Solve the problem by assuming the weight of each member can be represented as a vertical force, half of which is applied at each end of the member.

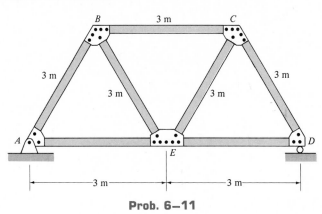

Prob. 6–11

6–13. The trussed building bent is subjected to a loading of 600 lb. Approximate each joint as a pin and determine the force in each member. State whether the members are in tension or compression. *Hint:* Since the load and truss are symmetrical, only half the truss has to be analyzed.

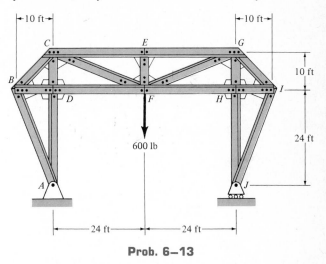

Prob. 6–13

*6–12. The *bridge truss* is subjected to the loadings shown. Determine the force in each member and indicate if the members are in tension or compression. Approximate each joint as a pin.

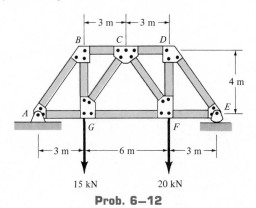

Prob. 6–12

6–14. For the given loading, determine the force in members *BC*, *CJ*, and *JI* of the *Howe roof truss*. Indicate whether the members are in tension or compression. Also indicate all the zero-force members.

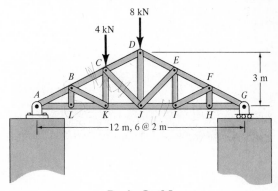

Prob. 6–14

227

6–15. Determine the force in each member of the truss expressed in terms of the load **P** and geometry given.

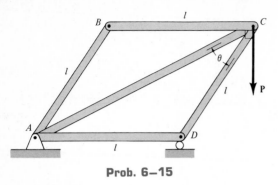

Prob. 6–15

*6–16. Determine the force in members *GR* and *OM* of the truss and indicate whether the members are in tension or compression. Also, indicate all zero-force members.

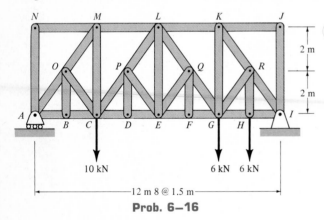

Prob. 6–16

6–17. The roof truss is subjected to the loadings shown. Determine the force in each member and indicate if the members are in tension or compression. Approximate each joint as a pin.

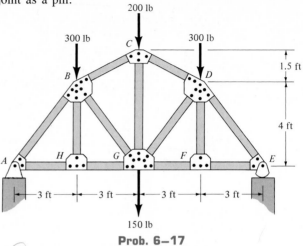

Prob. 6–17

6–18. The *Fink truss* supports the wind loads shown. Determine the force in each member and indicate if the members are in tension or compression. Approximate each joint as a pin.

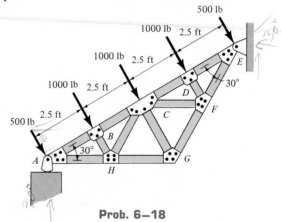

Prob. 6–18

6–19. Determine the force in members *CD* and *CM* of the *Baltimore bridge truss* and indicate whether the members are in tension or compression. Also, indicate all zero-force members.

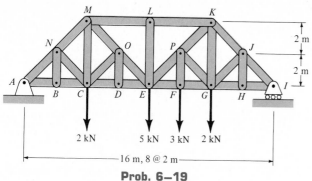

2 kN   5 kN   3 kN   2 kN

16 m, 8 @ 2 m

Prob. 6–19

6–21. Determine the force in each member of the truss. State whether the members are in tension or compression.

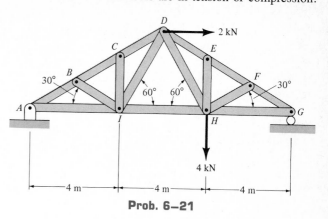

4 kN

4 m   4 m   4 m

Prob. 6–21

***6–20.** Determine the force in each member of the truss and indicate whether the members are in tension or compression. *Hint:* Start the analysis at joint *B* by summing the forces along member *BC* to determine the force in *BC*. Then analyze joint *C*, etc.

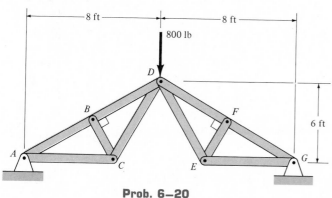

8 ft   8 ft

800 lb

6 ft

Prob. 6–20

The Method of Sections 6.4

The *method of sections* is used to determine the loadings acting within a body. It is based on the principle that if a body is in equilibrium, then any part of the body is also in equilibrium. To apply this method, one passes an *imaginary section* through the body, thus cutting it into two parts. When a free-body diagram of one of the parts is drawn, the loads acting at the section must be *included* on the free-body diagram. One then applies the equations of equilibrium to the part in order to determine the loading at the section. For example,

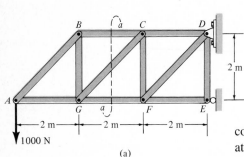

(a)

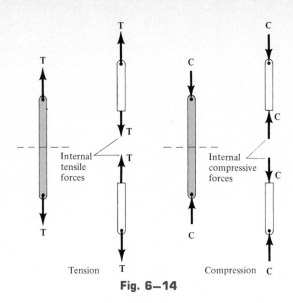

Fig. 6—14

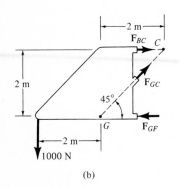

(b)

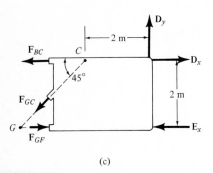

(c)

Fig. 6—15

consider the truss members shown on the left in Fig. 6–14. The internal loads at the section indicated by the dashed line can be obtained using one of the free-body diagrams shown on the right. Clearly, it can be seen that equilibrium requires that the member in tension be subjected to a "pull" **T** at the section, whereas the member in compression is subjected to a "push" **C.**

The method of sections can also be used to "cut" or section several members of an entire truss. If either of the two parts of the truss is isolated as a free-body diagram, we can then apply the equations of equilibrium to that part to determine the member forces at the "cut section." Since only *three* independent equilibrium equations ($\Sigma F_x = 0$, $\Sigma F_y = 0$, $\Sigma M_O = 0$) can be applied to the isolated part of the truss, one should try to select a section that, in general, passes through not more than *three* members in which the forces are unknown. For example, consider the truss in Fig. 6–15a. If the force in member *GC* is to be determined, section *aa* would be appropriate. The free-body diagrams of the two parts are shown in Fig. 6–15b and c. In particular, note that the line of action of each cut member force is specified from the *geometry* of the truss, since the force in a member passes along its axis. Also, the member forces acting on one part of the truss are equal but opposite to those acting on the other part—Newton's third law. As noted above, members assumed to be in *tension* (*BC* and *GC*) are subjected to a "pull," whereas the member in *compression* (*GF*) is subjected to a "push."

The three unknown member forces $\mathbf{F}_{BC}$, $\mathbf{F}_{GC}$, and $\mathbf{F}_{GF}$ can be obtained by applying the three equilibrium equations to the free-body diagram in Fig. 6–15b. If, however, the free-body diagram in Fig. 6–15c is considered, the three support reactions $\mathbf{D}_x$, $\mathbf{D}_y$, and $\mathbf{E}_x$ will have to be determined *first*. Why? (This, of course, is done in the usual manner by considering a free-body diagram of the *entire truss.*) When applying the equilibrium equations, one should consider ways of writing the equations so as to yield a *direct solution* for each of the unknowns, rather than having to solve simultaneous equations. For example, summing moments about *C* in Fig. 6–15b would yield a direct solution for $\mathbf{F}_{GF}$ since $\mathbf{F}_{BC}$ and $\mathbf{F}_{GC}$ create zero moment about C. Likewise, $\mathbf{F}_{BC}$ can be directly obtained by summing moments about G. Finally, $\mathbf{F}_{GC}$ can be found directly from a force summation in the vertical direction since $\mathbf{F}_{GF}$ and $\mathbf{F}_{BC}$ have no vertical components. This ability to *directly determine* the force in a particular truss member is one of the main advantages of using the method of sections.*

As in the method of joints, there are two ways in which one can determine the correct sense of an unknown member force.

1. *Always assume* that the unknown member forces at the cut section are in *tension*, i.e., "pulling" on the member. By doing this, the numerical solution of the equilibrium equations will yield *positive scalars for members in tension and negative scalars for members in compression.*

2. The correct sense of an unknown member force can in many cases be determined "by inspection." For example, $\mathbf{F}_{BC}$ is a tensile force as represented in Fig. 6–15b, since moment equilibrium about G requires that $\mathbf{F}_{BC}$ create a moment opposite to that of the 1000-N force. Also, $\mathbf{F}_{GC}$ is tensile since its vertical component must balance the 1000-N force acting downward. In more complicated cases, the sense of an unknown member force may be *assumed*. If the solution yields a *negative* scalar, it indicates that the force's sense is *opposite* to that shown on the free-body diagram. This is the method we will use in the example problems which follow.

PROCEDURE FOR ANALYSIS

The following procedure provides a means for applying the method of sections to determine the forces in the members of a truss.

Free-Body Diagram. Make a decision as to how to "cut" or section the truss through the members where forces are to be determined. Before isolating the appropriate section, it may first be necessary to determine the truss's *external* reactions, so that the three equilibrium equations are used *only* to solve for member forces at the cut section. Draw the free-body diagram of that part of the sectioned truss which has the least number of forces acting on it. Use one of the two methods described above for establishing the sense of an unknown member force.

Equations of Equilibrium. Try to apply the three equations of equilibrium such that simultaneous solution of equations is avoided. In this regard, moments should be summed about a point that lies at the intersection of the lines of action of two unknown forces, so that the third unknown force is determined directly from the moment equation. If two of the unknown forces are *parallel*, forces may be summed *perpendicular* to the direction of these unknowns to determine *directly* the third unknown force.

The following examples numerically illustrate these concepts.

---

*By comparison, if the method of joints were used to determine, say, the force in member GC, it would be necessary to analyze joints A, B, and G in sequence.

Example 6–5

Determine the force in members GE, GC, and BC of the bridge truss shown in Fig. 6–16a. Indicate whether the members are in tension or compression.

Solution

Section aa in Fig. 6–16a has been chosen since it cuts through the *three* members whose forces are to be determined. In order to use the method of sections, however, it is *first* necessary to determine the external reactions at A or D. Why? A free-body diagram of the entire truss is shown in Fig. 6–16b. Applying the equations of equilibrium, we have

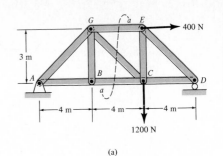

(a)

$$\overset{+}{\rightarrow}\Sigma F_x = 0; \qquad 400\ \text{N} - A_x = 0 \qquad A_x = 400\ \text{N}$$

$$\zeta + \Sigma M_A = 0; \qquad -1200\ \text{N}(8\ \text{m}) - 400\ \text{N}(3\ \text{m}) + D_y(12\ \text{m}) = 0$$

$$D_y = 900\ \text{N}$$

$$+\uparrow\Sigma F_y = 0; \qquad A_y - 1200\ \text{N} + 900\ \text{N} = 0 \qquad A_y = 300\ \text{N}$$

Free-Body Diagrams. The free-body diagrams of the sectioned truss are shown in Fig. 6–16c and d. For the analysis the free-body diagram in Fig. 6–16c will be used since it involves the least number of forces.

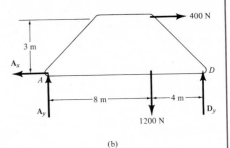

(b)

Equations of Equilibrium. Summing moments about point G eliminates $\mathbf{F}_{GE}$ and $\mathbf{F}_{GC}$ and yields a direct solution for F_{BC}.

$$\zeta + \Sigma M_G = 0; \qquad -300\ \text{N}(4\ \text{m}) - 400\ \text{N}(3\ \text{m}) + F_{BC}(3\ \text{m}) = 0$$

$$F_{BC} = 800\ \text{N} \quad (\text{T}) \qquad\qquad\qquad Ans.$$

In the same manner, by summing moments about point C we obtain a direct solution for F_{GE}.

$$\zeta + \Sigma M_C = 0; \qquad -300\ \text{N}(8\ \text{m}) + F_{GE}(3\ \text{m}) = 0$$

$$F_{GE} = 800\ \text{N} \quad (\text{C}) \qquad\qquad\qquad Ans.$$

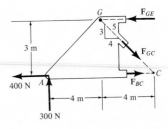

(c)

Since $\mathbf{F}_{BC}$ and $\mathbf{F}_{GE}$ have no vertical components, summing forces in the y direction directly yields F_{GC}, i.e.,

$$+\uparrow\Sigma F_y = 0; \quad 300\ \text{N} - \tfrac{3}{5}F_{GC} = 0 \qquad F_{GC} = 500\ \text{N} \quad (\text{T}) \qquad Ans.$$

Obtain these results by applying the equations of equilibrium to the free-body diagram shown in Fig. 6–16d.

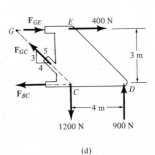

(d)

Fig. 6–16

Example 6–6

Determine the force in member CF of the truss shown in Fig. 6–17a. Indicate whether the member is in tension or compression. The reactions at the supports are shown in the figure.

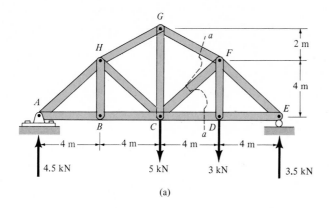

(a)

Fig. 6–17

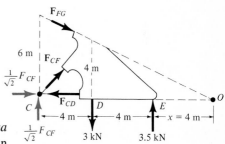

(b)

Solution

Free-Body Diagram. The truss is sectioned along the dashed line aa shown in Fig. 6–17a. This section will "expose" the internal force in member CF as "external" on a free-body diagram of either the right or left portion of the truss. The free-body diagram of the right portion, which is the easiest to analyze, is shown in Fig. 6–17b. There are three unknowns, F_{FG}, F_{CF}, and F_{CD}.

Equations of Equilibrium. The most direct method for solving this problem requires application of the moment equation about a point that eliminates two of the unknown forces. Hence, to obtain $\mathbf{F}_{CF}$, we will eliminate $\mathbf{F}_{FG}$ and $\mathbf{F}_{CD}$ by summing moments about point O, Fig. 6–17b. Note that the location of point O measured from E is determined from proportional triangles, i.e., $4/(4 + x) = 6/(8 + x)$, $x = 4$ m. Or, in another manner, the slope of member GF has a drop of 2 m to a horizontal distance of $CD = 4$ m. Since FD is 4 m, Fig. 6–17a, then from D to O, Fig. 6–17b, the distance must be 8 m.

An easy way to determine the moment of $\mathbf{F}_{CF}$ about point O is to resolve $\mathbf{F}_{CF}$ into its two rectangular components and then use the principle of transmissibility to move $\mathbf{F}_{CF}$ to point C. We have

$$\zeta + \Sigma M_O = 0; \quad -\frac{1}{\sqrt{2}} F_{CF}(12 \text{ m}) + (3 \text{ kN})(8 \text{ m}) - (3.5 \text{ kN})(4 \text{ m}) = 0$$

$$F_{CF} = 1.18 \text{ kN} \quad (C) \qquad\qquad Ans.$$

Example 6–7

Determine the force in member EB of the truss shown in Fig. 6–18a. Indicate whether the member is in tension or compression. The reactions of the supports are shown in the figure.

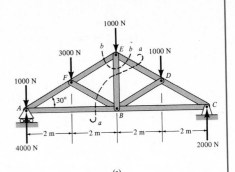

(a)

Solution

Free-Body Diagrams. By the method of sections, any imaginary vertical section that cuts through EB, Fig. 6–18a, will also have to cut through three other members for which the forces are unknown. For example, section aa cuts through ED, EB, FB, and AB. If a free-body diagram of the left side of this section is considered, Fig. 6–18b, it is possible to determine $\mathbf{F}_{ED}$ by summing moments about B to eliminate the other three unknowns; however, $\mathbf{F}_{EB}$ cannot be determined from the remaining two equilibrium equations. One possible way of determining $\mathbf{F}_{EB}$ is to first determine $\mathbf{F}_{ED}$ from section aa, then use this result on section bb, Fig. 6–18a, which is shown in Fig. 6–18c. Here the force system is concurrent and our sectioned free-body diagram is the same as the free-body diagram for the pin at E (method of joints).

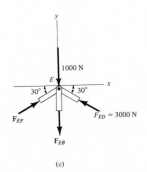

(b)

Equations of Equilibrium. In order to determine the moment of $\mathbf{F}_{ED}$ about point B, Fig. 6–18b, we will resolve it into its rectangular components and, by the principle of transmissibility, extend this force to point C as shown. The moments of 1000 N, F_{AB}, F_{FB}, F_{EB}, and $F_{ED} \cos 30°$ are all zero about B. Therefore,

$$\zeta + \Sigma M_B = 0; \quad 1000(4) + 3000(2) - 4000(4) + F_{ED} \sin 30°(4)$$

$$F_{ED} = 3000 \text{ N} \quad (C)$$

Considering now the free-body diagram of section bb, Fig. 6–18c, we have

$$\xrightarrow{+} \Sigma F_x = 0; \qquad F_{EF} \cos 30° - 3000 \cos 30° = 0$$

$$F_{EF} = 3000 \text{ N} \quad (C)$$

$$+ \uparrow \Sigma F_y = 0; \qquad 2(3000 \sin 30°) - 1000 - F_{EB} = 0$$

$$F_{EB} = 2000 \text{ N} \quad (T) \qquad \qquad \textit{Ans.}$$

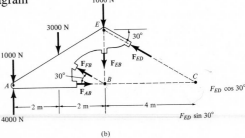

(c)

Fig. 6–18

Problems

6–22. Determine the force in members *BC*, *HC*, and *HG* of the bridge truss, and indicate whether the members are in tension or compression.

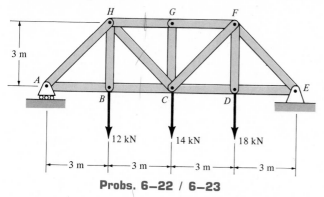

Probs. 6–22 / 6–23

6–23. Determine the force in members *GF*, *CF*, and *CD* of the bridge truss, and indicate whether the members are in tension or compression.

*6–24. The *Howe bridge truss* is subjected to the loading shown. Determine the force in members *HD*, *CD*, and *GD*, and indicate whether the members are in tension or compression.

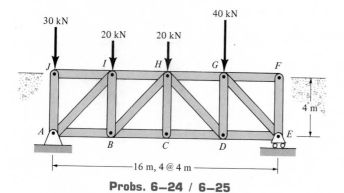

Probs. 6–24 / 6–25

6–25. The *Howe bridge truss* is subjected to the loading shown. Determine the force in members *HI*, *HB*, and *BC*, and indicate whether the members are in tension or compression.

6–26. Determine the force in members *CD*, *CJ*, *KJ*, and *DJ* of the truss which serves to support the deck of a bridge. State if the members are in tension or compression.

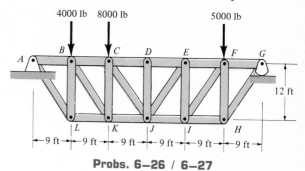

Probs. 6–26 / 6–27

6–27. Determine the force in members *EI* and *JI* of the truss which serves to support the deck of a bridge. State if the members are in tension or compression.

*6–28. Determine the force in members *BC*, *FC*, and *FE*. State if the members are in tension or compression.

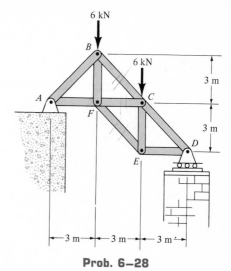

Prob. 6–28

235

6–29. Determine the force in members *CE*, *FE*, and *CD*. Indicate if the members are in tension or compression. *Hint:* The force acting at the pin *G* is directed along the bar *GD*. Why?

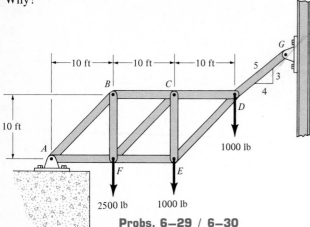

Probs. 6–29 / 6–30

6–30. Determine the force in members *BC*, *FC*, and *FE*. Indicate if the members are in tension or compression. *Hint:* The force acting at the pin *G* is directed along the bar *GD*. Why?

6–31. The skewed truss carries the load shown. Determine the force in members *CB*, *BE*, and *EF* and indicate if the members are in tension or compression. Assume that all joints are pinned.

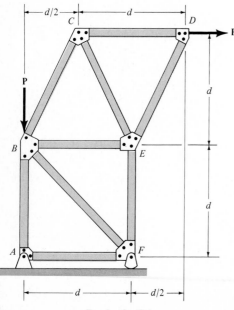

Prob. 6–31

236

*6–32. Determine the force developed in members *GB* and *GF* of the bridge truss, and indicate whether the members are in tension or compression.

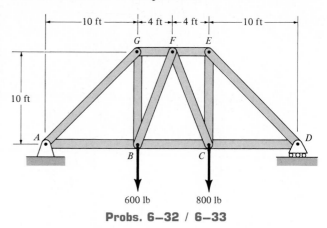

Probs. 6–32 / 6–33

6–33. Determine the force developed in members *FC* and *BC* of the bridge truss, and indicate whether the members are in tension or compression.

6–34. Compute the force in members *BG*, *HG*, and *BC* of the truss and indicate whether the members are in tension or compression.

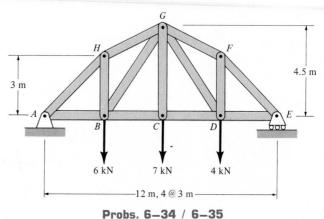

Probs. 6–34 / 6–35

6–35. Compute the force in members *CD*, *GD*, and *GC* of the truss and indicate whether the members are in tension or compression.

*6–36. Determine the force in members *CD*, *CF*, and *FG* of the symmetrical truss. Indicate if the members are in tension or compression.

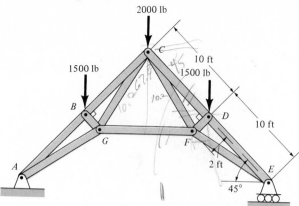

Prob. 6–36

6–37. Determine the force in members *KJ*, *NJ*, *ND*, and *CD* of the *K* truss. Indicate if the members are in tension or compression. *Hint:* Use sections *aa* and *bb*.

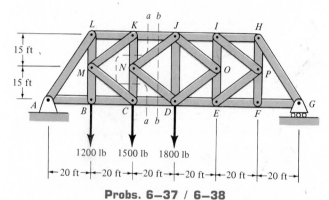

Probs. 6–37 / 6–38

6–38. Determine the force in members *JI* and *DE* of the *K* truss. Indicate if the members are in tension or compression.

6–39. The *Howe roof truss* supports the vertical loading shown. Determine the force in members *KJ*, *CD*, and *KD*, and indicate whether the members are in tension or compression.

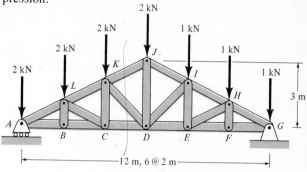

Probs. 6–39 / 6–40

*6–40. The *Howe roof truss* supports the vertical loading shown. Determine the force in members *JI* and *JD*, and indicate whether the members are in tension or compression.

6–41. Determine the force in members *DE*, *DL*, and *ML* of the roof truss. Indicate if the members are in tension or compression.

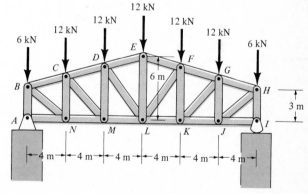

Probs. 6–41 / 6–42

6–42. Determine the force in members *EF* and *EL* of the roof truss. Indicate if the members are in tension or compression.

237

6–43. The tower truss is subjected to the loads shown. Determine the force in BC, BF, and FG. Use a single equilibrium equation for the calculation of each force. Indicate if the members are in tension or compression. The left side, $ABCD$, stands vertical.

*6–44. The tower truss is subjected to the loads shown. Determine the force in members BG and CF. Indicate if the members are in tension or compression. The left side $ABCD$ stands vertical.

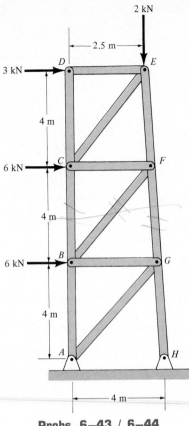

Probs. 6–43 / 6–44

Space Trusses 6.5 *

A *space truss* consists of members joined together at their ends to form a stable three-dimensional structure. The simplest element of a space truss is a *tetrahedron,* formed by connecting six members together, as shown in Fig. 6–19. Any additional members added to this basic element would be redundant in supporting the force **P.** A *simple space truss* can be built from this basic tetrahedral element by adding three additional members and a joint forming a system of multiconnected tetrahedrons.

Assumptions for Design.   The members of a space truss may be treated as two-force members provided the external loading is applied at the joints and the joints consist of ball-and-socket connections. These assumptions are justified if the welded or riveted connections of the joined members intersect at a common point and the weight of the members can be neglected. In cases where the weight of a member is to be included in the analysis, it is generally satisfactory to apply it as a vertical force, half of its magnitude applied at each end of the member.

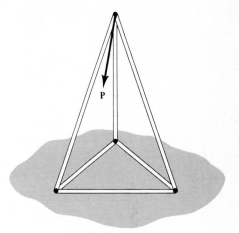

Fig. 6–19

PROCEDURE FOR ANALYSIS

Either the method of sections or the method of joints can be used to determine the forces developed in the members of a simple space truss.

Method of Sections.   If only a *few* member forces are to be determined, the method of sections may be used. When an imaginary section is passed through a truss, and the truss is separated into two parts, the force system acting on one of the parts must satisfy the *six* scalar equilibrium equations: $\Sigma F_x = 0$, $\Sigma F_y = 0$, $\Sigma F_z = 0$, $\Sigma M_x = 0$, $\Sigma M_y = 0$, $\Sigma M_z = 0$ (Eqs. 5–6). By proper choice of the section and axes for summing forces and moments, many of the unknown member forces in a space truss can be computed *directly,* using a single equilibrium equation.

Method of Joints.   Generally, if the forces in *all* the members of the truss must be determined, the method of joints is most suitable for the analysis. When using the method of joints, it is necessary to solve the three scalar equilibrium equations $\Sigma F_x = 0$, $\Sigma F_y = 0$, $\Sigma F_z = 0$ at each joint. The solution of many simultaneous equations can be avoided if the force analysis begins at a joint having at least one known force and at most three unknown forces. If the three-dimensional geometry of the force system at the joint is hard to visualize, it is recommended that a Cartesian vector analysis be used in the solution. The following example numerically illustrates this procedure.

Example 6–8

Determine the forces acting in the members of the space truss shown in Fig. 6–20a. Indicate whether the members are in tension or compression.

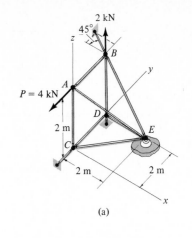

(a)

Solution

Since there are one known force and three unknown forces acting at joint A, the force analysis of the truss will begin at joint A.

Joint A (Fig. 6–20b).   Expressing each force that acts on the free-body diagram of joint A in vector notation yields

$$\mathbf{P} = \{-4\mathbf{j}\} \text{ kN}, \qquad \mathbf{F}_{AB} = F_{AB}\mathbf{j}, \qquad \mathbf{F}_{AC} = -F_{AC}\mathbf{k},$$

$$\mathbf{F}_{AE} = F_{AE}\left(\frac{\mathbf{r}_{AE}}{r_{AE}}\right) = F_{AE}(0.577\mathbf{i} + 0.577\mathbf{j} - 0.577\mathbf{k})$$

For equilibrium,

$$\Sigma\mathbf{F} = 0; \qquad \mathbf{P} + \mathbf{F}_{AB} + \mathbf{F}_{AC} + \mathbf{F}_{AE} = 0$$

$$-4\mathbf{j} + F_{AB}\mathbf{j} - F_{AC}\mathbf{k} + 0.577F_{AE}\mathbf{i} + 0.577F_{AE}\mathbf{j} - 0.577F_{AE}\mathbf{k} = 0$$

$$\Sigma F_x = 0; \qquad 0.577F_{AE} = 0$$
$$\Sigma F_y = 0; \qquad -4 + F_{AB} + 0.577F_{AE} = 0$$
$$\Sigma F_z = 0; \qquad -F_{AC} - 0.577F_{AE} = 0$$
$$F_{AC} = F_{AE} = 0 \qquad\qquad\qquad \textit{Ans.}$$
$$F_{AB} = 4 \text{ kN} \quad (T) \qquad\qquad\qquad \textit{Ans.}$$

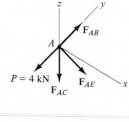

(b)

Since F_{AB} is known, joint B may be analyzed next.

Joint B (Fig. 6–20c)

$$\Sigma F_x = 0; \qquad -R_B \cos 45° + 0.707F_{BE} = 0$$
$$\Sigma F_y = 0; \qquad -4 + R_B \sin 45° = 0$$
$$\Sigma F_z = 0; \qquad 2 + F_{BD} - 0.707F_{BE} = 0$$
$$R_B = F_{BE} = 5.66 \text{ kN} \quad (T), \qquad F_{BD} = 2 \text{ kN} \quad (C) \qquad \textit{Ans.}$$

The *scalar* equations of equilibrium may be applied directly to the force systems on the free-body diagrams of joints D and C, since the force components are easily determined. Show that

$$F_{DE} = F_{DC} = F_{CE} = 0 \qquad\qquad \textit{Ans.}$$

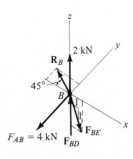

(c)

Fig. 6–20

Problems

6–45. Determine the force in each member of the space truss and indicate whether the members are in tension or compression. The truss is supported by ball-and-socket joints at D, C, and E. *Hint:* The support reaction at E acts along member EB. Why?

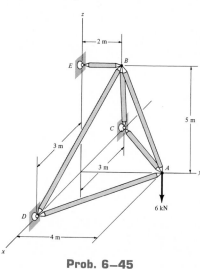

Prob. 6–45

6–46. The tetrahedral truss rests on roller supports at points A, B, and C. Determine the force in each member and indicate if the members are in tension or compression.

Prob. 6–46

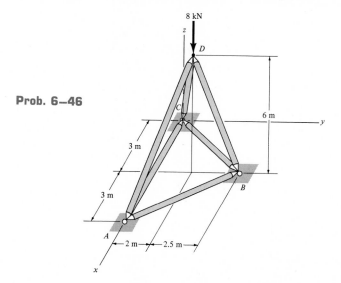

6–47. The space truss is used to support vertical forces at joints B, C, and D. Determine the force in each of the members and indicate whether the members are in tension or compression.

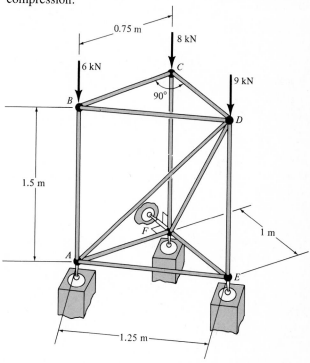

Prob. 6–47

6–48. Determine the force in each of the members of the space truss and indicate whether the members are in tension or compression. The truss is supported by a ball-and-socket joint at A and short links at B and C.

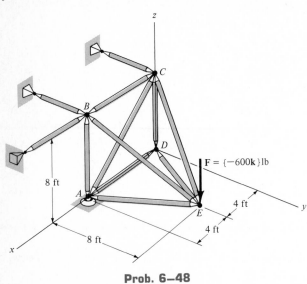

F = {−600k}lb

Prob. 6–48

6–49. The space truss is supported by a ball-and-socket joint at A and short links at D, E, and F. Determine the force in each member and indicate if the members are in tension or compression.

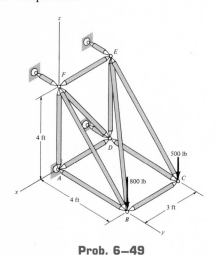

Prob. 6–49

6–50. The *tower truss* is subjected to a horizontal force of $P = \{900\,j\}$ N at its apex. Determine the force in members CD and DB and state whether the members are in tension or compression. The truss is supported by short links at A, G, and F.

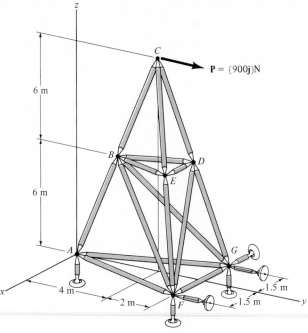

P = {900j}N

Prob. 6–50

Frames and Machines 6.6

Frames and machines are two common types of structures which are often composed of pin-connected *multiforce members*, i.e., members that are subjected to more than two forces. *Frames* are generally stationary and are used to support loads, whereas *machines* contain moving parts and are designed to transmit and alter the effect of forces. Provided a frame or machine is properly constrained and contains no more supports or members than are necessary to prevent collapse, then the forces acting at the joints and supports can be determined by applying the equations of equilibrium to each member. Once the forces at the joints are obtained, it is then possible to *design* the size of the members, connections, and supports using the theory of strength of materials.

Free-Body Diagrams.   In order to determine the forces acting at the joints and supports of a frame or machine, the structure must be disassembled and the free-body diagrams of its parts must be drawn. In this regard, the following important points *must* be observed:

1. Isolate each part by drawing its *outlined shape*. Then show all the forces and/or couple moments that act on the part. Make sure to *label* or *identify* each known and unknown force and couple moment and indicate any dimensions used for taking moments. Most often the equations of equilibrium are easier to apply if the forces are represented by their rectangular components. As usual, the sense of an unknown force or couple moment can be assumed.
2. Identify the two-force members in the structure, and represent their free-body diagrams as having two equal but opposite forces acting at their points of application. The line of action of the forces is defined by the line joining the two points where the forces act (see Sec. 5.4). *This step is particularly important in order to avoid solving an unnecessary number of equilibrium equations.* (See Example 6–13.)
3. Forces common to any two *contacting* members act with equal magnitudes but opposite sense on the respective members. If the two members are treated as a *"system" of connected members*, then these forces are *"internal"* and are *not shown* on the *free-body diagram of the system;* however, if the free-body diagram of *each member* is drawn, the forces are *"external"* and *must* be shown on the free-body diagrams.

The following examples graphically illustrate application of these points in drawing the free-body diagrams of a dismembered frame or machine. Each example should be thoroughly understood. In all cases, the weights of the members are neglected.

Example 6–9

For the frame shown in Fig. 6–21a, draw the free-body diagram of (a) each member, (b) the pin at B, and (c) the two members connected together.

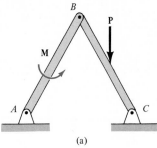

(a)

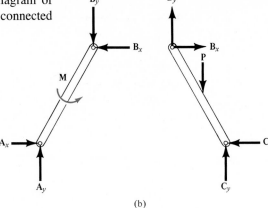

(b)

Solution

Part (a). By inspection the members are *not* two-force members. Instead, as shown on the free-body diagrams, Fig. 6–21b, BC is subjected to *three forces*, namely, the *resultants* of the two components of reaction at pins B and C and the external force $\mathbf{P}$. Likewise, AB is subjected to the *resultant* pin-reactive forces at A and B and the external couple $\mathbf{M}$.

Part (b). It can be seen in Fig. 6–21a that the pin at B is subjected to only *two forces*, i.e., the force of member BC on the pin and the force of member AB on the pin. For *equilibrium* these forces and therefore their respective components must be equal but opposite, Fig. 6–21c.* Notice carefully how Newton's third law is applied between the pin and its contacting members, i.e., the effect of the pin on the two members, Fig. 6–21b, and the equal but opposite effect of the two members on the pin, Fig. 6–21c. Realize also that $\mathbf{B}_x$ and $\mathbf{B}_y$ shown equal but opposite in Fig. 6–21b on members AB and BC is *not* the effect of Newton's third law; instead, it results from the *equilibrium* of the pin, Fig. 6–21c.

Part (c). The free-body diagram of both members connected together, yet removed from the supporting pins at A and C, is shown in Fig. 6–21d. Note that the force components $\mathbf{B}_x$ and $\mathbf{B}_y$ are *not shown* on this diagram since they form equal but opposite collinear pairs of *internal* forces (Fig. 6–21b) and therefore cancel out.† Also note that to be consistent when later applying the equilibrium equations, the unknown force components at A and C must act in the *same sense* as those shown in Fig. 6–21b.

(c)

(d)

Fig. 6–21

---

*The length of the pin is considered *short*, so that any moment created by the force interactions on the pin will be neglected.

†This is similar to not including internal forces exerted between adjacent particles of a rigid body when drawing the free-body diagram of the entire rigid body.

Example 6–10

For the frame shown in Fig. 6–22a, draw the free-body diagrams of (a) each of the three members, and (b) members *ABC* and *BD* together.

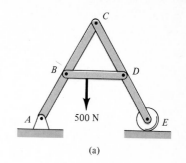

Solution

Part (a). By inspection, none of the three members of the frame are two-force members. Instead, each is subjected to three forces. The free-body diagrams are shown in Fig. 6–22b. Notice that equal but opposite force reactions occur at *B*, *C*, and *D*. Draw free-body diagrams of the pins at *B*, *C*, and *D* and show why this is so.

(a)

Fig. 6–22a

Part (b). The free-body diagram of *ABC* and *BD* together is shown in Fig. 6–22c. Since the entire frame is in equilibrium, the force system on these two members also satisfies the equilibrium equations. Why not show the force components B_x and B_y on this diagram?

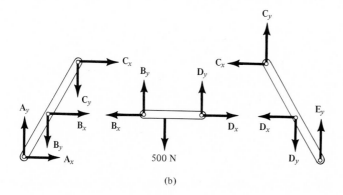

(b)

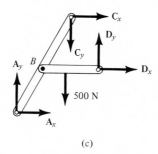

(c)

Fig. 6–22b and c

Example 6–11

For the frame shown in Fig. 6–23, draw the free-body diagrams of (a) the entire frame and (b) each of its parts.

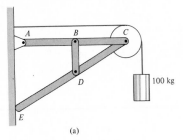

(a)

Fig. 6–23a

Solution

Part (a). The free-body diagram for the entire frame *including* the pulley, cord, and 100-kg mass is shown in Fig. 6–23b. Since the cord is continuous, it maintains a constant tension of 981 N throughout its length. Consequently, the 981-N force shown acting to the left represents the effect of the wall on the cord.

Part (b). By inspection, member *BD* is a two-force member, and therefore equal but opposite collinear forces act at its end points. The free-body diagrams are shown in Fig. 6–23c. Each of these diagrams should be carefully studied. In particular, note that the pulley is held in equilibrium by the two 981-N force components which the pin at *C* exerts on the pulley. The free-body diagram of the pin at *C* has been included here since *three force interactions* occur *on* this pin. The force components C_x and C_y represent the effect of member *ABC*, the force components C'_x and C'_y represent the effect of member *EDC*, and finally the two 981-N force components represent the effect of the pulley. Writing the equilibrium equations $\Sigma F_x = 0$ and $\Sigma F_y = 0$ for the pin at *C* will thus enable one to obtain two equations that relate the unknown force components C_x, C'_x and C_y, C'_y. Note that if only *two force interactions* occur at a pin connection, such as pin *D* in Fig. 6–23c, the force components acting must always be equal in magnitude but with opposite sense, and thus it is generally not necessary to consider its free-body diagram for the analysis.

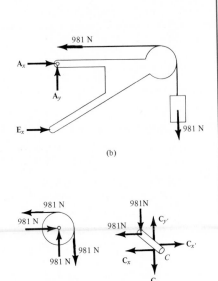

(b)

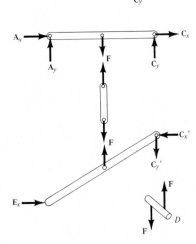

(c)

Fig. 6–23b and c

Example 6–12

Draw the free-body diagram of each part of the smooth piston and link mechanism shown in Fig. 6–24a.

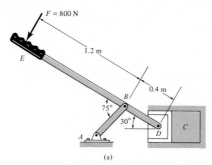

(a)

Fig. 6–24a

Solution

By inspection member *AB* is a two-force member. The free-body diagrams of the parts are shown in Fig. 6–24b. Since the pins at *B* and *D* *connect only two parts together,* the forces there are shown as equal but opposite on the separate free-body diagrams of their connected members. In particular, four components of force act on the piston: D_x and D_y represent the effect of the pin (or lever *EBD*), N_w is the *resultant force* of the wall, and **P** is the resultant compressive force caused by the material within the cylinder.

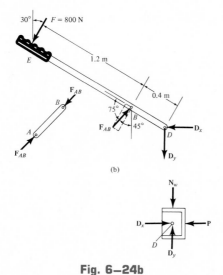

(b)

Fig. 6–24b

247

Equations of Equilibrium.   Provided the structure (frame or machine) is properly supported and contains no more supports or members than are necessary to prevent its collapse, then the unknown forces at the supports and connections can be determined from the equations of equilibrium. If the structure lies in the x-y plane, then for each free-body diagram drawn the loading must satisfy $\Sigma F_x = 0$, $\Sigma F_y = 0$, and $\Sigma M_O = 0$. The selection of the free-body diagrams used for the analysis is *completely arbitrary*. They may represent each of the members of the structure, a portion of the structure, or its entirety. For example, consider finding the six components of the pin reactions at A, B, and C for the frame shown in Fig. 6–25a. If the frame is dismembered, Fig. 6–25b, these unknowns can be determined by applying the three equations of equilibrium to each of the two members (total of six equations). The free-body diagram of the *entire frame* can also be used for part of the analysis, Fig. 6–25c. Hence, if so desired, all six unknowns can be determined by applying the three equilibrium equations to the entire frame, Fig. 6–25c, and also to either one of its members. Furthermore, the answers can be checked in part by applying the three equations of equilibrium to the remaining ''second'' member. In general, then, this problem can be solved by writing *at most* six equilibrium equations using free-body diagrams of the members and/or the combination of connected members. Any more than six equations written would *not* be unique from the original six and would only serve to check the results.

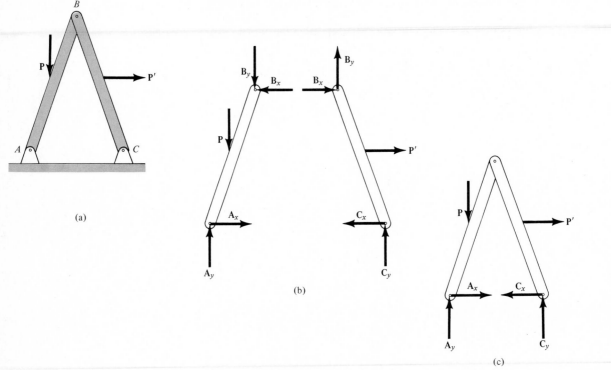

Fig. 6–25

PROCEDURE FOR ANALYSIS

The following procedure provides a method for determining the *joint reactions* of frames and machines (structures) composed of multiforce members.

Free-Body Diagrams. Draw the free-body diagram of the entire structure, a portion of the structure, or each of its members. The choice should be made so that it leads to the most direct solution to the problem.

Forces common to two members which are in contact act with equal magnitude but opposite sense on the respective free-body diagrams of the members. Recall that all *two-force members,* regardless of their shape, have equal but opposite collinear forces acting at the ends of the member. The unknown forces acting at the joints of multiforce members should be represented by their rectangular components. In many cases it is possible to tell by inspection the proper sense of the unknown forces; however, if this seems difficult, the sense can be assumed.

Equations of Equilibrium. Count the total number of unknowns to make sure that an equivalent number of equilibrium equations can be written for solution. Recall that in general three equilibrium equations can be written for each rigid body represented in two dimensions. Many times, the solution for the unknowns will be straightforward if moments are summed about a point that lies at the intersection of the lines of action of as many unknown forces as possible. If after obtaining the solution an unknown force magnitude is found to be negative, it means the sense of the force is the reverse of that shown on the free-body diagrams.

The examples that follow illustrate this procedure. All these examples should be *thoroughly understood* before proceeding to solve the problems.

249

Example 6–13

Determine the horizontal and vertical components of force which the pin at C exerts on member CB of the frame in Fig. 6–26a.

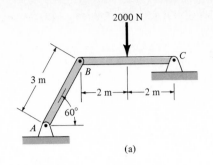

(a)

Solution I

Free-Body Diagrams. By inspection it can be seen that AB is a two-force member. The free-body diagrams of the two members are shown in Fig. 6–26b.

Equations of Equilibrium. The *three unknowns*, C_x, C_y, and F_{AB}, can be determined by applying the three equations of equilibrium to member CB. We have

$\zeta + \Sigma M_C = 0;$ $2000(2) - (F_{AB} \sin 60°)(4) = 0$ $F_{AB} = 1154.7$ N

$\xrightarrow{+} \Sigma F_x = 0;$ $1154.7 \cos 60° - C_x = 0$ $C_x = 577$ N   *Ans.*

$+ \uparrow \Sigma F_y = 0;$ $1154.7 \sin 60° - 2000 + C_y = 0$ $C_y = 1000$ N   *Ans.*

Solution II

Free-Body Diagrams. If one does not recognize that AB is a two-force member, then more work is involved in solving this problem. The free-body diagrams are shown in Fig. 6–26c.

Equations of Equilibrium. The *six unknowns*, A_x, A_y, B_x, B_y, C_x, C_y, are determined by applying the three equations of equilibrium to each member. We have

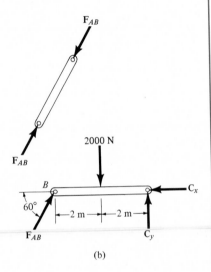

(b)

   Member AB

$\zeta + \Sigma M_A = 0;$ $B_x(3 \sin 60°) - B_y(3 \cos 60°) = 0$ (1)

$\xrightarrow{+} \Sigma F_x = 0;$ $A_x - B_x = 0$ (2)

$+ \uparrow \Sigma F_y = 0;$ $A_y - B_y = 0$ (3)

   Member BC

$\zeta + \Sigma M_C = 0;$ $2000(2) - B_y(4) = 0$ (4)

$\xrightarrow{+} \Sigma F_x = 0;$ $B_x - C_x = 0$ (5)

$+ \uparrow \Sigma F_y = 0;$ $B_y - 2000 + C_y = 0$ (6)

The results for C_x and C_y can be determined by solving these equations in the following sequence: (4), (1), (5), then (6). The results are:

$$B_y = 1000 \text{ N}$$
$$B_x = 577 \text{ N}$$
$$C_x = 577 \text{ N} \qquad \textit{Ans.}$$
$$C_y = 1000 \text{ N} \qquad \textit{Ans.}$$

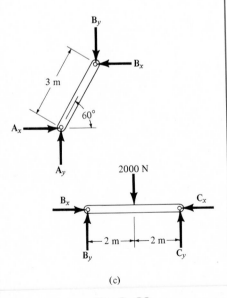

(c)

Fig. 6–26

By comparison, Solution I is simpler since the requirement that $\mathbf{F}_{AB}$ in Fig. 6–26b be equal, opposite, and collinear at the ends of member AB automatically satisfies Eqs. (1), (2), and (3) above and therefore eliminates the need to write these equations. *As a result, always spot the two-force members before starting the analysis!*

Example 6–14

The beam shown in Fig. 6–27a is pin-connected at B. Determine the reactions at its supports. Neglect its weight and thickness.

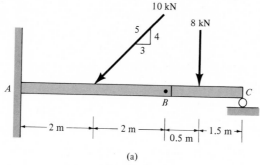

(a)

Fig. 6–27a

Solution

Free-Body Diagrams. The free-body diagrams of the two beam segments are shown in Fig. 6–27b.

Equations of Equilibrium
  Segment BC

$\xrightarrow{+} \Sigma F_x = 0;$ $B_x = 0$

$\downdownarrows + \Sigma M_B = 0;$ $-8 \text{ kN}(0.5 \text{ m}) + C_y(2 \text{ m}) = 0$

$+\uparrow \Sigma F_y = 0;$ $B_y - 8 \text{ kN} + C_y = 0$

  Segment AB

$\xrightarrow{+} \Sigma F_x = 0;$ $A_x - 10 \text{ kN}(\frac{3}{5}) + B_x = 0$

$\downdownarrows + \Sigma M_A = 0;$ $M_A - 10 \text{ kN}(\frac{4}{5})(2 \text{ m}) - B_y(4 \text{ m}) = 0$

$+\uparrow \Sigma F_y = 0;$ $A_y - 10 \text{ kN}(\frac{4}{5}) - B_y = 0$

Solving each of these equations successively, using previously calculated results, we obtain

        $A_x = 6 \text{ kN}$ $A_y = 14 \text{ kN}$ $M_A = 40 \text{ kN} \cdot \text{m}$ *Ans.*

        $B_x = 0$ $B_y = 6 \text{ kN}$

        $C_y = 2 \text{ kN}$ *Ans.*

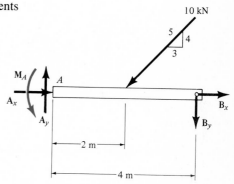

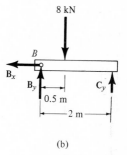

(b)

Fig. 6–27b

Example 6–15

Determine the horizontal and vertical components of force which the pin at C exerts on member $ABCD$ of the frame shown in Fig. 6–28a.

Solution

Free-Body Diagrams. By inspection, the three components of reaction at the supports can be determined from a free-body diagram of the entire frame, Fig. 6–28b. Also, the free-body diagram of each frame member is shown in Fig. 6–28c. Notice that member BE is a two-force member. As shown by the colored dashed lines, the forces at B, C, and E have equal magnitudes but opposite directions on the separate free-body diagrams.

Equations of Equilibrium. The six unknowns A_x, A_y, T_B, C_x, C_y and D_x will be determined from the equations of equilibrium applied to the entire frame and then to member CEF. We have

Entire Frame

$$\zeta + \Sigma M_A = 0; \quad -981 \text{ N}(2 \text{ m}) + D_x(2.8 \text{ m}) = 0 \quad D_x = 700.7 \text{ N}$$
$$\Rightarrow \Sigma F_x = 0; \quad A_x - 700.7 \text{ N} = 0 \quad A_x = 700.7 \text{ N}$$
$$+\uparrow \Sigma F_y = 0; \quad A_y - 981 \text{ N} = 0 \quad A_y = 981 \text{ N}$$

Member CEF

$$\zeta + \Sigma M_C = 0; \quad -981 \text{ N}(2 \text{ m}) - (T_B \sin 45°)(1.6 \text{ m}) = 0$$
$$T_B = -1734.2 \text{ N}$$
$$\Rightarrow \Sigma F_x = 0; \quad -C_x - (-1734.2 \cos 45° \text{ N}) = 0$$
$$C_x = 1226.2 \text{ N} \qquad \textit{Ans.}$$
$$+\uparrow \Sigma F_y = 0; \quad C_y - (-1734.2 \sin 45° \text{ N}) - 981 \text{ N} = 0$$
$$C_y = -245.2 \text{ N} \qquad \textit{Ans.}$$

Since the magnitudes of forces T_B and C_y were calculated as negative quantities, they were assumed to be acting in the wrong sense on the free-body diagrams, Fig. 6–28c. The correct sense of these forces might have been determined "by inspection" *before* applying the equations of equilibrium to member CEF. As shown in Fig. 6–28c, moment equilibrium about point E on member CEF indicates that C_y must actually act *downward* to counteract the moment created by the 981-N force about point E. Similarly, summing moments about point C, it is seen that the vertical component of force T_B must actually act *upward*. The above calculations can be checked in part by applying the three equilibrium equations to member $ABCD$, Fig. 6–28c.

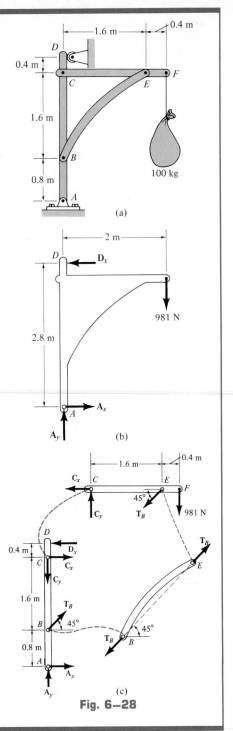

(a)

(b)

(c)

Fig. 6–28

Example 6–16

The smooth disk shown in Fig. 6–29a is pinned at D and has a weight of 20 lb. Neglecting the weights of the other members, determine the horizontal and vertical components of reaction at pins B and D.

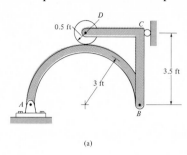

(a)

Fig. 6–29

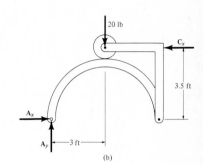

(b)

Solution

Free-Body Diagrams. By inspection, the three components of reaction at the supports can be determined from a free-body diagram of the entire frame, Fig. 5–29b. Also, free-body diagrams of the members are shown in Fig. 6–29c.

Equations of Equilibrium. The eight unknowns can of course be obtained by applying the eight equilibrium equations to each member—three to member AB, three to member BCD, and two to the disk. (Moment equilibrium is automatically satisfied for the disk.) If this is done, however, all the results can be obtained only from a simultaneous solution of some of the equations. (Try it and find out.) To avoid this situation, it is best to first determine the three support reactions on the *entire* frame; then, using these results, the remaining five equilibrium equations can be applied to two other parts in order to solve successively for the other unknowns.

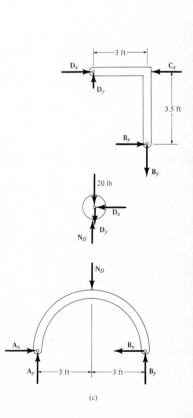

(c)

Entire Frame

$$\zeta+\Sigma M_A = 0; \quad -20 \text{ lb}(3 \text{ ft}) + C_x(3.5 \text{ ft}) = 0 \quad C_x = 17.1 \text{ lb}$$
$$\Rightarrow\Sigma F_x = 0; \quad A_x - 17.1 \text{ lb} = 0 \quad A_x = 17.1 \text{ lb}$$
$$+\uparrow\Sigma F_y = 0; \quad A_y - 20 \text{ lb} = 0 \quad A_y = 20 \text{ lb}$$

Member AB

$$\Rightarrow\Sigma F_x = 0; \quad 17.1 \text{ lb} - B_x = 0 \quad B_x = 17.1 \text{ lb} \qquad \textit{Ans.}$$
$$\zeta+\Sigma M_B = 0; \quad -20 \text{ lb}(6 \text{ ft}) + N_D(3 \text{ ft}) = 0 \quad N_D = 40 \text{ lb}$$
$$+\uparrow\Sigma F_y = 0; \quad 20 \text{ lb} - 40 \text{ lb} + B_y = 0 \quad B_y = 20 \text{ lb} \qquad \textit{Ans.}$$

Disk

$$\pm\Sigma F_x = 0; \quad D_x = 0 \qquad \textit{Ans.}$$
$$+\uparrow\Sigma F_y = 0; \quad 40 \text{ lb} - 20 \text{ lb} - D_y = 0 \quad D_y = 20 \text{ lb} \qquad \textit{Ans.}$$

Example 6–17

Determine the tension in the cables and also the force **P** required to support the 600-N force using the frictionless pulley system shown in Fig. 6–30*a*.

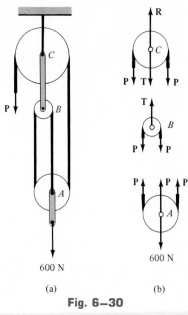

Fig. 6–30

(a)   (b)

Solution

Free-Body Diagrams. A free-body diagram of each pulley *including* its pin and a portion of the contacting cable is shown in Fig. 6–30*b*. Since the cable is *continuous* and the pulleys are frictionless, the cable has a *constant tension* of P N acting throughout its length (see Example 5–6). The link connection between pulleys B and C is a two-force member and therefore it has an unknown tension of T N acting on it. Notice that the *principle of action, equal but opposite reaction* must be carefully observed for forces **P** and **T** when the *separate* free-body diagrams are drawn.

Equations of Equilibrium. The three unknowns are obtained as follows:

Pulley A

$$+\uparrow \Sigma F_y = 0; \qquad 3P - 600 \text{ N} = 0 \qquad P = 200 \text{ N} \qquad\qquad Ans.$$

Pulley B

$$+\uparrow \Sigma F_y = 0; \qquad T - 2P = 0 \qquad T = 400 \text{ N} \qquad\qquad Ans.$$

Pulley C

$$+\uparrow \Sigma F_y = 0; \qquad R - 2P - T = 0 \qquad R = 800 \text{ N} \qquad\qquad Ans.$$

Example 6–18

Determine the tension in the cables and also the force **P** required to support the 600-N force using the frictionless pulley system shown in Fig. 6–31*a*.

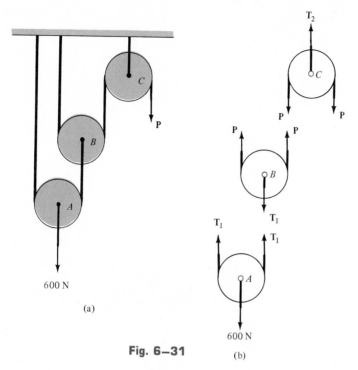

(a)

600 N

Fig. 6–31   (b)

Solution

Free-Body Diagrams. A free-body diagram of each pulley *including* its pin and a portion of its contacting cable is shown in Fig. 6–31*b*. Observe how the principle of action, equal but opposite reaction is applied to forces T_1 and **P** on the separate free-body diagrams.

Equations of Equilibrium. The three unknowns are obtained as follows:

Pulley A

$$+\uparrow \Sigma F_y = 0; \qquad 2T_1 - 600\text{ N} = 0 \qquad T_1 = 300\text{ N} \qquad\qquad Ans.$$

Pulley B

$$+\uparrow \Sigma F_y = 0; \qquad 2P - T_1 = 0 \qquad P = 150\text{ N} \qquad\qquad Ans.$$

Pulley C

$$+\uparrow \Sigma F_y = 0; \qquad T_2 - 2P = 0 \qquad T_2 = 300\text{ N} \qquad\qquad Ans.$$

Example 6–19

A man having a weight of 150 lb supports himself by means of the cable and pulley system shown in Fig. 6–32a. If the seat has a weight of 15 lb, determine the equilibrium force that he must exert on the cable at A and the force he exerts on the seat. Neglect the weight of the cables and pulleys.

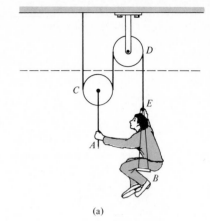
(a)

Solution I

Free-Body Diagrams. The free-body diagrams of the man, seat, and pulley C are shown in Fig. 6–32b. The *two* cables are subjected to tensions T_A and T_E, respectively. The man is subjected to three forces: his weight, the tension T_A of cable AC, and the reaction N_s of the seat.

Equations of Equilibrium. The three unknowns are obtained as follows:

Man

$$+\uparrow \Sigma F_y = 0; \qquad T_A + N_s - 150 \text{ lb} = 0 \qquad (1)$$

Seat

$$+\uparrow \Sigma F_y = 0; \qquad T_E - N_s - 15 \text{ lb} = 0 \qquad (2)$$

Pulley C

$$+\uparrow \Sigma F_y = 0; \qquad 2T_E - T_A = 0 \qquad (3)$$

The magnitude of force T_E can be determined by adding Eqs. (1) and (2) to eliminate N_s and then using Eq. (3). The other unknowns are then obtained by resubstitution of T_E.

$$T_A = 110 \text{ lb} \qquad \qquad \textit{Ans.}$$
$$T_E = 55 \text{ lb}$$
$$N_s = 40 \text{ lb} \qquad \qquad \textit{Ans.}$$

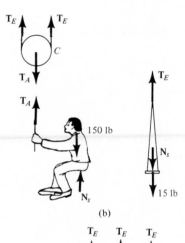
(b)

Solution II

Free-Body Diagram. By using the dashed section shown in Fig. 6–32a, the man, pulley, and seat can be considered as a *single system,* Fig. 6–32c. Here N_s and T_A are *internal* forces and hence are not included on the "combined" free-body diagram.

Equations of Equilibrium. Applying $\Sigma F_y = 0$ yields a *direct* solution for T_E.

$$+\uparrow \Sigma F_y = 0; \quad 3T_E - 15 \text{ lb} - 150 \text{ lb} = 0 \qquad T_E = 55 \text{ lb}$$

The other unknowns can be obtained from Eqs. (2) and (3).

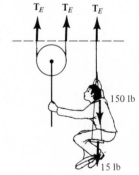
(c)

Fig. 6–32

Example 6–20

Determine the horizontal and vertical force components acting at the pin connections B and C of the frame shown in Fig. 6–33a.

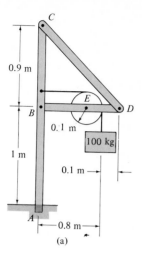

(a)

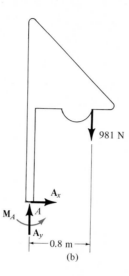

981 N

M_A A A_x

A_y

(b)

Fig. 6–33

Solution

Free-Body Diagrams. The free-body diagrams for members CD, BED, and the pulley are shown in Fig. 6–33b. Notice that member CD is a two-force member. Furthermore, the pulley is held in equilibrium by 981-N force components exerted on it by the pin at point E. By inspection, both force and moment equilibrium of the pulley are satisfied.

Equations of Equilibrium. The three unknowns will be determined by applying the equilibrium equations to member BED. We have

$\zeta + \Sigma M_B = 0;$ $-981 \text{ N}(0.7 \text{ m}) + T_C \sin 45° \ (0.9 \text{ m}) = 0$

$T_C = 1079.2 \text{ N}$ *Ans.*

$\xrightarrow{+} \Sigma F_x = 0;$ $B_x - 981 \text{ N} - 1079.2 \cos 45° \text{ N} = 0$

$B_x = 1744.0 \text{ N}$ *Ans.*

$+ \uparrow \Sigma F_y = 0;$ $1079.2 \sin 45° \text{ N} - 981 \text{ N} + B_y = 0$

$B_y = 218.0 \text{ N}$ *Ans.*

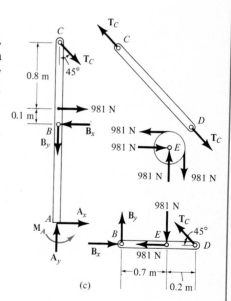

(c)

Example 6–21

The 100-kg block is held in equilibrium by means of the pulley and continuous cable system shown in Fig. 6–34a. If the cable is attached to the pin at B, compute the forces which this pin exerts on each of its connecting members.

Solution

Free-Body Diagrams. A free-body diagram of each member of the frame is shown in Fig. 6–34b. By inspection, members AB and CB are two-force members. Furthermore, the cable must be subjected to a force of 490.5 N in order to hold pulley D in equilibrium. A free-body diagram of the pin at B is needed, since *four interactions* occur at this pin. These are caused by the attached cable (490.5 N), member AB ($\mathbf{F}_{AB}$), member CB ($\mathbf{F}_{CB}$), and pulley B ($\mathbf{B}_x$ and $\mathbf{B}_y$).

Equations of Equilibrium. Applying the equations of force equilibrium to pulley B, we have

$$\xrightarrow{+} \Sigma F_x = 0; \quad B_x - 490.5 \cos 45° \text{ N} = 0 \quad B_x = 346.8 \text{ N} \qquad \textit{Ans.}$$
$$+ \uparrow \Sigma F_y = 0; \quad B_y - 490.5 \sin 45° \text{ N} - 490.5 \text{ N} = 0$$
$$B_y = 837.3 \text{ N} \qquad \textit{Ans.}$$

Using these results, equilibrium of the pin requires that

$$+ \uparrow \Sigma F_y = 0; \quad \tfrac{4}{5} F_{CB} - 837.3 \text{ N} - 490.5 \text{ N} = 0 \quad F_{CB} = 1659.7 \text{ N} \quad \textit{Ans.}$$
$$\xrightarrow{+} \Sigma F_x = 0; \quad F_{AB} - \tfrac{3}{5}(1659.7 \text{ N}) - 346.8 \text{ N} = 0 \quad F_{AB} = 1342.6 \text{ N} \quad \textit{Ans.}$$

It may be noted that the two-force member CB is subjected to bending as caused by the force $\mathbf{F}_{CB}$. From the standpoint of design, it would be better to make this member *straight* (from C to B) so that the force $\mathbf{F}_{CB}$ would only create tension in the member.

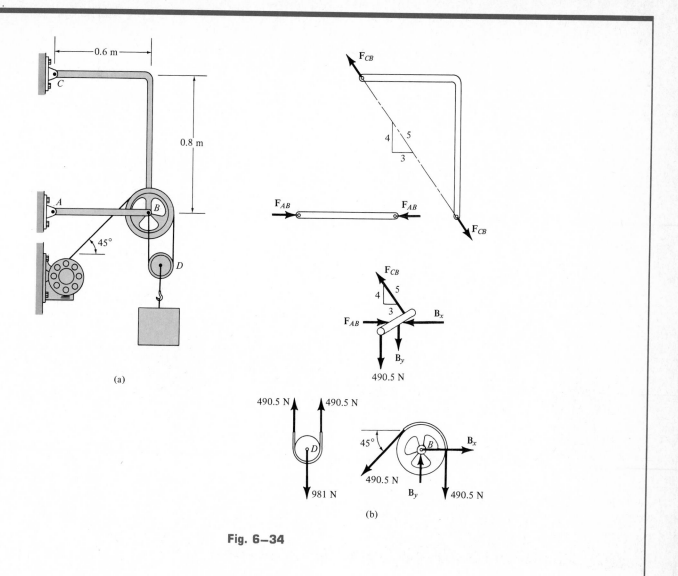

(a)

(b)

Fig. 6—34

Before solving the following problems, it is suggested that a brief review be made of all the previous examples. This may be done by covering each solution and trying to locate the two-force members, drawing the free-body diagrams, and conceiving ways of applying the equations of equilibrium to obtain the solution.

Problems

6–51. Draw the specified free-body diagram in each of the following problems. Identify all two-force members and determine the total number of unknown force and couple-moment magnitudes and directions. Neglect the weights of the members unless otherwise stated. Although not shown on each photo, assume the geometry (size and angles) is known.

a. The ground exerts a vertical load of 2.5 kN on the pad P of the outrigger when it is used to support the crane in a stable position. If all lettered points represent pin connections, draw the free-body diagrams for the arm ABC and the hydraulic cylinder AD.

Prob. 6–51a

b. The portion CB of the screw for the pipe vice is subjected to a compression of 300 N. Draw free-body diagrams of the smooth pipe, which is in contact with the vice at B, D, and E, the screw ACB, and the vice frame without the screw and pipe. The frame is fixed connected to its base at F.

Prob. 6–51b

c. Draw the free-body diagram for the boom ABC together with the load G_2 of the lift truck. The boom has a mass of 350 kg and mass center at G_1. The load has a mass of 200 kg and mass center at G_2. The boom is pin connected to the frame of the truck at C and to a hydraulic cylinder BD. In the position shown, the hydraulic cylinder EF is not operative and therefore does not contribute a significant force on the boom.

Prob. 6–51c

d. Sand in the shovel of the front-end loader has a weight of 350 lb and center of gravity at G. If all lettered points represent pin connections, draw the free-body diagrams for the shovel and members FH, FEB, and $ABCD$.

Prob. 6–51d

e. The screw of the pipe cutter exerts a downward vertical force of 7 N at *A* on the arm *BAC*. The pipe contacts the vice at the rollers *D* and *E* and at the cutting wheel *B*. Draw separate free-body diagrams of the pipe and the arm *BAC*, which is pin connected to the frame at *C*.

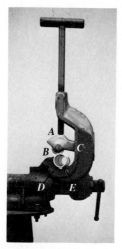

Prob. 6–51e

g. The soil held in the bucket of the backhoe has a mass of 30 kg and mass center at *G*. All lettered points represent pin connections. In particular, point *K* lies behind the hydraulic cylinder *LI* and is on member *FIJ*. Draw free-body diagrams of the bucket, stick *CDEFH*, and pin *B*.

Prob. 6–51g

f. The region *HI* of the screw in the gear puller is subjected to a compressive force of 30 lb. Pin connections exist at points *A*, *B*, *C*, and *D*. The horizontal bar is in smooth contact with the arms at *E* and *F*. Draw free-body diagrams for the bar *EACF* and arm *EBG*.

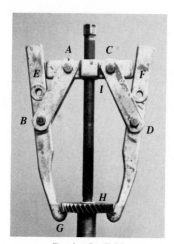

Prob. 6–51f

h. The man in the basket *G* of the lift truck weighs 200 lb. All lettered points represent pin connections. Draw the free-body diagrams of boom *ABCDE*, arm together with basket *EFG*, and pin *H*.

Prob. 6–51h

261

*6–52. Determine the force **P** needed to suspend the 100-lb weight. Each pulley has a weight of 10 lb. Also, what are the cord reactions at *A* and *B*?

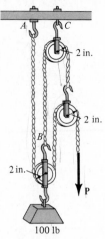

2 in.

2 in.

2 in.

100 lb

Prob. 6–52

6–53. Determine the force **P** needed to suspend the 40-lb weight. Also, what are the reactions at *A* and *B*?

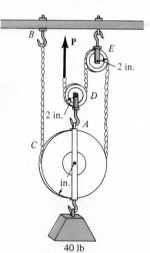

2 in.

2 in.

in.

40 lb

Probs. 6–53 / 6–54

6–54. Determine the force **P** needed to suspend the 40-lb weight. The pulleys *D* and *E* each weigh 10 lb and the large pulley *C* weighs 25 lb. Also, what are the reactions at *A* and *B*?

6–55. Determine the force **P** needed to support the 20-kg mass using the *Spanish Burton rig*. Also, what are the reactions at the supporting hooks *A*, *B*, and *C*?

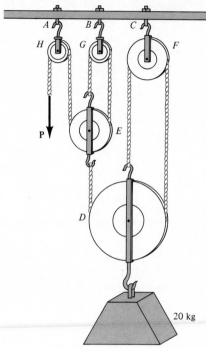

20 kg

Probs. 6–55 / 6–56

*6–56. Determine the force **P** needed to support the 50-kg mass using the *Spanish Burton rig*. The pulleys have a mass of $m_D = 10$ kg, $m_E = m_F = 5$ kg, and $m_G = m_H = 2$ kg. Also, what are the reactions at the supporting hooks *A*, *B*, and *C*?

262

6-57. Compute the tension **T** in the cord, and determine the angle θ that the pulley-supporting link AB makes with the vertical. Neglect the mass of the pulleys and the link. The block has a weight of 200 lb and the cord is attached to the pin at B. The pulleys have radii of $r_1 = 2$ in. and $r_2 = 1$ in.

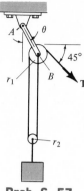

Prob. 6-57

6-58. A force of 8 lb is applied to the handles of the pliers. Determine the force developed on the smooth bolt B and the reaction that the pin A exerts on its attached members.

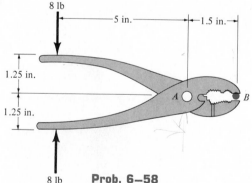

Prob. 6-58

6-59. Determine the horizontal and vertical components of force at pin A.

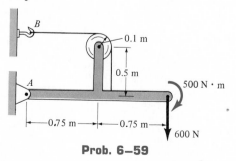

Prob. 6-59

*6-60. The compound beam is pin-supported at B and supported by a rocker at A and C. There is a hinge (pin) at D. Determine the reactions at the supports.

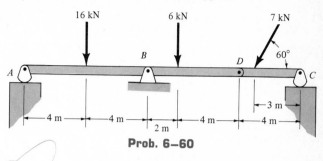

Prob. 6-60

6-61. The compound beam is supported by a rocker at B and is fixed to the wall at A. If it is hinged (pinned) together at C, determine the reactions at the supports.

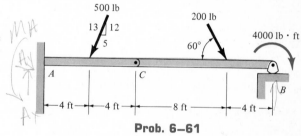

Prob. 6-61

6-62. The truck exerts the three forces shown on the girders of the bridge. Determine the reactions at the supports when it is in the position shown. The girders are connected together by a short vertical link DC.

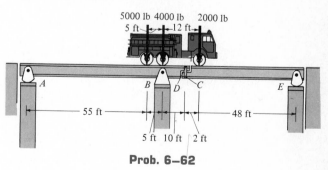

Prob. 6-62

263

6–63. The compound beam is fixed supported at C and supported by rockers at A and B. If there are hinges (pins) at D and E, determine the components of reaction at the supports.

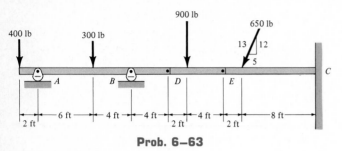

Prob. 6–63

6–65. Determine the horizontal and vertical components of reaction at the pins A, B, and C, and the reactions at the fixed support D of the three-member frame.

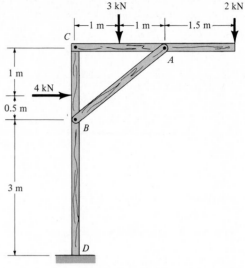

Prob. 6–65

*6–64. Determine the compressive force exerted on the specimen by a vertical load of 50 N applied to the toggle press.

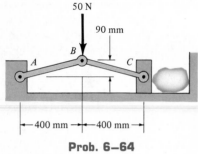

Prob. 6–64

6–66. Determine the horizontal and vertical components of force acting at the pins A and B necessary for equilibrium.

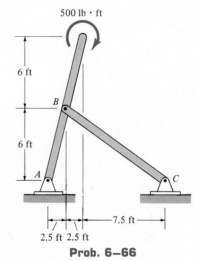

Prob. 6–66

6–67. The truck rests on the scale, which consists of a series of compound levers. If a mass of $m = 15$ kg is placed on the pan P and it is required that $x = 480$ mm to balance the "beam" ABC, determine the mass of the truck. There are pins at all points lettered. Take $EF = 0.2$ m, $FD = 3$ m, $KG = GH = 7$ m, and $HI = KJ = 0.2$ m. Is it necessary for the truck to be symmetrically placed on the scale? Explain.

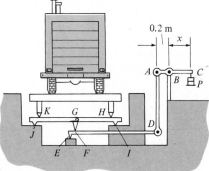

Prob. 6–67

*6–68. The clamp has a rated load capacity of 1500 lb. Determine the compressive force this creates in the portion AB of the screw and the magnitude of force exerted at the pin C. The screw is pin-connected at its end B and passes through the pin-connected (swivel) block at A.

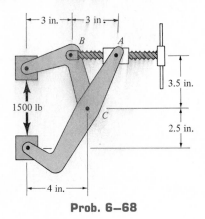

Prob. 6–68

6–69. Determine the force which the smooth roller at C exerts on beam AB. Also, what are the horizontal and vertical components of reaction at pin A? Neglect the weight of the frame and roller.

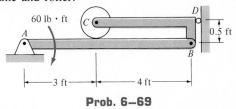

Prob. 6–69

6–70. Determine the horizontal and vertical components of force which the pins exert on member ABD.

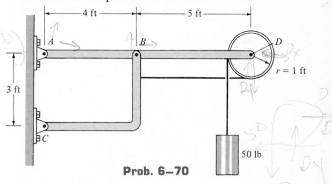

Prob. 6–70

6–71. Determine the horizontal and vertical components of force which the pins exert on member BCE of the frame.

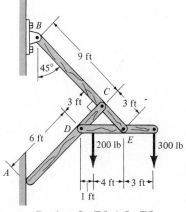

Probs. 6–71 / 6–72

*6–72. Determine the horizontal and vertical components of force which the pins at D and E exert on member DE of the frame.

265

6–73. Determine the horizontal and vertical components of force at C which member ABC exerts on member CEF.

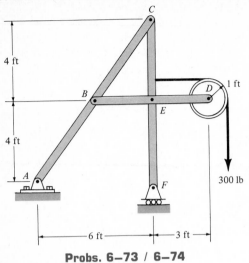

Probs. 6–73 / 6–74

6–74. Determine the horizontal and vertical components of force which the connecting pins at B, E, and D exert on member BED.

6–75. Multiple levers used in truck scales, discussed in Prob. 6–67, can be made *shorter* and more effective by using a *compound arrangement* such as shown for the pan scale. If the mass on the pan is 4 kg, determine the reactions at pins A, B, and C and the distance x of the 25-g mass to keep the scale in balance.

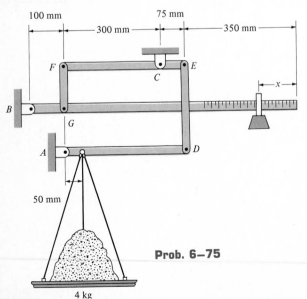

Prob. 6–75

***6–76.** Using a series of levers is more efficient than using a single lever. For example, the 8000-lb truck is balanced on a scale using either (a) three levers or (b) one lever as shown. In case (a) determine the force P required for equilibrium of the truck and in case (b), using the same force P, determine the length l of the single lever necessary for balancing the truck.

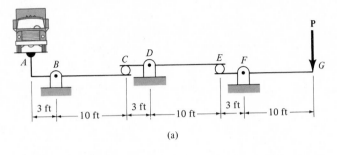

(a)

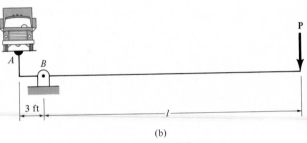

(b)

Prob. 6–76

6–77. A man having a weight of 175 lb attempts to lift himself using one of the two methods shown. Determine the total force he must exert on bar AB in each case and the normal reaction he exerts on the platform at C. Neglect the weight of the platform.

(a)   **Probs. 6–77 / 6–78**   (b)

6–78. A man having a weight of 175 lb attempts to lift himself using one of the two methods shown. Determine the total force he must exert on bar AB in each case and the normal reaction he exerts on the platform at C. The platform has a weight of 30 lb.

6–79. Determine the horizontal and vertical components of force at pin B and the normal force the pin at C exerts on the slot. Also, determine the moment and horizontal and vertical reactions of force at A. The block F has a weight of 50 lb and is supported by a cord which passes over a smooth peg (or pulley) of negligible size at E and is attached to the frame at D.

Prob. 6–79

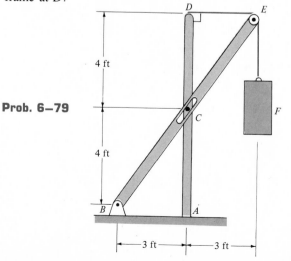

***6–80.** The flatbed trailer has a weight of 7000 lb and center of gravity at G_T. It is pin-connected to the cab at D. The cab has a weight of 6000 lb and center of gravity at G_C. Determine the range of values x for the position of the 2000-lb load L so that when it is placed over the rear axle, no axle is subjected to more than 5500 lb. The load has a center of gravity at G_L.

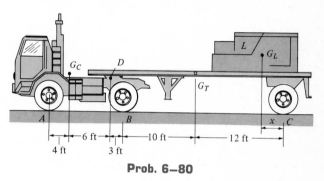

Prob. 6–80

6–81. The bucket of the backhoe and its contents have a weight of 1200 lb and a center of gravity at G. Determine the forces in the hydraulic cylinder AB and in links AC and AD in order to hold the load in the position shown. The bucket is pinned at E. *Hint: AB, AC,* and *AD* are all two-force members.

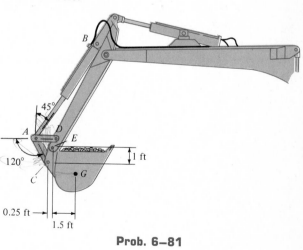

Prob. 6–81

267

6–82. The engine hoist is used to support the 200-kg engine. Determine the force acting in the hydraulic cylinder *AB*, the horizontal and vertical components of force at the pin *C*, and the reactions at the fixed support *D*.

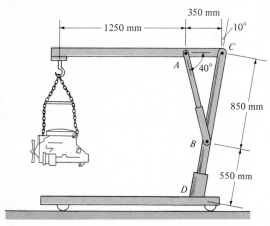

Prob. 6–82

6–83. The man has a weight of 150 lb and stands on the uniform plank having a weight of 40 lb. Determine the force he exerts on the plank if he pulls with just enough force to lift the plank off the support at *B*. The plank rests on the smooth surface at *A*.

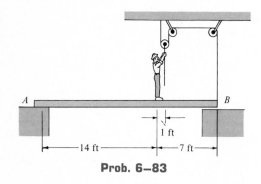

Prob. 6–83

*6–84. The machine shown is used for forming metal plates. It consists of two toggles *ABC* and *DEF*, which are operated by the hydraulic cylinder *H*. The toggles push the movable bar *G* forward, pressing the plate *p* into the cavity. If the force which the plate exerts on the head is $P = 12$ kN, determine the force **F** in the hydraulic cylinder when $\theta = 30°$.

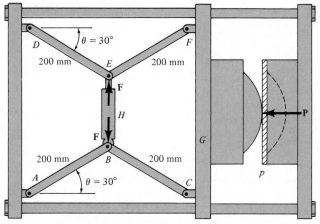

Prob. 6–84

6–85. The two-member frame is used to support the 150-lb weight. Determine the horizontal and vertical components of force which the pins exert on each member.

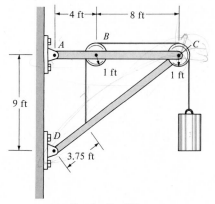

Prob. 6–85

6–86. The front hood of most automobiles does not pivot on single hinges. Instead, the hood is guided by a linkage such as that shown, so that when the hood is raised, the rear edge of the hood is moved forward a short distance. If the hood has a weight of 25 lb, with center of gravity at G, determine the stretch in the spring AB necessary to hold the hood in equilibrium when it is in the position shown. The hood is symmetric so that half its weight (12.5 lb) is supported by each side spring.

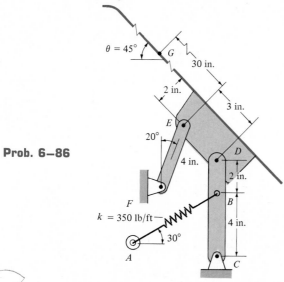

Prob. 6–86

6–87. The two-member frame supports the loading shown. Determine the force of the roller at B on member AC and the horizontal and vertical components of force which the pin at C exerts on member CB. The roller does not contact member CB.

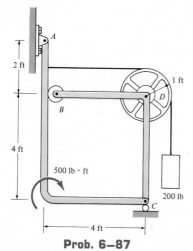

Prob. 6–87

*6–88. The piston C moves vertically between the two smooth walls. If the spring has a stiffness of $k = 15$ lb/in. and is unstretched when $\theta = 0°$, determine the couple $\mathbf{M}$ that must be applied to AB to hold the mechanism in equilibrium when $\theta = 30°$.

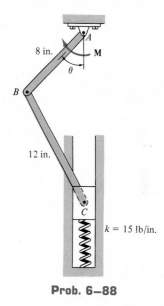

Prob. 6–88

6–89. Determine the horizontal force $\mathbf{F}$ required to maintain equilibrium of the slider mechanism when $\theta = 60°$.

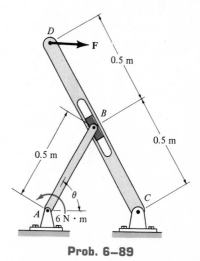

Prob. 6–89

6–90. The compound shears is used to cut metal parts. Determine the vertical cutting force exerted on the specimen S, if a force of F = 20 lb is applied at the grip G. The lobe CDE is in smooth contact with the head of the shear blade at E.

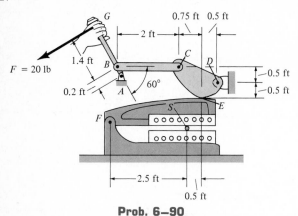

Prob. 6–90

6–91. The tongs consist of two jaws pinned to links at A, B, C, and D. Determine the horizontal and vertical components of force exerted on the 500-lb stone at F and G in order to lift it.

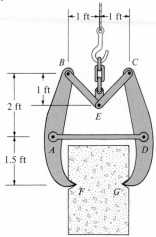

Prob. 6–91

*6–92. The toggle clamp is subjected to a force F at the handle. Determine the corresponding clamping force P acting at the grip.

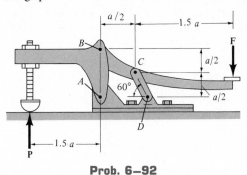

Prob. 6–92

6–93. Determine the force that the jaws J of the metal cutters exert on the smooth cable C if 100-N vertical forces are applied to the handles. The jaws are pinned at E and A, and D and B.

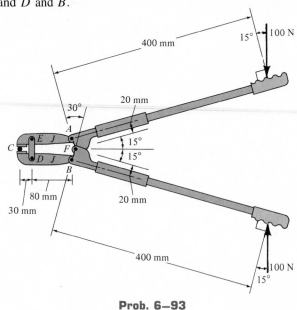

Prob. 6–93

6–94. A 150-lb man stands at the end of the uniform 50-lb beam and pulls on the cord such that the spring at *C* compresses 0.48 ft. If the beam is held in the horizontal position, determine the force he exerts on the beam at *A* and the horizontal and vertical components of force at the pin *B*.

6–95. The frame supports the 80-lb weight. Determine the horizontal and vertical components of force which the pins exert on member *ABCD*. Note that the pin at *C* is attached to member *ABCD* and passes through the smooth slot in member *ECF*.

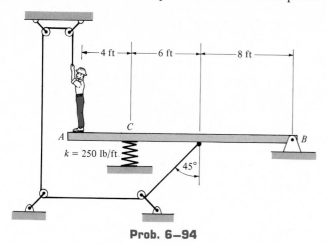

Prob. 6–94

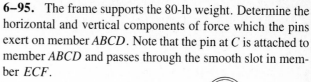

Prob. 6–95

*6–96.** The hydraulic, truck-mounted crane is used to lift a beam which has a mass of 1 Mg. Determine the horizontal and vertical components of force acting at each of the joints when the crane is in the position shown. Assume that all joints are pin-connected.

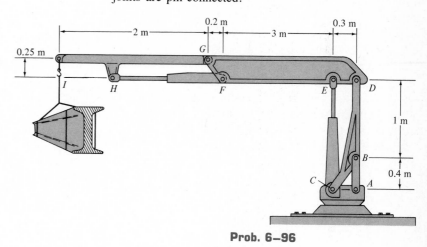

Prob. 6–96

6–97. The clamshell bucket is often used for dredging. When cable *B* is pulled, the buckets remain in the closed position shown, whereas, if cable *A* is pulled, while *B* is slack, the buckets will separate. The mechanism is *symmetrical*. If block *F* has a weight of 150 lb, and one of the buckets weighs 550 lb, with center of gravity at *G*, determine the horizontal force at the jaws *H* and the components of reaction at the pin *C* when the bucket is supported by cable *B* while the cable *A* is slack. No load is held within the bucket.

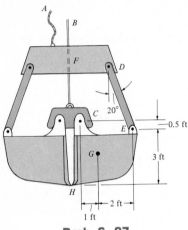

Prob. 6–97

6–98. A vertical force of 40 N is applied to the handle of the sector press. The handle of the press is fixed to gear *G*, which in turn is in mesh with the sector gear *C*. Note that *AB* is pinned at its ends to the gear *C* and the underside of the table *EF*, which is allowed to move vertically due to the smooth guides at *E* and *F*. If the gears exert tangential forces between them, determine the compressive force developed on the specimen *S* when a vertical force of 40 N is applied to the handle of the press.

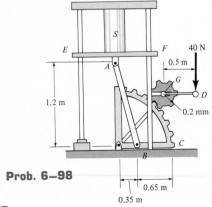

Prob. 6–98

6–99. The spring mechanism is used as a shock absorber for a load applied to the drawbar *AB*. Determine the equilibrium length of each spring when the 80-N force is applied. Each spring has an unloaded length of 200 mm, and the drawbar slides along the smooth guide posts *CG* and *EF*. The bottom springs pass around the guide posts and the ends of all springs are attached to their respective members.

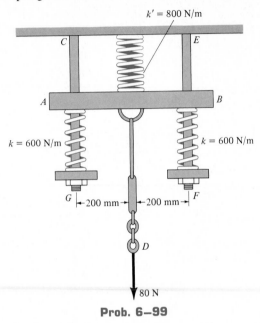

Prob. 6–99

■*6–100. The spring has an unstretched length of 0.3 m. Determine the angle θ for equilibrium if the uniform links each have a mass of 5 kg.

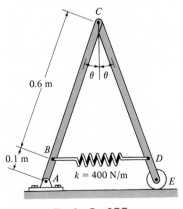

Prob. 6–100

6–101. The tractor shovel shown carries a 500-kg load of soil, having a center of gravity at *G*. Compute the forces developed in the hydraulic cylinders *IJ* and *BC* due to this loading.

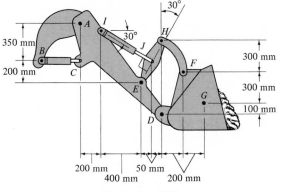

Prob. 6–101

6–102. If each of the three links of the mechanism has a weight of 25 lb, determine the angle θ for equilibrium. The spring, which always remains horizontal, is unstretched when $\theta = 0°$.

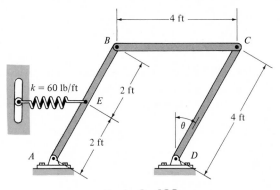

Prob. 6–102

6–103. The winch is used to lift a scaffold by applying a force **P** perpendicular to the handle. The handle is pinned at *A* and rotates the drum *D* by means of the driving pawl *BF*. Determine the minimum magnitude of force **P** needed to lift a scaffold that requires a cable tension of 6 kN. If the force **P** and pawl *BF* are released, what force does the loose locking pawl *CE* develop on the drum to support the load? *Hint:* The pawls *BF* and *CE* are two-force members.

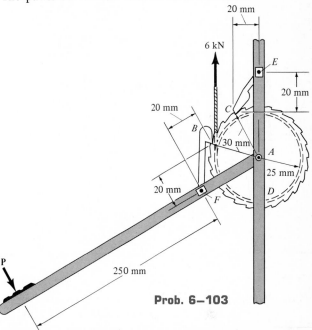

Prob. 6–103

*6–104. The four-member "A" frame supports a vertical force of 600 N. If it is assumed that the supports at *A* and *E* are smooth collars, *G* is a pin, and all other joints are ball-and-sockets, determine the *x, y, z* force components which member *BD* exerts on *EDC* and *FG*.

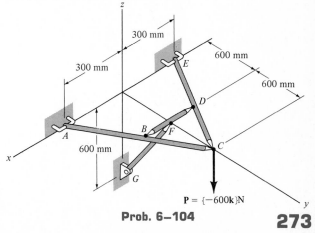

$\mathbf{P} = \{-600\mathbf{k}\}\mathrm{N}$

Prob. 6–104

6–105. The two-member frame is connected at its ends using ball-and-socket joints. Determine the components of reaction at the supports A and C and the tension in the cable ED.

6–107. The structure is subjected to the loadings shown. Member AB is supported by a ball-and-socket at A and fits through a snug hole at B. Member CD is supported by a pin at C. Determine the x, y, z components of reaction at A and C.

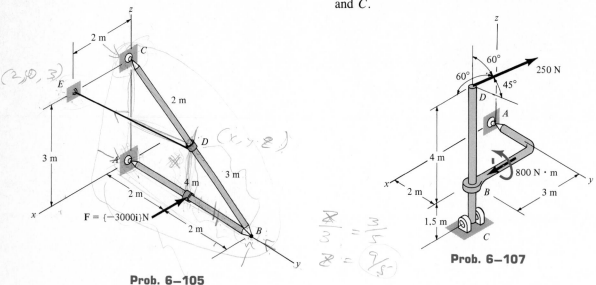

Prob. 6–105

Prob. 6–107

6–106. The tripod rests on the smooth surface at $A, B,$ and C. Its three legs, each having the same length, are held in position by the struts $DE, DF,$ and FE. If the members are assumed to be connected by ball-and-socket joints, determine the force in each strut for equilibrium.

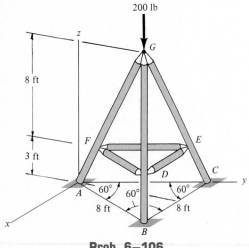

Prob. 6–106

Review Problems

*6–108. Determine the force in each member of the truss and indicate whether the members are in tension or compression.

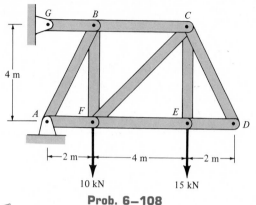

10 kN   15 kN

Prob. 6–108

6–109. Determine the clamping force exerted on the smooth pipe at *B* if a force of 20 lb is applied to the handles of the pliers. The pliers are pinned together at *A*.

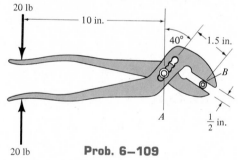

Prob. 6–109

6–110. Determine the force in each member of the truss. Indicate whether the members are in tension or compression.

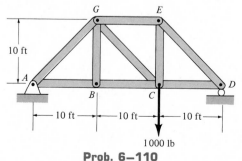

Prob. 6–110

6–111. Determine the horizontal and vertical components of force that the pins at *A* and *C* exert on the two-bar mechanism.

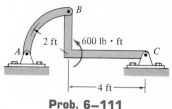

Prob. 6–111

*6–112. The three pin-connected members shown in the *top view* support a *downward* force of 60 lb at *G*. If only vertical forces are supported at the connections *B*, *C*, *E* and pad supports *A*, *D*, *F*, determine the reaction at each pad.

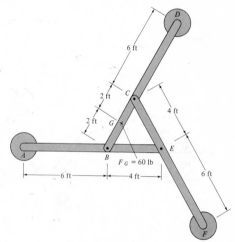

Prob. 6–112

275

6–113. Determine the force developed in each of the members of the space truss. Indicate whether the members are in tension or compression. The crate has a weight of 150 lb.

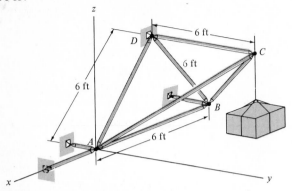

Prob. 6–113

6–117. Determine the horizontal and vertical components of force that pins A, B, and C exert on their connecting members.

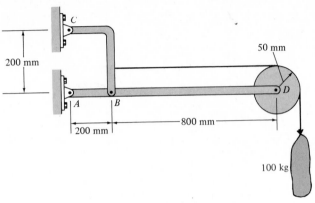

Prob. 6–117

6–114. Determine the force in member BC of the truss. Indicate whether the member is in tension or compression.

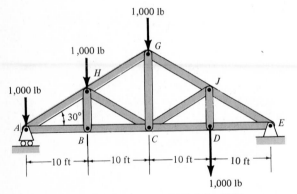

Probs. 6–114 / 6–115 / 6–116

6–115. Determine the force in member GJ of the truss. Indicate whether the member is in tension or compression.

*6–116. Determine the force in member GC of the truss. Indicate whether the member is in tension or compression.

6–118. The *scissors truss* is used to support the roof load. Determine the force in members BF and FD, and indicate whether the members are in tension or compression.

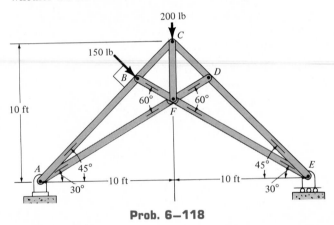

Prob. 6–118

276

6–119. Two space trusses are used to equally support the uniform 150-kg sign. Determine the force developed in members *AB*, *AC*, and *BC* of truss *ABCD*, and indicate whether these members are in tension or compression. Horizontal short links support the truss at joints *B*, *C*, and *D*.

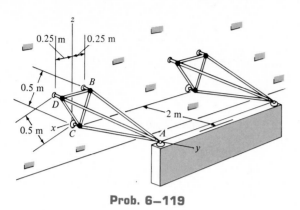

Prob. 6–119

*6–120. If a force of *P* = 6 lb is applied perpendicular to the handle of the mechanism, determine the magnitude of force **F** for equilibrium. The members are pin-connected at *A*, *B*, *C*, and *D*.

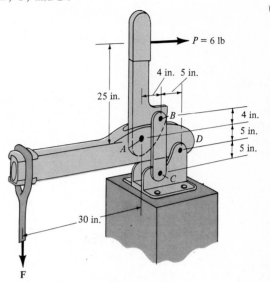

Prob. 6–120

Center of Gravity and Centroid

7

In this chapter we will extend the concepts developed in Sec. 4.8 and discuss the method used to determine the location of the center of gravity and center of mass for a system of discrete particles, and then we will expand its application to include a body of arbitrary shape. The same method of analysis will also be used to determine the geometric center, or centroid, of lines, areas, and volumes. Once the centroid has been located, we will then show how to obtain the area and volume of a surface of revolution and determine the resultants of distributed loadings.

Center of Gravity and Center of Mass for a System of Particles

7.1

Center of Gravity. Consider the system of n particles fixed within a region of space as shown in Fig. 7–1. The weights of the particles comprise a system of parallel forces which can be replaced by a single (equivalent) resultant weight and a defined point of application. This point is called the *center of gravity G*. To find its $\bar{x}, \bar{y}, \bar{z}$ coordinates, one must use the principles outlined in Sec. 4.8. This requires that the resultant weight be equal

to the total weight of all the particles; i.e., $W = \sum\limits_{i=1}^{n} W_i$. The sum of the

moments of the weights of all particles in the system is then equal to the moment of the resultant weight about the respective x, y, and z axes.*

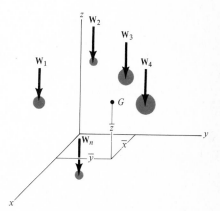

---

*If the weights of the particles are all parallel to the z axis, as shown in Fig. 7–1, then they will create no moment about this axis. However, to determine $\bar{z}$ it may be helpful to imagine the coordinate system and the particles fixed in it as being rotated $90°$ about either the x or y axis. This way one can "visualize" how the z-axis moments are developed.

Fig. 7–1

For example, if moments are summed about the y axis, we have

$$\overline{x}W = \widetilde{x}_1W_1 + \widetilde{x}_2W_2 + \cdots + \widetilde{x}_nW_n$$

where $\widetilde{x}_1$, etc. represent the x coordinate positions of the particles. This calculation and the two others like it can be written symbolically as

$$\overline{x} = \frac{\Sigma\widetilde{x}W}{\Sigma W} \qquad \overline{y} = \frac{\Sigma\widetilde{y}W}{\Sigma W} \qquad \overline{z} = \frac{\Sigma\widetilde{z}W}{\Sigma W} \qquad (7\text{--}1)$$

Here $\widetilde{x}, \widetilde{y}, \widetilde{z}$ represent the coordinate locations of each particle.

Center of Mass. To study problems concerning the motion of *matter* under the influence of force, i.e., dynamics, it is necessary to locate a point called the *center of mass*. Provided the acceleration of gravity g for every particle is constant, then $W = mg$. Substituting into Eqs. 7–1 and canceling g from both the numerator and denominator yields

$$\overline{x} = \frac{\Sigma\widetilde{x}m}{\Sigma m} \qquad \overline{y} = \frac{\Sigma\widetilde{y}m}{\Sigma m} \qquad \overline{z} = \frac{\Sigma\widetilde{z}m}{\Sigma m} \qquad (7\text{--}2)$$

By comparison then, the location of the center of gravity coincides with that of the center of mass.* Recall, however, that particles have "weight" only when under the influence of a gravitational attraction, whereas the center of mass is independent of gravity. For example, it would be meaningless to define the center of gravity of a system of particles representing the planets of our solar system, while the center of mass of this system is important.

7.2 Center of Gravity, Center of Mass, and Centroid for a Body

Center of Gravity. When the principles used to determine Eqs. 7–1 are applied to a system of particles composing a rigid body, one obtains the same form as these equations except each particle located at $(\widetilde{x}, \widetilde{y}, \widetilde{z})$ is thought to have a *differential weight dW*. Consequently, *integration* is required rather than a discrete summation of the terms. The resulting equations are

$$\overline{x} = \frac{\int \widetilde{x}\, dW}{\int dW} \qquad \overline{y} = \frac{\int \widetilde{y}\, dW}{\int dW} \qquad \overline{z} = \frac{\int \widetilde{z}\, dW}{\int dW} \qquad (7\text{--}3)$$

*This is not true in the exact sense since the weights are not parallel to each other, rather they are all *concurrent* at the earth's center. Furthermore, the acceleration of gravity g is actually different for each particle since it depends upon the distance from the earth's center to the particle. For all practical purposes, however, both of these effects can be neglected.

In order to properly use these equations, the differential weight dW must be expressed in terms of the associated volume dV. If γ represents the specific weight of the body, measured as a weight per unit volume, then $dW = \gamma\, dV$ and therefore

$$\bar{x} = \frac{\int \tilde{x}\gamma\, dV}{\int \gamma\, dV} \qquad \bar{y} = \frac{\int \tilde{y}\gamma\, dV}{\int \gamma\, dV} \qquad \bar{z} = \frac{\int \tilde{z}\gamma\, dV}{\int \gamma\, dV} \qquad (7\text{--}4)$$

Here integration must be performed throughout the entire volume of the body.

Center of Mass. The *density* ρ, or mass per unit volume, is related to γ by the equation $\gamma = \rho g$, where g is the acceleration of gravity. Substituting this relationship into Eqs. 7–4 and canceling g from both the numerators and denominators yields similar equations (with ρ replacing γ) that can be used to determine the body's *center of mass*.

Centroid. The *centroid* is a point which defines the *geometrical center* of an object. Its location can be determined from formulas similar to those used to determine the body's center of gravity or center of mass. For example, if the material composing a body is uniform or *homogeneous,* the *density or specific weight* will be *constant* throughout the body, and therefore this term will factor out of the integrals and *cancel* from both the numerators and denominators of Eqs. 7–4. The resulting formulas define the centroid of the body since they are independent of the body's weight and instead depend only on the body's volume.

Volume. Using a differential element dV, the coordinates for the centroid C of the volume of space occupied by an object, Fig. 7–2, are

$$\bar{x} = \frac{\int \tilde{x}\, dV}{\int dV} \qquad \bar{y} = \frac{\int \tilde{y}\, dV}{\int dV} \qquad \bar{z} = \frac{\int \tilde{z}\, dV}{\int dV} \qquad (7\text{--}5)$$

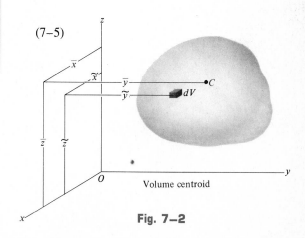

Volume centroid

Fig. 7–2

281

Area. In a similar manner, the centroid for the surface area of an object, such as a plate or shell, Fig. 7–3, can be found by subdividing the area into differential elements dA and computing the "moments" of these area elements about the coordinate axes. Doing this gives the coordinates for the centroid of the area, namely,

$$\bar{x} = \frac{\int \tilde{x}\, dA}{\int dA} \qquad \bar{y} = \frac{\int \tilde{y}\, dA}{\int dA} \qquad \bar{z} = \frac{\int \tilde{z}\, dA}{\int dA} \qquad (7\text{–}6)$$

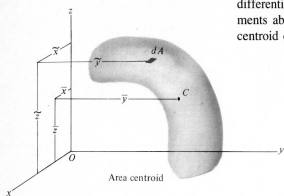

Area centroid

Fig. 7–3

Line. If the geometry of the object, such as a thin rod or wire, takes the form of a line, Fig. 7–4, the manner of finding its centroid is identical to the procedure outlined above. The results are

$$\bar{x} = \frac{\int \tilde{x}\, dL}{\int dL} \qquad \bar{y} = \frac{\int \tilde{y}\, dL}{\int dL} \qquad \bar{z} = \frac{\int \tilde{z}\, dL}{\int dL} \qquad (7\text{–}7)$$

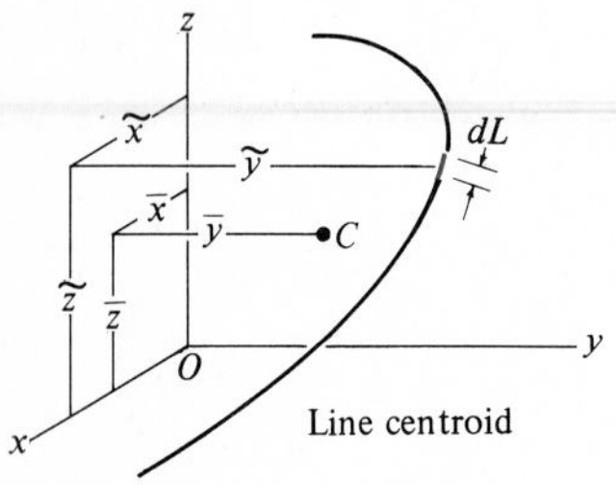

Notice that in all of the above cases the location of C does not necessarily have to be within the object; rather, it can be located off the object in space. Also, the *centroids* of some shapes may be partially or completely specified by using *symmetry conditions*. In cases where the shape has an axis of symmetry, the centroid of the shape will lie along that axis. For example, the centroid C for the line shown in Fig. 7–5 must lie along the y axis, since for

Line centroid

Fig. 7–4

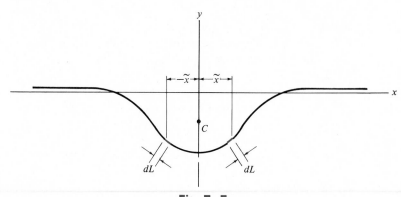

Fig. 7–5

every elemental length dL at a distance $+\widetilde{x}$ to the right of the y axis, there is an identical element at a distance $-\widetilde{x}$ to the left. The total moment for all the elements about the axis of symmetry will therefore cancel; i.e., $\int \widetilde{x}\,dL = 0$ (Eq. 7–7) so that $\bar{x} = 0$. In cases where a shape has two or three axes of symmetry, it follows that the centroid lies at the intersection of these axes, Fig. 7–6.

It is important to remember that the terms $\widetilde{x}, \widetilde{y}, \widetilde{z}$ in Eqs. 7–4 to 7–7 refer to the "moment arms" or perpendicular distances from the coordinate planes to the *center of gravity or centroid of the differential element* used in the equations. If possible, this differential element should be chosen such that it is of differential size or thickness in only *one direction*. When this is done only a single integration is required to cover the entire region.

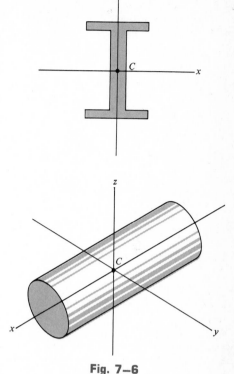

Fig. 7–6

PROCEDURE FOR ANALYSIS

The following procedure provides a method for determining the center of gravity or centroid of an object using a single integration.

Differential Element. Specify the coordinate axes and choose an appropriate differential element for integration. For lines this element dL is represented as a differential line segment; for areas the element dA is generally a rectangle, having a finite height and differential width; and for volumes the element dV is either a circular disk, having a finite radius and differential thickness, or a shell, having a finite length and radius and differential thickness. Locate the element so that it intersects the boundary of the line, area, or volume at an *arbitrary point* (x, y, z).

Size and Moment Arms. Express the length dL, area dA, or volume dV of the element in terms of the coordinates used to define the boundary of the object. Determine the perpendicular distances (moment arms) from the coordinate planes to the centroid or center of gravity of the element. For an x, y, z coordinate system, these dimensions are represented by the coordinates $\widetilde{x}, \widetilde{y}, \widetilde{z}$.

Integrations. Substitute the data computed above into the appropriate equations (Eqs. 7–4 to 7–7) and perform the integrations.* Note that integration can be accomplished when the function in the integrand is expressed in terms of the *same variable as the differential thickness of the element*. The limits of the integral are then defined from the two extreme locations of the element's differential thickness, so that when the elements are "summed" or the integration performed, the entire region is covered.

The following examples numerically illustrate this procedure.

*Formulas for integration are given in Appendix A.

Example 7–1

Locate the centroid of the thin rod bent into the shape of a quarter circle having a radius r as shown in Fig. 7–7.

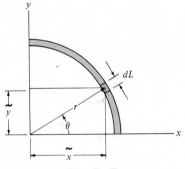

Fig. 7–7

Solution

Differential Element. The boundary of the rod can easily be defined using polar coordinates. The "arc" element intersects the curve at the arbitrary point (r, θ), Fig. 7–7.

Length and Moment Arms. The length of the element is $dL = r\, d\theta$, and the centroid is located at $\widetilde{x} = r \cos \theta$, $\widetilde{y} = r \sin \theta$.

Integrations. Using Eqs. 7–7 and integrating with respect to θ yields

$$\bar{x} = \frac{\displaystyle\int \widetilde{x}\, dL}{\displaystyle\int dL} = \frac{\displaystyle\int_0^{\pi/2} r \cos \theta\, (r\, d\theta)}{\displaystyle\int_0^{\pi/2} r\, d\theta} = \frac{r^2}{\dfrac{\pi}{2} r} = \frac{2r}{\pi} \qquad Ans.$$

$$\bar{y} = \frac{\displaystyle\int \widetilde{y}\, dL}{\displaystyle\int dL} = \frac{\displaystyle\int_0^{\pi/2} r \sin \theta\, (r\, d\theta)}{\displaystyle\int_0^{\pi/2} r\, d\theta} = \frac{r^2}{\dfrac{\pi}{2} r} = \frac{2r}{\pi} \qquad Ans.$$

Actually, by inspection only $\bar{x}$ has to be calculated, since $\bar{x} = \bar{y}$ due to symmetry.

Example 7–2

Locate the centroid of the triangular area shown in Fig. 7–8a.

Solution I

Differential Element. A rectangular differential element of thickness dx is shown in Fig. 7–8a. The element intersects the curve at the arbitrary point (x, y).

Area and Moment Arms. The area of the element is $dA = y\,dx$, and the centroid is located at $\widetilde{x} = x$, $\widetilde{y} = \tfrac{1}{2}y$.

Integrations. Using Eqs. 7–6 and integrating with respect to x yields

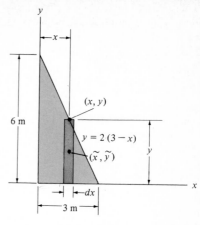

(a)

$$\bar{x} = \frac{\int_A \widetilde{x}\,dA}{\int_A dA} = \frac{\int_0^3 xy\,dx}{\int_0^3 y\,dx} = \frac{\int_0^3 x\,2(3-x)\,dx}{\int_0^3 2(3-x)\,dx} = \frac{\left.\tfrac{6}{2}x^2 - \tfrac{2}{3}x^3\right|_0^3}{\left.6x - \tfrac{2}{2}x^2\right|_0^3} = 1\ \text{m} \quad Ans.$$

$$\bar{y} = \frac{\int_A \widetilde{y}\,dA}{\int_A dA} = \frac{\int_0^3 (\tfrac{1}{2}y)\,y\,dx}{\int_0^3 y\,dx} = \frac{\int_0^3 \tfrac{4}{2}(3-x)^2\,dx}{\int_0^3 2(3-x)\,dx} = \frac{\left.2(9x - \tfrac{6}{2}x^2 + \tfrac{1}{3}x^3)\right|_0^3}{\left.6x - \tfrac{2}{2}x^2\right|_0^3} = 2\ \text{m} \quad Ans.$$

Solution II

Differential Element. A rectangular differential element of thickness dy is shown in Fig. 7–8b. As before, the element intersects the curve at the arbitrary point (x, y).

Area and Moment Arms. The area of the element is $dA = x\,dy$, and the centroid is located at $\widetilde{x} = \tfrac{1}{2}x$, $\widetilde{y} = y$.

Integrations. Using Eqs. 7–6, and integrating with respect to y yields

$$\bar{x} = \frac{\int_A \widetilde{x}\,dA}{\int_A dA} = \frac{\int_0^6 (\tfrac{1}{2}x)x\,dy}{\int_0^6 x\,dy} = \frac{\int_0^6 \tfrac{1}{2}(3 - \tfrac{1}{2}y)^2\,dy}{\int_0^6 (3 - \tfrac{1}{2}y)\,dy} = \frac{\left.\tfrac{1}{2}(9y - \tfrac{3}{2}y^2 + \tfrac{1}{12}y^3)\right|_0^6}{\left.3y - \tfrac{1}{4}y^2\right|_0^6} = 1\ \text{m} \quad Ans.$$

$$\bar{y} = \frac{\int_A \widetilde{y}\,dA}{\int_A dA} = \frac{\int_0^6 yx\,dy}{\int_0^6 x\,dy} = \frac{\int_0^6 y(3 - \tfrac{1}{2}y)\,dy}{\int_0^6 (3 - \tfrac{1}{2}y)\,dy} = \frac{\left.\tfrac{3}{2}y^2 - \tfrac{1}{6}y^3\right|_0^6}{\left.3y - \tfrac{1}{4}y^2\right|_0^6} = 2\ \text{m} \quad Ans.$$

(b)

Fig. 7–8

Example 7–3

Locate the centroid of the area shown in Fig. 7–9a.

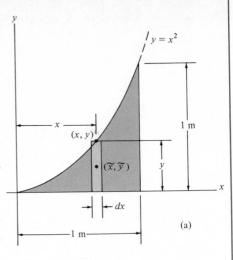

(a)

Solution I

Differential Element. A differential element of thickness dx is shown in Fig. 7–9a. The element intersects the curve at the arbitrary point (x, y).

Area and Moment Arms. The area of the element is $dA = y\, dx$, and the centroid is located at $\widetilde{x} = x$, $\widetilde{y} = y/2$.

Integrations. Using Eqs. 7–6 and integrating with respect to x yields

$$\bar{x} = \frac{\displaystyle\int_A \widetilde{x}\, dA}{\displaystyle\int_A dA} = \frac{\displaystyle\int_0^1 xy\, dx}{\displaystyle\int_0^1 y\, dx} = \frac{\displaystyle\int_0^1 x^3\, dx}{\displaystyle\int_0^1 x^2\, dx} = \frac{0.250}{0.333} = 0.75 \text{ m} \qquad Ans.$$

$$\bar{y} = \frac{\displaystyle\int_A \widetilde{y}\, dA}{\displaystyle\int_A dA} = \frac{\displaystyle\int_0^1 \left(\frac{y}{2}\right) y\, dx}{\displaystyle\int_0^1 y\, dx} = \frac{\displaystyle\int_0^1 \left(\frac{x^2}{2}\right) x^2\, dx}{\displaystyle\int_0^1 x^2\, dx} = \frac{0.100}{0.333} = 0.3 \text{ m} \quad Ans.$$

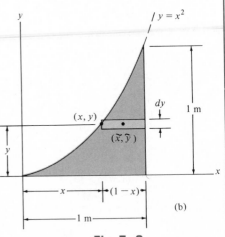

(b)

Fig. 7–9

Solution II

Differential Element. The differential element of thickness dy is shown in Fig. 7–9b. Again the element intersects the curve at the arbitrary point (x, y).

Area and Moment Arms. The area of the element is $dA = (1 - x)\, dy$, and the centroid is located at

$$\widetilde{x} = x + \left(\frac{1 - x}{2}\right) = \frac{1 + x}{2}, \qquad \widetilde{y} = y$$

Integrations. Using Eqs. 7–6, and integrating with respect to y, we obtain

$$\bar{x} = \frac{\displaystyle\int_A \widetilde{x}\, dA}{\displaystyle\int_A dA} = \frac{\displaystyle\int_0^1 \left(\frac{1 + x}{2}\right)(1 - x)\, dy}{\displaystyle\int_0^1 (1 - x)\, dy} = \frac{\frac{1}{2}\displaystyle\int_0^1 (1 - y)\, dy}{\displaystyle\int_0^1 (1 - \sqrt{y})\, dy} = \frac{0.250}{0.333} = 0.75 \text{ m} \quad Ans.$$

$$\bar{y} = \frac{\displaystyle\int_A \widetilde{y}\, dA}{\displaystyle\int_A dA} = \frac{\displaystyle\int_0^1 y(1 - x)\, dy}{\displaystyle\int_0^1 (1 - x)\, dy} = \frac{\displaystyle\int_0^1 (y - y^{3/2})\, dy}{\displaystyle\int_0^1 (1 - \sqrt{y})\, dy} = \frac{0.100}{0.333} = 0.3 \text{ m} \quad Ans.$$

Example 7–4

Locate the $\bar{x}$ centroid of the shaded area bounded by the two curves $y = x$ and $y = x^2$, Fig. 7–10.

Solution I

Differential Element. A differential element of thickness dx is shown in Fig. 7–10a. The element intersects the curves at *arbitrary* points (x, y_1) and (x, y_2).

Area and Moment Arm. The area of the element is $dA = (y_2 - y_1)\,dx$, and the centroid is located at $\tilde{x} = x$.

Integrations. Using Eq. 7–6, we have

$$\bar{x} = \frac{\displaystyle\int_A x^2\,dA}{\displaystyle\int_A dA} = \frac{\displaystyle\int_0^1 x(y_2 - y_1)\,dx}{\displaystyle\int_0^1 (y_2 - y_1)\,dx} = \frac{\displaystyle\int_0^1 x(x - x^2)\,dx}{\displaystyle\int_0^1 (x - x^2)\,dx} = \frac{\frac{1}{12}}{\frac{1}{6}} = 0.5 \text{ ft.} \quad Ans.$$

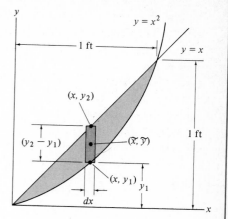

(a)

Solution II

Differential Element. A differential element having a thickness dy and intersecting the curves at the arbitrary points (x_2, y) and (x_1, y) is shown in Fig. 7–10b.

Area and Moment Arm. The area of the element is $dA = (x_1 - x_2)\,dy$, and the centroid is located at

$$\tilde{x} = x_2 + \frac{x_1 - x_2}{2} = \frac{x_1 + x_2}{2}$$

Integrations. Using Eq. 7–6, we have

$$\bar{x} = \frac{\displaystyle\int_A \tilde{x}\,dA}{\displaystyle\int_A dA} = \frac{\displaystyle\int_0^1 \left(\frac{x_1 + x_2}{2}\right)(x_1 - x_2)\,dy}{\displaystyle\int_0^1 (x_1 - x_2)\,dy} = \frac{\displaystyle\int_0^1 \left(\frac{\sqrt{y} + y}{2}\right)(\sqrt{y} - y)\,dy}{\displaystyle\int_0^1 (\sqrt{y} - y)\,dy}$$

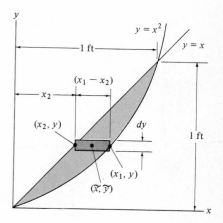

(b)

Fig. 7–10

$$= \frac{\dfrac{1}{2}\displaystyle\int_0^1 (y - y^2)\,dy}{\displaystyle\int_0^1 (\sqrt{y} - y)\,dy} = \frac{\frac{1}{12}}{\frac{1}{6}} = 0.5 \text{ ft} \qquad\qquad Ans.$$

287

Example 7–5

Locate the $\bar{y}$ centroid for the paraboloid of revolution, which is generated by revolving the shaded area shown in Fig. 7–11a about the y axis.

Solution I

Differential Element. An element having the shape of a *thin disk* is chosen, Fig. 7–11a. This element has a radius of $r = z$ and a thickness of dy. In this "disk" method of analysis, the element of planar area, dA, is always taken *perpendicular* to the axis of revolution. Here the element intersects the generating curve at the arbitrary point $(0, y, z)$.

Volume and Moment Arm. The volume of the element is $dV = (\pi z^2)\,dy$, and the centroid is located at $\tilde{y} = y$.

Integrations. Using the second of Eqs. 7–5 and integrating with respect to y yields

$$\bar{y} = \frac{\displaystyle\int_V \tilde{y}\,dV}{\displaystyle\int_V dV} = \frac{\displaystyle\int_0^{100} y(\pi z^2)\,dy}{\displaystyle\int_0^{100} (\pi z^2)\,dy} = \frac{100\pi \displaystyle\int_0^{100} y^2\,dy}{100\pi \displaystyle\int_0^{100} y\,dy} = 66.7 \text{ mm} \quad Ans.$$

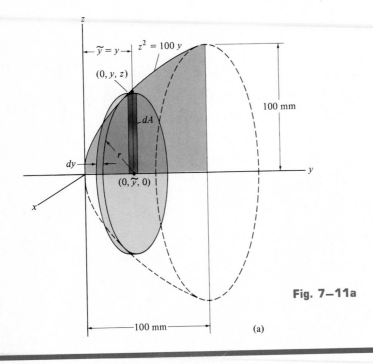

Fig. 7–11a

(a)

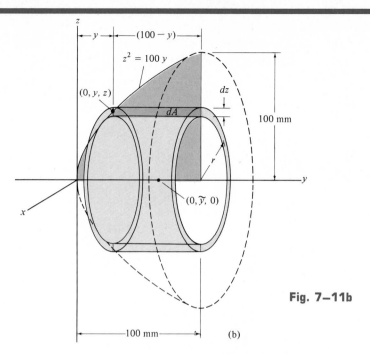

Fig. 7–11b

(b)

Solution II

Differential Element. As shown in Fig. 7–11*b*, the volume element can be chosen in the form of a *thin cylindrical shell,* where the shell radius is $r = z$ and the thickness is dz. In this "shell" method of analysis, the element of planar area, dA, is always taken *parallel* to the axis of revolution. Here the element intersects the generating curve at point $(0, y, z)$.

Volume and Moment Arm. The volume of the element is $dV = 2\pi r\, dA = 2\pi z(100 - y)\, dz$, and the centroid is located at $\widetilde{y} = y + (100 - y)/2 = (100 + y)/2$.

Integrations. Using the second of Eqs. 7–5 and integrating with respect to z yields

$$\overline{y} = \frac{\int_V \widetilde{y}\, dV}{\int_V dV} = \frac{\int_0^{100} \left(\dfrac{100 + y}{2}\right) 2\pi z(100 - y)\, dz}{\int_0^{100} 2\pi z(100 - y)\, dz}$$

$$= \frac{\pi \int_0^{100} z(10^4 - 10^{-4}z^4)\, dz}{2\pi \int_0^{100} z(100 - 10^{-2}z^2)\, dz} = 66.7 \text{ mm} \qquad Ans.$$

289

Example 7–6

Determine the location of the center of mass of the right circular cylinder shown in Fig. 7–12a. The density of the cylinder varies directly with its distance from the base, such that $\rho = 200z \ \mathrm{kg/m^3}$.

Solution

For reasons of material symmetry

$$\bar{x} = \bar{y} = 0 \qquad Ans.$$

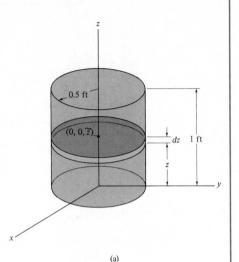

(a)

Differential Element. A disk element of thickness dz is chosen for integration, Fig. 7–12a, since the *density of the entire element is constant* for a given value of z. The element is located along the z axis at the arbitrary point $(0, 0, z)$.

Volume and Moment Arm. The volume of the element is $dV = \pi(0.5)^2 \, dz$. The centroid is located at $\tilde{z} = z$.

Integrations. Using an equation similar to the third of Eqs. 7–4 and integrating with respect to z, noting that $\rho = 200z$, we have

$$\bar{z} = \frac{\displaystyle\int_V \tilde{z}\rho \, dV}{\displaystyle\int_V \rho \, dV} = \frac{\displaystyle\int_0^1 z(200z)\pi(0.5)^2 \, dz}{\displaystyle\int_0^1 (200z)\pi(0.5)^2 \, dz}$$

$$= \frac{\displaystyle\int_0^1 z^2 \, dz}{\displaystyle\int_0^1 z \, dz} = 0.667 \ \mathrm{m} \qquad Ans.$$

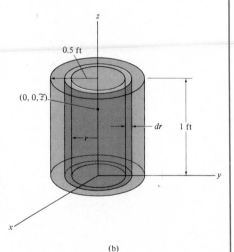

(b)

Fig. 7–12

Note: It is not possible to use a shell element for integration such as shown in Fig. 7–12b, since the density of the material composing the shell would *vary* along the shell's height, and hence the location of $\tilde{z}$ for the element would not be easily determined.

Problems

7–1. Locate the centroid $\bar{x}$ of the circular rod. Express the answer in terms of the radius r and semiarc angle α.

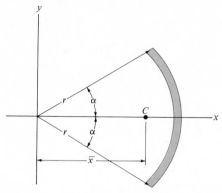

Prob. 7–1

7–2. Determine the distance $\bar{x}$ to the center of mass of the homogeneous rod bent into the shape shown. If the rod has a mass per unit length of 0.5 kg/m, determine the reactions at the fixed support O.

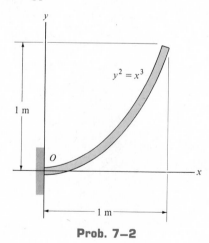

Prob. 7–2

7–3. Locate the center of gravity $\bar{x}$ of the homogeneous rod bent in the form of a parabola. If the rod has a weight per unit length of 0.5 lb/ft, determine the vertical reaction at A and the x and y components of reaction at B.

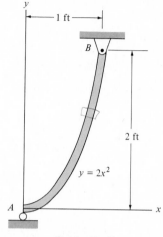

Prob. 7–3

■*7–4. Locate the center of gravity $\bar{x}$, $\bar{y}$ of the homogeneous rod bent into the shape of the arch shown. What is its weight if it has a mass of 3 kg/m? Solve the problem by evaluating the integrals using Simpson's rule.

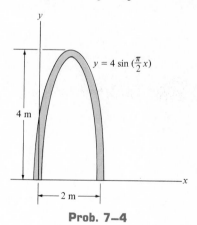

Prob. 7–4

291

7–5. Locate the centroid of the shaded area.

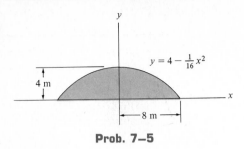

$y = 4 - \frac{1}{16}x^2$

4 m

8 m

Prob. 7–5

7–6. Locate the centroid of the shaded area.

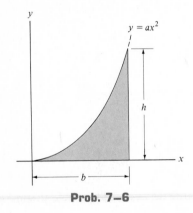

$y = ax^2$

h

b

Prob. 7–6

7–7. Locate the centroid of the parabolic area.

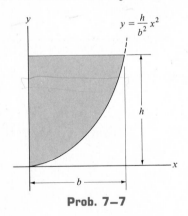

$y = \frac{h}{b^2}x^2$

h

b

Prob. 7–7

*7–8. The plate has a thickness of 0.25 ft and a specific weight of $\gamma = 180$ lb/ft³. Determine the location of its center of gravity. Also, compute the tension in each of the cords used to support it.

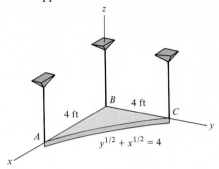

z

B 4 ft

4 ft

C

A

$y^{1/2} + x^{1/2} = 4$

y

x

Prob. 7–8

7–9. Locate the centroid $\bar{x}$ of the shaded area.

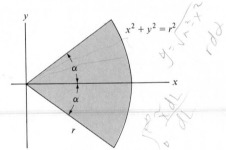

$x^2 + y^2 = r^2$

α

α

r

Prob. 7–9

7–10. Locate the centroid of the semicircular area.

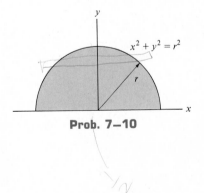

$x^2 + y^2 = r^2$

r

Prob. 7–10

292

7–11. The steel plate is 0.3 m thick and has a density of 7850 kg/m³. Determine the location of its center of gravity. Also compute the reactions at the pin and roller support.

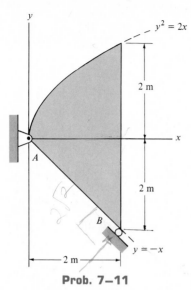

Prob. 7–11

*7–12. Locate the centroid of the shaded area. *Hint:* Choose elements of thickness dy and length $[(2 - y) - y^2]$.

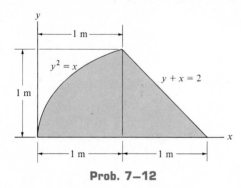

Prob. 7–12

■7–13. Locate the centroid $\bar{x}$ of the shaded area. Solve the problem by evaluating the integrals using Simpson's rule.

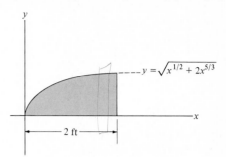

Probs. 7–13 / 7–14

■7–14. Locate the centroid $\bar{y}$ of the shaded area. Solve the problem by evaluating the integrals using Simpson's rule.

■7–15. Locate the centroid $\bar{x}$ of the shaded area. Solve the problem by evaluating the integrals using Simpson's rule.

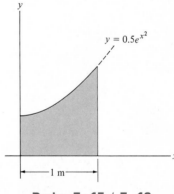

Probs. 7–15 / 7–16

■*7–16. Locate the centroid $\bar{y}$ of the shaded area. Solve the problem by evaluating the integrals using Simpson's rule.

7–17. Locate the centroid $\bar{z}$ of the thin conical shell. *Suggestion:* Use ring elements having a center at $(0, 0, z)$, radius y, and thickness $dL = \sqrt{(dy)^2 + (dz)^2}$.

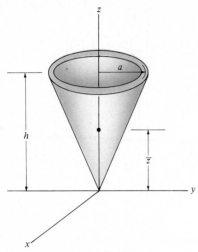

Prob. 7–17

7–18. Locate the center of gravity of the thin homogeneous hemispherical shell. *Suggestion:* Choose a ring element having a center at $(x, 0, 0)$, radius z, and thickness $dL = \sqrt{(dx)^2 + (dz)^2}$.

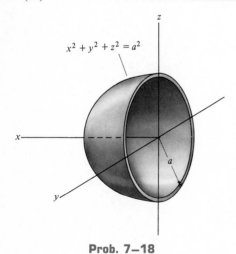

Prob. 7–18

7–19. Locate the centroid of the bell-shaped volume formed by revolving the shaded area about the y axis.

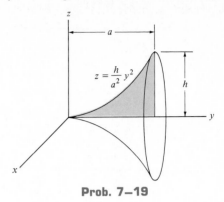

Prob. 7–19

*7–20. Locate the centroid of the cone formed by rotating the shaded area about the z axis.

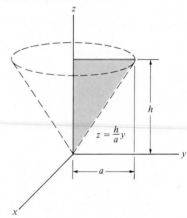

Prob. 7–20

7–21. Determine the distance $\bar{z}$ to the centroid of the spherical segment.

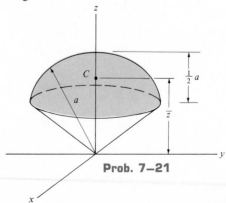

Prob. 7–21

7-22. Locate the center of gravity of the wood object generated by rotating the shaded area about the x axis. The density of wood is 560 kg/m³. Also, determine the reactions at the supports. There is a ball-and-socket joint at A, and there are rollers at B and C.

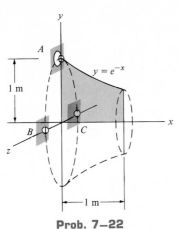

$y = e^{-x}$

1 m

1 m

Prob. 7-22

7-23. Locate the center of gravity of the hemisphere. The density of the material varies linearly from zero at the origin O to ρ_O at the surface. *Suggestion:* Choose a hemispherical shell element for integration and use the result of Prob. 7-18.

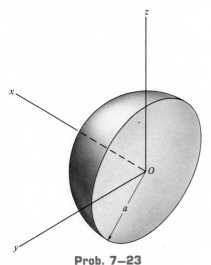

Prob. 7-23

***7-24.** Determine the location $\bar{z}$ of the centroid for the tetrahedron. *Suggestion:* Choose a triangular "plate" element parallel to the x-y plane and of thickness dz.

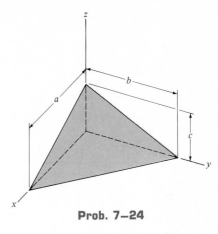

Prob. 7-24

7.3 Composite Bodies

In many cases, a body can be sectioned or divided into several parts having simpler shapes. Provided the weight and location of the center of gravity of each of these "composite parts" are known, one can eliminate the need for integration to determine the center of gravity for the entire body. Following the procedure outlined in Sec. 7.1 yields formulas analogous to Eqs. 7–1, since there is a finite number of weights to be accounted for. Thus, the necessary formulas for finding the center of gravity for a three-dimensional body become

$$\bar{x} = \frac{\Sigma \tilde{x} W}{\Sigma W} \qquad \bar{y} = \frac{\Sigma \tilde{y} W}{\Sigma W} \qquad \bar{z} = \frac{\Sigma \tilde{z} W}{\Sigma W} \qquad (7\text{–}8)$$

where $\tilde{x}$, $\tilde{y}$, and $\tilde{z}$ represent the *algebraic distances* from the center of gravity of each composite part to the origin of the coordinates, and ΣW represents the sum of the weights of the composite parts or simply the *total weight* of the body.

When the body has a *constant density,* the center of gravity *coincides* with the centroid of the body. The centroid for composite lines, areas, and volumes can be found using relations analogous to Eqs. 7–8; however, the W's are replaced by L's, A's, and V's, respectively. Centroids for common shapes of lines, areas, shells, and volumes are given in the table on the inside back cover of this book.

PROCEDURE FOR ANALYSIS

The following procedure provides a method for determining the center of gravity of a body or the centroid of a composite geometrical object represented by a line, area, or volume.

Composite Parts. Using a sketch, divide the body or object into a finite number of composite parts. If a *hole,* or geometric region having no material, is located within a composite part, the hole is considered as an *additional* composite part having *negative* weight or size.

Moment Arms. Establish the coordinate axes on the sketch and determine the location $\tilde{x}$, $\tilde{y}$, $\tilde{z}$ of the center of gravity or centroid of each part.

Summations. Determine $\bar{x}$, $\bar{y}$, $\bar{z}$ by applying the center of gravity equations, Eqs. 7–8, or the analogous centroid equations. If an object is *symmetrical* about an axis, recall that the centroid of the object lies on the axis.

If desired, the calculations can be arranged in tabular form, as indicated in the following three examples.

Example 7-7

Locate the centroid of the wire shown in Fig. 7-13a.

Solution

Composite Parts. The wire is divided into three segments as shown in Fig. 7-13b.

Moment Arms. The location of the centroid for each piece is determined and indicated on the sketch. In particular, the centroid of segment ① is determined either by integration or using the table on the inside back cover.

Summations. The calculations are tabulated as follows:

Segment	L (mm)	$\tilde{x}$ (mm)	$\tilde{y}$ (mm)	$\tilde{z}$ (mm)	$\tilde{x}L$ (mm²)	$\tilde{y}L$ (mm²)	$\tilde{z}L$ (mm²)
1	$\pi(60) = 188.5$	60	−38.2	0	11 310	−7200	0
2	40	0	20	0	0	800	0
3	20	0	40	−10	0	800	−200
	$\Sigma L = 248.5$				$\Sigma \tilde{x}L = 11\,310$	$\Sigma \tilde{y}L = -5600$	$\Sigma \tilde{z}L = -200$

Thus,

$$\bar{x} = \frac{\Sigma \tilde{x}L}{\Sigma L} = \frac{11\,310}{248.5} = 45.5 \text{ mm} \qquad Ans.$$

$$\bar{y} = \frac{\Sigma \tilde{y}L}{\Sigma L} = \frac{-5600}{248.5} = -22.5 \text{ mm} \qquad Ans.$$

$$\bar{z} = \frac{\Sigma \tilde{z}L}{\Sigma L} = \frac{-200}{248.5} = -0.8 \text{ mm} \qquad Ans.$$

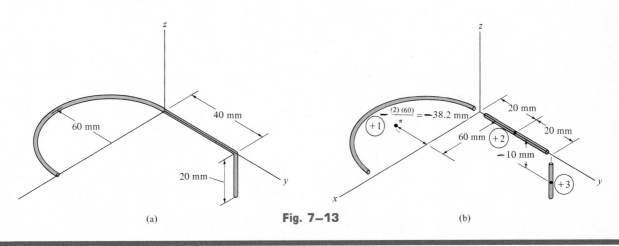

Fig. 7-13

(a) (b)

297

Example 7–8

Locate the centroid of the plate area shown in Fig. 7–14*a*.

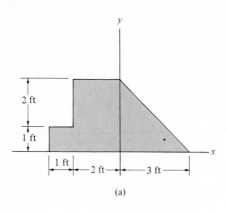

(a)

Fig. 7–14

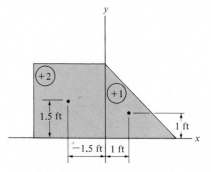

Solution

Composite Parts. The plate is divided into three segments as shown in Fig. 7–14*b*. Here the area of the small rectangle ③ is considered "negative" since it must be subtracted from the larger one ②.

Moment Arms. The centroid of each segment is located as indicated in the figure. Note that the $\widetilde{x}$ coordinates of ② and ③ are *negative*.

Summations. Taking the data from Fig. 7–14*b*, the calculations are tabulated as follows:

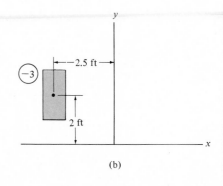

(b)

Segment	A (ft²)	$\widetilde{x}$ (ft)	$\widetilde{y}$ (ft)	$\widetilde{x}A$ (ft³)	$\widetilde{y}A$ (ft³)
1	$\frac{1}{2}(3)(3) = 4.5$	1	1	4.5	4.5
2	$(3)(3) = 9$	-1.5	1.5	-13.5	13.5
3	$-(2)(1) = -2$	-2.5	2	5	-4
	$\Sigma A = 11.5$			$\Sigma \widetilde{x}A = -4$	$\Sigma \widetilde{y}A = 14$

Thus,

$$\bar{x} = \frac{\Sigma \widetilde{x}A}{\Sigma A} = \frac{-4}{11.5} = -0.348 \text{ ft} \qquad Ans.$$

$$\bar{y} = \frac{\Sigma \widetilde{y}A}{\Sigma A} = \frac{14}{11.5} = 1.22 \text{ ft} \qquad Ans.$$

Example 7–9

Locate the center of mass of the composite assembly shown in Fig. 7–15a. The conical frustum has a density of $\rho_c = 8$ Mg/m³ and the hemisphere has a density of $\rho_h = 4$ Mg/m³.

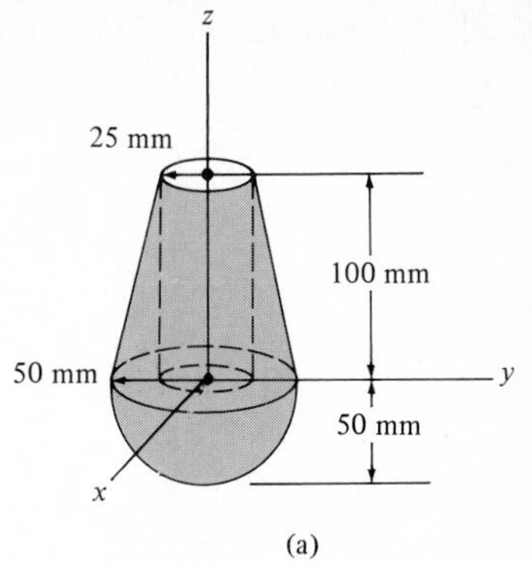

Solution

Composite Parts. The assembly can be thought of as consisting of four segments as shown in Fig. 7–15b. For the calculations, segments 3 and 4 must be considered as "negative" volumes in order that the four pieces, when added together, yield the total composite shape shown in Fig. 7–15a.

Moment Arm. Using the table on the inside back cover, the computations for the centroid $\tilde{z}$ of each piece are shown in the figure.

Summations. Because of *symmetry*, note that

$$\bar{x} = \bar{y} = 0 \qquad \textit{Ans.}$$

Since $W = mg$ and g is constant, the third of Eqs. 7–8 becomes $\bar{z} = \Sigma \tilde{z}m/\Sigma m$. The mass of each piece can be computed from $m = \rho V$ and used for the calculations. Also, 1 Mg/m³ $= 10^{-6}$ kg/mm³, so that

Segment	m (kg)	$\tilde{z}$ (mm)	$\tilde{z}m$ (kg · mm)
1	$8(10^{-6})(\frac{1}{3})\pi(50)^2(200) = 4.189$	50	209.440
2	$4(10^{-6})(\frac{2}{3})\pi(50)^3 = 1.047$	-18.75	-19.635
3	$-8(10^{-6})(\frac{1}{3})\pi(25)^2(100) = -0.524$	$100 + 25 = 125$	-65.450
4	$-8(10^{-6})\pi(25)^2(100) = -1.571$	50	-78.540
	$\Sigma m = 3.141$		$\Sigma \tilde{z}m = 45.815$

Thus,

$$\bar{z} = \frac{\Sigma \tilde{z}m}{\Sigma m} = \frac{45.815}{3.141} = 14.6 \text{ mm} \qquad \textit{Ans.}$$

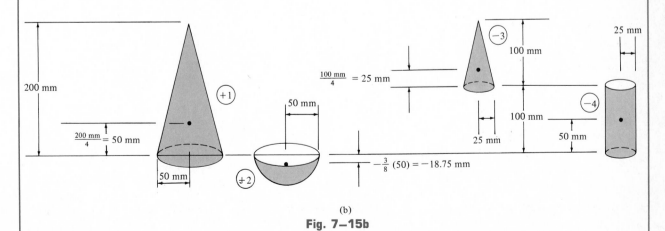

(b)

Fig. 7–15b

(Fig. 7–15a panel, right side)

25 mm
100 mm
50 mm
50 mm
(a)

Fig. 7–15a

Problems

7–25. The "roll-formed" member has a cross section as shown. Determine the location $\bar{y}$ of the centroid C. Neglect the thickness of the material and any slight bends at the corners.

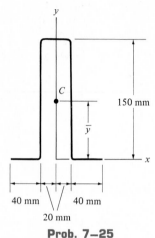

Prob. 7–25

7–26. The steel and aluminum plate assembly is bolted together and fastened to the wall. Each plate has a width of 200 mm and thickness of 20 mm. If the mass density of A and B is $\rho_s = 7.85$ Mg/m^3, and for C, $\rho_{al} = 2.71$ Mg/m^3, determine the location $\bar{x}$ of the center of mass. Neglect the size of the bolts.

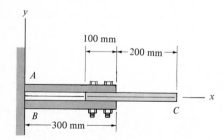

Prob. 7–26

7–27. Determine the location $(\bar{x}, \bar{y})$ of the center of gravity of the homogeneous wire bent in the form of a triangle. Neglect any slight bends at the corners. If the wire is lifted using a thread T attached to it at C, determine the angle of tilt AB makes with the horizontal when the wire is in equilibrium.

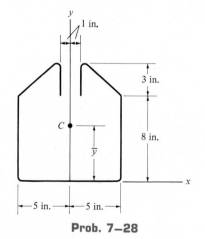

Prob. 7–27

*7–28. Determine the location $\bar{y}$ of the centroid for the cold-formed metal strut having the cross section shown. Neglect the thickness of the material and slight bends at the corners.

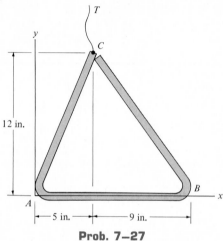

Prob. 7–28

7–29. The three members of the frame each have a weight per unit length of 4 lb/ft. Locate the position $(\bar{x}, \bar{y})$ of the center of gravity. Neglect the size of the pins at the joints and the thickness of the members. Also, calculate the reactions at the fixed support A.

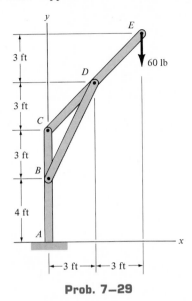

Prob. 7–29

7–30. The truss is made from five members, each having a length of 4 m and a mass of 7 kg/m. If the mass of the gusset plates at the joints and the thickness of the members can be neglected, determine the distance d to where the hoisting cable must be attached, so that the truss does not tip (rotate) when it is lifted.

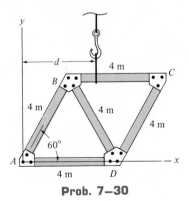

Prob. 7–30

7–31. The lamp assembly consists of two 2-ft-long uniform conduit pipes each having a weight of 1.6 lb, the uniform ballast B having a weight of 28 lb, and the light fixture having a weight of 33 lb and a center of gravity at G_l. Determine the location $\bar{y}$ of the center of gravity G of the entire assembly, measured from the ceiling. Also compute the tensions in the two supporting chains CE and DF.

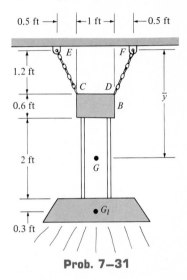

Prob. 7–31

*7–32. Determine the location $(\bar{x}, \bar{y})$ of the center of gravity of the three-wheeler. The location of the center of gravity of each component and its weight are tabulated in the figure. If the three-wheeler is symmetrical with respect to the x-y plane, determine the normal reactions it exerts on the ground.

1. Rear wheels	18 lb
2. Mechanical components	85 lb
3. Frame	120 lb
4. Front wheel	8 lb

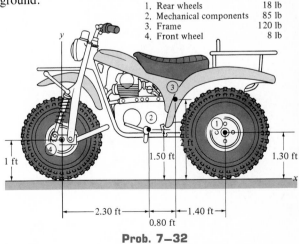

Prob. 7–32

301

7–33. Determine the location $(\bar{x}, \bar{y})$ of the center of gravity of the helicopter. The locations of the centers of gravity of the various components and their weights are tabulated in the figure.

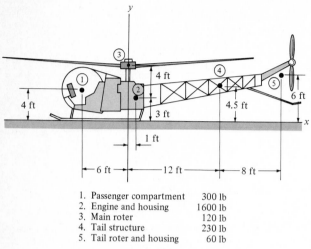

1. Passenger compartment	300 lb
2. Engine and housing	1600 lb
3. Main roter	120 lb
4. Tail structure	230 lb
5. Tail roter and housing	60 lb

Prob. 7–33

7–34. Determine the location $(\bar{x}, \bar{y})$ of the centroid C of the cross-sectional area for the structural member constructed from two equal-sized channels welded together as shown. Assume all corners are square. Neglect the size of the welds.

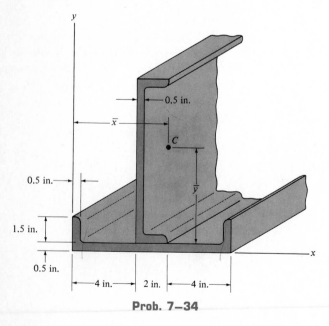

Prob. 7–34

7–35. Determine the weight and location $(\bar{x}, \bar{y})$ of the center of gravity G of the concrete retaining wall. The wall has a length of 10 ft, and concrete has a specific gravity of $\gamma = 150$ lb/ft^3.

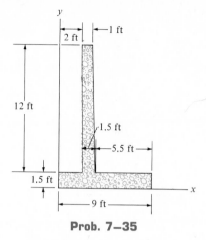

Prob. 7–35

*7–36. Determine the location $\bar{y}$ of the centroid of the beam's cross section built up from a channel and a wide-flange beam.

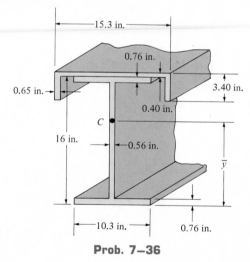

Prob. 7–36

7–37. Determine the location $\bar{y}$ of the centroid C for a beam having the cross-sectional area shown. The beam is symmetric with respect to the y axis.

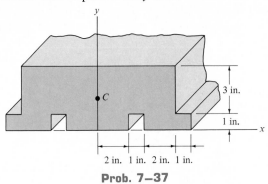

Prob. 7–37

7–38. Determine the distance $\bar{y}$ to the centroid of the trapezoidal area in terms of the dimensions shown.

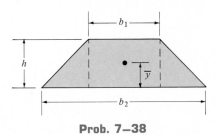

Prob. 7–38

7–39. Locate the centroid $C(\bar{x}, \bar{y})$ of the channel's cross-sectional area.

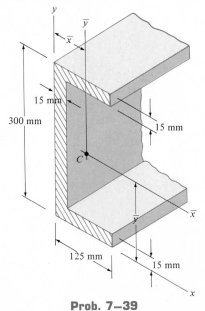

Prob. 7–39

*7–40.** Locate the centroid $\bar{y}$ of the concrete beam having the tapered cross section shown.

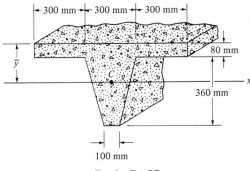

Prob. 7–40

7–41. The arched structure is 3 ft thick and is supported at C by a pin and bears against the smooth wall at B. If the specific weight of the material is $\gamma = 90 \text{ lb/ft}^3$, determine the location $(\bar{x}, \bar{y})$ of its center of gravity and the reactions at its supports.

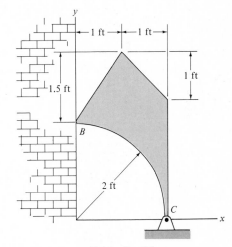

Prob. 7–41

303

7–42. The strut was formed in a cold-rolling mill. Determine the location $\bar{y}$ of the centroid of its cross-sectional area.

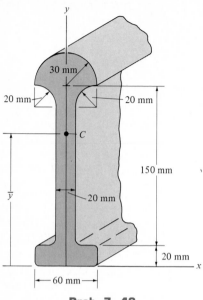

Prob. 7–42

7–43. Use the result of Prob. 7–9 and determine the distance $\bar{x}$ to the centroid C of the shaded area which is part of a circle having a radius of r.

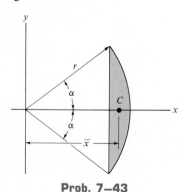

Prob. 7–43

*7–44. Locate the center of gravity $(\bar{x}, \bar{y}, \bar{z})$ of the sheet-metal bracket if the material is homogeneous and has a constant thickness. If the bracket is resting on the horizontal x-y plane as shown, determine the maximum angle of tilt θ which it can have before it falls over, i.e., begins to rotate about the y axis.

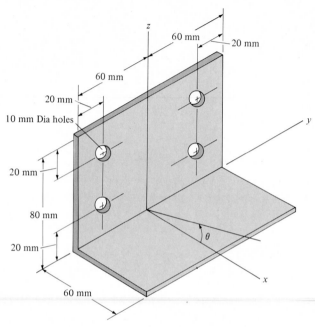

Prob. 7–44

7–45. A wooden buoy is made from two cones having a common radius of 1.5 ft. Locate the centroid $\bar{z}$.

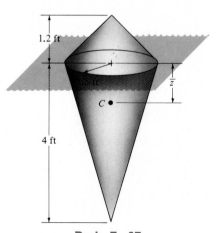

Prob. 7–45

7–46. Determine the location $\bar{z}$ of the center of mass. The material has a mass density of $\rho = 3$ Mg/m³. There is a hole bored through the center.

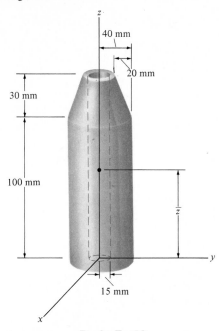

Prob. 7–46

7–47. The wooden table is made from a square board having a weight of 15 lb. Each of the legs weighs 2 lb and is 3 ft long. Determine how high its center of gravity is from the floor. Also, what is the greatest angle measured from the horizontal, through which its top surface can be tilted on two of its legs before it begins to overturn? Neglect the thickness of each leg.

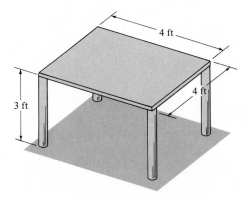

Prob. 7–47

*7–48. The container for a water tank is made from a cylindrical and hemispherical shell. Determine the position $\bar{z}$ of the center of gravity G of the water when the tank is full, i.e., when the water level is at the top of the cylinder, $z = 0$.

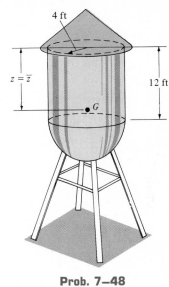

Prob. 7–48

7–49. Determine the distance h to which a hole must be bored into the cylinder so that the center of mass G of the assembly is located at $\bar{x} = 64$ mm. The material has a density of 8 Mg/m³.

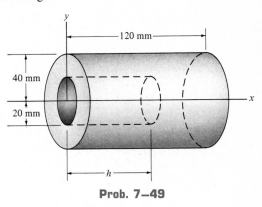

Prob. 7–49

7–50. Each of the three homogeneous plates welded to the rod has a density of $6 \ Mg/m^3$ and a thickness of 10 mm. Determine the length l of plate C and the angle of placement, θ, so that the center of mass of the assembly lies on the y axis. Plates A and B lie in the x-y and z-y planes, respectively.

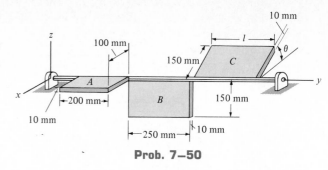

Prob. 7–50

★7.4 Theorems of Pappus and Guldinus

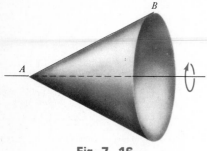

Fig. 7–16

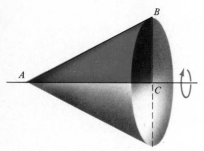

Fig. 7–17

The two *theorems of Pappus and Guldinus,* which were first developed by Pappus of Alexandria during the third century A.D. and then restated at a later time by the Swiss mathematician Paul Guldin or Guldinus (1577–1643), are used to find the surface area and volume of any object of revolution.

A *surface area of revolution* is generated by revolving a *plane curve* about a nonintersecting fixed axis in the plane of the curve; whereas a *volume of revolution* is generated by revolving a *plane area* about a nonintersecting fixed axis in the plane of the area. For example, if the *line AB* shown in Fig. 7–16 is rotated about a fixed axis, it generates the *surface area* of a cone; if the *triangular area ABC* shown in Fig. 7–17 is rotated about the axis, it generates the *volume* of a cone.

The statements and proofs of the theorems of Pappus and Guldinus follow. The proofs require that the generating curves and areas do *not* cross the axis about which they are rotated; otherwise, two sections on either side of the axis would generate areas or volumes having opposite signs and hence cancel each other.

Surface Area. *The area of a surface of revolution equals the product of the length of the generating curve and the distance traveled by the centroid of the curve in generating the surface area.*

Proof. When a differential length dL of the curve shown in Fig. 7–18 is revolved about an axis through a distance $2\pi r$, it generates a ring having a surface area $dA = 2\pi r \ dL$. The entire surface area, generated by revolving the entire curve about the axis, is therefore $A = 2\pi \int_L r \ dL$. Since $\int_L r \ dL = \bar{r}L$,

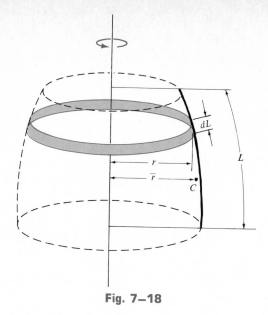

Fig. 7–18

where $\bar{r}$ locates the centroid C of the generating curve L, the area becomes $A = 2\pi\bar{r}L$. In general, though,

$$A = \theta\bar{r}L \qquad (7\text{--}9)$$

where A = surface area of revolution
  θ = angle of revolution measured in radians, $\theta \leqslant 2\pi$
  $\bar{r}$ = perpendicular distance from the axis of revolution to the centroid of the generating curve
  L = length of the generating curve

Volume. *The volume of a surface of revolution equals the product of the generating area and the distance traveled by the centroid of the area in generating the volume.*

  Proof. When the differential area dA shown in Fig. 7–19 is revolved about an axis through a distance $2\pi r$, it generates a ring having a volume $dV = 2\pi r\, dA$. The entire volume, generated by revolving A about the axis, is therefore $V = 2\pi \int_A r\, dA$. Since $\int_A r\, dA = \bar{r}A$, where $\bar{r}$ locates the centroid C of the generating area A, the volume becomes $V = 2\pi\bar{r}A$. In general, though,

$$V = \theta\bar{r}A \qquad (7\text{--}10)$$

where V = volume of revolution
  θ = angle of revolution measured in radians, $\theta \leqslant 2\pi$
  $\bar{r}$ = perpendicular distance from the axis of revolution to the centroid of the generating area
  A = generating area

The following examples illustrate application of the above two theorems.

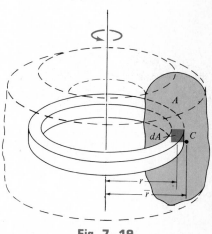

Fig. 7–19

307

Example 7–10

Show that the surface area of a sphere is $A = 4\pi R^2$ and its volume is $V = \frac{4}{3}\pi R^3$.

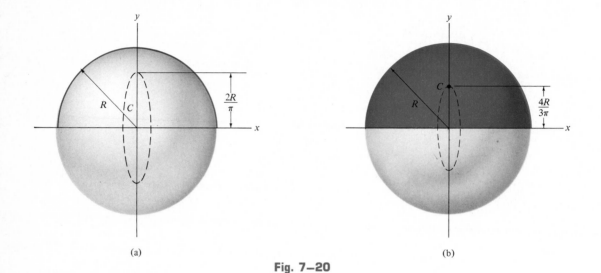

(a)   (b)

Fig. 7–20

Solution

The sphere in Fig. 7–20a is generated by rotating a semicircular *arc* about the x axis. Using the table on the inside back cover, it is seen that the centroid of this arc is located at a distance $\bar{y} = 2R/\pi$ from the x axis of rotation. Since the centroid moves through an angle of $\theta = 2\pi$ rad in generating the sphere, applying Eq. 7–9 gives the surface area of the sphere as:

$$A = \theta\bar{y}L; \qquad A = 2\pi\left(\frac{2R}{\pi}\right)\pi R = 4\pi R^2 \qquad \textit{Ans.}$$

In a similar manner, the volume of the sphere is generated by rotating the semicircular *area* in Fig. 7–20b about the x axis. Using the table on the inside back cover to locate the centroid, and applying Eq. 7–10, we have

$$V = \theta\bar{r}A; \qquad V = 2\pi\left(\frac{4R}{3\pi}\right)\left(\frac{1}{2}\pi R^2\right) = \frac{4}{3}\pi R^3 \qquad \textit{Ans.}$$

Example 7–11

Determine the mass of concrete needed to construct the arched beam shown in Fig. 7–21. The density of concrete is $\rho_c = 2.1 \text{ Mg/m}^3$. Also, determine the surface area of the beam.

Solution

The mass of the arch may be determined by using the density ρ_c provided the volume of concrete is known. The cross-sectional area of the beam is composed of two rectangles having centroids at points C_1 and C_2 as shown in Fig. 7–21b. The volume generated by each of these areas is equal to the product of the cross-sectional area and the distance traveled by its centroid. In generating the arch, the centroids move through an angle of $\theta = \pi(90°/180°) = 1.57$ radians. Point C_1 acts at a distance of $d_1 = 4.5 - 0.1 = 4.4$ m from the center of rotation, point O; whereas C_2 acts at $d_2 = 4 + 0.15 = 4.15$ m from O. Applying Eq. 7–10, we obtain the total volume:

$$V = \theta \Sigma \bar{x} A$$
$$= (1.57 \text{ rad})[(4.4 \text{ m})(1 \text{ m})(0.2 \text{ m}) + (4.15)(0.15 \text{ m})(0.3 \text{ m})]$$
$$= 1.676 \text{ m}^3$$

The required mass of concrete is then

$$m = \rho_c V = (2.1 \text{ Mg/m}^3)(1.676 \text{ m}^3) = 3.52 \text{ Mg} \qquad \textit{Ans.}$$

The surface area of the arch is determined by dividing the perimeter into eight line segments, Fig. 7–21c, and then applying Eq. 7–9. Thus,

$$A = \theta \Sigma \bar{x} L = 1.57 \text{ rad}[4 \text{ m}(0.15 \text{ m}) + 2(4.15 \text{ m})(0.3 \text{ m}) +$$
$$(4.3 \text{ m})(1 \text{ m} - 0.15 \text{ m}) + 2(4.4 \text{ m})(0.2 \text{ m}) + (4.5 \text{ m})(1 \text{ m})]$$
$$= 20.4 \text{ m}^2 \qquad \textit{Ans.}$$

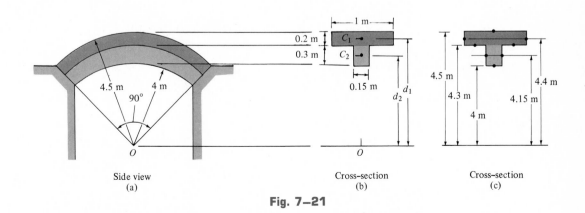

Side view
(a)

Cross-section
(b)

Cross-section
(c)

Fig. 7–21

Problems

7–51. Using integration, determine the area and the centroidal distance $\bar{y}$ of the shaded area. Then, using the second theorem of Pappus–Guldinus, determine the volume of a paraboloid formed by revolving the area about the x axis.

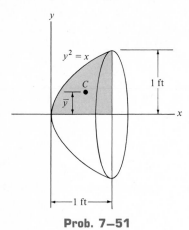

Prob. 7–51

*7–52. Using integration, compute both the area and the centroidal distance $\bar{x}$ of the shaded region. Then, using the second theorem of Pappus–Guldinus, compute the volume of the solid generated by revolving the shaded area about the y axis.

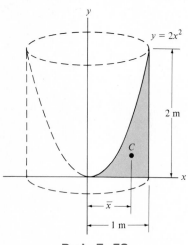

Prob. 7–52

■7–53. Using Simpson's rule, determine the area and the centroidal distance y of the shaded area. Then using the second theorem of Pappus–Guldinus, determine the volume formed by revolving the area about the x axis.

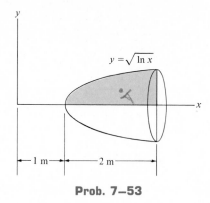

Prob. 7–53

7–54. The anchor ring is made of steel having a specific weight of 490 lb/ft^3. Determine its weight. The cross section is circular as shown.

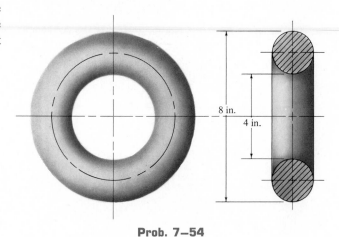

Prob. 7–54

7–55. A circular sea wall is made of concrete. Determine the total weight of the wall if the concrete has a density of $\rho = 150$ lb/ft^3.

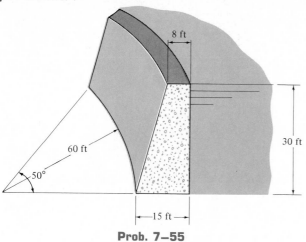

Prob. 7–55

*7–56. Determine the volume of concrete needed to construct the curb.

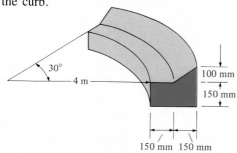

Prob. 7–56

7–57. The circular tube and flange was formed in a die by impacting an aluminum slug with a punch. Determine the amount of aluminum necessary to make it. The tube is a full circular part. Its cross section is shown in the figure.

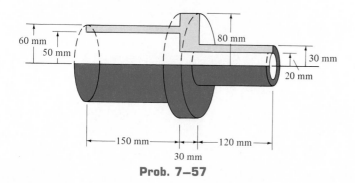

Prob. 7–57

7–58. The *rim* of a flywheel has the cross section *A-A* shown. Determine the volume of material needed for its construction.

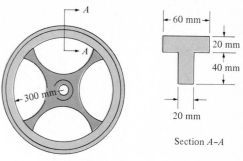

Section *A–A*

Prob. 7–58

7–59. Sand is piled between two walls as shown. Assume the pile to be a quarter section of a cone and that 26 percent of this volume is voids (air space). Use the second theorem of Pappus–Guldinus to determine the volume of sand.

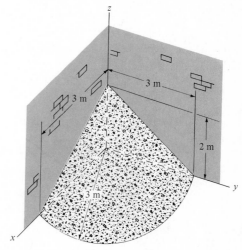

Prob. 7–59

311

*7–60. The full circular aluminum housing is used in an automotive brake system. Half the cross section is shown in the figure. Determine its weight if aluminum has a specific weight of 169 lb/ft³.

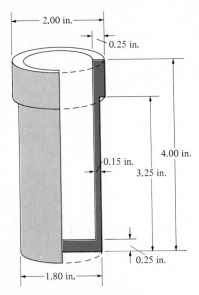

Prob. 7–60

7–61. Half the cross section of the steel housing is shown in the figure. There are six 10-mm-diameter bolt holes used around its rim. Determine its mass. The density of steel is 7.85 Mg/m³. The housing is a full circular part.

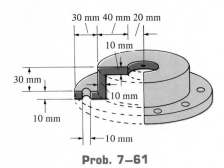

Prob. 7–61

7–62. The tank is fabricated from a hemisphere and cylindrical shell. Determine the vertical reactions that each of the four symmetrically placed legs exerts on the floor if the tank contains water which is 12 ft deep in the tank. The specific gravity of water is 62.4 lb/ft³. Neglect the weight of the tank.

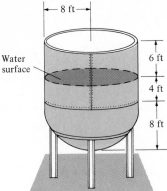

Probs. 7–62 / 7–63

7–63. Determine the approximate amount of paint needed to cover the outside surface of the tank. Assume that a gallon of paint covers 400 ft².

*7–64. The hopper is filled to its top with coal. Determine the volume of coal if the voids (air space) are 35 percent of the volume of the hopper.

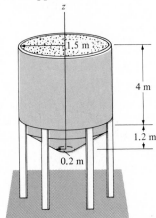

Probs. 7–64 / 7–65

7–65. If the hopper is made of steel plate having a thickness of 10 mm, determine its weight when empty. Steel has a density of 7.85 Mg/m³.

■**7–66.** The circular footing *A* having a diameter of 7 ft is used to anchor the circular leg *B* of an electrical transmission tower and thereby prevent the tower from overturning due to the wind. When computing the vertical uplifting strength **T** of the anchor, it is generally assumed that the weight of the soil within an inverted frustum of a cone with sides sloping outward at 30° with the vertical will resist the upward force of **T.** If the soil has a specific weight of 90 lb/ft³, and the footing and leg *B* are made of concrete having a specific weight of 150 lb/ft³, determine the approximate depth *h* to which it must be buried in order to resist a force of *T* = 25 kip on the leg of the tower.

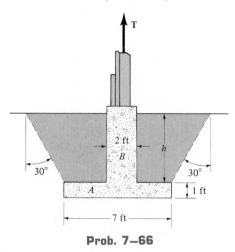

Prob. 7–66

7–67. Determine the height *h* to which liquid should be poured into the conical cup so that it contacts half the surface area on the inside of the cup.

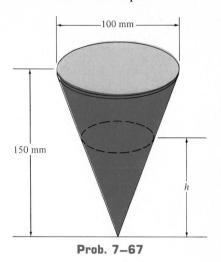

Prob. 7–67

Resultant of a Distributed Force System **7.5**

In Chapter 4 we considered ways of simplifying a system of concentrated forces which act on a body. In many practical situations, however, the body may be subjected to loadings distributed over its surface. Common examples include the effects of wind and hydrostatic pressure. The effects caused by these loadings can be studied in a simple manner if we replace them by their resultants. Here we will use the methods of Sec. 4.8 and show how to compute the resultant force of a distributed loading and specify its line of action. Two special cases will be considered.

Pressure Distribution over a Flat Surface. When a body supports a loading over its surface, its distribution is described by a *loading function* which defines the intensity of the load measured as a *force per unit area* or *pressure* acting on the body's surface. For example, the loading distribution or pressure

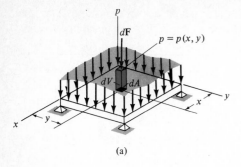

(a)

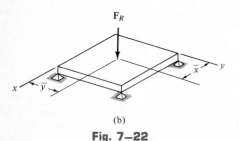

(b)

Fig. 7—22

acting on the plate shown in Fig. 7–22a is described by the loading function $p = p(x, y)$ Pa.* Knowing this function, we can determine the *magnitude* of the infinitesimal force dF acting on the differential area dA m² of the plate, located at the arbitrary point (x, y). This force magnitude is simply $dF = [p(x, y) \text{ N/m}^2](dA \text{ m}^2) = [p(x, y) \, dA]$ N. The entire loading on the plate is therefore represented as a system of *parallel forces* infinite in number and each acting on separate differential areas dA. This system of parallel forces will now be simplified to a single resultant force $\mathbf{F}_R$ acting through a unique point $(\bar{x}, \bar{y})$ on the plate, Fig. 7–22b.

Magnitude of Resultant Force. To determine the *magnitude* of $\mathbf{F}_R$, it is necessary to sum each of the differential forces dF acting over the plate's *entire surface area A*. This sum may be expressed mathematically as an integral.

$$F_R = \Sigma F; \qquad \boxed{F_R = \int p(x, y) \, dA} \qquad (7\text{–}11)$$

Note that $p(x, y) \, dA = dV$, the colored differential *volume element* shown in Fig. 7–22a. Therefore, the result indicates that the *magnitude of the resultant force is equal to the total volume under the distributed-loading diagram*.

Location of Resultant Force. The location $(\bar{x}, \bar{y})$ of $\mathbf{F}_R$ is determined by setting the moments of $\mathbf{F}_R$ equal to the moments of all the forces dF about the respective x and y axes. From Fig. 7–22a and 7–22b, we get

$$\bar{x} = \frac{\displaystyle\int xp(x, y) \, dA}{\displaystyle\int p(x, y) \, dA} \qquad \bar{y} = \frac{\displaystyle\int yp(x, y) \, dA}{\displaystyle\int p(x, y) \, dA} \qquad (7\text{–}12)$$

Setting $dV = p(x, y) \, dA$, we can also write

$$\bar{x} = \frac{\displaystyle\int x \, dV}{\displaystyle\int dV} \qquad \bar{y} = \frac{\displaystyle\int y \, dV}{\displaystyle\int dV} \qquad (7\text{–}13)$$

Hence, it can be seen that the *line of action of the resultant force passes through the geometric center or centroid of the volume under the distributed loading diagram*.

Linear Distribution of Load along a Straight Line. In many cases a pressure distribution acting on a flat surface will be symmetric about an axis of the surface upon which it acts. An example of such a loading is shown in Fig. 7–23a. Here the loading function $p = p(x)$ Pa is only a function of x since the pressure is uniform along the y axis. If we multiply $p = p(x)$ by the width a m of the plate, we obtain $w = [p(x) \text{ N/m}^2](a \text{ m}) = w(x)$ N/m. This loading function, shown in Fig. 7–23b, is a measure of load distribution along a *line*

*A pascal, Pa = 1 N/m².

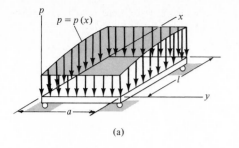

(a)

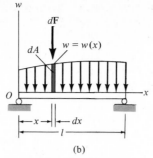

(b)

Fig. 7–23

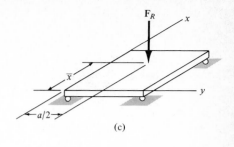

(c)

directed parallel to the x axis. As noted, it is measured as a force per unit length, rather than a force per unit area. Consequently, the load intensity diagram for $w = w(x)$ represents a system of coplanar parallel forces, Fig. 7–23b. Each infinitesimal force $d\mathbf{F}$ acts over an element of length dx such that at any point x, $dF = (w(x)\ \text{N/m})(dx\ \text{m}) = w(x)\ dx\ \text{N}$. Using the same method as above, we will now simplify the loading to a single resultant force and specify its location $\bar{x}$.

Magnitude of Resultant Force. The magnitude of $\mathbf{F}_R$ is determined from the sum

$$F_R = \int_L w(x)\ dx \qquad (7\text{–}14)$$

Note that $w(x)\ dx = dA$, where dA is the colored differential area shown in Fig. 7–23b. Thus we can state *the magnitude of the resultant force is equal to the total area under the distributed-loading diagram.*

Location of the Resultant Force. The location $\bar{x}$ of the line of action of $\mathbf{F}_R$ is determined by equating the moments of the force resultant and the force distribution about point O, Fig. 7–23b. This yields

$$\bar{x} = \frac{\displaystyle\int xw(x)\ dx}{\displaystyle\int w(x)\ dx} \qquad (7\text{–}15)$$

Setting $dA = w(x)\ dx$, we can also write

$$\bar{x} = \frac{\displaystyle\int x\ dA}{\displaystyle\int dA} \qquad (7\text{–}16)$$

This equation locates the $\bar{x}$ coordinate for the geometric center or centroid for the area. Hence, *the line of action of the resultant force passes through the geometric center or centroid of the area under the distributed-loading diagram.* Once $\bar{x}$ is determined, by symmetry, $\mathbf{F}_R$ passes through point $(\bar{x}, a/2)$ on the surface of the plate, Fig. 7–23c.

315

Example 7–12

In each case, determine the magnitude and location of the resultant of the distributed load acting on the beams in Fig. 7–24.

Solution

Uniform Loading. As indicated $w = 400$ lb/ft, which is constant over the entire beam, Fig. 7–24a. This loading forms a rectangle, the area of which is equal to the resultant force, Fig. 7–24b, i.e.,

$$F_R = (400 \text{ lb/ft})(10 \text{ ft}) = 4000 \text{ lb} \qquad Ans.$$

The location of F_R passes through the center or centroid C of the area, so that

$$\bar{x} = 5 \text{ ft} \qquad Ans.$$

Triangular Loading. Here the loading varies uniformly in intensity from 0 to 600 N/m, Fig. 7–24c. These values can be verified by substitution of $x = 0$ and $x = 6$ m into the loading function $w = 100x$ N/m. The area of this triangular loading is equal to F_R, Fig. 7–24d. From the table on the inside back cover, $A = \frac{1}{2}bh$, so that

$$F_R = \tfrac{1}{2}(6 \text{ m})(600 \text{ N/m}) = 1800 \text{ N} \qquad Ans.$$

The line of action of F_R passes through the centroid C of the triangle. Using the table on the inside back cover, this point lies at a distance of one third the length of the beam, measured from the right side. Hence,

$$\bar{x} = 6 \text{ m} - \tfrac{1}{3}(6 \text{ m}) = 4 \text{ m} \qquad Ans.$$

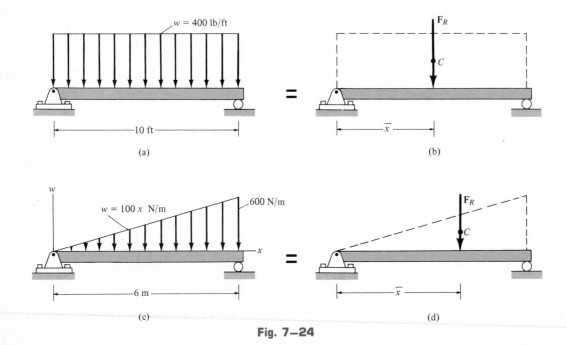

Fig. 7–24

Example 7–13

Determine the magnitude and location of the resultant of the distributed load acting on the beam shown in Fig. 7–25a.

Solution

The area of the loading diagram is a *trapezoid,* and therefore the solution can be obtained directly from the area and centroid formulas for a trapezoid listed on the inside back cover. Since these formulas are not easily remembered, instead we will solve this problem by using "composite" areas. In this regard, we can divide the trapezoidal loading into a rectangular and triangular loading as shown in Fig. 7–25b. The magnitude of the force represented by each of these loadings is equal to its associated *area,*

$$F_1 = \tfrac{1}{2}(9 \text{ ft})(50 \text{ lb/ft}) = 225 \text{ lb}$$
$$F_2 = (9 \text{ ft})(50 \text{ lb/ft}) = 450 \text{ lb}$$

The lines of action of these parallel forces act through the *centroid* of their associated areas and therefore intersect the beam at

$$\bar{x}_1 = \tfrac{1}{3}(9 \text{ ft}) = 3 \text{ ft}$$
$$\bar{x}_2 = \tfrac{1}{2}(9 \text{ ft}) = 4.5 \text{ ft}$$

The two parallel forces $\mathbf{F}_1$ and $\mathbf{F}_2$ can be reduced to a single resultant $\mathbf{F}_R$. The magnitude of $\mathbf{F}_R$ is

$$+\downarrow F_R = \Sigma F; \qquad F_R = 225 + 450 = 675 \text{ lb} \qquad \textit{Ans.}$$

With reference to point A, Fig. 7–25b and c, we can define the location of $\mathbf{F}_R$. We require

$$\curvearrowright + M_{R_A} = \Sigma M_A; \qquad \bar{x}(675) = 3(225) + 4.5(450)$$
$$\bar{x} = 4.0 \text{ ft} \qquad \textit{Ans.}$$

Note: The trapezoidal area in Fig. 7–25a can also be divided into two triangular areas as shown in Fig. 7–25d. In this case

$$F_1 = \tfrac{1}{2}(9 \text{ ft})(100 \text{ lb/ft}) = 450 \text{ lb}$$
$$F_2 = \tfrac{1}{2}(9 \text{ ft})(50 \text{ lb/ft}) = 225 \text{ lb}$$

and

$$\bar{x}_1 = \tfrac{1}{3}(9 \text{ ft}) = 3 \text{ ft}$$
$$\bar{x}_2 = \tfrac{1}{3}(9 \text{ ft}) = 3 \text{ ft}$$

Using these results, show that $F_R = 675$ lb and $\bar{x} = 4.0$ ft as obtained above.

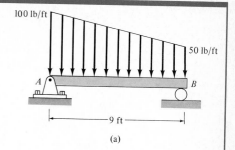

(a)

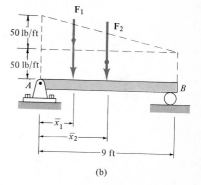

(b)

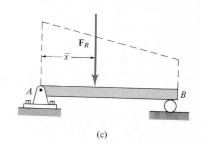

(c)

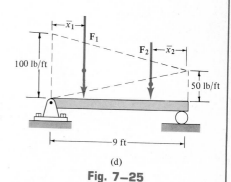

(d)

Fig. 7–25

317

Example 7–14

Determine the magnitude and location of the resultant force acting on the beam in Fig. 7–26a.

Solution

As shown, the colored differential area element $dA = w \, dx = 60x^2 \, dx$ will be used for integration. Applying Eq. 7–14, by summing these elements from $x = 0$ to $x = 2$ m, we obtain the resultant force $\mathbf{F}_R$,

$$F_R = \Sigma F;$$

$$F_R = \int_A dA = \int_0^2 60x^2 \, dx = 60 \left[\frac{x^3}{3} \right]_0^2 = 60 \left[\frac{2^3}{3} - \frac{0^3}{3} \right]$$

$$= 160 \text{ N} \qquad\qquad\qquad\qquad\qquad\qquad\qquad\qquad \textit{Ans.}$$

Since the element of area dA is located an arbitrary distance x from O, the location $\bar{x}$ of $\mathbf{F}_R$ *measured from* O, Fig. 7–26b, is determined from Eq. 7–16.

$$\bar{x} = \frac{\displaystyle\int_A x \, dA}{\displaystyle\int_A dA} = \frac{\displaystyle\int_0^2 x(60 \, x^2) \, dx}{160} = \frac{60 \left[\dfrac{x^4}{4} \right]_0^2}{160} = \frac{60 \left[\dfrac{2^4}{4} - \dfrac{0^4}{4} \right]}{160}$$

$$= 1.5 \text{ m} \qquad\qquad\qquad\qquad\qquad\qquad\qquad\qquad \textit{Ans.}$$

These results may be checked by using the table on the inside back cover, where it is shown that for a semiparabolic area of height a, length b, and shape shown in Fig. 7–26a,

$$A = \frac{ab}{3} = \frac{240(2)}{3} = 160 \text{ N} \quad \text{and} \quad \bar{x} = \frac{3}{4}a = \frac{3}{4}(2) = 1.5 \text{ m}$$

As an exercise, try obtaining the results for the triangular loading in Example 7–12 using Eqs. 7–14 and 7–15.

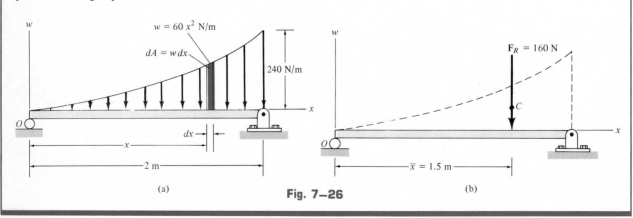

(a)

Fig. 7–26

(b)

Example 7–15

A distributed loading of $p = 800x$ Pa acts over the top surface of the beam shown in Fig. 7–27a. Determine the magnitude and location of the resultant force.

Solution

The loading function $p = 800x$ Pa indicates that the load intensity varies uniformly from $p = 0$ at $x = 0$ to $p = 7200$ Pa at $x = 9$ m. Since the intensity is uniform along the width of the beam, the loading may be viewed in two dimensions as shown in Fig. 7–27b. Here

$$w = (800 \; x \; \text{N/m}^2)(0.2 \; \text{m})$$
$$= (160x) \; \text{N/m}$$

At $x = 9$ m, note that $w = 1440$ N/m. Thus the magnitude of the resultant force is

$$F_R = \tfrac{1}{2}(9 \; \text{m})(1440 \; \text{N/m}) = 6480 \; \text{N} = 6.48 \; \text{kN} \qquad Ans.$$

The *line of action* of $\mathbf{F}_R$ passes through the *centroid C* of the triangle. Hence,

$$\bar{x} = 9 \; \text{m} - \tfrac{1}{3}(9 \; \text{m}) = 6 \; \text{m} \qquad Ans.$$

The results are shown in Fig. 7–27c.

We may also view the resultant $\mathbf{F}_R$ as *acting* through the *centroid* of the *volume* of the loading diagram $p = p(x)$ in Fig. 7–27a. Hence $\mathbf{F}_R$ intersects the *x-y* plane at the point (6 m, 0) on the *x* axis. Furthermore, the *magnitude* of $\mathbf{F}_R$ is equal to the *volume* under the loading curve, i.e.,

$$F_R = V = \tfrac{1}{2}(7200 \; \text{N/m}^2)(9 \; \text{m})(0.2 \; \text{m}) = 6.48 \; \text{kN} \qquad Ans.$$

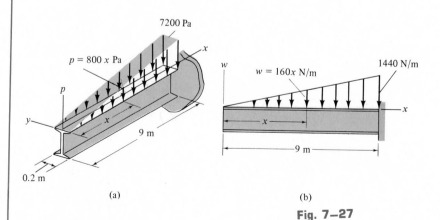

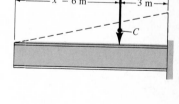

(a)

(b)

(c)

Fig. 7–27

Problems

*7–68. Replace the loading by an equivalent force and couple moment acting at point O.

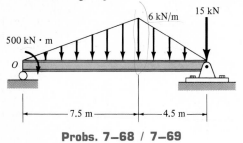

Probs. 7–68 / 7–69

7–69. Replace the loading by a single resultant force, and specify the location of the force on the beam measured from point O.

7–70. Replace the loading by a single resultant force and specify its location on the beam, measured from point O.

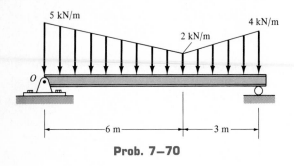

Prob. 7–70

7–71. Determine the length b of the triangular load and its position a on the beam such that the resultant force is zero and the resultant couple moment acting on the beam is 8 kN · m clockwise.

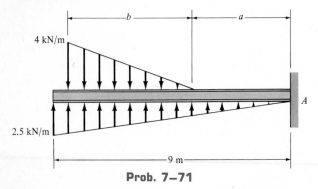

Prob. 7–71

*7–72. Determine the reactions at the wall A and the rocker C. The two members are connected by a pin at B.

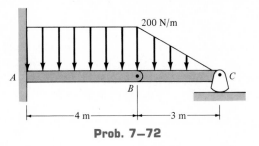

Prob. 7–72

7–73. If the maximum intensity of the distributed load acting on the beam is $w = 4$ kN/m, determine the reactions at the pin A and roller B for equilibrium.

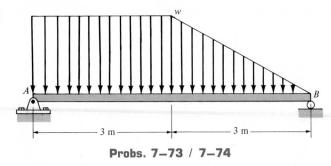

Probs. 7–73 / 7–74

7–74. If either the pin at A or the roller at B can support a load no greater than 6 kN, determine the maximum intensity of the distributed load w kN/m, so that failure of a support does not occur.

7–75. The cantilever footing is used to support a wall near its edge *A* such that it causes a uniform soil pressure under the footing. Determine the uniform distribution of force, w_A and w_B, measured in lb/ft at pads *A* and *B*, necessary to balance the wall forces of 1.6 kip and 4 kip as shown.

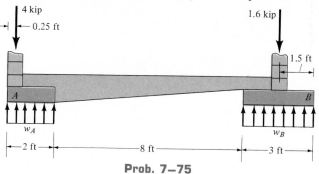

Prob. 7–75

7–78. Determine the horizontal and vertical components of force which the pins exert on member *ABC*.

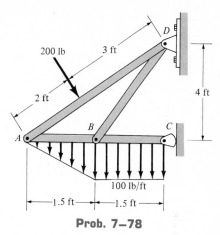

Prob. 7–78

*7–76. A cantilever beam, having an extended length of 3 m, is subjected to a vertical force $F = 800$ N acting at its end. Assuming that the wall resists this load with linearly varying distributed loads over the 0.15-m length of the beam, determine the intensities w_1 and w_2 for equilibrium.

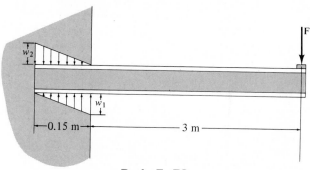

Prob. 7–76

7–79. Determine the horizontal and vertical components of reaction at the pin connections *A*, *B*, *C*, *D*, and *E* of the three-member frame.

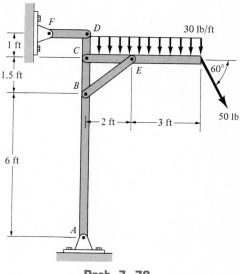

Prob. 7–79

7–77. The beam is subjected to the two concentrated loads as shown. Assuming that the foundation exerts a linearly varying load distribution on its bottom, determine the load intensities w_1 and w_2 for equilibrium (a) in terms of the parameters shown; (b) set $P = 500$ lb, $L = 12$ ft.

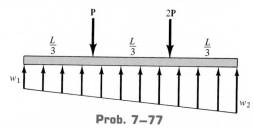

Prob. 7–77

321

*7–80. The wall footing is used to support the load of 12,000 lb. Determine the intensities w_1 and w_2 of the distributed loading acting on the base of the footing for equilibrium.

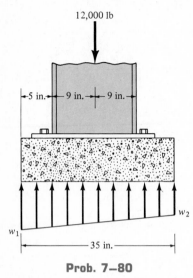

Prob. 7–80

7–81. The wind loading on the chemical distillation tower varies in the manner shown. Determine the resultant force of the wind and specify its location on the tower, measured from its base, $y = 0$.

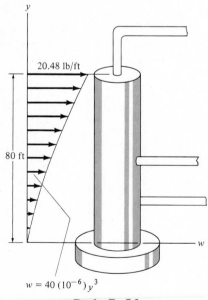

Prob. 7–81

7–82. Determine the magnitude and location of the resultant force acting on the beam.

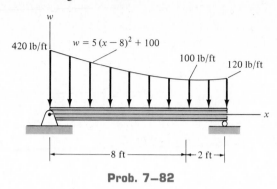

Prob. 7–82

■7–83. Determine the reactions on the beam at A and B due to the distributed loading. Evaluate the integrals using Simpson's rule.

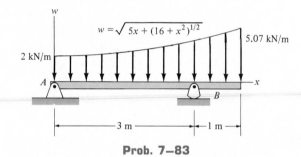

Prob. 7–83

*7–84. Determine the magnitude and location of the resultant force acting on the beam.

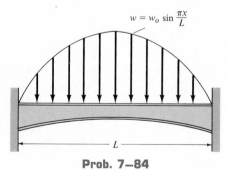

Prob. 7–84

7–85. The variation of pressure on the interior wall of a deep rectangular silo can be approximated by $p = 20(1 - e^{-0.1z})$ lb/ft^2, where z is in feet. Simplify the load to a single resultant force and specify its location measured down from the top point O. The wall of the silo has a width of 2 ft.

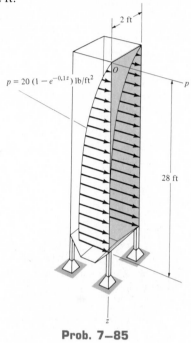

2 ft

$p = 20(1 - e^{-0.1z})$ lb/ft^2

O

p

28 ft

Prob. 7–85

7–86. The rectangular bin is filled with coal, which creates a pressure distribution along wall A that varies as shown, i.e., $p = 4z^3$ lb/ft^2, where z is in feet. Compute the resultant force created by the coal, and specify its location measured from the top surface of the coal.

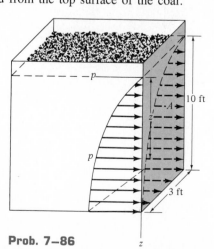

p

A

10 ft

z

p

3 ft

Prob. 7–86

z

7–87. The loading acting on a square plate generates a pressure distribution that is parabolic. Determine the magnitude and location of the resultant force. Also, what are the reactions at the rollers B and C and the ball-and-socket joint A? Neglect the weight of the plate.

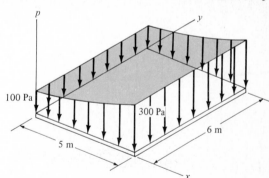

p

C

$p = 30 y^2$ Pa

y

480 Pa

B

A

4 m

2 m

4 m

x

Prob. 7–87

*7–88. The pressure loading on the plate is described by the function $p = [-240/(x + 1) + 340]$ Pa. Determine the magnitude of the resultant force and the coordinates $(\bar{x}, \bar{y})$ where the line of action of the force intersects the plate.

p

y

100 Pa

300 Pa

6 m

5 m

x

Prob. 7–88

323

*7.6 Fluid Pressure Acting on a Submerged Surface

In Sec. 7.5 we discussed the method used to simplify a distributed loading which is uniform along an axis of a flat surface. In this section we will not only generalize this method to include surfaces which are curved and have a variable shape, but apply it to the surface of a body which is submerged in a fluid.

Pressure Loading. According to Pascal's law, a fluid at rest creates a pressure p at a point that is the *same* in *all* directions. The magnitude of p, measured as a force per unit area, depends upon the specific weight γ or mass density ρ of the fluid and the depth z of the point from the fluid surface. The relationship can be expressed mathematically as

$$p = \gamma z = \rho g z \tag{7-17}$$

where g is the acceleration of gravity. Equation 7–17 is only valid for fluids that are assumed *incompressible,* as in the case of most liquids.* Gases are compressible fluids, and since their density changes significantly with both pressure and temperature, Eq. 7–17 cannot be used.

Consider now what effect the pressure of a liquid has at points A, B, and C, located on the top surface of the submerged plate shown in Fig. 7–28. Since

*Note that for water, $\gamma = 62.4 \ \text{lb/ft}^3$, or $\gamma = 9810 \ \text{N/m}^3$ since $\rho = 1000 \ \text{kg/m}^3$.

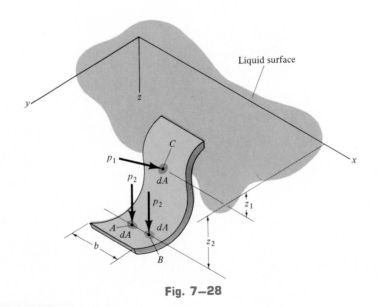

Fig. 7–28

points A and B are both at depth z_2 from the liquid surface, the *pressure* at these points has a magnitude of $p_2 = \gamma z_2$. Likewise, point C is at depth z_1; hence, $p_1 = \gamma z_1$. In all cases, the pressure acts *normal* to the surface area dA located at the point, Fig. 7–28.

Flat Plate of Constant Width. A flat rectangular plate of constant width, which is submerged in a liquid having a specific weight γ, is shown in Fig. 7–29a. The plane of the plate makes an angle θ with the horizontal, such that its top edge is located at a depth z_1 from the liquid surface and its bottom edge is located at a depth z_2. Since pressure varies linearly with depth, Eq. 7–17, the distribution of pressure over the plate's surface is represented by a trapezoidal volume of loading. The magnitude of the *resultant force* $\mathbf{F}_R$ is equal to the *volume* of this loading diagram and has a *line of action* that passes through the volume's centroid C. Note that $\mathbf{F}_R$ does *not* act at the center of the plate; rather, it acts at point P, called the *center of pressure*.

Since the plate has a *uniform width,* the loading distribution may also be viewed in two dimensions, Fig. 7–29b. Here the loading intensity is measured as force/length and varies linearly from $w_1 = bp_1 = b\gamma z_1$ to $w_2 = bp_2 = b\gamma z_2$. The magnitude of $\mathbf{F}_R$ in this case equals the trapezoidal *area* and $\mathbf{F}_R$ has a *line of action* that passes through the area's *centroid C*. For numerical applications, the area and location of the centroid for a trapezoid are tabulated on the inside back cover.

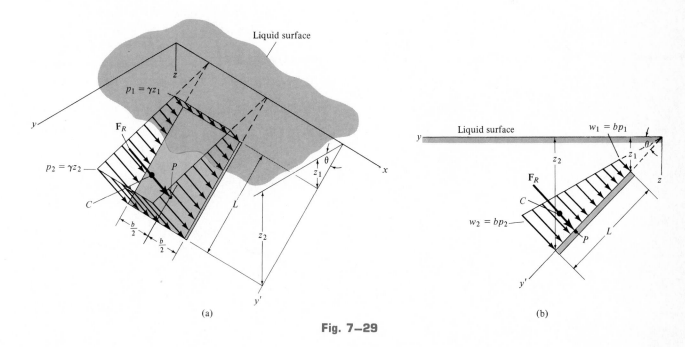

(a)

(b)

Fig. 7–29

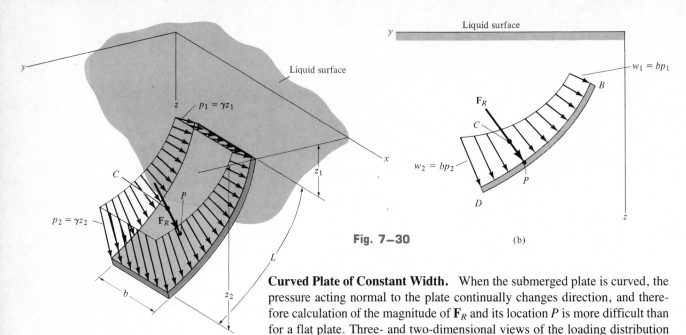

Liquid surface

$p_1 = \gamma z_1$

$p_2 = \gamma z_2$

$\mathbf{F}_R$

z_1

z_2

L

b

(a)

Liquid surface

$w_1 = bp_1$

B

$\mathbf{F}_R$

C

$w_2 = bp_2$

P

D

Fig. 7–30

(b)

Curved Plate of Constant Width. When the submerged plate is curved, the pressure acting normal to the plate continually changes direction, and therefore calculation of the magnitude of $\mathbf{F}_R$ and its location P is more difficult than for a flat plate. Three- and two-dimensional views of the loading distribution are shown in Fig. 7–30a and b, respectively. Here integration can be used to determine both F_R and the location of the centroid C or center of pressure P.

A simpler method exists, however, for calculating the magnitude of $\mathbf{F}_R$ and its location along a curved (or flat) plate having a *constant width*. This method requires separate calculations for the horizontal and vertical *components* of $\mathbf{F}_R$. For example, the distributed loading acting on the top surface of the curved plate DB in Fig. 7–30b can be represented by the *equivalent loading* shown in Fig. 7–31. Here it is seen that DB supports a horizontal force along its *vertical projection AD*. This force, $\mathbf{F}_{AD}$, has a magnitude that equals the area under the trapezoid and acts through the centroid C_{AD} of this area. The

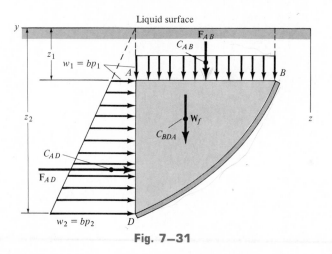

Liquid surface

z_1

$w_1 = bp_1$

$\mathbf{F}_{AB}$

C_{AB}

A

B

z_2

C_{AD}

$\mathbf{F}_{AD}$

C_{BDA}

$\mathbf{W}_f$

$w_2 = bp_2$

D

Fig. 7–31

distributed loading along the *horizontal projection AB* is constant, since all points lying in this plane are at the same depth from the surface of the liquid. The magnitude of $\mathbf{F}_{AB}$ is simply the area of the rectangle. This force acts through the centroid C_{AB} (or midpoint) of AB. In addition to $\mathbf{F}_{AB}$, the curved surface DB must also support the downward *weight of fluid* $\mathbf{W}_f$ contained within the block BDA. This force has a magnitude of $W_f = (\gamma b)(Area_{BDA})$ and acts through the centroid of BDA. Summing the three coplanar forces shown in Fig. 7–31 yields $\mathbf{F}_R = \Sigma\mathbf{F} = \mathbf{F}_{AD} + \mathbf{F}_{AB} + \mathbf{W}_f$. The center of pressure P is determined by applying the principle of moments to this force system and its resultant about a convenient reference point. The final results for $\mathbf{F}_R$ and its location P will be equivalent to those shown in Fig. 7–30.

Flat Plate of Variable Width. The pressure distribution acting across the top face of a submerged plate having a variable width is shown in Fig. 7–32. The resultant force of this loading equals the volume described by the plate area as its base and linear varying pressure distribution as its altitude. The shaded element shown in Fig. 7–32 may be used if integration is chosen to determine this volume. The element consists of a strip of area $dA = x\,dy'$ located at a depth z below the liquid surface. Since a uniform pressure $p = \gamma z$ (force/area) acts on dA, the magnitude of the differential force $d\mathbf{F}$ is equal to $dF = dV = p\,dA = \gamma z(x\,dy')$. Integrating over the entire volume yields

$$F_R = \int dV = V \qquad (7\text{–}18)$$

The centroid of V defines the point through which $\mathbf{F}_R$ acts. The center of pressure, which lies on the surface of the plate just below C, has coordinates $P(\bar{x}, \bar{y}')$ defined by the equations

$$\bar{x} = \frac{\int \tilde{x}\,dV}{\int dV} \qquad \bar{y}' = \frac{\int \tilde{y}'\,dV}{\int dV}$$

This point should *not* be mistaken for the centroid of the plate *area*.

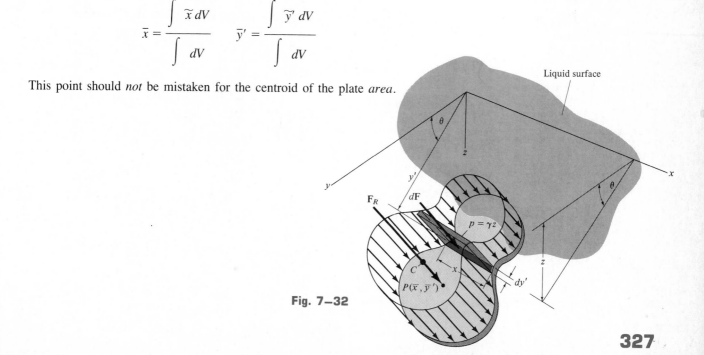

Fig. 7–32

Example 7–16

Determine the magnitude and location of the resultant hydrostatic force acting on the submerged plate AB shown in Fig. 7–33a. The plate has a width of 1.5 m; $\rho_w = 1000$ kg/m³.

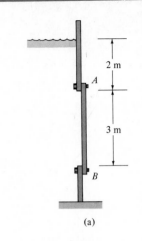

(a)

Solution

Since the plate has a constant width, the distributed loading can be viewed in two dimensions as shown in Fig. 7–33b. The intensity of the load at A and B is computed as

$$w_A = b\rho_w g z_A = (1.5\ \text{m})(1000\ \text{kg/m}^3)(9.81\ \text{m/s}^2)(2\ \text{m})$$
$$= 29.4\ \text{kN/m}$$
$$w_B = b\rho_w g z_B = (1.5\ \text{m})(1000\ \text{kg/m}^3)(9.81\ \text{m/s}^2)(5\ \text{m})$$
$$= 73.6\ \text{kN/m}$$

The magnitude of the resultant force $\mathbf{F}_R$ created by the distributed load is

$$F_R = (\text{area of trapezoid})$$
$$= \tfrac{1}{2}(3)(29.4 + 73.6) = 154.5\ \text{kN} \qquad Ans.$$

This force acts through the centroid of the area,

$$h = \frac{1}{3}\left(\frac{2(29.4) + 73.6}{29.4 + 73.6}\right)(3) = 1.29\ \text{m} \qquad Ans.$$

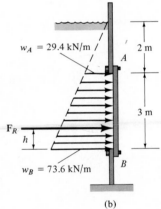

(b)

measured upward from B, Fig. 7–33b.

The same results can be obtained by considering two components of $\mathbf{F}_R$ defined by the triangle and rectangle shown in Fig. 7–33c. Each force acts through its associated centroid and has a magnitude of

$$F_{Re} = (29.4\ \text{kN/m})(3\ \text{m}) = 88.2\ \text{kN}$$
$$F_t = \tfrac{1}{2}(44.2\ \text{kN/m})(3\ \text{m}) = 66.3\ \text{kN}$$

Hence,

$$F_R = F_{Re} + F_t = 88.2 + 66.3 = 154.5\ \text{kN} \qquad Ans.$$

The location of $\mathbf{F}_R$ is determined by summing moments about B, Fig. 7–33b and c, i.e.,

$$\curvearrowleft + (M_R)_B = \Sigma M_B; \quad (154.5)h = 88.2(1.5) + 66.3(1)$$
$$h = 1.29\ \text{m} \qquad Ans.$$

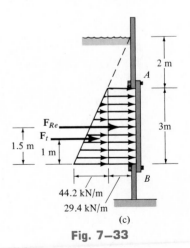

(c)

Fig. 7–33

Example 7–17

Determine the magnitude of the resultant hydrostatic force acting on the surface of a sea wall shaped in the form of a parabola as shown in Fig. 7–34a. The wall is 5 m long; $\rho_w = 1020$ kg/m^3.

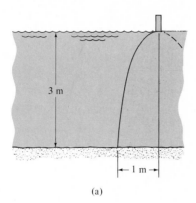

(a)

Fig. 7–34

Solution

The horizontal and vertical components of the resultant force will be calculated, Fig. 7–34b. Since

$$w_B = b\rho_w g z_B = 5\ \text{m}(1020\ \text{kg/m}^3)(9.81\ \text{m/s}^2)(3\ \text{m}) = 150.0\ \text{kN/m}$$

then

$$F_x = \tfrac{1}{2}(3\ \text{m})(150.0\ \text{kN/m}) = 225.0\ \text{kN}$$

The area of the parabolic sector ABC can be determined using the table on the inside back cover. Hence, the weight of water within this region is

$$F_y = (\rho_w g b)(Area_{ABC})$$
$$= (1020\ \text{kg/m}^3)(9.81\ \text{m/s}^2)5\ \text{m}[\tfrac{1}{3}(1\ \text{m})(3\ \text{m})] = 50.0\ \text{kN}$$

The resultant force is therefore

$$F_R = \sqrt{F_x^2 + F_y^2} = \sqrt{(225.0)^2 + (50.0)^2}$$
$$= 230.5\ \text{kN} \qquad\qquad Ans.$$

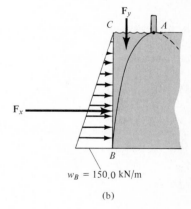

(b)

Example 7–18

Determine the magnitude and location of the resultant force acting on the triangular end plates of the water trough shown in Fig. 7–35a; $\rho_w = 1000$ kg/m^3.

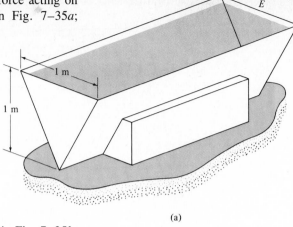

(a)

Solution

The pressure loading acting on the end plate E is shown in Fig. 7–35b. The magnitude of the resultant force **F** is equal to the volume of this loading diagram. Choosing the differential volume element shown in the figure, we get

$$dF = dV = p\, dA = \rho_w gz(2x\, dz) = 19\,620zx\, dz$$

The equation of line AB is

$$x = 0.5(1 - z)$$

Hence, substituting and integrating with respect to z from $z = 0$ to $z = 1$ m yields

$$F = V = \int dV = \int_0^1 (19\,620)z[0.5(1 - z)]\, dz$$

$$= 9810 \int_0^1 (z - z^2)\, dz = 1635 \text{ N}$$

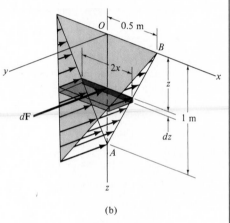

(b)

Fig. 7–35

or

$$F = 1.64 \text{ kN} \qquad \textit{Ans.}$$

This resultant passes through the *centroid of the volume*. Because of symmetry,

$$\bar{x} = 0 \qquad \textit{Ans.}$$

Since $\tilde{z} = z$ for the volume element, then

$$\bar{z} = \frac{\int \tilde{z}\, dV}{\int dV} = \frac{\int_0^1 z(19\,620)z[0.5(1 - z)]\, dz}{1635} = \frac{9810 \int_0^1 (z^2 - z^3)\, dz}{1635}$$

$$= 0.5 \text{ m} \qquad \textit{Ans.}$$

Problems

7–89. The tank is filled with water to a depth of 2 m. Determine the resultant force the water exerts on one of the sides of the tank. If oil instead of water is placed in the tank, to what depth should it reach so that it creates the same resultant force? $\rho_o = 900$ kg/m^3 and $\rho_w = 1000$ kg/m^3.

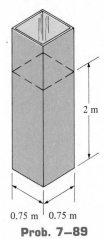

Prob. 7–89

7–90. When the tide water A subsides, the tide gate automatically swings open to drain the marsh B. For the condition of high tide shown, determine the horizontal resultant forces developed at the hinge C and stop block D. The length of the gate is 6 m and its height is 4 m. $\rho_w = 1$ Mg/m^3.

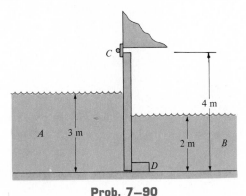

Prob. 7–90

7–91. The concrete dam is designed so that its face AB has a gradual slope into the water as shown. Because of this, the frictional force at the base BD of the dam is increased due to the hydrostatic force of the water acting on the dam. Calculate the hydrostatic force acting on the face AB of the dam. The dam is 60 ft wide. $\gamma_w = 62.4$ lb/ft^3.

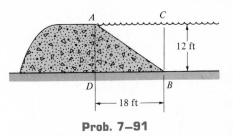

Prob. 7–91

*7–92. The tank is used to store a liquid having a density of 80 lb/ft^3. If it is filled to the top, determine the magnitude of force the liquid exerts on each of its two sides $ABDC$ and $BDFE$.

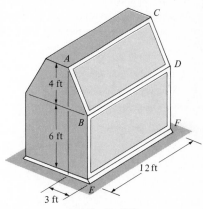

Prob. 7–92

331

7-93. The concrete "gravity" dam is held in place by its own weight. If the density of concrete is $\rho_c = 2.5$ Mg/m³, and water has a density of $\rho_w = 1.0$ Mg/m³, determine the smallest width d that will prevent the dam from overturning about its end A. The dam has a width of 8 m.

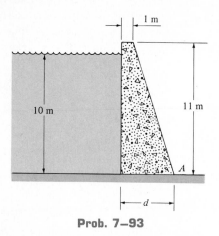

Prob. 7-93

7-94. The quarter circular concrete "gravity" dam is held in place by its own weight. If water has a density of $\rho_w = 1.0$ Mg/m³, determine the smallest possible density of the material composing the dam so that the dam will be prevented from overturning about its end A.

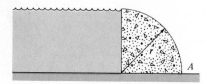

Prob. 7-94

7-95. The storage tank contains oil having a density of $\rho_o = 0.90$ Mg/m³. If the tank is 1.5 m wide, calculate the resultant force acting on the inclined side AB of the tank, caused by the oil, and specify its location along AB, measured from A.

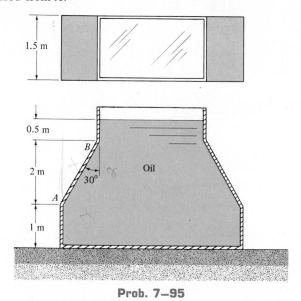

Prob. 7-95

*7-96. The structure shown is used for temporary storage of oil at sea for later loading into ships. When it is empty the water level is at A (sea level). As oil is loaded, the water is displaced through exit ports at D. If the riser EC is filled with oil, i.e., to level C, determine the height h of the oil level above sea level. $\rho_o = 900$ kg/m³ and $\rho_w = 1020$ kg/m³.

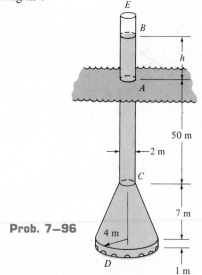

Prob. 7-96

7–97. If the vessel in Prob. 7–96 is filled with oil, i.e., until it reaches a depth of 58 m below sea level, how high h will the oil level extend above sea level?

7–98. The arched wall AB is shaped in the form of a quarter circle. If it is 8 m long, determine the horizontal and vertical components of the resultant force caused by the water acting on the wall. $\rho_w = 1000$ kg/m^3.

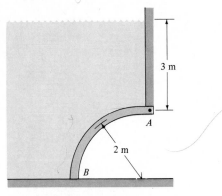

3 m

2 m

A

B

Prob. 7–98

*7–100. The tank is filled to the top ($y = 0.5$ m) with water having a density of $\rho_w = 1.0$ Mg/m^3. Determine the resultant force of the water pressure acting on the flat end plate C of the tank, and specify its location measured from the top of the tank.

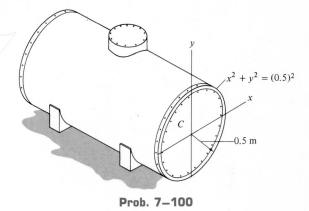

y

$x^2 + y^2 = (0.5)^2$

x

C

0.5 m

Prob. 7–100

7–99. The semicircular tunnel passes under a river which is 9 m deep. Determine the resultant hydrostatic force acting per meter of length along the length of the tunnel. The tunnel is 6 m wide; $\rho_w = 1.0$ Mg/m^3.

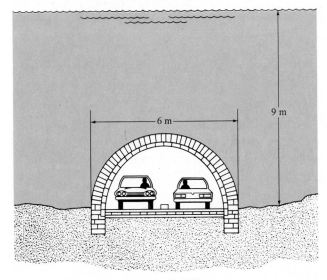

6 m

9 m

Prob. 7–99

Review Problems

7–101. Locate the center of gravity of the two-block assembly. The densities of materials A and B are $\rho_A = 150$ lb/ft^3 and $\rho_B = 400$ lb/ft^3, respectively.

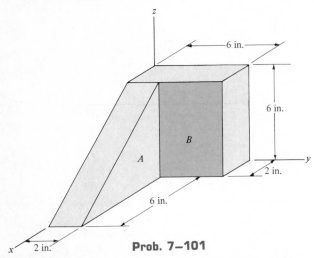

Prob. 7–101

7–102. Locate the center of gravity $\bar{y}$ of the volume generated by revolving the shaded area about the y axis. The material is homogeneous.

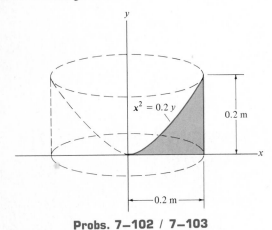

Probs. 7–102 / 7–103

7–103. Using integration, compute the area and the centroidal distance $\bar{x}$ for the shaded area. Then using the second theorem of Pappus–Guldinus, compute the volume of the solid generated by revolving the shaded area about the y axis.

*7–104. Determine the distance $\bar{y}$ to the centroidal axis $\bar{x}\bar{x}$ of the beam's cross-sectional area. Neglect the size of the corner welds at A and B for the calculation.

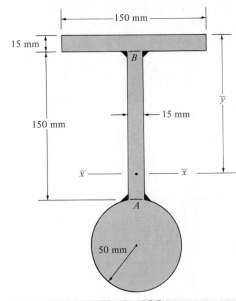

Prob. 7–104

7–105. Determine the reactions at the supports.

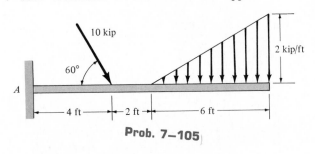

Prob. 7–105

7–106. Locate the center of gravity of the volume generated by revolving the shaded area about the x axis. The material is homogeneous.

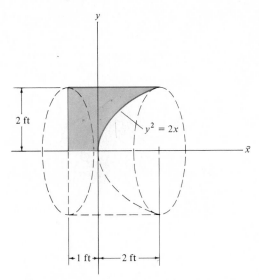

Probs. 7–106 / 7–107

7–107. Using integration, compute the area and the centroidal distance $\bar{y}$ for the shaded area. Then using the second theorem of Pappus–Guldinus, compute the volume of the solid generated by revolving the shaded area about the x axis.

*7–108. Determine the components of reaction at the fixed support A for the beam loaded as shown.

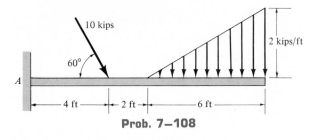

Prob. 7–108

7–109. Determine the resultant forces at pins B and C of the four-member frame.

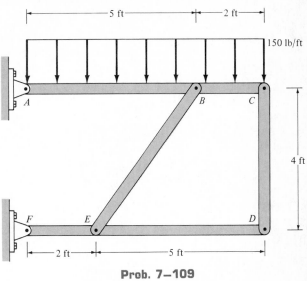

Prob. 7–109

7–110. Determine the magnitude of the resultant hydrostatic force acting per meter of length on the sea wall; $\rho_w = 1.0 \, \text{Mg/m}^3$.

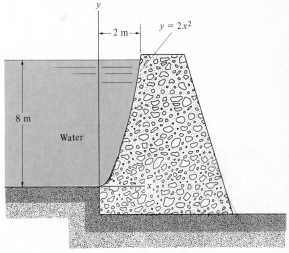

Prob. 7–110

335

7–111. Locate the distance $\bar{x}$ to the centroid C of the shaded area.

*7–112. Locate the distance $\bar{y}$ to the centroid C of the shaded area.

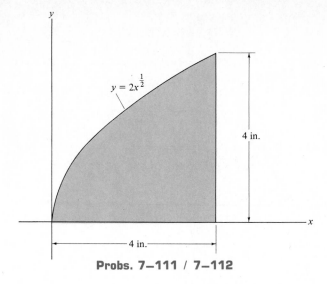

$$y = 2x^{\frac{1}{2}}$$

4 in.

4 in.

Probs. 7–111 / 7–112

Internal Forces

8

In Chapter 6 we discussed application of the equations of equilibrium to determine the forces acting at the *connections* of a structural member. Once these forces are obtained we can extend the equilibrium analysis to compute the member's *internal loadings*. In this chapter we will develop a technique for finding the internal loading at specific *points* within a structural member, and then we will generalize this method to find the point-to-point variation of loading along the axis of a beam. A graph showing this variation of internal load will allow us to find the critical point where the *maximum* internal loading occurs. This information is important for the proper design of a structure using the theory of mechanics of materials. The analysis of cables will be treated in the last part of the chapter.

Internal Forces Developed in Structural Members   8.1

The design of a structural member requires an investigation of the forces acting *within* the member which are necessary to balance the forces acting external to it. The *method of sections* can be used to determine these *internal* forces. This requires that a "cut" or section be made perpendicular to the axis of the member at the point where the internal loading is to be determined. A free-body diagram of either segment of the "cut" member is isolated and the internal loads are then determined from the equations of equilibrium. The loadings obtained in this manner will actually represent the *resultant* of a distribution of force per unit area, called *stress,* which acts over the member's cross-sectional area at the cut section. Using the theory of mechanics of materials, one can relate this stress distribution to the resultant internal loadings and thereby develop a formula which can be used for the design of the member.

In order to illustrate the procedure for finding the internal loadings in a structural member, consider the "simply supported" beam shown in Fig. 8–1a which is subjected to the forces $\mathbf{F}_1$ and $\mathbf{F}_2$. The *support reactions* $\mathbf{A}_x$, $\mathbf{A}_y$, and $\mathbf{B}_y$ can be determined by applying the equations of equilibrium, using the free-body diagram of the *entire beam*, Fig. 8–1b. To determine the *internal loadings* at point C, for example, it is necessary to pass an imaginary section through the beam, cutting it into two segments at that point, Fig. 8–1a. Doing this "exposes" the internal forces as *external* on the free-body diagram of each segment, Fig. 8–1c. Since both segments (AC and BC) were prevented from translating *and* rotating *before* the beam was sectioned, equilibrium of each section is maintained if rectangular force components $\mathbf{A}_C$ and $\mathbf{V}_C$ and a resultant couple moment $\mathbf{M}_C$ are developed at the cut section. Note that these loadings must be equal in magnitude and opposite in direction on each of the segments (Newton's third law). The magnitude of each unknown can be determined by applying the three equations of equilibrium to either segment AC or CB. A *direct solution* for $\mathbf{A}_C$ is obtained by applying $\Sigma F_x = 0$; $\mathbf{V}_C$ is obtained directly from $\Sigma F_y = 0$; and $\mathbf{M}_C$ is determined by summing moments about point C, $\Sigma M_C = 0$, in order to eliminate the moments of the unknowns $\mathbf{A}_C$ and $\mathbf{V}_C$.

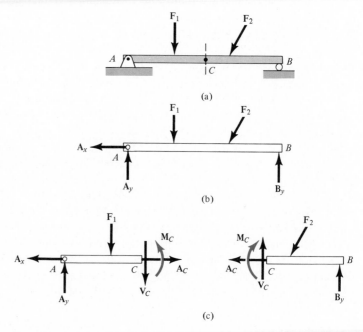

(a)

(b)

(c)

Fig. 8–1

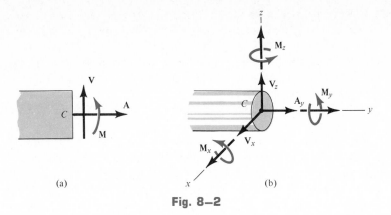

Fig. 8–2

In structural mechanics, the force components **A,** acting normal to the beam at the cut section, and **V,** acting tangent to the section, are termed the *axial force* and *shear force,* respectively. The couple **M** is referred to as the *bending moment,* Fig. 8–2a. In three dimensions, a general internal force and moment resultant will act at the section. The x, y, z components of these loadings are shown in Fig. 8–2b. Here $\mathbf{A}_y$ is the *axial force,* $\mathbf{V}_x$ and $\mathbf{V}_z$ are *shear force components,* $\mathbf{M}_y$ is a *torsional or twisting moment,* and $\mathbf{M}_x$ and $\mathbf{M}_z$ are *bending moment components.* For structural applications, these loadings are often represented at the geometric center or centroid (C) of the section's cross-sectional area. In general, the magnitude for each loading will be different at various points along the axis of a structural member. In all cases, however, the method of sections can be used to determine their values.

Free-Body Diagrams. Since frames and machines are composed of *multi-force members,* each of their members will generally be subjected to axial, shear, and bending loadings. For example, consider the frame shown in Fig. 8–3a. If section *aa* is passed through the frame to determine the internal loadings at points F, G, and H, the resulting free-body diagram of the top portion of this section is shown in Fig. 8–3b. At each point where a member is sectioned there is an unknown axial force, shear force, and bending moment. As a result, we cannot apply the three equations of equilibrium to this section in order to obtain these *nine unknowns.** Instead, to solve this problem one has to *first dismember* the frame and determine the reactions at the connections of the members using the techniques of Sec. 6.6. Once this is done, one can section *each member* at its appropriate point and then apply the three equations of equilibrium to determine **A, V,** and **M.** For example, the free-body diagram of segment DG, Fig. 8–3c, can be used to determine the internal loadings at G provided the pin reactions $\mathbf{D}_x$ and $\mathbf{D}_y$ are known.

*Recall that this method of analysis worked well for trusses since truss members are *straight two-force members* which only support an axial load.

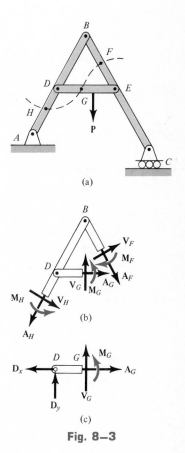

(a)

(b)

(c)

Fig. 8–3

PROCEDURE FOR ANALYSIS

The following procedure provides a means for applying the method of sections to determine the internal loadings at a specific location in a member.

Support Reactions.   Before the member is "cut" or sectioned, it may first be necessary to determine the member's support reactions, so that the equilibrium equations are used only to solve for the internal loadings when the member is sectioned. If the member is part of a frame or machine, the reactions at its connections are determined using the methods of Sec. 6.6.

Free-Body Diagram.   Keep all distributed loadings, couple moments, torques, and forces acting on the member in their *exact location,* then pass an imaginary section through the member, perpendicular to its axis at the point where the internal loading is to be determined. Draw a free-body diagram of one of the "cut" segments on either side of the section and indicate the *x, y, z* components of the force and moment resultants at the section. In particular, if the member is subjected to a *coplanar* system of forces, only **A, V,** and **M** act at the section. In many cases it may be possible to tell by inspection the proper sense of the unknown loadings; however, if this seems difficult, the sense can be assumed.

Equations of Equilibrium.   Apply the equations of equilibrium to obtain the unknown internal loadings. In most cases, moments should be summed at the section about axes passing through the *centroid* of the member's cross-sectional area, in order to eliminate the unknown axial and shear forces and thereby obtain direct solutions for the moment components. If the solution of the equilibrium equations yields a negative scalar, the assumed sense of the quantity is opposite to that shown on the free-body diagram.

The following examples numerically illustrate this procedure.

Example 8–1

The bar is fixed at its end and is loaded as shown in Fig. 8–4a. Determine the internal axial force at points B and C.

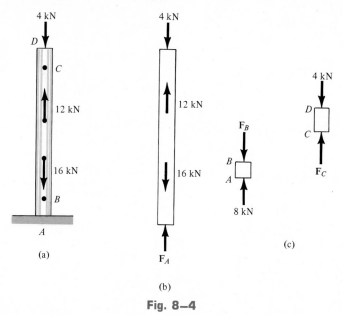

(a)

(b)

(c)

Fig. 8–4

Solution

Support Reactions. A free-body diagram of the entire bar is shown in Fig. 8–4b. By inspection, only an axial force $\mathbf{F}_A$ acts at the fixed support since the loads are symmetrically applied along the bar's axis.

$$+ \uparrow \Sigma F_z = 0; \quad F_A - 16 \text{ kN} + 12 \text{ kN} - 4 \text{ kN} = 0 \qquad F_A = 8 \text{ kN}$$

Free-Body Diagrams. The internal forces at B and C will be found using the free-body diagrams of the sectioned bar shown in Fig. 8–4c. In particular, segments AB and DC will be chosen here since they contain the *least* number of forces.

Equations of Equilibrium
  Segment AB

$$+ \uparrow \Sigma F_z = 0; \qquad 8 \text{ kN} - F_B = 0 \qquad F_B = 8 \text{ kN} \qquad\qquad Ans.$$

  Segment DC

$$+ \uparrow \Sigma F_z = 0; \qquad F_C - 4 \text{ kN} = 0 \qquad F_C = 4 \text{ kN} \qquad\qquad Ans.$$

Try working this problem in the following manner: Determine F_B from segment BD. (Note that this approach does *not require* solution for the support reaction at A.) Using the result for F_B, isolate segment BC to determine F_C.

Example 8–2

The circular shaft is subjected to three concentrated torques as shown in Fig. 8–5a. Determine the internal torques at points B and C.

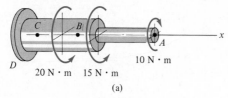

(a)

Fig. 8–5a

Solution

Support Reactions. Since the shaft is subjected only to collinear torques, a torque reaction occurs at the support, Fig. 8–5b. Using the right-hand rule to define the positive directions of the torques, we require

$\Sigma M_x = 0;$
$$-10 \text{ N} \cdot \text{m} + 15 \text{ N} \cdot \text{m} + 20 \text{ N} \cdot \text{m} - T_D = 0 \qquad T_D = 25 \text{ N} \cdot \text{m}$$

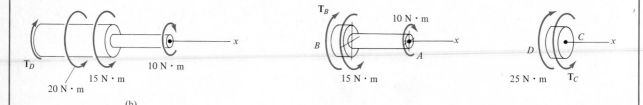

(b)

(c)

Fig. 8–5b and c

Free-Body Diagrams. The internal torques at B and C will be found using the free-body diagrams of the shaft segments AB and CD shown in Fig. 8–5c.

Equations of Equilibrium. Applying the equation of moment equilibrium along the shaft's axis, we have

Segment AB

$\Sigma M_x = 0; \quad -10 \text{ N} \cdot \text{m} + 15 \text{ N} \cdot \text{m} - T_B = 0 \qquad T_B = 5 \text{ N} \cdot \text{m} \qquad$ *Ans.*

Segment CD

$\Sigma M_x = 0; \qquad\qquad T_C - 25 \text{ N} \cdot \text{m} \qquad T_C = 25 \text{ N} \cdot \text{m} \qquad\qquad$ *Ans.*

Try to solve for T_C by using segment CA. Note that this approach does not require a solution for the support reaction at D.

Example 8–3

The three smooth plates of the joint shown in Fig. 8–6*a* are connected by two bolts. Determine the shear force in each bolt at section *bb* between the plates and the resultant axial force supported by the top plate at section *aa*. Assume that each bolt carries an equal amount of shear force.

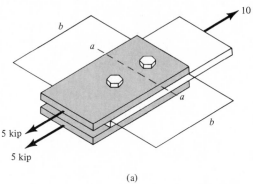

5 kip

5 kip

10 kip

Fig. 8–6

(a)

Solution

Support Reactions. The free-body diagrams and equilibrium requirements for the top *or* bottom plate, the middle plate, and a bolt are shown in Fig. 8–6*b*.

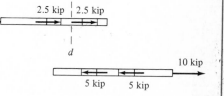

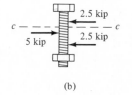

(b)

Free-Body Diagrams. The free-body diagrams of sections *bb* and *aa* are shown in Fig. 8–6*c*.

Equations of Equilibrium. Applying the equations of force equilibrium, we have

$$\xrightarrow{+} \Sigma F_x = 0; \qquad 5 \text{ kip} - 2V_b = 0 \qquad V_b = 2.5 \text{ kip} \qquad \textit{Ans.}$$

$$\xrightarrow{+} \Sigma F_x = 0; \quad 5 \text{ kip} - 2.5 \text{ kip} - A_p = 0 \qquad A_p = 2.5 \text{ kip} \qquad \textit{Ans.}$$

Note: It is also possible to solve this problem in another way. For example, show that sections *cc* through the bolt and *dd* through the plate, Fig. 8–6*b*, yield the results previously obtained.

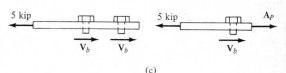

(c)

Example 8–4

Determine the internal axial force, shear force, and bending moment acting just to the left, point B, and just to the right, point C, of the 6-kN concentrated force shown on the beam in Fig. 8–7a.

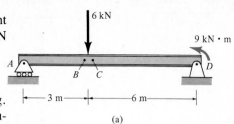

(a)

Solution

Support Reactions. The free-body diagram of the beam is shown in Fig. 8–7b. When computing the external reactions, note that the 9-kN · m couple is a free vector and can therefore act anywhere on the beam. Here we will only determine A_y since segments AB and AC will be used for the analysis.

$$\zeta + \Sigma M_D = 0; \quad 9 \text{ kN} \cdot \text{m} + (6 \text{ kN})(6 \text{ m}) - A_y(9 \text{ m}) = 0$$
$$A_y = 5 \text{ kN}$$

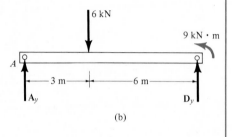

(b)

Free-Body Diagrams. The free-body diagrams of the left segments AB and AC of the beam are shown in Fig. 8–7c and d. Note that the 9-kN · m couple moment is *not included* on these diagrams. Although it is a free vector, it must be kept in its original position until *after* the section is made and the appropriate body isolated; only then will the true loading situation be represented.

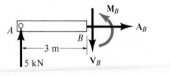

(c)

Equations of Equilibrium

Segment AB

$\xrightarrow{+} \Sigma F_x = 0;$	$A_B = 0$	*Ans.*
$+ \uparrow \Sigma F_y = 0;$	$5 \text{ kN} - V_B = 0 \qquad V_B = 5 \text{ kN}$	*Ans.*
$\zeta + \Sigma M_B = 0;$	$-(5 \text{ kN})(3 \text{ m}) + M_B = 0 \qquad M_B = 15 \text{ kN} \cdot \text{m}$	*Ans.*

Segment AC

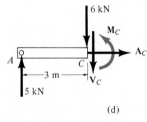

$\xrightarrow{+} \Sigma F_x = 0;$	$A_C = 0$	*Ans.*
$+ \uparrow \Sigma F_y = 0;$	$5 \text{ kN} - 6 \text{ kN} - V_C = 0 \qquad V_C = -1 \text{ kN}$	*Ans.*
$\zeta + \Sigma M_C = 0;$	$-(5 \text{ kN})(3 \text{ m}) + M_C = 0 \qquad M_C = 15 \text{ kN} \cdot \text{m}$	*Ans.*

(d)

Fig. 8–7

Note that the moment arm for the 5-kN force in both cases is approximately 3 m, since B and C are "almost" coincident.

Example 8–5

Determine the internal axial force, shear force, and bending moment acting at point B of the two-member frame shown in Fig. 8–8a.

Solution

Support Reactions. A free-body diagram of each member is shown in Fig. 8–8b. Since CD is a two-force member, the equations of equilibrium have to be applied only to member AC.

$\zeta + \Sigma M_A = 0;$ $-400 \text{ lb}(4 \text{ ft}) + (\frac{3}{5})F_{DC}(8 \text{ ft}) = 0$ $F_{DC} = 333.3 \text{ lb}$

$\overset{+}{\rightarrow}\Sigma F_x = 0;$ $-A_x + (\frac{4}{5})(333.3 \text{ lb}) = 0$ $A_x = 266.7 \text{ lb}$

$+\uparrow \Sigma F_y = 0;$ $A_y - 400 \text{ lb} + \frac{3}{5}(333.3 \text{ lb}) = 0$ $A_y = 200 \text{ lb}$

Free-Body Diagrams. Passing an imaginary section perpendicular to the axis of member AC through point B yields the free-body diagrams of segments AB and BC shown in Fig. 8–8c. Notice that it is important when constructing these diagrams to keep the distributed loading exactly as it is until *after* the section is made. Only then can it be simplified to a single resultant force. Why? Also, notice that $\mathbf{A}_B$, $\mathbf{V}_B$, and $\mathbf{M}_B$ act with equal magnitude but opposite direction on each segment—Newton's third law.

Equations of Equilibrium. Applying the equations of equilibrium to segment AB, we have

$\overset{+}{\rightarrow}\Sigma F_x = 0;$ $A_B - 266.7 \text{ lb} = 0$ $A_B = 266.7 \text{ lb}$ *Ans.*

$+\uparrow \Sigma F_y = 0;$ $200 \text{ lb} - 200 \text{ lb} - V_B = 0$ $V_B = 0$ *Ans.*

$\zeta + \Sigma M_B = 0;$ $M_B - 200 \text{ lb}(4 \text{ ft}) + 200 \text{ lb}(2 \text{ ft}) = 0$

                    $M_B = 400 \text{ lb} \cdot \text{ft}$ *Ans.*

As an exercise, try to obtain the foregoing results using segment BC.

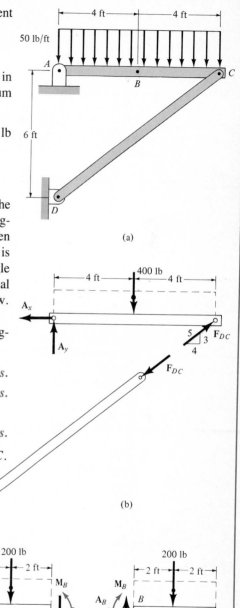

(a)

(b)

(c)

Fig. 8–8

Example 8–6

Determine the axial force, shear force, and bending moment acting at point E of the frame loaded as shown in Fig. 8–9a.

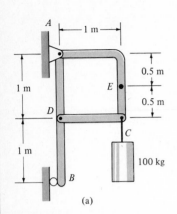

(a)

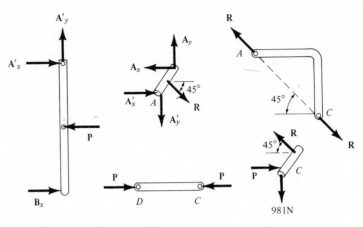

(b)

Fig. 8–9

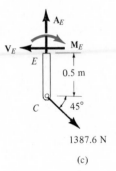

(c)

Solution

Support Reactions. The free-body diagram of each member of the frame is shown in Fig. 8–9b. Note that members AC and CD are two-force members. The pins at A and C are also included here since there are *three interactions* at each pin. At pin A A_x and A_y represent the effect of the support on the pin, A_x' and A_y' represent the effect of member ADB on the pin, and $\mathbf{R}$ represents the effect of member AC on the pin. Identify the three forces acting on pin C.

Applying the force equilibrium equation in the vertical direction to pin C gives

$$+\uparrow \Sigma F_y = 0; \quad R \sin 45° - 981 \text{ N} = 0 \quad R = 1387.6 \text{ N}$$

Since we wish to determine the external loadings at E, it is not necessary to determine $\mathbf{P}$.

Free-Body Diagram. The free-body diagram of section CE is shown in Fig. 8–9c.

Equations of Equilibrium

$$\xrightarrow{+}\Sigma F_x = 0; \quad 1387.6 \cos 45° \text{ N} - V_E = 0 \qquad V_E = 981 \text{ N} \qquad Ans.$$

$$+\uparrow \Sigma F_y = 0; \quad -1387.6 \sin 45° \text{ N} + A_E = 0 \qquad A_E = 981 \text{ N} \qquad Ans.$$

$$\downarrow + \Sigma M_E = 0; \quad 1387.6 \cos 45° \text{ N}(0.5 \text{ m}) - M_E = 0 \qquad M_E = 490.5 \text{ N} \cdot \text{m} \quad Ans.$$

Example 8–7

A force of $\mathbf{F} = \{-3\mathbf{i} + 7\mathbf{j} - 4\mathbf{k}\}$ kN acts at the corner of a beam extended from a fixed wall as shown in Fig. 8–10a. Determine the internal loadings at a section passing through point A.

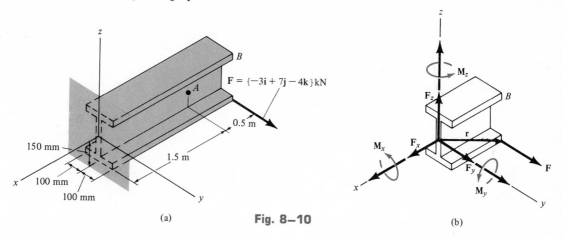

(a)   **Fig. 8–10**   (b)

Solution

This problem can be solved by considering a section of the beam that will *not* involve the support reactions.

Free-Body Diagram. The free-body diagram of segment AB is shown in Fig. 8–10b. The components of the resultant force $\mathbf{F}_A$ and moment $\mathbf{M}_A$ pass through the *centroid* of the cross-sectional area at A.

Equations of Equilibrium

$$\Sigma \mathbf{F} = 0; \qquad \mathbf{F}_A - 3\mathbf{i} + 7\mathbf{j} - 4\mathbf{k} = 0$$
$$\mathbf{F}_A = \{3\mathbf{i} - 7\mathbf{j} + 4\mathbf{k}\} \text{ kN} \qquad \qquad Ans.$$

$$\Sigma \mathbf{M}_A = 0; \quad \mathbf{M}_A + \mathbf{r} \times \mathbf{F} = \mathbf{M}_A + \begin{vmatrix} \mathbf{i} & \mathbf{j} & \mathbf{k} \\ -0.5 & 0.1 & -0.15 \\ -3 & 7 & -4 \end{vmatrix} = 0$$

$$\mathbf{M}_A = \{-0.650\mathbf{i} + 1.55\mathbf{j} + 3.20\mathbf{k}\} \text{ kN} \cdot \text{m} \qquad Ans.$$

Note that $\mathbf{F}_x = \{3\mathbf{i}\}$ kN represents the axial force A, whereas $\mathbf{F}_y = \{-7\mathbf{j}\}$ kN and $\mathbf{F}_z = \{4\mathbf{k}\}$ kN are components of the shear force $V = \sqrt{F_y^2 + F_z^2}$. Also, the torsional moment is $\mathbf{M}_x = \{-0.65\mathbf{i}\}$ kN · m, and the bending moment is determined from the components $\mathbf{M}_y = \{1.55\mathbf{j}\}$ kN · m and $\mathbf{M}_z = \{3.20\mathbf{k}\}$ kN · m; i.e., $M_b = \sqrt{(M_y)^2 + (M_z)^2}$.

Problems

8–1. Draw the specified free-body diagram in each of the following problems. Neglect the weights of the members unless otherwise stated. Although not shown on each photo, assume the geometry (size and angles) is known.

a. The torque wrench is subjected to a downward force of 15 lb applied perpendicular to the handle at A. Draw the free-body diagram of the section AB of the wrench and indicate the internal axial force, shear force and moment at point B. If section AC is considered, which of the internal loadings increases from B to C? Note that this increase is the reason why the handle was tapered to a thicker cross-sectional area at C than at B.

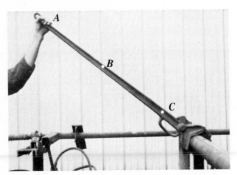

Prob. 8–1a

b. The beam AF has been designed to support a roof which will transmit a vertical force of 2 kN to the beam at points B, C, D, and E. Also, a vertical force of 1 kN will be applied at its end, F. It is assumed to be supported by a pin connection at A and roller at H. Draw a free-body diagram of segments AG and GF of the beam.

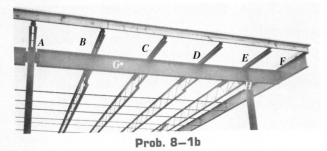

Prob. 8–1b

c. The overhead crane AB supports a load L of 300 lb. If the crane (beam) has a weight of 80 lb/ft and is supported on wheels (rollers) at its ends A and B, draw the free-body diagram of segment CB of the crane.

Prob. 8–1c

d. The bar AB is used to lift a bundle of pipes P which develop vertical forces of 200 lb in each of the cables attached at its ends A and B. Assume that each of the slender rods AC and CB is pin connected to the bar and hook at its ends. Draw a free-body diagram of segment AD of the bar.

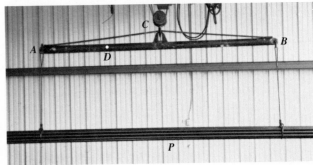

Prob. 8–1d

8–2. Draw a free-body diagram of bolts A and B used for the connection. Then determine the maximum shear force in B and the tensile force in A. Neglect the effects of friction in the calculation.

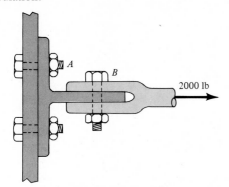

Prob. 8–2

8–3. The joint is held together using a single bolt as shown. Draw a free-body diagram of the bolt and then make appropriate sections through the bolt in order to determine the shear force between (a) plates A and B, and (b) plates B and C.

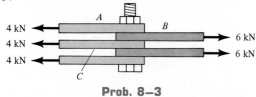

Prob. 8–3

***8–4.** The pliers are used to grip the pipe. If a force of 20 lb is applied to the handles, determine the shear force developed in the pin at A between the two parts C and D. Assume the jaws of the pliers exert only normal forces on the pipe.

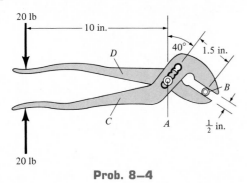

Prob. 8–4

8–5. The beam is supported by a bolt at C and a cable AB. The cable is attached to the beam using a clevis, as shown in the adjacent sketch. If $w = 2$ kN/m, determine the shear force in the pin at A.

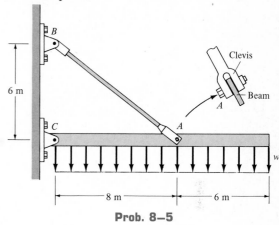

Prob. 8–5

8–6. Determine the maximum intensity of the distributed loading w that can be placed on the beam if the bolt at A can support a shear force of 25 kN before it fails.

8–7. The column is fixed to the floor and is subjected to the loads shown. Determine the internal axial force, shear force, and moment at points A and B.

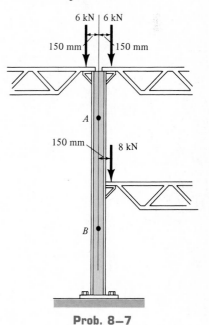

Prob. 8–7

349

*8–8. The forces act on the shaft shown. Determine the internal axial force at points A, B, and C.

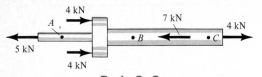

Prob. 8–8

8–9. The rod is subjected to the forces shown. Determine the internal axial force at points A, B, and C.

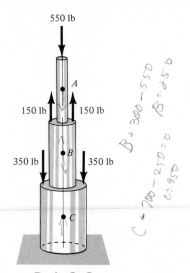

Prob. 8–9

8–10. The shaft is supported by the two smooth bearings A and B. The four pulleys attached to the shaft are used to transmit power to adjacent machinery. If the torques applied to the pulleys are as shown, determine the internal torques at points C, D, and E.

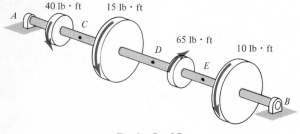

Prob. 8–10

8–11. The shaft is supported by smooth bearings at A and B and subjected to the torques shown. Determine the internal torques at points C, D, and E.

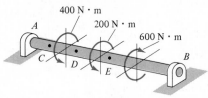

Prob. 8–11

*8–12. Three torques act on the shaft as shown. Determine the internal torques at points A, B, C, and D.

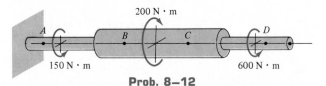

Prob. 8–12

8–13. The beam-column is fixed to the floor and supports the load shown. Determine the internal axial force, shear force, and moment at points A and B due to this loading.

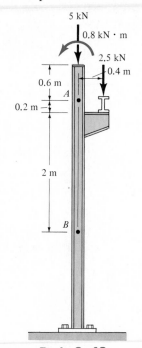

Prob. 8–13

8–14. Determine the axial force, shear force, and moment at point C.

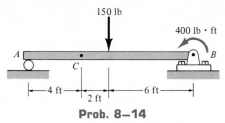

150 lb

400 lb · ft

A

C

B

4 ft

2 ft

6 ft

Prob. 8–14

8–15. The jack AB is used to straighten the bent beam DE using the arrangement shown. If the axial compressive force in the jack is 5000 lb, determine the internal moment developed at point C of the top beam. Neglect the weights of the beams.

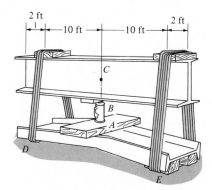

2 ft

10 ft

10 ft

2 ft

C

B

A

D

E

Probs. 8–15 / 8–16

*8–16. Solve Prob. 8–15 assuming that each beam has a uniform weight of 150 lb/ft.

8–17. Determine the axial force, shear force, and moment at point A if the clamp exerts a compressive force of $F = 85$ lb on board B. Force **F** acts along the central axis of the screw.

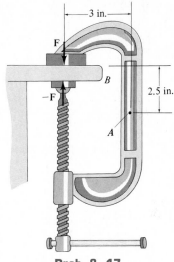

3 in.

F

B

2.5 in.

–F

A

Prob. 8–17

8–18. Determine the internal axial force, shear force, and moment at point D.

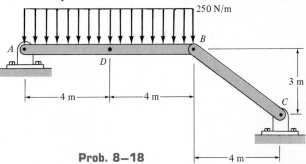

250 N/m

A

D

B

4 m

4 m

3 m

C

4 m

Prob. 8–18

8–19. Determine the axial force, shear force, and moment acting at point C and at point D, which is located just to the right of the roller support at B.

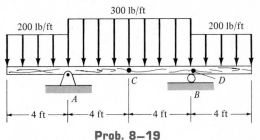

300 lb/ft

200 lb/ft

200 lb/ft

C

D

A

B

4 ft

4 ft

4 ft

4 ft

Prob. 8–19

*8–20. Determine the shear force and moment in the beam at points C and D.

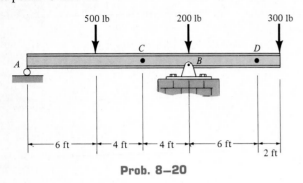

Prob. 8–20

8–21. Determine the axial force, shear force, and moment at points B and C of the beam.

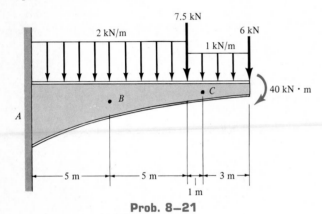

Prob. 8–21

8–22. The carriage of the crane exerts forces of 2000 lb on the beam as shown. Determine the moment developed in the beam at point C when the carriage is in the position shown. Neglect the weight of the beam and carriage.

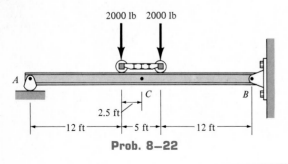

Prob. 8–22

8–23. The beam weighs 280 lb/ft. Determine the internal axial force, shear force, and moment at point C.

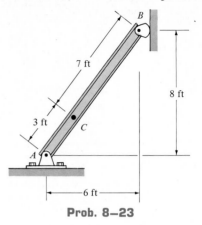

Prob. 8–23

*8–24. The member supports the loads at the rigid joint B. Determine the internal axial force, shear force, and moment at point D.

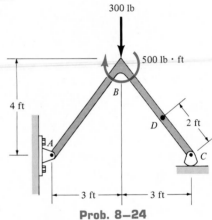

Prob. 8–24

8–25. Determine the internal axial force, shear force, and moment at points C and D of the beam.

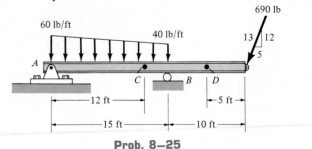

Prob. 8–25

■8–26. Determine the internal shear force and moment acting at point C of the beam. Use Simpson's rule to evaluate the integrals used for the calculation.

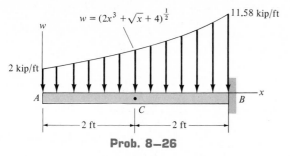

$w = (2x^3 + \sqrt{x} + 4)^{\frac{1}{2}}$

w

11.58 kip/ft

2 kip/ft

A C B x

|← 2 ft →|← 2 ft →|

Prob. 8–26

8–27. Determine the internal axial force, shear force, and moment acting at points D and E of the two-member frame.

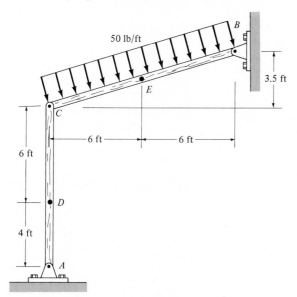

B

50 lb/ft

E

3.5 ft

C

|← 6 ft →|← 6 ft →|

6 ft

D

4 ft

A

Prob. 8–27

*8–28. Determine the internal axial force, shear force, and moment at point D of the two-member frame.

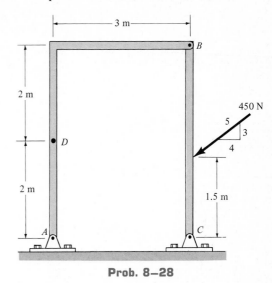

|← 3 m →|

B

2 m

D

450 N

5

3

4

2 m

1.5 m

A C

Prob. 8–28

8–29. Determine the internal axial force, shear force, and moment at point C.

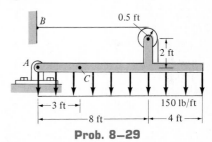

B 0.5 ft

2 ft

A

C

150 lb/ft

|← 3 ft →|

|← 8 ft →|← 4 ft →|

Prob. 8–29

8–30. Determine the internal axial force, shear force, and moment at point C.

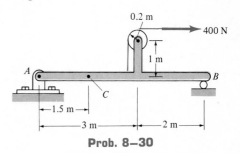

0.2 m

400 N

1 m

A

B

C

|← 1.5 m →|

|← 3 m →|← 2 m →|

Prob. 8–30

8-31. Determine the internal axial force, shear force, and moment at points E and D. There is a pin or hinge at B.

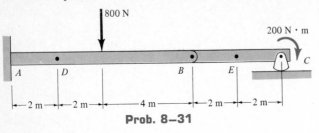

Prob. 8-31

*8-32. Determine the ratio of a/b for which the shear force will be zero at the midpoint C of the beam.

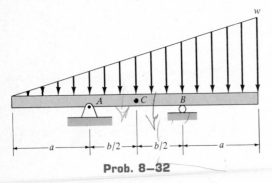

Prob. 8-32

8-33. Determine the internal axial force, shear force, and moment acting at points A and B of the smooth hook.

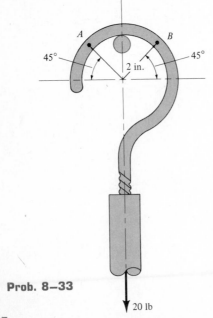

Prob. 8-33

8-34. Two forces act on the column. Determine the resultant internal force and moment components acting in the x, y, z directions at a horizontal section taken through point O. Express the results as Cartesian force and moment vectors.

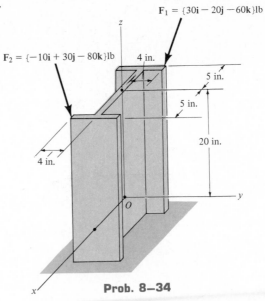

Prob. 8-34

8-35. Two forces act at the corners of the concrete block. If the block has a uniform density of 150 lb/ft^3, determine the internal force and moment components acting in the x, y, and z directions at a horizontal section taken through point O.

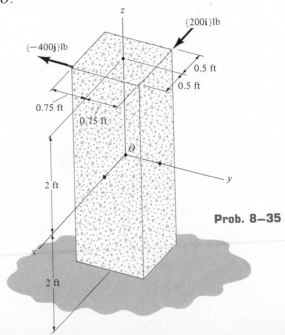

Prob. 8-35

354

*8–36. A vertical force of 60 N is applied to the handle of the pipe wrench as shown. Determine the x, y, z components of internal force and moment at point A.

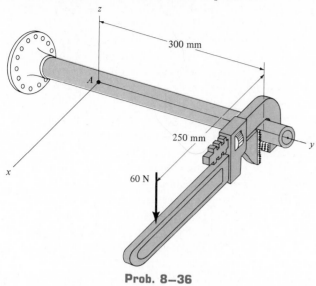

Prob. 8–36

8–37. Determine the x, y, z components of internal force and moment at point C in the pipe assembly. Neglect the weight of the pipe.

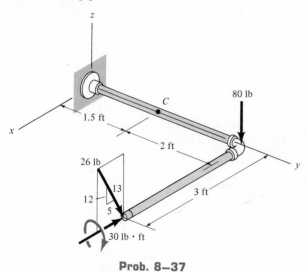

Prob. 8–37

8–38. Determine the internal loadings in the rod at point F. *Hint:* See Example 5–18 for part of the analysis.

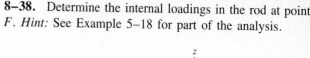

Prob. 8–38

355

*8.2 Shear and Moment Diagrams for a Beam

Beams are structural members which are designed to support loadings applied perpendicular to their axis. In general, beams are long, straight bars, having a constant cross-sectional area. Often beams are classified as to how they are supported. For example, a *simply supported beam* is pinned at one end and roller supported at the other, Fig. 8–11, whereas a *cantilevered beam* is fixed at one end and free at the other. The actual design of a beam requires a detailed knowledge of the *variation* of the internal shear force V and bending moment M acting at *each point* along the axis of the beam. After the force and bending-moment analysis is complete, one can then use the theory of strength of materials to determine the beam's required cross-sectional area.

The *variations* of V and M as a function of the position x of an *arbitrary point* along the beam's axis can be obtained by using the method of sections discussed in Sec. 8.1. Here, however, it is necessary to locate the imaginary section at an arbitrary distance x from the end of the beam rather than at a specified point. If the results are plotted, the graphical variations of V and M as a function of x are termed the *shear diagram* and *bending-moment diagram*, respectively.

In general, the internal shear and bending-moment functions will be discontinuous, or their slope will be discontinuous at points where a distributed load changes or where concentrated forces or couples are applied. Because of this, shear and bending-moment functions must be determined for *each segment* of the beam located between any two discontinuities of loading. For example, sections located at x_1, x_2, and x_3 will have to be used to describe the variation of V and M throughout the length of the beam in Fig. 8–11. These functions will be valid *only* within regions from O to a for x_1, from a to b for x_2, and from b to l for x_3.

The internal axial force will not be considered in the following discussion for two reasons. In most cases, the loads applied to a beam act perpendicular to the beam's axis and hence produce only an internal shear force and bending moment. For design purposes, the beam's resistance to shear, and particularly to bending, is more important than its ability to resist axial force.

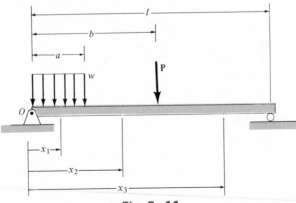

Fig. 8–11

Sign Convention. Before presenting a method for determining the shear and bending moment as functions of x and later plotting these functions (shear and bending-moment diagrams), it is first necessary to establish a *sign convention* so as to define "positive" and "negative" shear force and bending moment acting in the beam. [This is analogous to assigning coordinate directions x positive to the right and y positive upward, when plotting a function $y = f(x)$.] Although the choice of a sign convention is arbitrary, here we will choose the one used in the majority of books on mechanics. It is illustrated in Fig. 8–12a. On the *left-hand face* (L.H.F.) of a beam segment, the internal shear force **V** acts downward and the internal moment **M** acts counterclockwise. In accordance with Newton's third law, an equal and opposite shear force and bending moment must act on the *right-hand face* (R.H.F.) of a segment. Perhaps an easy way to remember this sign convention is to isolate a small beam segment and note that *positive shear tends to rotate the segment clockwise*, Fig. 8–12b, and a *positive bending moment tends to bend the segment, if it were elastic, concave upward*, Fig. 8–12c.

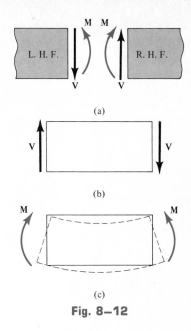

(a)

(b)

(c)

Fig. 8–12

PROCEDURE FOR ANALYSIS

The following procedure provides a method for constructing the shear and bending-moment diagrams for a beam.

Support Reactions. Determine all the reactive couples and forces acting of the beam and resolve the forces into components acting perpendicular and parallel to the beam's axis.

Shear and Moment Functions. Specify separate coordinates x having an origin at the beam's *left end* and extending to regions of the beam between concentrated forces and/or couples and where there is no discontinuity of distributed loading. Section the beam perpendicular to its axis at each distance x and from the free-body diagram of one of the segments, determine the unknowns V and M at the cut section as functions of x. On the free-body diagram, **V** and **M** should be shown acting in their *positive sense*, in accordance with the sign conventions given in Fig. 8–12. V is obtained from $\Sigma F_y = 0$ and M is obtained by summing moments about point S located at the cut section, $\Sigma M_S = 0$.

Shear and Moment Diagrams. Plot the shear diagram (V versus x) and the moment diagram (M versus x). If computed values of the functions describing V and M are *positive*, the values are plotted above the x axis, whereas negative values are plotted below the axis. Generally, it is convenient to plot the shear and bending-moment diagrams directly below the free-body diagram of the beam.

The following examples numerically illustrate this procedure.

Example 8-8

Draw the shear and bending-moment diagrams for the beam shown in Fig. 8-13a.

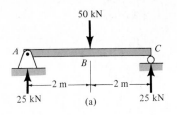

(a)

Solution

Support Reactions. The support reactions have been computed, Fig. 8-13a.

Shear and Moment Functions. The beam is sectioned at an arbitrary distance x from point A, extending within the region AB, and the free-body diagram of the left segment is shown in Fig. 8-13b. The unknowns $\mathbf{V}$ and $\mathbf{M}$ are indicated acting in the *positive sense* on the right-hand face of the segment according to the established sign convention. Why? Applying the equilibrium equations yields

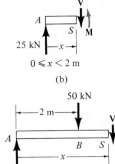

(b)

$$+\uparrow \Sigma F_y = 0; \qquad\qquad V = 25 \text{ kN} \qquad\qquad (1)$$
$$\wr+\Sigma M_S = 0; \qquad\qquad M = 25x \text{ kN} \cdot \text{m} \qquad\qquad (2)$$

A free-body diagram for a left segment of the beam extending a distance x within the region BC is shown in Fig. 8-13c. As always, $\mathbf{V}$ and $\mathbf{M}$ are shown acting in the positive sense. Hence,

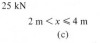

(c)

$$+\uparrow \Sigma F_y = 0; \qquad\qquad 25 - 50 - V = 0$$
$$V = -25 \text{ kN} \qquad\qquad (3)$$
$$\wr+\Sigma M_S = 0; \qquad\qquad M + 50(x - 2) - 25(x) = 0$$
$$M = (100 - 25x) \text{ kN} \cdot \text{m} \qquad\qquad (4)$$

Shear and Moment Diagrams. When Eqs. (1) to (4) are plotted within the regions in which they are valid, the shear and bending-moment diagrams shown in Fig. 8-13d are obtained. The shear diagram indicates that the internal shear force is always 25 kN (positive) along beam segment AB. Just to the right of point B, the shear force changes sign and remains at a constant value of -25 kN for segment BC. The moment diagram starts at zero, increases linearly to point C at $x = 2$ m, where $M_{\max} = 25$ kN(2 m) $= 50$ kN $\cdot$ m, and thereafter decreases back to zero.

It is seen in Fig. 8-13d that the graph of the shear and moment diagrams is discontinuous at points of concentrated force, i.e., points A, B, and C. For this reason, as stated earlier, it is necessary to express both the shear and bending-moment functions separately for regions between concentrated loads. It should be realized, however, that all loading discontinuities are mathematical, arising from the *idealization of a concentrated force and couple*. Physically, all loads are applied over a finite area, and if this load variation could be accounted for, all shear and bending-moment diagrams would actually be continuous over the beam's entire length.

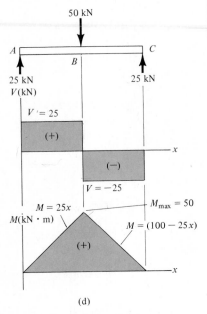

(d)

Fig. 8-13

Example 8–9

Draw the shear and bending-moment diagrams for the beam shown in Fig. 8–14a.

Solution

Support Reactions. The support reactions have been computed, Fig. 8–14a.

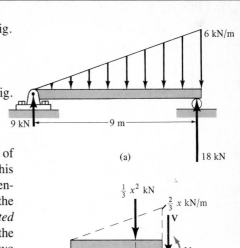

(a)

Shear and Moment Functions. A free-body diagram for a left segment of the beam is shown in Fig. 8–14b. The distributed loading acting on this segment, which has an intensity of $\frac{2}{3}x$ at its end, is replaced by a concentrated force *after* the segment is isolated as a free-body diagram. Since the distributed load now has a length x, the *magnitude* of the *concentrated force* is equal to $\frac{1}{2}(x)(\frac{2}{3}x) = \frac{1}{3}x^2$. This force *acts through the centroid* of the distributed loading area, a distance $\frac{1}{3}x$ from the right end. Applying the two equations of equilibrium yields*

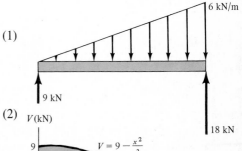

(b)

$$+\uparrow \Sigma F_y = 0; \qquad 9 - \frac{1}{3}x^2 - V = 0$$

$$V = \left(9 - \frac{x^2}{3}\right) \text{kN} \qquad (1)$$

$$\zeta + \Sigma M_S = 0; \qquad M + \frac{1}{3}x^2\left(\frac{x}{3}\right) - 9x = 0$$

$$M = \left(9x - \frac{x^3}{9}\right) \text{kN} \cdot \text{m} \qquad (2)$$

Shear and Moment Diagrams. The shear and bending-moment diagrams shown in Fig. 8–14c are obtained by plotting Eqs. (1) and (2).

The point of *zero shear* can be found using Eq. (1):

$$V = 9 - \frac{x^2}{3} = 0$$

$$x = 5.20 \text{ m}$$

This value of x happens to represent the point on the beam where the *maximum moment* occurs (see Sec. 8.3). Using Eq. (2), we have

$$M_{\max} = \left(9(5.20) - \frac{(5.20)^3}{9}\right) \text{kN} \cdot \text{m}$$

$$= 31.18 \text{ kN} \cdot \text{m}$$

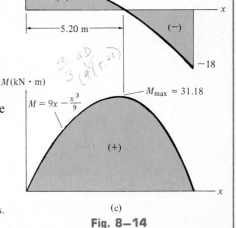

(c)

Fig. 8–14

*Note from Eqs. (2) and (1) that $dM/dx = V$ and $dV/dx = -w$. This is a consequence of Eqs. 8–1 and 8–2, which will be presented in the next section.

359

Problems

8–39. Draw the shear and bending-moment diagrams for the beam (a) in terms of the parameters shown; (b) set $P = 800$ lb, $a = 8$ ft, $b = 4$ ft.

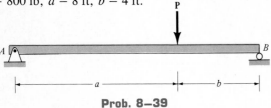

Prob. 8–39

*8–40. Draw the shear and bending-moment diagrams for the cantilever beam.

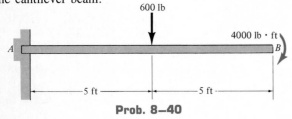

Prob. 8–40

8–41. Draw the shear and bending-moment diagrams for the beam.

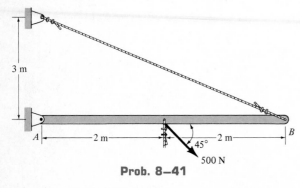

Prob. 8–41

8–42. Draw the shear and bending-moment diagrams for the beam $ABCD$ (a) in terms of the parameters shown; (b) set $P = 400$ lb, $a = 4$ ft.

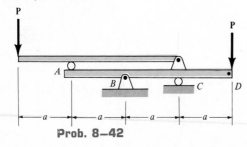

Prob. 8–42

8–43. Draw the shear and bending-moment diagrams for the beam ABC.

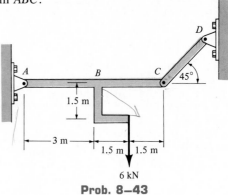

Prob. 8–43

*8–44. Draw the shear and bending-moment diagrams for the beam (a) in terms of the parameters shown; (b) set $M = 500$ N $\cdot$ m, $L = 8$ m.

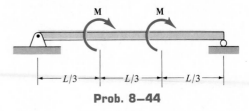

Prob. 8–44

8–45. Draw the shear and bending-moment diagrams for the beam (a) in terms of the parameters shown; (b) set $P = 800$ lb, $a = 5$ ft, $L = 12$ ft.

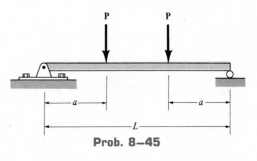

Prob. 8–45

8–46. Draw the shear and bending-moment diagrams for the beam.

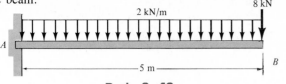

Prob. 8–46

8–47. Draw the shear and bending-moment diagrams for the beam (a) in terms of the parameters shown; (b) set $w = 800$ lb/ft, $L = 12$ ft.

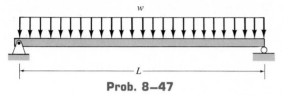

Prob. 8–47

*8–48. Draw the shear and bending-moment diagrams for the beam.

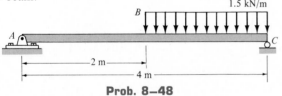

Prob. 8–48

8–49. Draw the shear and bending-moment diagrams for the beam.

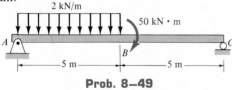

Prob. 8–49

8–50. Draw the shear and bending-moment diagrams for beam ABC. Note that there is a pin at B. Solve the problem for (a) the parameters shown; (b) set $w = 5$ kN/m, $L = 10$ m.

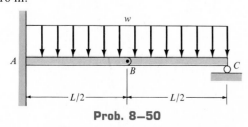

Prob. 8–50

8–51. The traveling crane consists of a 5-m-long beam having a uniform mass of 20 kg/m. The chain hoist and its supported load exert a force of 8 kN on the beam when $x = 2$ m. Draw the shear and moment diagrams for the beam. The guide wheels at the ends A and B exert only vertical reactions on the beam. Neglect the size of the trolley at C.

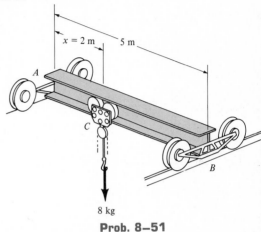

Prob. 8–51

*8–52. The wall is subjected to a soil pressure which is assumed to vary as shown. If it is fixed supported at A, determine the internal moment as a function of x and draw the shear and bending-moment diagrams.

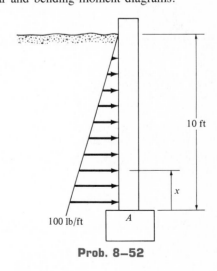

Prob. 8–52

8–53. Draw the shear and bending-moment diagrams for the beam (a) in terms of the parameters shown; (b) set $w = 250$ lb/ft, $L = 12$ ft.

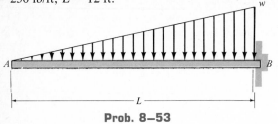

Prob. 8–53

8–54. Draw the shear and bending-moment diagrams for the beam (a) in terms of the parameters shown; (b) set $w = 250$ lb/ft, $L = 12$ ft.

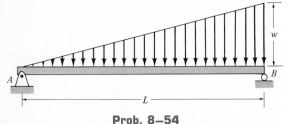

Prob. 8–54

8–55. Draw the shear and bending-moment diagrams for the beam.

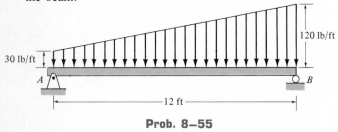

Prob. 8–55

***8–56.** Draw the shear and bending-moment diagrams for the beam.

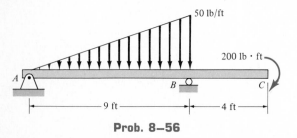

Prob. 8–56

8–57. Draw the shear and bending-moment diagrams for the beam.

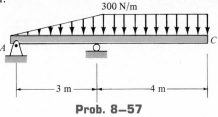

Prob. 8–57

8–58. The beam has a depth of 2 ft and is subjected to a uniform distributed loading of 50 lb/ft which acts at an angle of 30° from the vertical as shown. Determine the axial force, shear force, and moment in the beam as a function of x. *Hint:* The moment loading is to be determined from a point along the centerline of the beam (x axis).

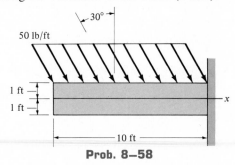

Prob. 8–58

8–59. The jib crane carries a load of 8 kN on the trolley T, which rolls along the bottom flange of the beam CD. Determine the position x of the trolley that will create the greatest bending moment in the column at E. What is this moment? As a result of end constraints, 0.3 m $\leq x \leq 3.9$ m.

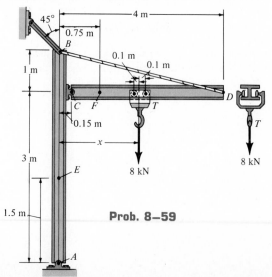

Prob. 8–59

*8–60. Determine the position x of the trolley in Prob. 8–59 that will create the greatest shear force at F. What is this shear force?

8–61. Determine the internal axial force, shear force, and the bending moment as a function of $0° \leqslant \theta \leqslant 180°$ and $0 \leqslant x \leqslant 2$ ft for the beam loaded as shown. For what angle θ is the internal moment equal to zero?

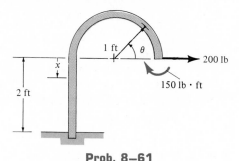

Prob. 8–61

8–62. Determine the internal axial force, shear force, and bending moment in the ring as a function of θ.

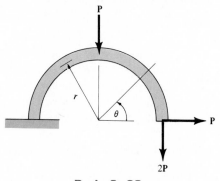

Prob. 8–62

8–63. The cantilevered beam is made of material having a specific weight of γ. Determine the internal shear and moment in the beam as a function of x.

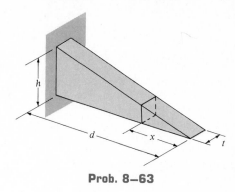

Prob. 8–63

*8–64. The beam will fail when the maximum moment developed in it is M_O. Determine the position x of the concentrated force $\mathbf{P}$ and its smallest magnitude that will cause the beam to fail.

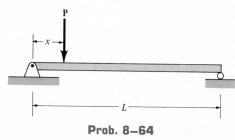

Prob. 8–64

8–65. Determine the distance a as a fraction of the beam's length L for locating the roller support so that the moment in the beam at B is zero.

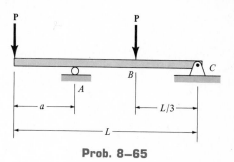

Prob. 8–65

8–66. Determine the distance a between the supports in terms of the beam's length L so that the bending moment in the *symmetric* beam is zero at the beam's center. The intensity of the distributed load at the center of the beam is w_o.

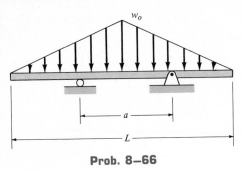

Prob. 8–66

8–67. Express the internal shear and moment components acting in the rod as a function of y, where $0 \leqslant y \leqslant 4$ ft.

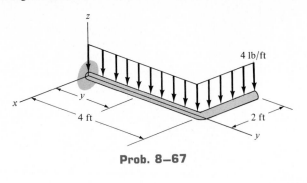

Prob. 8–67

★ # 8.3 Relations Between Distributed Load, Shear, and Moment

In cases where a beam is subjected to several concentrated forces, couples, and distributed loads, the method of constructing the shear and bending-moment diagrams discussed in Sec. 8.2 may become quite tedious. In this section a simpler method for constructing these diagrams is discussed—a method based upon differential relations that exist between the load, shear, and bending moment.

Consider the beam AD shown in Fig. 8–15a, which is subjected to an arbitrary distributed loading $w = w(x)$ and a series of concentrated forces and couples. In the following discussion, the *distributed load* will be considered *positive* when the *loading acts downward* as shown. The free-body diagram for a small segment of the beam having a length Δx is shown in Fig. 8–15b. Since this segment has been chosen at a point x along the beam which is *not* subjected to a concentrated force or couple, any results obtained will not apply at points of concentrated loading. The internal shear force and bending moment shown on the free-body diagram are assumed to act in the *positive sense* according to the established sign convention, Fig. 8–15. Note that both the shear force and moment acting on the right-hand face must be increased by a small, finite amount in order to keep the segment in equilibrium. The distributed loading has been replaced by a resultant force $w(x) \Delta x$ that acts at a fractional distance $\varepsilon(\Delta x)$ from the right end, where $0 < \varepsilon < 1$ [for example, if $w(x)$ is *uniform*, $\varepsilon = \frac{1}{2}$]. Applying the equations of equilibrium, we have

$+\uparrow\Sigma F_y = 0;$ $V - w(x)\,\Delta x - (V + \Delta V) = 0$

$$\Delta V = -w(x)\,\Delta x$$

$\downarrow+\Sigma M_O = 0;$ $-V\,\Delta x - M + w(x)\,\Delta x[\varepsilon(\Delta x)] + (M + \Delta M) = 0$

$$\Delta M = V\,\Delta x - w(x)\varepsilon(\Delta x)^2$$

Dividing by Δx and taking the limit as $\Delta x \rightarrow 0$ the above equations become

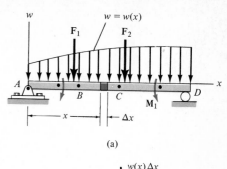

(a)

$$\frac{dV}{dx} = -w(x) \tag{8-1}$$

$$\frac{\text{Slope of}}{\text{Shear Diagram}} = \frac{\text{Distributed}}{\text{Load Intensity}}$$

and

$$\frac{dM}{dx} = V \tag{8-2}$$

$$\frac{\text{Slope of}}{\text{Moment Diagram}} = \text{Shear}$$

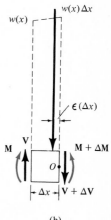

(b)

Fig. 8–15

These two equations provide a convenient means for plotting the shear and moment diagrams for a beam. At a specific point in a beam, Eq. 8–1 states that the intensity of the *distributed loading* is equal to the *slope* of the shear *diagram,* while Eq. 8–2 states that the shear force at the point is equal to the *slope* of the moment diagram. In particular, if the shear is equal to zero, $dM/dx = 0$, and therefore a point of zero shear corresponds to a point of maximum (or possibly minimum) moment.

Equations 8–1 and 8–2 may be rewritten in the form $dV = -w(x)\,dx$ and $dM = V\,dx$. Noting that $w(x)\,dx$ and $V\,dx$ represent the differential areas under the distributed-loading and shear diagrams, one can integrate these areas between two points B and C on the beam and write

$$\Delta V_{BC} = -\int w(x)\,dx$$

$$\frac{\text{Change}}{\text{in shear}} = \frac{-\text{area under}}{\text{loading curve}} \tag{8-3}$$

and

$$\Delta M_{BC} = \int V\,dx$$

$$\frac{\text{Change}}{\text{in moment}} = \frac{\text{area under}}{\text{shear diagram}} \tag{8-4}$$

Equation 8–3 states that the *area* under the distributed-loading curve between points B and C, Fig. 8–15a, is equal to the *change in shear* between these two points. Similarly, from Eq. 8–4, the area under the shear diagram drawn for the region from B to C is equal to the change in moment between B and C.

From the above derivations it should be clear that the above equations do not apply at points where a *concentrated* force or couple moment acts. These two special cases create *discontinuities* in the shear and moment diagrams, respectively. Consider, for example, the free-body diagrams of the beam segments shown in Fig. 8–16. From Fig. 8–16a, it is seen that force equilibrium requires the change in shear to be

$$+\uparrow \Sigma F_y = 0; \qquad\qquad \Delta V = -F \qquad\qquad (8\text{–}5)$$

Thus, when **F** acts *downward*, ΔV is negative so that the shear "jumps" *downward*, and when **F** acts *upward*, the jump (ΔV) is *upward*. Likewise, from Fig. 8–16b, letting $\Delta x \to 0$, moment equilibrium requires the change in moment to be

$$\zeta + \Sigma M_O = 0; \qquad\qquad \Delta M = M' \qquad\qquad (8\text{–}6)$$

In this case, if an applied moment **M'** is *clockwise*, then ΔM is positive so that the moment diagram shows a jump *upward*, and when **M'** acts *counterclockwise*, the jump (ΔM) is *downward*.

The following examples illustrate application of the above equations for the construction of the shear and moment diagrams. After working through these examples, it is recommended that Examples 8–8 and 8–9 be solved using this method.

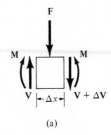

(a)

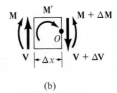

(b)

Fig. 8–16

Example 8–10

Draw the shear and bending-moment diagrams for the beam shown in Fig. 8–17a.

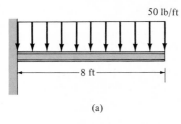

50 lb/ft

8 ft

(a)

Fig. 8–17

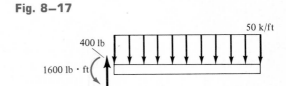

400 lb

50 k/ft

1600 lb · ft

(b)

Solution

Support Reactions (Fig. 8–17b). The reactions at the fixed support have been calculated and are shown on the free-body diagram of the beam.

Shear Diagram (Fig. 8–17c). The shear at the end points is plotted first. From the sign convention, Fig. 8–12, $V = +400$ at $x = 0$ (R.H.F.) and $V = 0$ at $x = 8$. Since $dV/dx = -w = -50$, a straight *negative* sloping line connects the end points.

Moment Diagram (Fig. 8–17d). From our sign convention, Fig. 8–12, the moments at the beam's end points, $M = -1600$ at $x = 0$ (R.H.F.) and $M = 0$ at $x = 8$, are plotted first. Successive values of shear on the shear diagram indicate the slope $dM/dx = V$ is always positive yet *linearly decreasing* from $dM/dx = 400$ at $x = 0$ to $dM/dx = 0$ at $x = 8$. Thus, a parabola having this characteristic connects the end points.

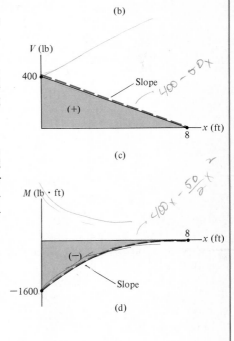

V (lb)

400

Slope

(+)

x (ft)

8

(c)

M (lb · ft)

8

x (ft)

(−)

Slope

−1600

(d)

Example 8–11

Draw the shear and bending-moment diagrams for the beam shown in Fig. 8–18a.

Solution

Support Reactions (Fig. 8–18b). The reactions at the fixed support have been calculated and are shown on the free-body diagram of the beam.

Shear Diagram (Fig. 8–18c). Using the established sign convention, Fig. 8–12, the shear at the ends of the beam is plotted first, i.e., $x = 0$, $V = +1080$; $x = 20$, $V = +600$.

Since the uniform distributed load is downward and *constant*, then the slope of the shear diagram is $dV/dx = -w = -40$ for $0 \leqslant x < 12$ as indicated.

The magnitude of shear at $x = 12$ is $+600$. This can be determined by first finding the area under the load diagram between $x = 0$ and $x = 12$, i.e., $\Delta V = -\int w(x)\, dx = -40(12) = -480$. Thus $V|_{x=12} = V|_{x=0} + (-480) = 1080 - 480 = 600$. Also, we can obtain this value by using the method of sections, Fig. 8–18e, where for equilibrium $V = +600$.

Since the load between $12 < x \leqslant 20$ is $w = 0$, then $dV/dx = 0$ as indicated. This brings the value of the shear to the required value of $V = 60$ at $x = 20$.

Moment Diagram (Fig. 8–18d). Again, using the established sign convention, Fig. 8–12, the moments at the ends of the beam are plotted first, i.e., $x = 0$, $M = -15,880$; $x = 20$, $M = -1000$.

Numerical values of shear change from $+1080$ to $+600$ for $0 \leqslant x < 12$. Each value of shear gives the slope of the moment diagram since $dM/dx = V$. As indicated, at $x = 0$, $dM/dx = +1080$ and at $x = 12$, $dM/dx = +600$. For $0 < x < 12$ specific values of the shear diagram are positive but linearly decreasing. Hence, the moment diagram is parabolic with a linear decreasing slope.

The magnitude of moment at $x = 12$ ft is -5800. This can be found from the area under the shear diagram, i.e., $\Delta M = \int V(dx) = 600(12) + \frac{1}{2}(1080 - 600)(12) = +10,080$, so that $M|_{x=12} = M|_{x=0} + 10,080 = -15,880 + 10,080 = -5800$. The more "basic" method of sections can also be used, where equilibrium at $x = 12$ requires $M = -5800$, Fig. 8–18e.

The moment diagram has a constant slope for $12 < x \leqslant 20$ since $dM/dx = V = +600$. This brings the value of $M = -1000$ at $x = 20$, as required.

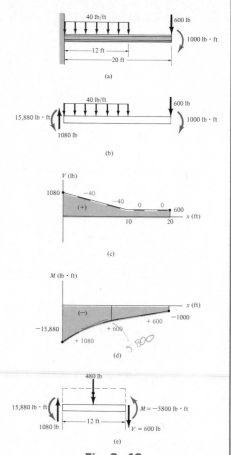

Fig. 8–18

Example 8–12

Draw the shear and moment diagrams for the beam in Fig. 8–19a.

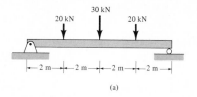

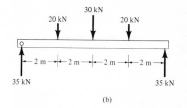

Fig. 8–19

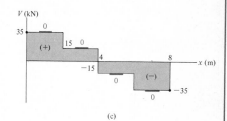

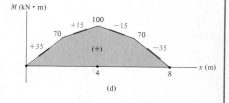

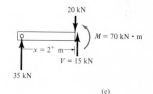

Solution

Support Reactions (Fig. 8–19b).   The reactions at the supports are shown on the free-body diagram.

Shear Diagram (Fig. 8–19b).   The end points $x = 0$, $V = +35$ and $x = 8$, $V = -35$ are plotted first.

Since there is no distributed load on the beam, the slope of the shear diagram throughout the beam's length is zero, i.e., $dV/dx = 0$. There is a discontinuity or "jump" of the shear diagram, however, at each concentrated force. From Eq. 8–5, $\Delta V = -F$, the change in shear is negative when the load acts downward and positive when the load acts upward. Stated another way, the "jump" follows the force, i.e., a downward force causes a downward jump, and vice versa. Thus, the 20-kN force at $x = 2$ m changes the shear force from 35 kN to 15 kN; the 30-kN force at $x = 4$ m changes the shear from 15 kN to -15 kN, etc. We can *also* obtain numerical values for the shear at a specified point in the beam by using the method of sections, as for example, $x = 2^{+}$ m, $V = 15$ kN in Fig. 8–19e.

Moment Diagram (Fig. 8–19d).   The end points $x = 0$, $M = 0$ and $x = 8$, $M = 0$ are plotted first.

Since the shear is constant in each region of the beam, the moment diagram has a corresponding constant positive or negative slope as indicated on the diagram. Numerical values of the moment at any point can be computed from the *area* under the shear diagram. For example, at $x = 2$ m, $\Delta M = \int V\,dx = 35(2) = 70$, so $M|_{x=2} = M|_{x=0} + 70 = 0 + 70 = 70$. Also, by the method of sections, we can determine the moment at a specified point, as, for example, $x = 2^{+}$ m, $M = 70$ kN · m, Fig. 8–19e.

369

Example 8–13

Sketch the shear and bending-moment diagrams for the beam shown in Fig. 8–20a.

Solution

Support Reactions. The reactions are calculated and indicated on the free-body diagram, Fig. 8–20b.

Shear Diagram. As in Example 8–12, the shear diagram can be constructed by "following the load" on the free-body diagram. In this regard, beginning at A, $V_A = +100$ lb. No load acts between A and C, so the shear remains constant, i.e., $dV/dx = w(x) = 0$. At C the 600-lb force is down, so the shear jumps down 600 lb, from 100 lb to -500 lb. Again the shear is constant (no load) and ends at -500 lb, point B. One might wonder why no jump or discontinuity in shear occurs at D, the point where the 4000-lb · ft moment is applied, Fig. 8–20a. Why this is not the case is indicated by the equilibrium conditions on the free-body diagram in Fig. 8–20f.

Moment Diagram. The moment at each end of the beam is zero. These two points are plotted first, Fig. 8–20d. The slope of the moment diagram from A to C is constant since $dM/dx = V = +100$. The value of the moment at C can be determined by the method of sections, Fig. 8–20e, or by computing the area under the shear diagram between A and C, i.e., $\Delta M_{AC} = M_C - M_A = (100 \text{ lb})(10 \text{ ft}) = 1000$ lb · ft. Since $M_A = 0$, then $M_C = 0 + 1000$ lb · ft $= 1000$ lb · ft. From C to D the slope is $dM/dx = V = -500$, Fig. 8–20c. The area under the shear diagram between points C and D is $\Delta M_{CD} = M_D - M_C = (-500 \text{ lb})(5 \text{ ft}) = -2500$ lb · ft. Thus $M_D = 1000 - 2500 = -1500$ lb · ft. A jump occurs at point D due to the concentrated couple moment of 4000 lb · ft. Using Eq. 8–6, the jump is *positive* since the couple moment is *clockwise*. Thus, at $x = 15^+$ ft, the moment is $M_D = -1500 + 4000 = 2500$ lb · ft. This value can *also* be determined by the method of sections, Fig. 8–20f. From point D the slope of $dM/dx = -500$ is maintained until the diagram closes to zero at B, Fig. 8–20d.

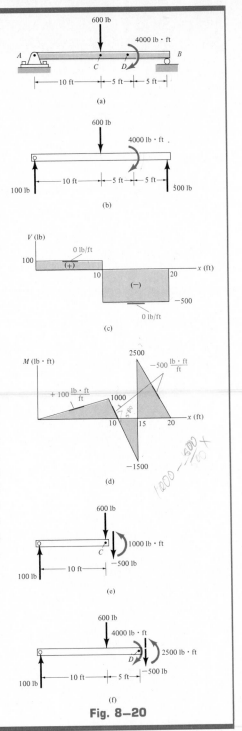

Fig. 8–20

Problems

*8–68. Draw the shear and bending-moment diagrams for the beam in Prob. 8–40.

8–69. Draw the shear and bending-moment diagrams for the beam in Prob. 8–46.

8–70. Draw the shear and bending-moment diagrams for the beam in Prob. 8–49.

8–71. Draw the shear and bending-moment diagrams for the beam in Prob. 8–54.

*8–72. Draw the shear and bending-moment diagrams for the beam in Prob. 8–56.

8–73. Draw the shear and bending-moment diagrams for the beam in Prob. 8–57.

8–74. The girder AB has a uniform weight of 500 lb/ft and is subjected to the concentrated forces caused by the floor beams. Draw the shear and bending-moment diagrams for the girder.

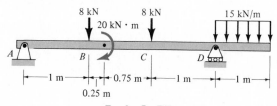

Prob. 8–74

8–75. Draw the shear and bending-moment diagrams for the beam.

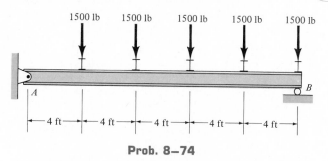

Prob. 8–75

*8–76. Draw the shear and bending-moment diagrams for the beam.

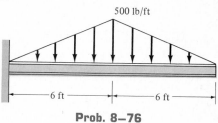

Prob. 8–76

8–77. Draw the shear and bending-moment diagrams for the beam.

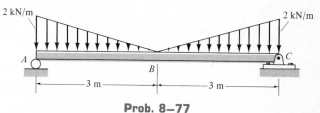

Prob. 8–77

8–78. Draw the shear and bending-moment diagrams for the beam.

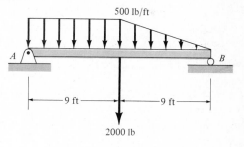

Prob. 8–78

8–79. Draw the shear and bending-moment diagrams for the beam.

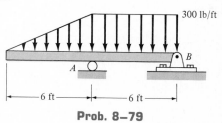

Prob. 8–79

*8–80. Draw the shear and bending-moment diagrams for the beam.

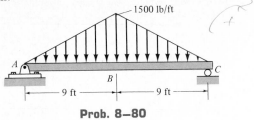

Prob. 8–80

8–81. Draw the shear and bending-moment diagrams for the beam.

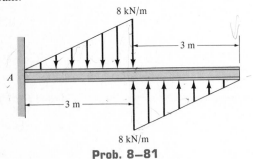

Prob. 8–81

8–82. Draw the shear and bending-moment diagrams for the column which is subjected to the loadings shown.

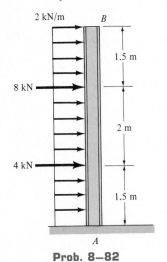

Prob. 8–82

8–83. Draw the shear and bending-moment diagrams for the beam.

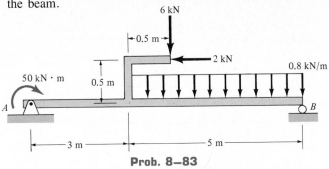

Prob. 8–83

*8–84. The beam consists of two segments pin-connected at B. Draw the shear and bending-moment diagrams for the beam.

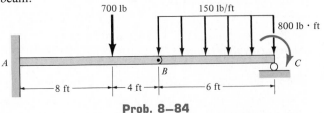

Prob. 8–84

8–85. The beam consists of three segments pin-connected at B and E. Draw the shear and bending-moment diagrams for the beam.

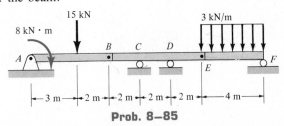

Prob. 8–85

8–86. The two segments of the girder are connected together by a short vertical link DC. Draw the shear and bending-moment diagrams for the girder.

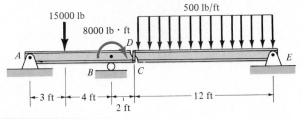

Prob. 8–86

Cables **8.4**[*]

Flexible cables and chains are often used in engineering structures for support and to transport loads from one member to another. When used for supporting suspension bridges and trolley wheels, cables form the main load-carrying element of the structure. In the force analysis of such systems, the weight of the cable itself may frequently be neglected; however, when cables are used for transmission lines and guys for radio antennas and derricks, the cable weight may become important and must be included in the structural analysis. Three cases will be considered in the analysis that follows: (1) a cable subjected to concentrated loads, (2) a cable subjected to a distributed load, and (3) a cable subjected to its own weight. Regardless of which loading conditions are present, provided the loading is coplanar with the cable, the requirements for equilibrium are formulated in an identical manner.

When deriving the necessary relations between the tensile force, sag, length, and span of a cable, we will make the assumption that the cable is *perfectly flexible* and *inextensible*. Due to flexibility, the cable offers no resistance to bending, and therefore, the tensile force acting in the cable is always tangent to the cable at points along its length. Being inextensible, the cable has a constant length both before and after the load is applied. As a result, once the load is applied, the geometry of the cable is fixed, and the cable or a segment of it can be treated as a rigid body.

Cable Subjected to Concentrated Loads. When a cable of negligible weight supports several concentrated loads, the cable takes the form of several straight-line segments, each of which is subjected to a constant tensile force. Consider, for example, the cable shown in Fig. 8–21, where the distances h and L_1, L_2, and L_3 and the loads P_1 and P_2 are known. The problem here is to determine the *nine unknowns* consisting of the tensions in each of the *three* segments, the four components of reaction at A and B, and the sags y_C and y_D at the *two* points C and D. To do this, we can write *two* equations of force equilibrium at each of points A, B, C, and D. This results in a total of *eight equations*.[*] To complete the solution, it will be necessary to know something about the geometry of the cable in order to obtain the necessary ninth equation. For example, if the cable's total *length L* is specified, then the Pythagorean theorem can be used to relate each of the three segmental lengths, written in terms of y_C, y_D, L_1, and L_2, to the total length L. Unfortunately, this type of problem cannot be easily solved by hand. Another possibility, however, is to specify one of the sags instead of the cable length. By doing this, the equilibrium equations are then sufficient for obtaining the force components and the remaining sag. Once the sag is obtained, the length of the cable can be determined by trigonometry. The following example illustrates a procedure for performing the equilibrium analysis for this type of problem.

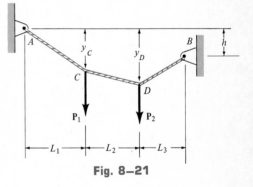

Fig. 8–21

*As will be shown in the following example, the eight equilibrium equations can *also* be written for the entire cable, or any part thereof. But *no more* than *eight* equations are available.

Example 8–14

Determine the tension in each segment of the cable shown in Fig. 8–22a.

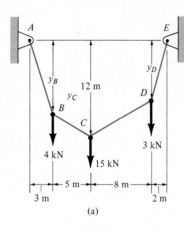

3 kN

4 kN

15 kN

5 m   8 m

3 m   2 m

(a)

Fig. 8–22a, b and c

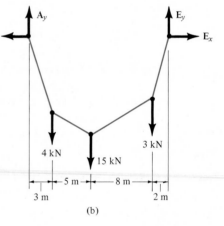

(b)

Solution

By inspection, there are four unknown external reactions (A_x, A_y, E_x, and E_y) and four unknown cable tensions, one in each cable segment. These eight unknowns along with the two unknown sags y_B and y_D can be determined from *ten* available equilibrium equations. One method is to apply these equations as force equilibrium ($\Sigma F_x = 0$, $\Sigma F_y = 0$) at each of the five points A through E. Here we will take a more direct approach. Consider the free-body diagram for the entire cable, Fig. 8–22b. Thus,

$\xrightarrow{+}\Sigma F_x = 0;$ $\qquad\qquad -A_x + E_x = 0$

$\zeta+\Sigma M_E = 0;$ $\quad -A_y(18) + 4(15) + 15(10) + 3(2) = 0 \qquad A_y = 12 \text{ kN}$

$+\uparrow\Sigma F_y = 0;$ $\qquad\quad 12 - 4 - 15 - 3 + E_y = 0 \qquad E_y = 10 \text{ kN}$

Since the sag $y_C = 12$ m is known, we will now consider the leftmost section, which cuts cable BC, Fig. 8–22c.

$\zeta+\Sigma M_C = 0;$ $\quad A_x(12) - 12(8) + 4(5) = 0 \qquad A_x = E_x = 6.33 \text{ kN}$

$\xrightarrow{+}\Sigma F_x = 0;$ $\qquad\qquad T_{BC} \cos\theta_{BC} - 6.33 = 0$

$+\uparrow\Sigma F_y = 0;$ $\qquad\quad 12 - 4 - T_{BC} \sin\theta_{BC} = 0$

$$\theta_{BC} = 51.6°$$

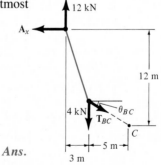

(c)

Thus,

$$T_{BC} = 10.2 \text{ kN} \qquad\qquad\qquad Ans.$$

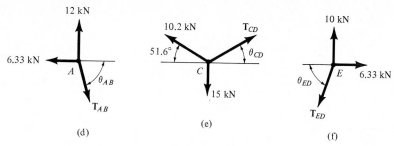

Fig. 8–22d, e and f

Proceeding now to analyze the equilibrium of points A, C, and E in sequence, we have

Point A (Fig. 8–22*d*)

$$\xrightarrow{+}\Sigma F_x = 0; \qquad T_{AB}\cos\theta_{AB} - 6.33 = 0$$
$$+\uparrow\Sigma F_y = 0; \qquad -T_{AB}\sin\theta_{AB} + 12 = 0$$
$$\theta_{AB} = 62.2°$$
$$T_{AB} = 13.6 \text{ kN} \qquad\qquad\qquad Ans.$$

Point C (Fig. 8–22*e*)

$$\xrightarrow{+}\Sigma F_x = 0; \qquad T_{CD}\cos\theta_{CD} - 10.2\cos 51.6° = 0$$
$$+\uparrow\Sigma F_y = 0; \qquad T_{CD}\sin\theta_{CD} + 10.2\sin 51.6° - 15 = 0$$
$$\theta_{CD} = 47.9°$$
$$T_{CD} = 9.44 \text{ kN} \qquad\qquad\qquad Ans.$$

Point E (Fig. 8–22*f*)

$$\xrightarrow{+}\Sigma F_x = 0; \qquad 6.33 - T_{ED}\cos\theta_{ED} = 0$$
$$+\uparrow\Sigma F_y = 0; \qquad 10 - T_{ED}\sin\theta_{ED} = 0$$
$$\theta_{ED} = 57.7°$$
$$T_{ED} = 11.8 \text{ kN} \qquad\qquad\qquad Ans.$$

Note that the maximum cable tension is in segment AB since this segment has the greatest slope (θ) and it is required that for any segment the horizontal component $T\cos\theta = A_x$ (a constant). Also, since the angles that the cable segments make with the horizontal have now been determined, one can determine the sags y_B and y_D, Fig. 8–22*a*, using trigonometry.

Cable Subjected to a Distributed Load. Consider the weightless cable shown in Fig. 8–23a, which is subjected to a loading function $w = w(x)$ *as measured in the x direction*. The free-body diagram of a small segment of the cable having a length Δs is shown in Fig. 8–23b. Since the tensile force in the cable changes continuously in both magnitude and direction along the length of the cable, this change is denoted on the free-body diagram by $\Delta\mathbf{T}$. The distributed load is represented by its resultant force $w(x)(\Delta x)$, which acts at a fractional distance $\varepsilon(\Delta x)$ from point O, where $0 < \varepsilon < 1$. Applying the equations of equilibrium yields

$\xrightarrow{+} \Sigma F_x = 0; \qquad -T\cos\theta + (T + \Delta T)\cos(\theta + \Delta\theta) = 0$

$+\uparrow \Sigma F_y = 0; \qquad -T\sin\theta - w(x)(\Delta x) + (T + \Delta T)\sin(\theta + \Delta\theta) = 0$

$\zeta + \Sigma M_O = 0; \qquad w(x)(\Delta x)\varepsilon(\Delta x) - T\cos\theta\,\Delta y + T\sin\theta\,\Delta x = 0$

Dividing each of the preceding equations by Δx and taking the limit as $\Delta x \to 0$, and hence $\Delta y \to 0$, $\Delta\theta \to 0$, and $\Delta T \to 0$, we obtain

$$\frac{d(T\cos\theta)}{dx} = 0 \tag{8–7}$$

$$\frac{d(T\sin\theta)}{dx} - w(x) = 0 \tag{8–8}$$

$$\frac{dy}{dx} = \tan\theta \tag{8–9}$$

Integrating Eq. 8–7, we have

$$T\cos\theta = \text{constant} = F_H \tag{8–10}$$

where F_H represents the horizontal component of tensile force at any point along the cable.

Integrating Eq. 8–8 gives

$$T\sin\theta = \int w(x)\,dx \tag{8–11}$$

Dividing Eq. 8–11 by Eq. 8–10 eliminates T. Then, using Eq. 8–9, we obtain

$$\tan\theta = \frac{dy}{dx} = \frac{1}{F_H}\int w(x)\,dx$$

Performing a second integration yields

$$y = \frac{1}{F_H}\int\left(\int w(x)\,dx\right)dx \tag{8–12}$$

This equation is used to determine the deflection curve for the cable, $y = f(x)$. The force F_H and the two constants, say C_1 and C_2, resulting from the integration of Eq. 8–12 are determined by applying the boundary conditions for the cable.

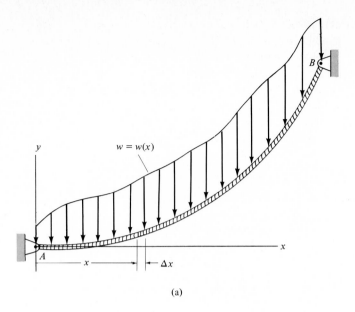

(a)

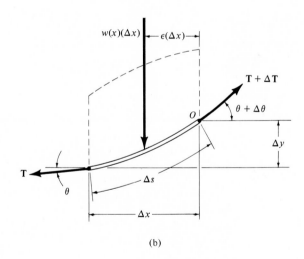

(b)

Fig. 8–23

Example 8–15

The cable of a suspension bridge supports half of the uniform road surface between the two columns at A and B, as shown in Fig. 8–24a. If this loading is w_o N/m, determine the maximum force developed in the cable and the required length of the cable. The span length L and sag h are known.

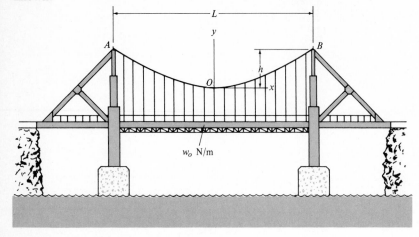

(a)

Fig. 8–24a

Solution

For reasons of symmetry, the origin of coordinates has been placed at the cable's center. The deflection curve of the cable can be found by using Eq. 8–12. Noting that $w(x) = w_o$, we have

$$y = \frac{1}{F_H} \int \left(\int w_o \, dx \right) dx$$

Integrating this equation twice gives

$$y = \frac{1}{F_H} \left(\frac{w_o x^2}{2} + C_1 x + C_2 \right) \tag{1}$$

The constants of integration may be determined by using the boundary conditions $y = 0$ at $x = 0$ and $dy/dx = 0$ at $x = 0$. Substituting into Eq. (1) yields $C_1 = C_2 = 0$. The deflection curve then becomes

$$y = \frac{w_o}{2F_H} x^2 \tag{2}$$

This is the equation of a *parabola*. The constant F_H may be obtained by using the boundary condition $y = h$ at $x = L/2$. Thus,

$$F_H = \frac{w_o L^2}{8h} \tag{3}$$

Therefore, Eq. (2) becomes

$$y = \frac{4h}{L^2}x^2 \qquad (4)$$

The maximum tension in the cable may be determined using Eq. 8–10. Since $T = F_H/\cos\theta$, for $0 \leqslant \theta < \pi/2$, T_{max} will occur when θ is *maximum*, i.e., at point B, Fig. 8–24a. From Eq. (2) this occurs when

$$\left.\frac{dy}{dx}\right|_{max} = \tan\theta_{max} = \left.\frac{w_o}{F_H}x\right|_{x=L/2}$$

or

$$\theta_{max} = \tan^{-1}\left(\frac{w_o L}{2F_H}\right) \qquad (5)$$

Therefore,

$$T_{max} = \frac{F_H}{\cos(\theta_{max})} \qquad (6)$$

Using the triangular relationship shown in Fig. 8–24b, which is based on Eq. (5), Eq. (6) may be written as

$$T_{max} = \frac{\sqrt{4F_H^2 + w_o^2 L^2}}{2}$$

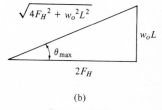

(b)

Fig. 8–24b

Substituting Eq. (3) into the above equation yields

$$T_{max} = \frac{w_o L}{2}\sqrt{1 + \left(\frac{L}{4h}\right)^2} \qquad Ans.$$

For a differential segment of cable length,

$$ds = \sqrt{(dx)^2 + (dy)^2} = \sqrt{1 + \left(\frac{dy}{dx}\right)^2}\,dx$$

Hence, the total length of the cable, $\mathcal{L}$, can be determined by integration, i.e.,

$$\mathcal{L} = \int ds = 2\int_0^{L/2}\sqrt{1 + \left(\frac{w_o x}{F_H}\right)^2}\,dx \qquad (7)$$

Integrating and substituting the limits yields

$$\mathcal{L} = \frac{L}{2}\sqrt{1 + \left(\frac{w_o L}{2F_H}\right)^2} + \frac{F_H}{w_o}\sinh^{-1}\left(\frac{w_o L}{2F_H}\right)$$

Using Eq. (3) and rearranging terms, the final result can be written as

$$\mathcal{L} = \frac{L}{2}\left[\sqrt{1 + \left(\frac{4h}{L}\right)^2} + \frac{L}{4h}\sinh^{-1}\left(\frac{4h}{L}\right)\right] \qquad Ans.$$

Cable Subjected to Its Own Weight. When the weight of the cable becomes important in the force analysis, the loading function along the cable becomes a function of the arc length s rather than the projected length x. A generalized loading function $w = w(s)$ acting along the cable is shown in Fig. 8–25a. The free-body diagram for a segment of the cable is shown in Fig. 8–25b. Applying the equilibrium equations to the force system on this diagram, one obtains relationships identical to those given by Eqs. 8–7 to 8–9, but with ds replacing dx. Therefore, it may be shown that

$$T \cos \theta = F_H$$

$$T \sin \theta = \int w(s) \, ds \tag{8–13}$$

$$\frac{dy}{dx} = \frac{1}{F_H} \int w(s) \, ds$$

To perform a direct integration of Eq. 8–13, it is necessary to replace dy/dx by ds/dx. Since

$$\frac{dy}{dx} = \sqrt{\left(\frac{ds}{dx}\right)^2 - 1}$$

we obtain

$$\frac{ds}{dx} = \left\{ 1 + \frac{1}{F_H^2} \left(\int w(s) \, ds \right)^2 \right\}^{1/2}$$

Separating the variables and integrating yields

$$x = \int \frac{ds}{\left\{ 1 + \dfrac{1}{F_H^2} \left(\displaystyle\int w(s) \, ds \right)^2 \right\}^{1/2}} \tag{8–14}$$

The two constants of integration, say C_1 and C_2, are found using the boundary conditions for the cable.

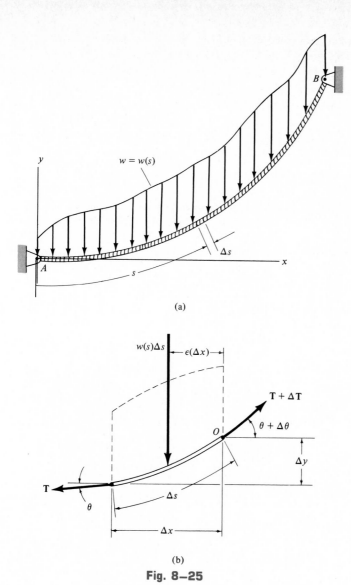

(a)

(b)

Fig. 8–25

Example 8–16

Determine the deflection curve, the length, and the maximum tension in the uniform cable shown in Fig. 8–26. The cable weighs $w_o = 5$ N/m.

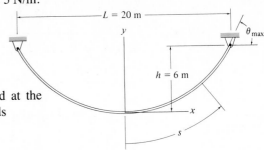

Fig. 8–26

Solution

For reasons of symmetry, the origin of coordinates is located at the center of the cable. Applying Eq. 8–14, where $w(s) = w_o$, yields

$$x = \int \frac{ds}{\left\{1 + \frac{1}{F_H^2}\left(\int w_o\, ds\right)^2\right\}^{1/2}}$$

Integrating the term under the integral sign in the denominator, we have

$$x = \int \frac{ds}{\left\{1 + \frac{1}{F_H^2}\left(w_o s + C_1\right)^2\right\}^{1/2}}$$

Substituting $u = (1/F_H)(w_o s + C_1)$ so that $du = (w_o/F_H)\, ds$, a second integration yields

$$x = \frac{F_H}{w_o}(\sinh^{-1} u + C_2)$$

or

$$x = \frac{F_H}{w_o}\left\{\sinh^{-1}\left[\frac{1}{F_H}(w_o s + C_1)\right] + C_2\right\} \qquad (1)$$

From Eq. 8–13,

$$\frac{dy}{dx} = \frac{1}{F_H}\int w_o\, ds$$

or

$$\frac{dy}{dx} = \frac{1}{F_H}(w_o s + C_1)$$

Since $dy/dx = 0$ at $s = 0$, $C_1 = 0$ and

$$\frac{dy}{dx} = \tan \theta = \frac{w_o s}{F_H} \qquad (2)$$

where θ is the angle of the cable at any point. The constant C_2 may be evaluated by using the condition $s = 0$ at $x = 0$ in Eq. (1), in which case $C_2 = 0$. To obtain the deflection curve, $y = f(x)$, solve for s in Eq. (1), which yields

$$s = \frac{F_H}{w_o}\sinh\left(\frac{w_o}{F_H}x\right) \qquad (3)$$

Now, substitute Eq. (3) into Eq. (2), in which case

$$\frac{dy}{dx} = \sinh\left(\frac{w_o}{F_H}x\right)$$

Hence

$$y = \frac{F_H}{w_o} \cosh\left(\frac{w_o}{F_H}x\right) + C_3 \qquad (4)$$

If the boundary condition $y = 0$ at $x = 0$ is applied, the constant $C_3 = -F_H/w_o$, and therefore the deflection curve becomes

$$y = \frac{F_H}{w_o}\left[\cosh\left(\frac{w_o}{F_H}x\right) - 1\right]$$

This equation defines the shape of a *catenary curve*. The constant F_H is obtained by using the boundary condition that $y = h$ at $x = L/2$, in which case

$$h = \frac{F_H}{w_o}\left[\cosh\left(\frac{w_o L}{2F_H}\right) - 1\right] \qquad (5)$$

Since $w_o = 5$ N/m, $h = 6$ m, and $L = 20$ m, Eqs. (4) and (5) become

$$y = \frac{F_H}{5\text{ N/m}}\left[\cosh\left(\frac{5\text{ N/m}}{F_H}x\right) - 1\right] \qquad (6)$$

$$6\text{ m} = \frac{F_H}{5\text{ N/m}}\left[\cosh\left(\frac{50\text{ N}}{F_H}\right) - 1\right] \qquad (7)$$

Equation (7) can be solved for F_H by using a trial-and-error procedure. The result is

$$F_H = 45.8\text{ N}$$

and therefore Eq. (6) becomes

$$y = 9.16[\cosh(0.109x) - 1]\text{ m} \qquad\qquad \textit{Ans.}$$

Using Eq. (3), with $x = 10$ m, the half-length of the cable is

$$\frac{\mathcal{L}}{2} = \frac{45.8\text{ N}}{5\text{ N/m}}\sinh\left[\frac{5\text{ N/m}}{45.8\text{ N}}(10\text{ m})\right] = 12.1\text{ m}$$

Hence,

$$\mathcal{L} = 24.2\text{ m} \qquad\qquad \textit{Ans.}$$

Since $T = F_H/\cos\theta$, the maximum tension occurs when θ is maximum, i.e., at $s = \mathcal{L}/2 = 12.1$ m. Using Eq. (2) yields

$$\left.\frac{dy}{dx}\right|_{\text{max}} = \tan\theta_{\text{max}} = \frac{5\text{ N/m}(12.1\text{ m})}{45.8\text{ N}} = 1.32$$

$$\theta_{\text{max}} = 52.9°$$

Thus,

$$T_{\text{max}} = \frac{45.8\text{ N}}{\cos 52.9°} = 75.9\text{ N} \qquad\qquad \textit{Ans.}$$

Problems

Neglect the weight of the cable in the following problems, *unless* specified.

8–87. Determine the tension in each segment of the cable and the cable's total length.

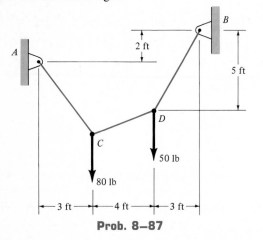

Prob. 8–87

*8–88. Cable *ABCD* supports the 4-kg lamp *E* and 6-kg lamp *F*. Determine the maximum tension in the cable and the sag of point *B*.

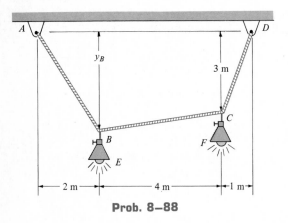

Prob. 8–88

8–89. The cable supports the loading shown. Determine the distance x_B point *B* acts from the wall.

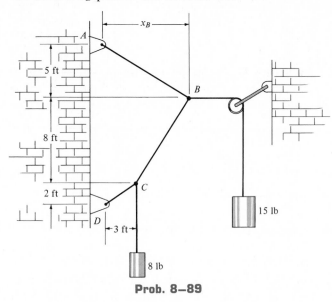

Prob. 8–89

8–90. The cable supports the three loads shown. Determine the sags y_B and y_D of points *B* and *D* and the tension in each segment of the cable.

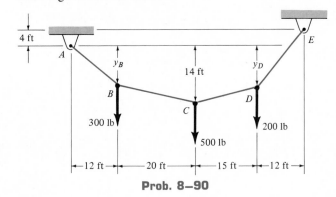

Prob. 8–90

8–91. Determine the force **P** needed to hold the cable in the position shown, i.e., so segment *CD* remains horizontal. Also, compute the sag y_D and the maximum tension in the cable.

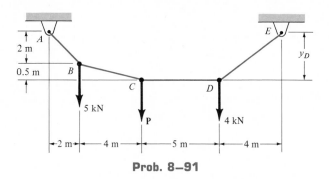

Prob. 8–91

***8–92.** If the side loading on the suspension bridge in Example 8–15 is $w_o = 0.75$ kN/m and $L = 190$ m, $h = 20$ m, determine the cable tension at point *B*.

8–93. The cable will break when the maximum tension reaches $T_{max} = 10$ kN. Determine the sag *h* to develop this maximum tension if the beam supports the uniform distributed load of $w = 700$ N/m.

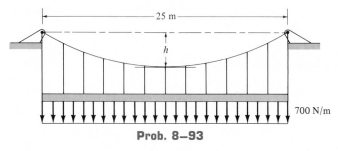

Prob. 8–93

8–94. Determine the maximum uniform loading w_o N/m that the cable can support if it is capable of sustaining a maximum tension of 40 kN before it will break.

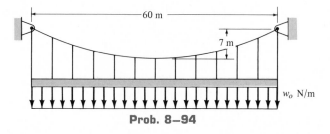

Prob. 8–94

8–95. The cable supports a pipe which weighs 50 lb/ft. Determine the tension in the cable at points *A*, *B*, and *C*.

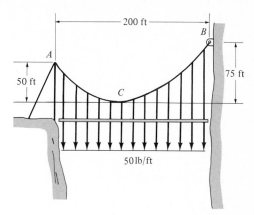

Prob. 8–95

***8–96.** The man picks up the 52-ft chain and holds it just high enough so it is completely off the ground. The chain has points of attachment *A* and *B* that are 50 ft apart. If the chain has a weight of 3 lb/ft, and the man weighs 150 lb, determine the force he exerts on the ground. Also, how high *h* must he lift the chain? *Hint:* The slopes at *A* and *B* are zero.

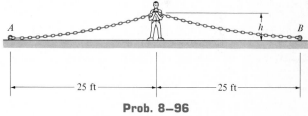

Prob. 8–96

8–97. A 100-lb cable is attached between two points at a distance of 50 ft apart having equal elevations. If the maximum tension developed in the cable is 75 lb, determine the length of the cable and the sag.

■**8–98.** The cable has a weight of 5 lb/ft and a length of 175 ft. Determine its minimum elevation d from the ground if it spans 150 ft. Also, compute the minimum tension in the cable.

50 ft d 50 ft

|←————— 150 ft —————→|

Prob. 8–98

■**8–99.** A 50-ft cable is suspended between two points a distance of 15 ft apart and at the same elevation. If the minimum tension in the cable is 200 lb, determine the total weight of the cable and the maximum tension developed in the cable.

■*8–100.** A 40-m-long chain has a total mass of 100 kg and is suspended between two points 10 m apart at the same elevation. Determine the maximum tension and the sag in the chain.

■**8–101.** The telephone wire has a mass of 500 g/m. If the cable sags 1.5 m, determine the maximum tension in the cable and its length between the poles.

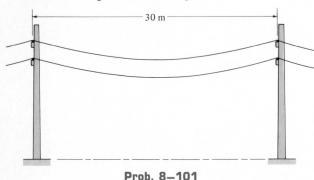

|←————— 30 m —————→|

Prob. 8–101

8–102. Show that the deflection curve of the cable discussed in Example 8–16 reduces to Eq. (4) in Example 8–15 when the *hyperbolic cosine function* is expanded in terms of a series and only the first two terms are retained. (The answer indicates that the *catenary* may be replaced by a *parabola* in the analysis of problems in which the sag is small. In this case, the cable weight is assumed to be uniformly distributed along the horizontal.)

8–103. A uniform cord is suspended between two points having the same elevation. Determine the sag-to-span ratio so that the maximum tension in the cord equals the cord's total weight.

■*8–104.** A cable has a weight of 2 lb/ft. If it can span 100 ft and has a sag of 12 ft, determine the length of the cable. The ends of the cable are supported from the same elevation.

■**8–105.** The chain has a weight of 3 lb/ft. Determine the tension at points A, B, and C necessary for equilibrium.

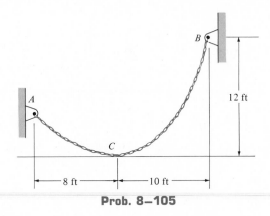

|←—— 8 ft ——→|←—— 10 ft ——→|

Prob. 8–105

8–106. The uniform beam weighs 500 lb and is held in the horizontal position by means of the cable AB, which has a weight of 5 lb/ft. If the slope of the cable at A is 30°, determine the length of the cable.

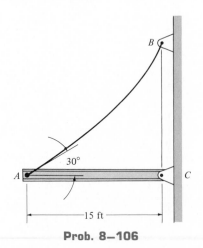

|←———— 15 ft ————→|

Prob. 8–106

8–107. A steel tape used for measurement in surveying has a length of 100.00 ft and a total weight of 2 lb. How much horizontal tension must be applied to the tape so that the distance marked on the ground is 99.90 ft? In practice the calculation should also include the effects of elastic stretching and temperature changes on the tape's length.

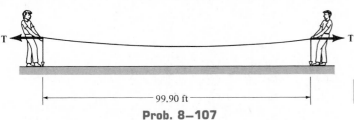

99.90 ft

Prob. 8–107

*8–108. Two identical cables, each having a weight of 8 N/m and length of 10 m, are attached to the top of the pole as shown. If the tension in the cables at C and D is 300 N, determine the distance from A to C and A to D.

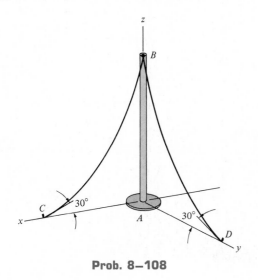

Prob. 8–108

8–109. The cable is subjected to the triangular loading. If the slope of the cable at point A is zero, determine the equation of the curve $y = f(x)$ which defines the cable shape AB, and the maximum tension developed in the cable.

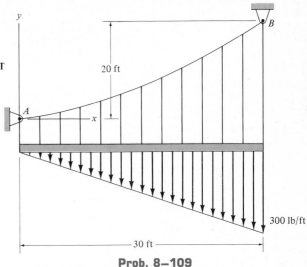

20 ft

30 ft

300 lb/ft

Prob. 8–109

8–110. The cable is subjected to the loading $w = (150 \cos (1.80x))$ lb/ft, where x is in ft and the argument for the cosine is in degrees. Determine the equation $y = f(x)$ which defines the cable shape AB and the maximum tension in the cable.

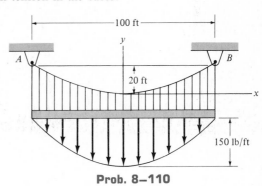

100 ft

20 ft

150 lb/ft

Prob. 8–110

387

8–111. The cable AB is subjected to a triangular loading. Neglecting the weight of the cable and assuming the angles with the tangents at points A and B are 45° and 80°, respectively, determine the deflection curve of the cable and the maximum tension developed in the cable.

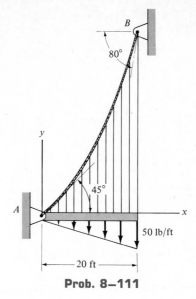

Prob. 8–111

Review Problems

*8–112. Draw the shear and bending-moment diagrams for the beam.

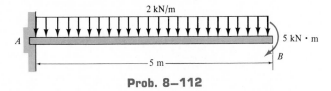

2 kN/m

A

5 kN · m

B

5 m

Prob. 8–112

8–113. Determine the internal axial force, shear force, and moment at points E and D. There is a pin or hinge at B.

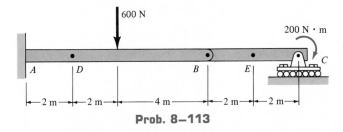

600 N

200 N · m

A   D   B   E   C

2 m   2 m   4 m   2 m   2 m

Prob. 8–113

8–114. Determine the internal axial force, shear force, and moment at point F of the frame.

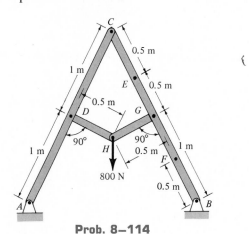

C

0.5 m

1 m

E

0.5 m

0.5 m

D   G

90°   90°

1 m   H   0.5 m   1 m

F

800 N

0.5 m

A   B

Prob. 8–114

8–115. Sketch the shear and bending-moment diagrams for the beam.

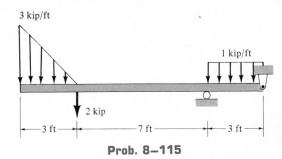

3 kip/ft

1 kip/ft

2 kip

3 ft   7 ft   3 ft

Prob. 8–115

*8–116. Determine the internal shear force and bending-moment in the beam for 3 ft $< x <$ 15 ft. Also, sketch the shear and bending-moment diagrams for the beam.

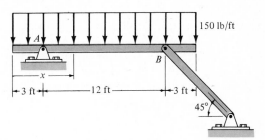

150 lb/ft

A

B

x

3 ft   12 ft   3 ft

45°

Prob. 8–116

8–117. Determine the x, y, z components of the internal axial force, shear force, and moment at point B of the rod.

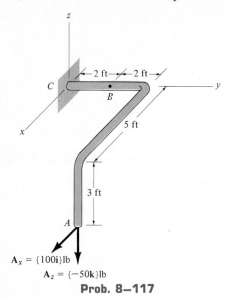

Prob. 8–117

8–118. Sketch the shear and bending-moment diagrams for the lathe spindle if it is subjected to the concentrated loads shown. The vertical equilibrium forces at bearings A and B are to be calculated.

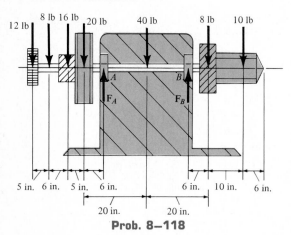

Prob. 8–118

■8–119. The cable has a weight of 8 lb/ft and a length of $\mathcal{L} = 150$ ft. Determine the sag h when it spans 100 ft. Also, compute the minimum tension in the cable.

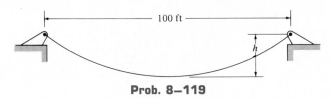

Prob. 8–119

***8–120.** Draw the shear and bending-moment diagrams for the beam.

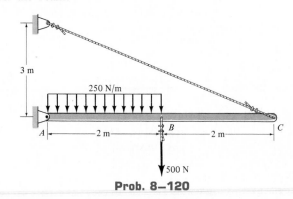

Prob. 8–120

Friction

In the previous chapters the surfaces of contact between two bodies were considered to be perfectly *smooth*. Because of this, the force of interaction between the bodies always acts *normal* to the surface at points of contact. In reality, however, all surfaces are *rough*, and depending upon the nature of the problem, the ability of a body to support a *tangential* as well as a *normal* force at its contacting surface must be considered. The tangential force is caused by friction, and in this chapter we will show how to analyze problems involving frictional forces. Specific application will include frictional forces on screws, bearings, disks, and belts. The analysis of rolling resistance is given in the last part of the chapter.

Characteristics of Dry Friction 9.1

Friction may be defined as a force of resistance acting on a body which prevents or retards slipping of the body relative to a second body with which it is in contact. This force always acts *tangent* to the surface at points of contact with other bodies and is directed so as to oppose the possible or existing motion of the body at these points.

In general, two types of friction can occur between surfaces. *Fluid friction* exists when the contacting surfaces are separated by a film of fluid (gas or liquid). The nature of fluid friction is studied in fluid mechanics since it depends upon knowledge of the velocity of the fluid and the fluid's ability to resist shearing force. In this book only the effects of *dry friction* will be presented. This type of friction is often called *Coulomb friction*, since its characteristics were extensively studied by C. A. Coulomb in 1781. Specifically, dry friction occurs between the contacting surfaces of bodies in the absence of a lubricating fluid.

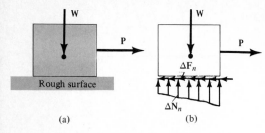

(a)

(b)

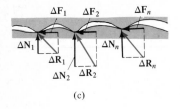

(c)

Theory of Dry Friction. The theory of dry friction can best be explained by considering what effects are caused by pulling horizontally on a block of uniform weight **W** which is resting on a rough horizontal surface, Fig. 9–1a. Throughout this discussion, it is necessary to consider the surfaces of contact to be *nonrigid or deformable*. As shown on the free-body diagram of the block, Fig. 9–1b, the floor exerts a *distribution* of both *normal force* ΔN_n and *frictional force* ΔF_n along the contacting surface. For equilibrium, the normal forces must act *upward* to balance the block's weight **W** and the frictional forces act to the left to prevent the applied force **P** from moving the block to the right. Close examination of the contacting surfaces between the floor and block reveals how these frictional and normal forces develop, Fig. 9–1c. It can be seen that many microscopic irregularities exist between the two surfaces and, as a result, reactive forces ΔR_n are developed at each of the protuberances.* These forces act at all points of contact and, as shown, each reactive force contributes both a frictional component ΔF_n and a normal component ΔN_n.

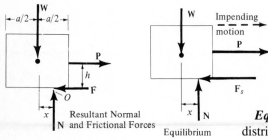

(d) Resultant Normal
N and Frictional Forces

(e) Equilibrium

Equilibrium. For simplicity in the following analysis, the effect of the distributed normal and frictional loadings will be indicated by their *resultants* N and F and then represented on the free-body diagram of the block as shown in Fig. 9–1d. Clearly, the distribution of ΔF_n in Fig. 9–1b indicates that **F** always acts *tangent to the contacting surface, opposite* to the direction of **P**. The normal force **N** is determined from the distribution of ΔN_n in Fig. 9–1b and is directed upward to balance the block's weight **W**. Notice that it acts a distance x to the right of the line of action of **W**, Fig. 9–1d. This location, which coincides with the centroid of the loading diagram in Fig. 9–1b, is necessary in order to balance the "tipping effect" caused by **P**. For example, if **P** is applied at a height h from the surface, Fig. 9–1d, then moment equilibrium about point O is satisfied if $Wx = Ph$ or $x = Ph/W$. In particular, note that the block will be on the verge of *tipping* if $x = a/2$.

Impending Motion. In cases where h is small or the surfaces of contact are rather "slippery," the frictional force **F** may *not* be great enough to balance the magnitude of **P**, and consequently the block will tend to slip *before* it can tip. In other words, as the magnitude of **P** is slowly increased, the magnitude of **F** correspondingly increases until it attains a certain *maxi-*

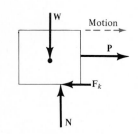

(f)

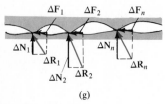

(g)

Fig. 9–1

*Besides mechanical interactions as explained here, a detailed treatment of the nature of frictional forces must also include the effects of temperature, density, cleanliness, and atomic or molecular attraction between the contacting surfaces. See D. Tabor, *Journal of Lubrication Technology*, 103, 169, 1981.

mum value F_s, called the *limiting static frictional force*, Fig. 9–1e. When this value is reached, the block is in *unstable equilibrium,* since any further increase in P will cause motion. Experimentally, it has been determined that the magnitude of the limiting static frictional force $\mathbf{F}_s$ is *directly proportional* to the magnitude of the resultant normal force $\mathbf{N}$. This may be expressed mathematically as

$$F_s = \mu_s N \qquad\qquad (9\text{--}1)$$

where the constant of proportionality, μ_s (mu "sub" s), is called the *coefficient of static friction.*

Typical values for μ_s, found in many engineering handbooks, are given in Table 9–1. Here it may be seen that, as in the case of aluminum on aluminum, it is possible for μ_s to be greater than 1. Furthermore, it should be noted that μ_s is dimensionless and depends only upon the characteristics of the two surfaces in contact. A wide range of values is given for each value of μ_s, since experimental testing was done under variable conditions of roughness and cleanliness of the contacting surfaces. For applications, therefore, it is important that both caution and judgment be exercised when selecting a coefficient of friction for a given set of conditions. When an exact calculation of F_s is required, the coefficient of friction should be determined directly by an experiment that involves the two materials to be used.

Motion. If the magnitude of $\mathbf{P}$ is increased so that it becomes greater than F_s, the frictional force at the contacting surfaces drops slightly to a smaller value F_k, called the *kinetic frictional force*. The block will *not* be held in equilibrium $(P > F_k)$; instead, it begins to slide with increasing speed, Fig. 9–1f. The drop made in the frictional force magnitude, from F_s (static) to F_k (kinetic), can be explained by again examining the surfaces of contact, Fig. 9–1g. Here it is seen that when $P > F_s$, then P has the capacity to shear off the peaks at the contact surfaces and cause the block to "lift" somewhat out of its settled position and "ride" on top of the peaks. Once the block begins to slide, high local temperatures at the points of contact cause momentary adhesion (welding) of these points. The continued shearing of these welds is the dominant mechanism creating friction. Since the resultant contact forces $\Delta\mathbf{R}_n$ are aligned slightly more in the vertical direction than before, Fig. 9–1c, they thereby contribute *smaller* frictional components, $\Delta\mathbf{F}_n$, as when the irregularities are meshed.

Experiments with sliding blocks indicate that the magnitude of the resultant frictional force $(\mathbf{F}_k)$ is directly proportional to the magnitude of the resultant normal force $\mathbf{N}$. This may be expressed mathematically as

$$F_k = \mu_k N \qquad\qquad (9\text{--}2)$$

where the constant of proportionality, μ_k, is called the *coefficient of kinetic friction*. Typical values for μ_k are approximately 25 per cent *smaller* than those listed in Table 9–1 for μ_s.

Table 9–1 Typical Values for μ_s

Contact Materials	Coefficient of Static Friction (μ_s)
Metal on ice	0.03–0.05
Wood on wood	0.30–0.70
Leather on wood	0.20–0.50
Leather on metal	0.30–0.60
Aluminum on aluminum	1.10–1.70

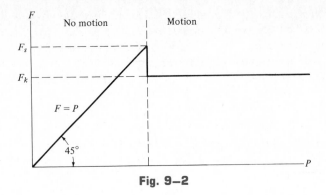

Fig. 9–2

The above effects can be summarized by reference to the graph in Fig. 9–2, which shows the variation of the frictional force **F** versus the applied load **P.** Here the frictional force is categorized in three different ways, namely, **F** is a *static-frictional force* if equilibrium is maintained, **F** is a *limiting static-frictional force* F_s when its magnitude reaches a maximum value needed to maintain equilibrium, and finally **F** is termed a *kinetic-frictional force* F_k when sliding occurs at the contacting surface. Notice also from the graph that for very large values of **P,** due to aerodynamic effects, F_k and likewise μ_k decrease somewhat.

Characteristics of Dry Friction. As a result of *experiments* that pertain to the foregoing discussion, the following rules which apply to bodies subjected to dry friction may be stated.

1. The frictional force acts *tangent* to the contacting surfaces in a direction *opposed* to the *relative motion* or tendency for motion of one surface against another.
2. The magnitude of the maximum static frictional force F_s that can be developed is independent of the area of contact provided the normal pressure is not very low nor great enough to severely deform or crush the contacting surfaces of the bodies.
3. The magnitude of the maximum static frictional force is generally greater than the magnitude of the kinetic frictional force for any two surfaces of contact. However, if one of the bodies is moving with a *very low velocity* over the surface of another, F_k becomes approximately equal to F_s, i.e., $\mu_s \approx \mu_k$.
4. When *slipping* at the point of contact is *about to occur,* the magnitude of the maximum static frictional force is proportional to the magnitude of the normal force at the point of contact, such that $F_s = \mu_s N$, Eq. 9–1.
5. When *slipping* at the point of contact is *occurring,* the magnitude of the kinetic frictional force is proportional to the magnitude of the normal force at the point of contact, such that $F_k = \mu_k N$, Eq. 9–2.

Angle of Friction. It should be observed that Eqs. 9–1 and 9–2 have a specific, yet *limited,* use in the solution of friction problems. In particular, the frictional force acting at a contacting surface is determined from $F_k = \mu_k N$ *only* if *relative motion* is occurring between the two surfaces. Furthermore, if two bodies are *stationary,* the magnitude of frictional force, F, *does not necessarily* equal $\mu_s N$; instead, F must satisfy the inequality $F \leqslant \mu_s N$. Only when *impending motion* occurs does F reach its upper limit, $F = F_s = \mu_s N$. This situation may be better understood by considering the block shown in Fig. 9–3a, which is acted upon by a force **P**. In this case consider $P = F_s$, so that the block is on the *verge of sliding.* For equilibrium, the normal force **N** and frictional force **F**$_s$ combine to create a resultant **R**$_s$. The angle ϕ_s that **R**$_s$ makes with **N** is called the *angle of static friction.* From the figure,

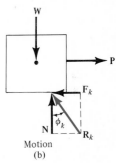

Impending motion
(a)

$$\phi_s = \tan^{-1}\left(\frac{F_s}{N}\right) = \tan^{-1}\left(\frac{\mu_s N}{N}\right) = \tan^{-1}\mu_s$$

Provided the block is *not in motion,* any horizontal force $P < F_s$ causes a resultant **R** which has a line of action directed at an angle ϕ from the vertical such that $\phi \leqslant \phi_s$. If **P** creates uniform *motion* of the block, then $P = F_k$. In this case, the resultant **R**$_k$ has a line of action defined by ϕ_k, Fig. 9–3b. This angle is referred to as the *angle of kinetic friction,* where

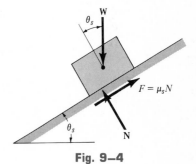

Motion
(b)

Fig. 9–3

$$\phi_k = \tan^{-1}\left(\frac{F_k}{N}\right) = \tan^{-1}\left(\frac{\mu_k N}{N}\right) = \tan^{-1}\mu_k$$

By comparison, $\phi_s > \phi_k$.

Angle of Repose. An experimental method which can be used to measure the coefficient of friction between two contacting surfaces consists of placing a block of one material having a known weight W on a plane made of the other material. The plane is inclined to the angle θ_s, at which point the block is on the *verge of sliding* and therefore $F_s = \mu_s N$. The free-body diagram of the block at this instant is shown in Fig. 9–4. Applying the force equations of equilibrium yields $N = W \cos \theta_s$ and $F_s = \mu N = W \sin \theta_s$ so that

$$\theta_s = \tan^{-1}\left(\frac{F_s}{N}\right) = \tan^{-1}\left(\frac{\mu_s N}{N}\right) = \tan^{-1}\mu_s$$

Fig. 9–4

The angle θ_s is referred to as the *angle of repose,* and by comparison it is equal to the angle of static friction ϕ_s. Once it is measured, the coefficient of static friction is obtained from $\mu_s = \tan \theta_s$.

Problems Involving Dry Friction 9.2

If a rigid body is in equilibrium when it is subjected to a system of forces that includes the effect of friction, the force system must not only satisfy the equations of equilibrium but *also* the laws that govern the frictional forces.

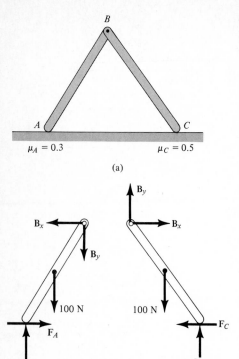

(a)

(b)

Fig. 9–5

Types of Frictional Problems. In general, there are three types of mechanics problems involving dry friction. They can easily be classified once the free-body diagrams are drawn and the total number of unknowns are identified and compared with the total number of available equilibrium equations. Each type of problem will now be explained and illustrated graphically by examples. In all cases the geometry and dimensions for the problem are assumed to be known.

Type 1. Problems in this category are strictly equilibrium problems which require *the total number of unknowns to be equal to the total number of available equilibrium equations*. Once the frictional forces are determined from the solution, however, their numerical values must be checked to be sure they satisfy the inequality $F \leq \mu_s N$; otherwise, slipping will occur and the body will not remain in equilibrium. A problem of this type is shown in Fig. 9–5a. Here we must determine the frictional forces at A and C to check if the equilibrium position of the bars can be maintained. If the bars are uniform and have known weights of 100 N each, then the free-body diagrams are shown in Fig. 9–5b. There are six unknown force components which can be determined *strictly* from the six equilibrium equations (three for each member). Once $\mathbf{F}_A$, $\mathbf{N}_A$, $\mathbf{F}_C$, and $\mathbf{N}_C$ are determined, then the bars will remain in equilibrium provided $F_A \leq 0.3\, N_A$ and $F_B \leq 0.5 N_B$ are satisfied.

Type 2. In this case *the total number of unknowns will equal the total number of available equilibrium equations plus the total number of available frictional equations*, $F = \mu N$. In particular, if *motion is impending* at the points of contact, then $F_s = \mu_s N$; whereas if the body is *slipping*, then $F_k = \mu_k N$. For example, consider the problem of finding the smallest angle θ at which the 100-N bar in Fig. 9–6a can be placed against the wall without slipping. The free-body diagram is shown in Fig. 9–6b. Here there are *five* unknowns: F_A, N_A, F_B, N_B, θ. For the solution there are *three* equilibrium equations and *two* frictional equations which apply at *both* points of contact, so that $F_A = 0.3N_A$ and $F_B = 0.4N_B$. (It should also be noted that the bar will not be in a state where motion impends *unless* the bar slips at *both* points A and B simultaneously.)

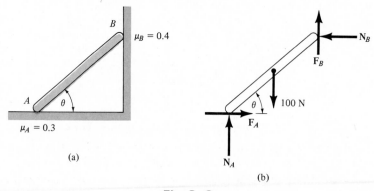

(a)

(b)

Fig. 9–6

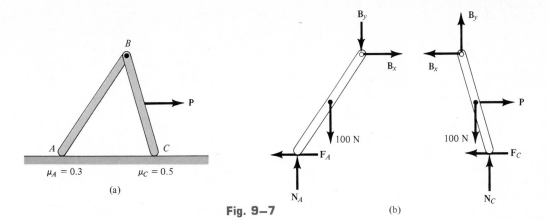

Fig. 9–7

(a)

(b)

Type 3. Here *the total number of unknowns will be less than the number of available equilibrium equations plus the total number of frictional equations or conditional equations for tipping.* As a result of this, several possibilities for motion or impending motion will exist and the problem will involve a determination of the kind of motion which actually occurs. For example, consider the two-member frame shown in Fig. 9–7a. Here we wish to determine the horizontal force **P** needed to cause movement of the frame. If each member has a weight of 100 N, then the free-body diagrams are as shown in Fig. 9–7b. There are *seven* unknowns: N_A, F_A, N_C, F_C, B_x, B_y, P. For a unique solution we must satisfy the *six* equilibrium equations (three for each member) and only *one* of two possible frictional equations. This means that as **P** increases its magnitude it will either cause slipping at A and no slipping at C, so that $F_A = 0.3\,N_A$ and $F_C \leqslant 0.5N_C$, or slipping occurs at C and no slipping at A, in which case $F_C = 0.5N_C$ and $F_A \leqslant 0.3N_A$. The actual situation can be determined by calculating P for each case and then choosing the case for which P is *smallest*. If in both cases the *same value* for P is calculated, then slipping at both points occurs simultaneously; i.e., the *seven unknowns* will satisfy *eight equations.* As a second example, consider a block having a width b, height h, and weight W which is resting on a rough surface, Fig. 9–8a. The force **P** needed to cause motion is to be determined. Inspection of the free-body diagram, Fig. 9–8b, indicates there are *four unknowns,* namely, P, F, N, and x. For a unique solution, however, we must satisfy the *three* equilibrium equations and either *one* friction equation or *one* conditional equation which requires the block not to tip. Hence two possibilities of motion exist. Either the block will *slip,* Fig. 9–8b, in which case $F = \mu_s N$ and the value obtained for x must satisfy $0 \leqslant x \leqslant b/2$, or the block will *tip,* Fig. 9–8c, in which case $x = b/2$ and the frictional force satisfies the inequality $F \leqslant \mu_s N$. The solution yielding the *smallest* value of P will define the type of motion the block undergoes. If it happens that the same value of P is calculated for both cases, then slipping and tipping will occur simultaneously, i.e., the *four unknowns* will satisfy *five equations.*

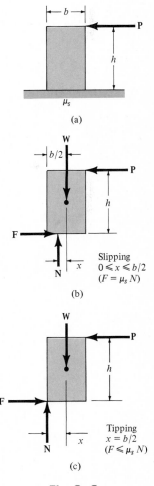

(a)

Slipping
$0 \leqslant x \leqslant b/2$
$(F = \mu_s N)$

(b)

Tipping
$x = b/2$
$(F \leqslant \mu_s N)$

(c)

Fig. 9–8

Equilibrium Versus Frictional Equations. It was stated earlier that the frictional force *always* acts so as to either oppose the relative motion or impede the motion of a body over its contacting surface. Realize, however, that we can *assume* the sense of the frictional force in problems which require F to be an "equilibrium force" and satisfy the inequality $F < \mu_s N$. The correct sense is made known *after* solving the equations of equilibrium for F. For example, if F is a negative scalar, the sense of **F** is the reverse of that which was assumed. This convenience of *assuming* the sense of **F** is possible because the equilibrium equations equate to zero the *components of vectors* acting in the *same direction*. In cases where the frictional equation $F_s = \mu_s N$ is used in the solution of a problem, however, the convenience of *assuming* the sense of **F** is *lost,* since the frictional equation relates only the *magnitudes* of two perpendicular vectors. Consequently, *F must always* be shown acting with its *correct sense* on the free-body diagram whenever the frictional equation is used for the solution of a problem.

PROCEDURE FOR ANALYSIS

The following procedure provides a method for solving equilibrium problems involving dry friction.

Free-Body Diagrams. Draw the necessary free-body diagrams and determine the number of unknowns or equations required for a complete solution. Unless stated in the problem, *always* show the frictional forces as *unknowns,* i.e., *do not* assume that $F = \mu N$. Recall that only three equations of coplanar equilibrium can be written for each body. Consequently, if there are more unknowns than equations of equilibrium, then it will be necessary to apply the frictional equation at some, if not all, points of contact to obtain the extra number of equations needed for a complete solution.

Equations of Friction and Equilibrium. Apply the equations of equilibrium and the necessary frictional equations (or conditional equations if tipping is involved) and solve for the unknowns. If the problem involves a three-dimensional force system such that it becomes difficult to obtain the force components or the necessary moment arms, apply the equations of equilibrium using Cartesian vectors.

The following example problems numerically illustrate this procedure.

Example 9–1

The uniform crate shown in Fig. 9–9a has a mass of 20 kg. If a force $P = 80$ N is applied to the crate, determine if the crate remains in equilibrium. The coefficient of static friction is $\mu_s = 0.3$.

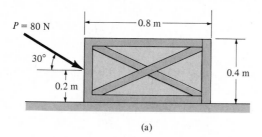

Fig. 9–9

(a)

Solution

Free-Body Diagram. As shown in Fig. 9–9b, the *resultant* normal force N_C must act a distance x from the crate's center line in order to counteract the tipping effect caused by **P.** There are *three unknowns:* F, N_C, and x, which can be determined strictly from the *three* equations of equilibrium.

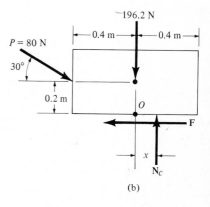

(b)

Equations of Equilibrium

$$\xrightarrow{+} \Sigma F_x = 0; \qquad 80 \cos 30° - F = 0$$
$$+ \uparrow \Sigma F_y = 0; \quad -80 \sin 30° + N_C - 196.2 = 0$$
$$\zeta + \Sigma M_O = 0; \quad 80 \sin 30°(0.4) - 80 \cos 30°(0.2) + N_C(x) = 0$$

Solving,

$$F = 69.3 \text{ N}$$
$$N_C = 236 \text{ N}$$
$$x = 0.0091 \text{ m}$$

Since all the answers are positive scalars, their sense shown on the free-body diagram is correct. In particular, $x = 0.0091$ m indicates the *resultant* normal force acts slightly to the *right* of the crate's center line. No tipping will occur since $x \leqslant 0.4$ m. Also, the *maximum* frictional force which can be developed at the surface of contact is $F_{max} = \mu_s N_C = 0.3(236 \text{ N}) = 70.8$ N. Since $F = 69.3$ N < 70.8 N, the crate will *not slip*, although it is very close to doing so.

Example 9–2

The pipe shown in Fig. 9–10a is gripped between two levers that are pinned together at C. If the coefficient of friction between the levers and the pipe is $\mu = 0.3$, determine the maximum angle θ at which the pipe can be gripped without slipping. Neglect the weight of the pipe.

Solution

Free-Body Diagram.   As shown in Fig. 9–10b, there are five unknowns: N_A, F_A, N_B, F_B, and θ. The *three* equations of equilibrium and *two* frictional equations at A and B apply. The frictional forces act toward C to prevent upward motion of the pipe.

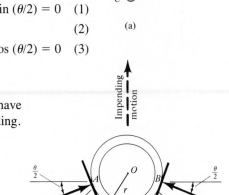

Equations of Friction and Equilibrium.   The frictional equations are

$$F_s = \mu_s N; \qquad\qquad F_A = \mu N_A$$
$$F_B = \mu N_B$$

Using these results, and applying the equations of equilibrium, yields

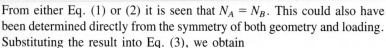

$\xrightarrow{+}\Sigma F_x = 0;$ $N_A \cos (\theta/2) + \mu N_A \sin (\theta/2) - N_B \cos (\theta/2) - \mu N_B \sin (\theta/2) = 0$ (1)

$\curvearrowleft + \Sigma M_O = 0;$ $\mu N_B(r) - \mu N_A(r) = 0$ (2)

$+\uparrow \Sigma F_y = 0;$ $N_A \sin (\theta/2) - \mu N_A \cos (\theta/2) + N_B \sin (\theta/2) - \mu N_B \cos (\theta/2) = 0$ (3)

(a)

From either Eq. (1) or (2) it is seen that $N_A = N_B$. This could also have been determined directly from the symmetry of both geometry and loading. Substituting the result into Eq. (3), we obtain

$$\sin (\theta/2) - \mu \cos (\theta/2) = 0$$

so that

$$\tan (\theta/2) = \frac{\sin (\theta/2)}{\cos (\theta/2)} = \mu = 0.3$$

$$\theta = 2 \tan^{-1} 0.3 = 33.4° \qquad\qquad Ans.$$

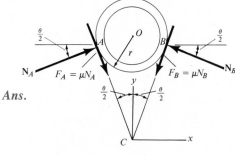

(b)

Fig. 9–10

Example 9–3

The uniform plank having a weight W and length l is supported at its ends A and B, where the coefficient of static friction is μ, Fig. 9–11a. Determine the greatest angle θ so the plank does not slip. Neglect the thickness of the plank for the calculation.

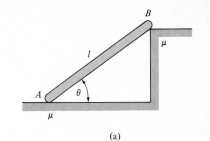

(a)

Solution

Free-Body Diagram. As shown in Fig. 9–11b, there are *five* unknowns: F_A, N_A, F_B, N_B, and θ. These can be determined from the *three* equilibrium equations and *two* frictional equations applied at points A and B. The frictional forces must be drawn with their correct sense so that they oppose the tendency for motion of the plank. Why?

Equations of Friction and Equilibrium. Writing the frictional equations,

$$F = \mu_s N; \qquad F_A = \mu N_A$$
$$F_B = \mu N_B$$

Using these results and applying the equations of equilibrium yields

$$\xrightarrow{+}\Sigma F_x = 0; \qquad \mu N_A + \mu N_B \cos\theta - N_B \sin\theta = 0 \qquad (1)$$
$$+\uparrow\Sigma F_y = 0; \quad N_A - W + N_B \cos\theta + \mu N_B \sin\theta = 0 \qquad (2)$$
$$\zeta + \Sigma M_G = 0; \quad -N_A\left(\frac{l}{2}\cos\theta\right) + \mu N_A\left(\frac{l}{2}\sin\theta\right) + N_B\left(\frac{l}{2}\right) = 0 \qquad (3)$$

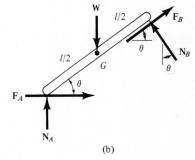

(b)

Fig. 9–11

Notice that moments were summed about the center of the plank G in order to eliminate **W.** We can now solve Eqs. (1) and (3), which reduce to

$$\mu N_A = N_B(\sin\theta - \mu\cos\theta)$$
$$N_B = N_A(\cos\theta - \mu\sin\theta)$$

Thus,

$$\mu N_A = N_A(\cos\theta - \mu\sin\theta)(\sin\theta - \mu\cos\theta)$$
$$\mu = \sin\theta\cos\theta - \mu\cos^2\theta - \mu\sin^2\theta + \mu^2\sin\theta\cos\theta$$
$$\mu = (1 + \mu^2)\sin\theta\cos\theta - \mu(\sin^2\theta + \cos^2\theta)$$

Since $\sin^2\theta + \cos^2\theta = 1$ and $\sin 2\theta = 2\sin\theta\cos\theta$, then

$$2\mu = \left(\frac{1 + \mu^2}{2}\right)\sin 2\theta$$

or

$$\theta = \frac{1}{2}\sin^{-1}\left(\frac{4\mu}{1 + \mu^2}\right) \qquad\qquad Ans.$$

Example 9-4

The homogeneous block shown in Fig. 9–12a has a weight of 20 lb and rests on the incline for which $\mu_s = 0.55$. Determine the largest angle of tilt, θ, of the plane before the block moves.

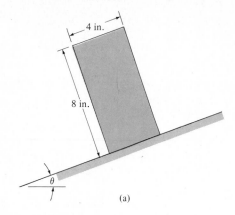

(a)   **Fig. 9–12**

Solution

Free-Body Diagram. As shown in Fig. 9–12b, the dimension x is used to locate the position of the resultant normal force $\mathbf{N}_B$ under the block. There are *four* unknowns, θ, N_B, F_B, and x. *Three* equations of equilibrium are available. The *fourth* equation is obtained by investigating the conditions for tipping or sliding of the block.

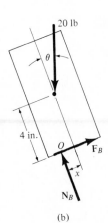

(b)

Equations of Equilibrium. Applying the equations of equilibrium yields

$$+\swarrow\Sigma F_x = 0; \qquad 20\sin\theta - F_B = 0 \qquad (1)$$
$$+\nwarrow\Sigma F_y = 0; \qquad N_B - 20\cos\theta = 0 \qquad (2)$$
$$\curvearrowright+\Sigma M_O = 0; \qquad -20\sin\theta(4) + 20\cos\theta(x) = 0 \qquad (3)$$

(Impending Motion of Block.) This requires use of the frictional equation

$$F_s = \mu_s N; \qquad F_B = 0.55 N_B \qquad (4)$$

Solving Eqs. (1) to (4) yields

$$N_B = 17.5\text{ lb} \qquad F_B = 9.64\text{ lb} \qquad \theta = 28.8° \qquad x = 2.2\text{ in.}$$

Since $x = 2.2$ in. > 2 in., the block will tip *before* sliding.

(Tipping of Block.) This requires

$$x = 2\text{ in.} \qquad (5)$$

Solving Eqs. (1) to (3) using Eq. (5) yields

$$N_B = 17.9\text{ lb} \qquad F_B = 8.94\text{ lb}$$
$$\theta = 26.6° \qquad\qquad Ans.$$

Example 9–5

Beam AB is subjected to a uniform load of 200 N/m and is supported at B by column BC, Fig. 9–13a. If the static coefficients of friction at B and C are $\mu_B = 0.2$ and $\mu_C = 0.5$, determine the force $\mathbf{P}$ needed to pull the column out from under the beam. Neglect the weight of the members.

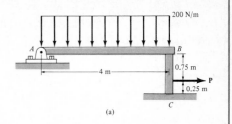
(a)

Solution

Free-Body Diagrams. The free-body diagram of beam AB is shown in Fig. 9–13b. Applying $\Sigma M_A = 0$, we obtain $N_B = 400$ N. This result is shown on the free-body diagram of the column, Fig. 9–13c. Referring to this member, the *four* unknowns F_B, P, F_C, and N_C are determined from the *three* equations of equilibrium and *one* frictional equation applied either at B or C.

Equations of Equilibrium and Friction. Applying the equations of equilibrium, we have

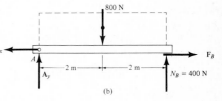
(b)

$$\xrightarrow{+}\Sigma F_x = 0; \qquad P - F_B - F_C = 0 \tag{1}$$
$$+\uparrow \Sigma F_y = 0; \qquad N_C - 400 = 0 \tag{2}$$
$$\curvearrowright+\Sigma M_C = 0; \qquad P(0.25) - F_B(1) = 0 \tag{3}$$

(Column Slips Only at B.) This requires $F_C \leqslant \mu_C N_C$ and

$$F_B = \mu_B N_B; \qquad F_B = 0.2(400) = 80 \text{ N}$$

Using this result and solving Eqs. (1) to (3), we obtain

$$P = 320 \text{ N}$$
$$F_C = 240 \text{ N}$$
$$N_C = 400 \text{ N}$$

Since $F_C = 240 > \mu_C N_C = 0.5(400) = 200$ N, the other case of movement must be investigated.

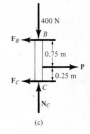
(c)

Fig. 9–13

(Column Slips Only at C.) Here $F_B \leqslant \mu_B N_B$ and

$$F_C = \mu_C N_C; \qquad F_C = 0.5 N_C \tag{4}$$

Solving Eqs. (1) to (4) yields

$$P = 267 \text{ N} \qquad\qquad Ans.$$
$$N_C = 400 \text{ N}$$
$$F_C = 200 \text{ N}$$
$$F_B = 66.7 \text{ N}$$

Obviously, this case of movement occurs first since it requires a *smaller* value for P.

Example 9–6

Determine the normal force that must be exerted on the 100-kg spool shown in Fig. 9–14a to push it up the 20° incline at constant velocity. The coefficients of static and kinetic friction at the points of contact are $(\mu_s)_A = 0.18$, $(\mu_k)_A = 0.15$ and $(\mu_s)_B = 0.45$, $(\mu_k)_B = 0.4$.

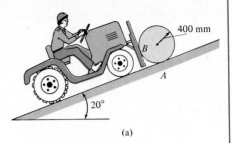

(a)

Solution

Free-Body Diagram. As shown in Fig. 9–14b, there are four unknowns N_A, F_A, N_B, and F_B acting on the spool. These can be determined from the *three* equations of equilibrium and *one* frictional equation, which applies either at A or B. If slipping only occurs at A, the spool will *slide* up the incline; whereas if slipping only occurs at B, the spool *rolls* up the incline. The problem requires determination of N_B.

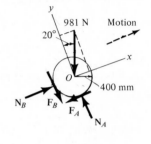

(b)

Fig. 9–14

Equations of Equilibrium and Friction

$$+ \nearrow \Sigma F_x = 0; \qquad -F_A + N_B - 981 \sin 20° = 0 \qquad (1)$$

$$+ \nwarrow \Sigma F_y = 0; \qquad N_A - F_B - 981 \cos 20° = 0 \qquad (2)$$

$$\gamma + \Sigma M_O = 0; \qquad -F_B(400 \text{ mm}) + F_A(400 \text{ mm}) = 0 \qquad (3)$$

(Spool Rolls up Incline.) In this case $F_A \leq 0.18 N_A$ and

$$(F_k)_B = (\mu_k)_B N_B; \qquad F_B = 0.40 N_B \qquad (4)$$

The direction of the frictional force at B must be correctly specified. Why? Since the spool is being forced up the plane, $\mathbf{F}_B$ acts downward to prevent the clockwise rolling motion of the spool, Fig. 9–14b. Solving Eqs. (1) to (4), we have

$$N_A = 1145.5 \text{ N} \qquad F_A = 223.7 \text{ N} \qquad N_B = 559.2 \text{ N} \qquad F_B = 223.7 \text{ N}$$

The assumption regarding no slipping at A should be checked.

$$F_A \leq (\mu_s)_A N_A; \qquad 223.7 \text{ N} \overset{?}{\leq} 0.18(1145.5) = 206.2 \text{ N}$$

The inequality does *not apply*, and therefore slipping occurs at A and not at B. Hence, the other case of motion must be investigated.

(Spool Slides up Incline.) In this case, $F_B \leq 0.45 N_B$ and

$$(F_k)_A = (\mu_k)_A N_A; \qquad F_A = 0.15 N_A \qquad (5)$$

Solving Eqs. (1) to (3) and (5) yields

$$N_A = 1084.5 \text{ N} \qquad F_A = 162.7 \text{ N} \qquad N_B = 498.2 \text{ N} \qquad F_B = 162.7 \text{ N}$$

The validity of the solution ($N_B = 498.2$ N) can be checked by testing the assumption that indeed no slipping occurs at B.

$$F_B \leq 0.45 N_B; \qquad 162.7 \text{ N} < 0.45(498.2 \text{ N}) = 224.2 \text{ N} \quad \text{(check)}$$

Problems

Assume that $\mu = \mu_s = \mu_k$ in the following problems where specified.

9–1. Draw the specified free-body diagram in each of the following problems and determine the number of unknown force magnitudes. Show all friction forces acting in their correct directions. If it is known that impending motion or slipping occurs, report the friction forces in the form $F = \mu N$, where μ is assumed to be known. Neglect the weights of the members unless otherwise stated. Although not shown on each photo, assume the geometry (size and angles) is known.

a. Draw a free-body diagram of the bicycle wheel which is pin connected to the frame at its center A. Although the frame is moving forward, the wheel is prevented from turning by application of the locked brakes at B. Sliding occurs on the road at C.

Prob. 9–1a

b. Draw the free-body diagrams of each of the three concrete pipes. Friction is developed at all the points of contact, A, B, C, and D. Each pipe has a uniform size and weight of 600 lb.

Prob. 9–1b

c. Draw a free-body diagram of the asphalt-roller which has a mass of 9 Mg, and center of mass at G, and is traveling forward at constant velocity. The engine powers the roller at A, whereas the roller at B is free to turn. Friction occurs at the points of contact, C and D, with the road.

Prob. 9–1c

d. The 20-kg plate is held in the jaws of the friction grip which contacts the plate at C and D. Draw the free-body diagrams of each of the two parts if they are pin connected at A. The chain (which can be considered as a smooth cable) is connected to one part at B and passes through a loop E formed in the other part.

Prob. 9–1d

9–2. The 5-kg cylinder is suspended from two equal-length cords. The end of each cord is attached to a ring of negligible mass, which passes along a horizontal shaft. If the coefficient of friction between each ring and the shaft is $\mu = 0.5$, determine the greatest distance d by which the rings can be separated and still support the cylinder.

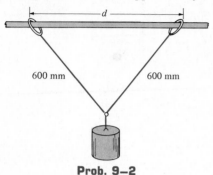

Prob. 9–2

9–3. The refrigerator has a weight of 180 lb and a center of gravity at G. Determine the force **P** required to move it. Also, determine the location of the resultant normal force measured from A. Take $\mu = 0.4$.

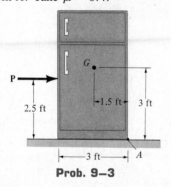

Prob. 9–3

*9–4. Determine the magnitude of force **P** needed to start moving the 40-kg crate. Also determine the location of the resultant normal force acting on the crate, measured from point A. The crate is symmetric with uniform mass. Take $\mu = 0.3$.

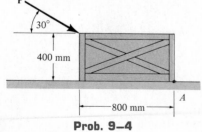

Prob. 9–4

9–5. Determine the horizontal force **P** needed to just start pushing the 300-lb crate up the plane. Take $\mu = 0.3$. Also, calculate the frictional force when $P = 500$ lb.

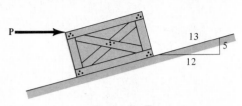

Prob. 9–5

9–6. The winch cable on a tow truck is subjected to a force of $T = 6$ kN when the cable is directed at $\theta = 60°$. Determine the magnitude of the brake frictional force **F** on the rear wheels B. The truck has a total mass of 4 Mg and mass center at G. Take $\mu = 0.5$.

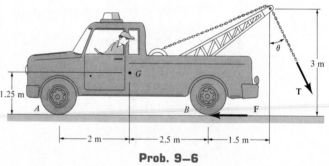

Prob. 9–6

9–7. The uniform dresser has a weight of 90 lb and rests on a tile floor for which $\mu_s = 0.25$. If the man pushes horizontally on it in the direction shown, determine the smallest magnitude of force **F** needed to move the dresser. Also, if the man has a weight of 150 lb, determine the smallest coefficient of friction between his shoes and the floor so that he does not slip.

Prob. 9–7

*9–8. The jeep and driver have a mass of 1.2 Mg. Determine the steepest slope θ which the jeep can climb at constant speed if (a) driving power is applied only to the rear wheels B, while the front wheels A are free to roll, and (b) driving power is applied to all four wheels. The coefficient of friction between the wheels and the ground is $\mu = 0.5$.

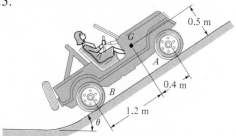

Prob. 9–8

9–9. The log has a coefficient of friction of $\mu = 0.3$ with the ground and a weight of 40 lb/ft. If a man can pull on the rope with a maximum force of 80 lb, determine the greatest length l of log he can drag.

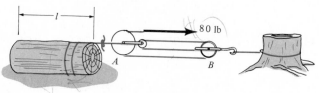

Prob. 9–9

9–10. One method of removing a fence post consists of using a wheel rim and chain arranged as shown. If the hub H of the wheel is smooth, whereas contact with the ground A is rough, determine the necessary friction force at A and the vertical force acting on the post which is effective in pulling it out. A tractor pulls on the chain with a horizontal force of $F = 200$ lb. Take $\mu_A = 0.8$.

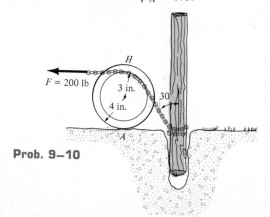

Prob. 9–10

9–11. Determine the maximum weight W the man can lift using the pulley system, without and then with the "leading block" or pulley at A. The man has a weight of 200 lb and the coefficient of friction between his feet and the ground is $\mu = 0.6$.

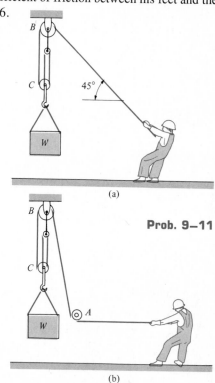

(a)

Prob. 9–11

(b)

*9–12. The board can be adjusted vertically by tilting it up and sliding the smooth pin A along the vertical guide G. When placed horizontally, the bottom C bears along the edge of the guide, where $\mu = 0.4$. Determine the largest dimension d which will support any applied force $\mathbf{F}$ without causing the board to slip downward.

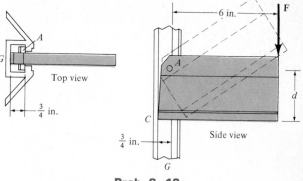

Prob. 9–12

407

9–13. The tractor has a weight of 4500 lb with center of gravity at G. The driving traction is developed at the rear wheels B, while the front wheels at A are free to roll. If the coefficient of friction between the wheels at B and the ground is $\mu = 0.5$, determine the largest load P it can tow without causing the wheels at B to slip or the front wheels at A to lift off the ground.

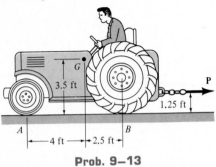

Prob. 9–13

9–14. The spool of wire having a mass of 150 kg rests on the ground A and against the wall at B. Determine the force **P** required to pull the wire horizontally off the spool. The coefficient of friction between the spool and its points of contact is $\mu = 0.25$.

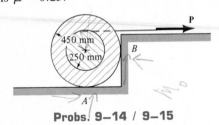

Probs. 9–14 / 9–15

9–15. The spool of wire having a mass of 150 kg rests on the ground A and against the wall at B. Determine all the forces acting on the spool if $P = 800$ N. The coefficient of friction between the spool and the ground at point A is $\mu = 0.35$. The wall at B is smooth.

*9–16. The left-end support of the beam acts as a collar so that the beam may be adjusted to any desired elevation. Contact with the vertical shaft is made at points A and B. If $\mu = 0.6$, determine the smallest distance x a load P may be placed on the beam without causing it to slip. Neglect the weight of the beam.

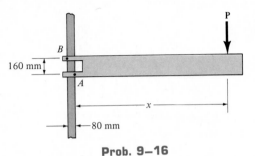

Prob. 9–16

9–17. Two boys, each weighing 60 lb, sit at the ends of a uniform board which has a weight of 30 lb. If the board rests at its center on a post having a coefficient of friction of $\mu = 0.6$ with the board, determine the greatest angle of tilt θ before slipping occurs. Neglect the size of the post and the thickness of the board in the calculation.

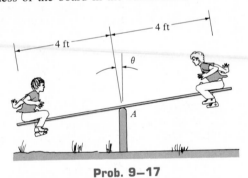

Prob. 9–17

9–18. The tow truck is used to pull the car out of the ditch using the winch at A. The cable force needed is 600 lb and the truck's back brakes are locked. The coefficient of friction between the wheels at B and the ground is $\mu = 0.5$, and the truck has a weight of 5000 lb and center of gravity at G. Determine if the car starts to move without tipping the truck or causing it to slide. The front wheels of the truck are free to roll.

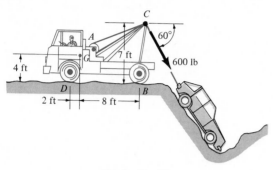

Prob. 9–18

9–19. Determine the greatest load W the boy can pull on the sled if the sled has a weight of 8 lb and $\mu_s = 0.1$ between the runners and the ice. The boy weighs 70 lb and $\mu_b = 0.4$ between his boots and the ice. Set $\theta = 30°$.

Probs. 9–19 / 9–20

*9–20. Determine the angle θ at which the cord must be directed so that the boy can pull the sled which supports a load of 400 lb. The sled weighs 8 lb and $\mu_s = 0.1$ between the runners and the ice. The boy weighs 70 lb and $\mu_b = 0.4$ between his boots and the ice.

9–21. The truck has a weight of 15,000 lb and a center of gravity at G. If it is traveling gradually off a road onto a steep shoulder, determine the greatest angle θ at which it can be parked so that it does not slide downward or tip over. Take $\mu_A = \mu_B = 0.4$. *Hint:* Draw a free-body diagram of the truck viewed from the back. For tipping require the normal force at B to be zero.

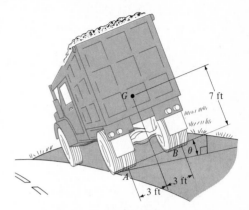

Prob. 9–21

9–22. The hoop of weight W is suspended by the peg at A. A horizontal force $\mathbf{P}$ is slowly applied at B. If the hoop begins to slip at A when $\theta = 30°$, determine the coefficient of friction between the hoop and the peg.

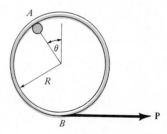

Prob. 9–22

9–23. Romeo has a mass of 80 kg and attempts a late-night rendezvous with Juliet using a ladder placed against the building as shown. If the coefficient of friction at A is $\mu_A = 0.5$, and $\mu_B = 0.6$, determine if he can stand within her reach, i.e., on the rung at C. Is it possible then for him to carry her down the ladder? Juliet has a mass of 65 kg. Neglect the weight of the ladder. Incidentally, Juliet is an engineering student who fully realizes the consequences of such an action.

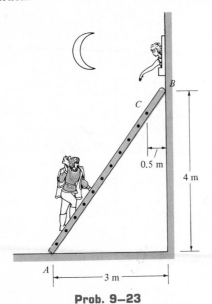

Prob. 9–23

*9–24. If a meter stick of uniform mass is placed on your index fingers A and B at the 200-mm and 900-mm position shown, determine what will happen as your hands are slowly drawn close together, i.e., at what mark on the meter stick will they meet? Take $\mu = 0.3$. The mass of the stick is 0.5 kg.

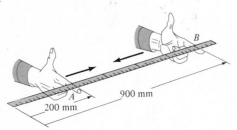

Prob. 9–24

9–25. If the coefficient of static friction between the drum and brake mechanism is $\mu_s = 0.4$, determine the horizontal and vertical components of reaction at pin O. Does the force $P = 150$ N prevent the drum from rotating? Neglect the weight and thickness of the brake. The drum has a mass of 25 kg.

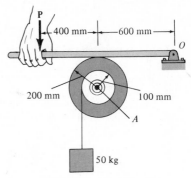

Probs. 9–25 / 9–26

9–26. Solve Prob. 9–25 if $P = 400$ N.

9–27. The disk has a mass of 50 kg and the block has a mass of 12 kg. If the coefficients of friction at the points of contact are $\mu_A = 0.5$ and $\mu_B = 0.2$, respectively, determine the smallest couple moment M which can be applied to the disk to cause motion. Neglect the mass of the connecting rod CD.

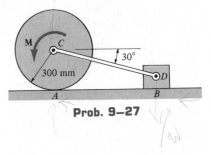

Prob. 9–27

*9–28. Three blocks A, B, and C have a weight of 10 lb, 6 lb, and 4 lb, respectively. They are resting on the incline for which $\mu_A = 0.15$, $\mu_B = 0.25$, and $\mu_C = 0.3$. Determine the incline angle θ for which impending motion of all the blocks occurs. Also find the stretch or compression in each of the connecting springs. The springs each have a stiffness of $k = 2$ lb/ft.

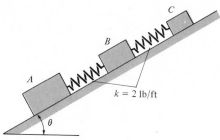

Prob. 9–28

9–29. The 80-lb boy stands on the beam and pulls on the cord with a force large enough to just cause him to slip. If $\mu_D = 0.4$ between his shoes and the beam, determine the reactions at A and B. The beam is uniform and has a weight of 100 lb. Neglect the size of the pulleys and the thickness of the beam.

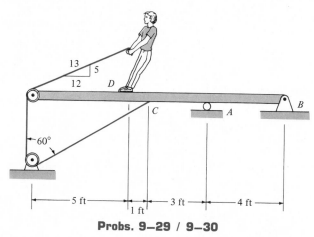

Probs. 9–29 / 9–30

9–30. The 80-lb boy stands on the beam and pulls with a force of 40 lb. If $\mu_D = 0.4$, determine the frictional force between his shoes and the beam and the reactions at A and B. The beam is uniform and has a weight of 100 lb. Neglect the size of the pulleys and the thickness of the beam.

9–31. The friction pawl is pinned at A and rests against the wheel at B. It allows freedom of movement when the wheel is rotating counterclockwise. Clockwise rotation is prevented since due to friction the pawl will then bind the wheel. If $\mu_B = 0.6$, determine the design angle θ which will prevent clockwise motion for any value of applied moment **M**. *Hint:* Neglecting its weight, the pawl is a two-force member.

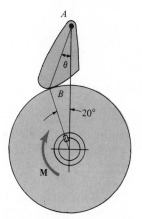

Prob. 9–31

■*9–32. The beam is supported by a pin at A and a roller at B which has negligible weight and a radius of 15 mm. If the coefficient of static friction is $\mu_B = \mu_C = 0.2$, determine the largest angle θ of the incline so that the roller does not slip for any force **P** applied to the beam.

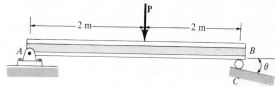

Prob. 9–32

■9–33. The empty drum has a weight W and center of gravity at G. If the coefficient of friction between it and the inclined plane is $\mu = 0.4$, determine the greatest angle θ of the plane without causing the disk to slip.

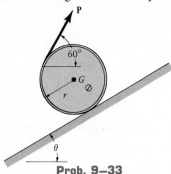

Prob. 9–33

9–34. The carton clamp on the lift fork has a coefficient of friction of $\mu = 0.5$ with any contacting carton made of cardboard. A cardboard carton has a coefficient of friction of $\mu' = 0.4$ with any other cardboard container. This being the case, compute the smallest horizontal force P the clamp must exert on the sides of a carton so that two cartons A and B each weighing 30 lb can be lifted. What smallest force P is required to lift three 30-lb cartons? The third carton C is placed between A and B.

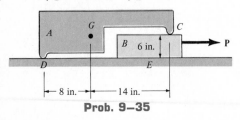

Prob. 9–34

9–35. Block B has a weight of 6 lb and A weighs 10 lb. If A has a center of gravity at G, determine the greatest horizontal force P which may be applied to B without causing movement of B. The coefficients of friction at C, D, and E are $\mu_C = 0.5$, $\mu_D = 0.3$, and $\mu_E = 0.6$.

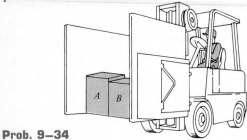

Prob. 9–35

*9–36. A man pushes against the stack of four uniform boxes each weighing 20 lb. If the coefficient of friction between each box is $\mu = 0.6$ and between the floor and a box $\mu' = 0.4$, determine the greatest horizontal force P that can be applied without causing any slipping or tipping.

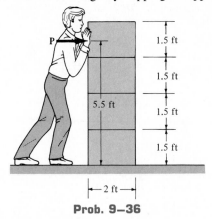

Prob. 9–36

9–37. Two blocks A and B, each having a mass of 10 kg, are connected by the linkage shown. If the coefficient of static friction at the contacting surfaces is $\mu_s = 0.5$, determine the largest vertical force P that may be applied to pin C of the linkage without causing the blocks to slip. Neglect the weight of the links.

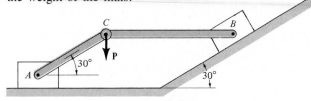

Prob. 9–37

9–38. The two plates are pin-connected at A and rest on the rough floor at B and C for which $\mu_B = \mu_C = 0.6$. Plate AB has a weight of 20 lb and AC weighs 75 lb. Determine the greatest angle θ between the plates so that they can remain in equilibrium and do not collapse.

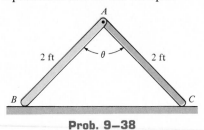

Prob. 9–38

9–39. A 45-kg disk rests on an inclined surface for which $\mu_s = 0.2$. Determine the maximum vertical force P that may be applied to link AB without causing the disk to slip at C.

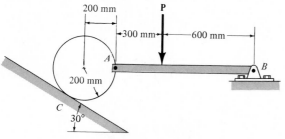

200 mm
300 mm
P
600 mm
A
200 mm
B
C
30°

Prob. 9–39

*9–40. The end C of the two-bar linkage rests on the top center of the 50-kg cylinder. If the coefficients of friction at C and E are $\mu_C = 0.6$ and $\mu_E = 0.3$, determine the largest vertical force P which can be applied at B without causing motion. Neglect the mass of the bars.

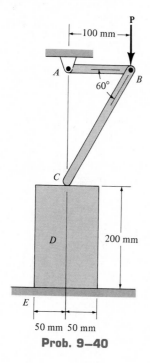

P
100 mm
A
60°
B
C
D
200 mm
E
50 mm 50 mm

Prob. 9–40

9–41. Two blocks A and B, each having a mass of 6 kg, are connected by the linkage shown. If the coefficient of static friction at the contacting surfaces is $\mu_B = 0.8$ and $\mu_A = 0.2$, determine the largest vertical force P that may be applied to pin C without causing the blocks to slip. Neglect the weight of the links.

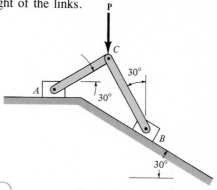

P
C
30°
A
30°
B
30°

Prob. 9–41

9–42. The wheel weighs 20 lb and rests on a surface for which $\mu_B = 0.2$. A cord wrapped around it is attached to the top of the 30-lb homogeneous block for which $\mu_D = 0.3$. Determine the smallest vertical force which can be applied tangentially to the wheel which will cause motion to impend.

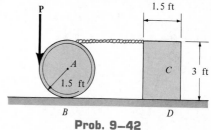

P
1.5 ft
A
1.5 ft
C
3 ft
B
D

Prob. 9–42

■9–43. The uniform rod has a weight W and rests on a smooth peg at A and against a wall at B for which $\mu = 0.25$. Determine the greatest angle θ for placement of the rod so that it does not slip.

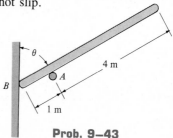

θ
B
A
4 m
1 m

Prob. 9–43

413

*9–44. The uniform 6-kg slender rod rests on the top center of the 3-kg block. If the coefficients of friction at the points of contact are $\mu_A = 0.4$, $\mu_B = 0.6$, and $\mu_C = 0.3$, determine the largest couple moment M which can be applied to the rod without causing motion of the rod.

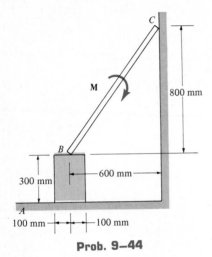

Prob. 9–44

9–45. The beam AB has a negligible mass and is subjected to a force of 200 N. It is supported at one end by a pin and at the other end by a spool having a mass of 40 kg. If a cable is wrapped around the inner core of the spool, determine the minimum cable force P needed to move the spool. The coefficients of friction at B and D are $\mu_B = 0.4$ and $\mu_D = 0.2$, respectively.

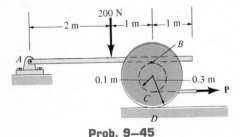

Prob. 9–45

9–46. The uniform 5-lb rod rests on the ground at B and against the wall at A. If the rod does not slip at B, determine the smallest coefficient of friction at A which will prevent slipping.

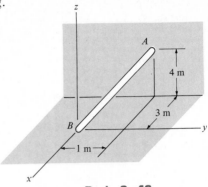

Prob. 9–46

9–47. The rod has a weight W and rests against the floor and wall for which the coefficients of friction are μ_A and μ_B, respectively. Determine the smallest value of θ for which the rod will not move. Solve the problem (a) using the parameters given, and (b) with $W = 20$ lb, $L = 8$ ft, $\mu_A = 0.2$ and $\mu_B = 0.4$.

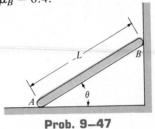

Prob. 9–47

*9–48. The semicylinder has a weight W and a center of gravity at G. The coefficient of friction at the ground is μ. If a horizontal force P is applied to its edge as shown, determine the angle of tilt θ before the cylinder slips.

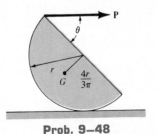

Prob. 9–48

9–49. Specify the angle θ and the magnitude P of the smallest force needed to tow the 120-kg crate up the incline. The coefficient of friction between the plane and the crate is $\mu = 0.5$. *Hint:* Using the force equilibrium equations, relate $P = f(\theta)$, then set $dP/d\theta = 0$ to determine θ.

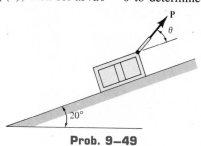

Prob. 9–49

■**9–50.** The uniform pole has a weight W and is lowered slowly from a vertical position $\theta = 90°$ toward the horizontal using a cable AB. If $\mu = 0.3$ at C, determine the angle θ at which the pole will start to slip.

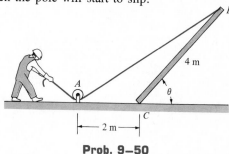

Prob. 9–50

Wedges **9.3**

A *wedge* is a simple machine which is often used to transform an applied force into much larger forces, directed at approximately right angles to the applied force. Also, wedges can be used to give small displacements or adjustments to heavy loads.

Consider, for example, the wedge shown in Fig. 9–15a, which is used to *lift* a block of weight W by applying a force P to the wedge. Free-body diagrams of the wedge and block are shown in Fig. 9–15b. Here we have excluded the weight of the wedge since it is usually *small* compared to the weight of the block. Also, note that the frictional forces F_1 and F_2 must oppose the motion of the wedge. Likewise, the frictional force F_3 of the wall on the block must act downward so as to oppose the block's upward motion. The locations of the resultant normal forces are not important in the force analysis since neither the block nor wedge will "tip." Hence the moment equilibrium equations will not be considered. There are seven unknowns consisting of the applied force P, needed to cause motion of the wedge, and six normal and frictional forces. The seven available equations consist of two force equilibrium equations ($\Sigma F_x = 0$, $\Sigma F_y = 0$) applied to the wedge and block (four equations total) and the frictional equation $F = \mu N$ applied at each surface of contact (three equations total).

Realize that if the block is to be *lowered*, the frictional forces will all act in a sense opposite to that shown in Fig. 9–15b. The applied force P will act to the right as shown if the coefficient of friction is small or the wedge angle θ is large. Otherwise, P would have to *pull* on the wedge in order to remove it. If P is *removed*, and friction forces hold the block in place, then the wedge is referred to as *self-locking*.

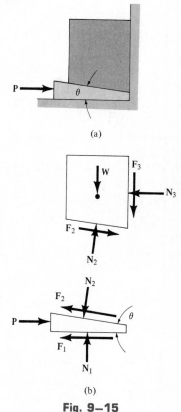

Fig. 9–15

415

Example 9–7

The uniform stone has a mass of 500 kg and is held in the horizontal position using a wedge as shown in Fig. 9–16a. If the coefficient of friction is $\mu = 0.3$ at all surfaces of contact, determine the force **P** used to remove the wedge. Is the wedge self-locking? Assume that the stone does not slip at A.

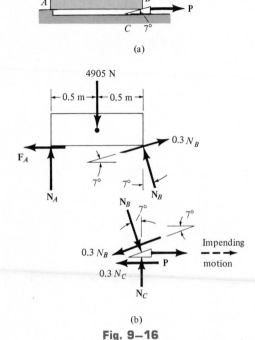

Fig. 9–16

Solution

Since the wedge is to be removed, slipping is about to occur at the surfaces of contact. Thus, $F = \mu N$, and the free-body diagrams are shown in Fig. 9–16b. Note that on the stone at A, $F_A < \mu N_A$, since slipping does not occur there. From the free-body diagram of the stone,

$$\zeta + \Sigma M_A = 0; \quad -4905(0.5) + (N_B \cos 7°)(1) + (0.3N_B \sin 7°)(1) = 0$$
$$N_B = 2383.1 \text{ N}$$

Using this result for the wedge, we have

$$\xrightarrow{+} \Sigma F_x = 0; \quad 2383.1 \sin 7° - 0.3(2383.1) \cos 7° + P - 0.3N_C = 0$$
$$+\uparrow \Sigma F_y = 0; \quad N_C - 2383.1 \cos 7° - 0.3(2383.1) \sin 7° = 0$$
$$N_C = 2452.5 \text{ N}$$
$$P = 1154.9 \text{ N} = 1.15 \text{ kN} \qquad \qquad Ans.$$

Since P is positive, indeed the wedge must be pulled out. Obviously, if P is zero, the wedge will remain in place (self-locking) and the frictional forces $\mathbf{F}_C$ and $\mathbf{F}_B$ developed at the points of contact will satisfy $F_B < \mu N_B$ and $F_C < \mu N_C$.

Frictional Forces on Screws 9.4*

In most cases screws are used as fasteners; however, in many types of machines, screws are incorporated to transmit power or motion from one part of the machine to another. A *square-threaded screw* is most commonly used for the latter purpose, especially when large forces are applied along the axis of the screw. This type of screw will be considered in the following analysis. The analysis of other types of screws, such as the V-thread, is based on the same principles.

A *screw* may be thought of simply as an inclined plane or wedge wrapped around a cylinder. A nut initially at position A on the screw shown in Fig. 9–17 will move up to B when rotated 360° around the screw. This rotation is equivalent to translating the nut up an inclined plane of height p and length $l = 2\pi r$, where r is the mean radius of the thread. The rise p for a single revolution is often referred to as the *pitch* of the screw, where the *pitch angle* is given by $\theta_p = \tan^{-1}(p/2\pi r)$.

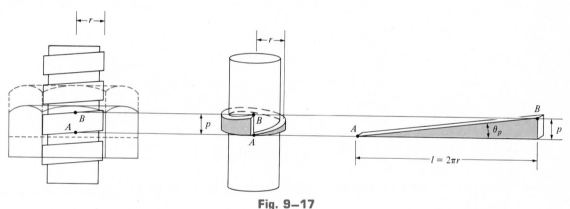

Fig. 9–17

Frictional Analysis. When a screw is subjected to large axial loads, the frictional forces developed at the thread become important in the analysis to determine the force needed to turn the screw. Consider, for example, the square-threaded jack screw shown in Fig. 9–18, which supports the vertical load **W** and twisting moment **M.*** If moments are summed about the axis of the screw, **M** can be thought of as being equivalent to the moment of a horizontal force **S** acting at the mean radius r of the thread, so that $M = Sr$. The reactive forces of the jack to loads S and W are actually distributed over the circumference of the screw thread in contact with the screw hole in the jack, that is, within region h shown in Fig. 9–18. For simplicity, this portion

*For applications, **M** is developed by applying a horizontal force **P** at a right angle to the end of a lever that would be fixed to the screw.

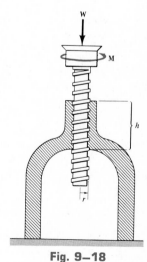

Fig. 9–18

417

Fig. 9–19

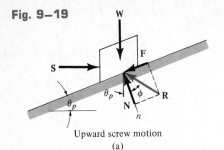

Upward screw motion
(a)

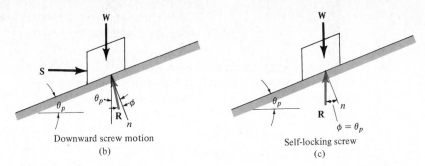

Downward screw motion
(b)

Self-locking screw
(c)

of thread can be imagined as being unwound from the screw and represented as a simple block resting on an inclined plane having the screw's pitch angle θ_p, Fig. 9–19a. The inclined plane represents the inside *supporting thread* of the jack base. The block is subjected to the total axial load **W** acting on the jack, the horizontal force **S** (which is related to the applied moment **M**), and the *resultant force* **R** exerted by the plane. As shown, force **R** has components acting normal, **N**, and tangent, **F**, to the contacting surfaces.

Upward Screw Motion. Provided M is great enough, the screw (and hence the block) is either on the verge of upward impending motion or motion is occurring. Under these conditions, **R** acts at an angle $(\phi + \theta_p)$ from the vertical as shown in the figure, where $\phi = \tan^{-1}(F/N) = \tan^{-1}(\mu N/N) = \tan^{-1}\mu$. Applying the two force equations of equilibrium to the block, we obtain

$$\xrightarrow{+}\Sigma F_x = 0; \qquad\qquad S - R\sin(\phi + \theta_p) = 0$$
$$+\uparrow\Sigma F_y = 0; \qquad\qquad R\cos(\phi + \theta_p) - W = 0$$

Eliminating R and solving for S, then substituting this value into the equation $M = Sr$, yields

$$M = Wr\tan(\phi + \theta_p) \qquad\qquad (9-3)$$

This equation gives the required value M necessary to cause upward impending motion of the screw when $\phi = \phi_s = \tan^{-1}\mu_s$ (the angle of static friction). If ϕ is replaced by $\phi_k = \tan^{-1}\mu_k$ (the angle of kinetic friction), Eq. 9–3 would give a smaller value M necessary to maintain uniform upward motion of the screw.

Downward Screw Motion. When the load **W** is to be *lowered*, the direction of **M** is reversed, in which case the angle ϕ (ϕ_s or ϕ_k) lies on the opposite side of the normal n to the plane supporting the block. This case is shown in Fig. 9–19b for $\phi < \theta_p$. Thus, Eq. 9–3 becomes

$$M = Wr\tan(\phi - \theta_p) \qquad\qquad (9-4)$$

Self-locking Screw. If the moment **M** is *removed*, the screw will remain *self-locking;* i.e., it will support the load **W** *by friction forces alone* provided $\phi > \theta_p$. However, if $\phi = \phi_s = \theta_p$, Fig. 9–19c, the screw will be on the verge of rotating downward. When $\theta_p > \phi_s$, a restraining moment is needed to prevent the screw from rotating downward.

Example 9–8

The turnbuckle shown in Fig. 9–20 has a square thread with a mean radius of 5 mm and a pitch of 2 mm. If the coefficient of friction between the screw and the turnbuckle is $\mu_s = 0.25$, determine the moment **M** that must be applied to draw the end screws closer together. Is the turnbuckle self-locking?

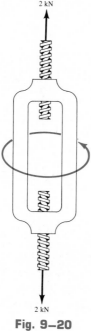

2 kN

2 kN

Fig. 9–20

Solution

The moment may be obtained by using Eq. 9–3. Since friction at two screws must be overcome, this requires

$$M = 2[Wr \tan (\phi_s + \theta_p)] \qquad (1)$$

Here, $W = 2000$ N, $r = 5$ mm, $\phi_s = \tan^{-1} \mu_s = \tan^{-1}(0.25) = 14.04°$, and $\theta_p = \tan^{-1}(p/2\pi r) = \tan^{-1}(2 \text{ mm}/[2\pi(5 \text{ mm})]) = 3.64°$. Substituting these values into Eq. (1) and solving gives

$$M = 2[(2000 \text{ N})(5 \text{ mm}) \tan (14.04° + 3.64°)]$$
$$M = 6375.1 \text{ N} \cdot \text{mm} = 6.38 \text{ N} \cdot \text{m} \qquad \textit{Ans.}$$

When the moment is *removed,* the turnbuckle will be self-locking; i.e., it will not unscrew, since $\phi_s > \theta_p$.

Problems

9–51. Determine the force **P** needed to lift the 100-lb load. Smooth rollers are placed between the wedges. The coefficient of friction between *A* and *C* and *B* and *D* is $\mu = 0.3$. Neglect the weight of each wedge.

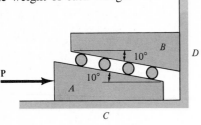

Prob. 9–51

*9–52. The wedge is used to level the floor of a building. For the floor loading shown, determine the horizontal force **P** that must be applied to move the wedge forward. The coefficient of friction between the wedge and the two surfaces of contact is $\mu = 0.25$. Neglect the weight of the wedge.

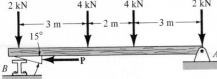

Prob. 9–52

9–53. The two blocks used in a measuring device have negligible weight. If the spring is compressed 5 in. when in the position shown, determine the smallest axial force *P* which the adjustment screw must exert on *B* in order to start the movement of *B* downward. The end of the screw is *smooth* and the coefficient of friction at all other points of contact is $\mu = 0.3$.

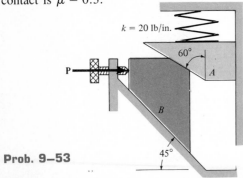

Prob. 9–53

9–54. The wedge blocks are used to hold the specimen in a tension testing machine. Determine the design angle θ of the wedges so that the specimen will not slip regardless of the applied load. Take $\mu_A = 0.1$ at *A* and $\mu_B = 0.6$ at *B*. Neglect the weight of the blocks.

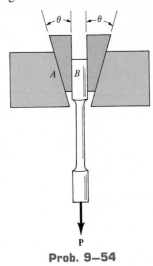

Prob. 9–54

9–55. The wedge has a negligible weight and a coefficient of static friction $\mu_s = 0.35$ with all contacting surfaces. Determine the angle θ so that it is "self-locking." This requires no slipping for any magnitude of the force **P** applied to the joint.

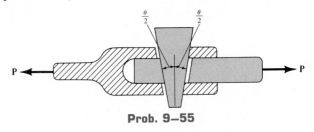

Prob. 9–55

*9–56. Each wedge has a weight of 400 lb. Determine how far the force **P** can compress the spring until wedge B slips on wedge A. What is the magnitude of **P** for this to occur?

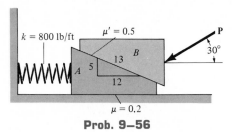

Prob. 9–56

9–57. The uniform roll of sheet metal has a mass of 120 kg and rests on a 30° wedge of negligible mass. If the coefficient of friction at all contacting surfaces is $\mu = 0.3$, determine the horizontal force P that must be applied to the wedge in order to lift the cylinder.

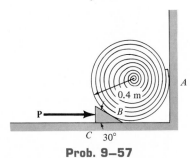

Prob. 9–57

9–58. The three stone blocks have weights of $W_A = 600$ lb, $W_B = 150$ lb, and $W_C = 500$ lb. Determine the smallest horizontal force P that must be applied to block C in order to move this block. The coefficient of friction between the blocks is $\mu = 0.3$ and between the floor and each block $\mu' = 0.5$.

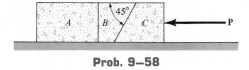

Prob. 9–58

9–59. Determine the largest weight of the wedge that can be placed between the 8-lb cylinder and the wall without upsetting equilibrium. The coefficient of friction at A and C is $\mu = 0.5$ and at B, $\mu' = 0.6$.

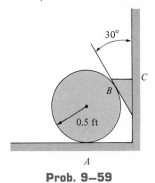

Prob. 9–59

*9–60. The square-threaded bolt is used to join two plates together. If the bolt has a mean diameter of $d = 0.75$ in. and a pitch of $p = 0.6$ in., determine the torque **M** required to tighten the nut so that it develops a tension in the bolt of $T = 30,000$ lb. The coefficient of static friction between the threads and the bolt is $\mu_s = 0.4$.

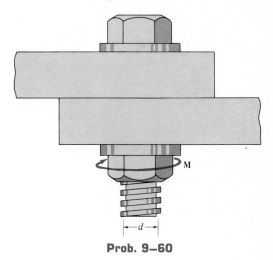

Prob. 9–60

9–61. Determine the horizontal force P applied perpendicular to the handle of the jack screw necessary to start lifting the 3-kN load. The square-threaded screw has a pitch of 5 mm and a mean diameter of 60 mm. The coefficient of friction for the screw is $\mu_s = 0.2$.

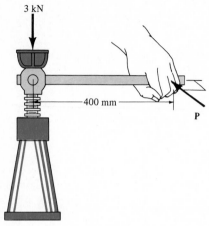

Prob. 9–61

9–62. Determine the clamping force on the block of wood if the screw of the "C" clamp is tightened with a twist of $M = 8$ N · m. The single square-threaded screw has a mean radius of 10 mm, a pitch of 3 mm, and the coefficient of friction is $\mu = 0.35$.

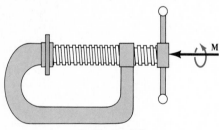

Prob. 9–62

9–63. The vice is used to grip the pipe. If a horizontal force of 25 lb is applied perpendicular to the end of the 10-in. handle, determine the compressive force F developed in the pipe. The square threads have a mean diameter of 1.5 in. and a pitch of 0.2 in. Take $\mu_s = 0.3$. How much force must be applied perpendicular to the handle to loosen the vice?

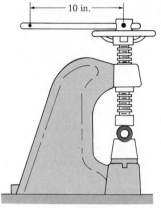

Prob. 9–63

*9–64. The shaft has a square-threaded screw with a pitch of $p = 9$ mm and a mean radius of 15 mm. If it is in contact with a plate gear having a mean radius of 20 mm, determine the resisting torque M on the plate gear which can be overcome if a torque of 7 N · m is applied to the shaft. The coefficient of friction at the screw is $\mu = 0.2$. Neglect friction of the bearings located at A and B.

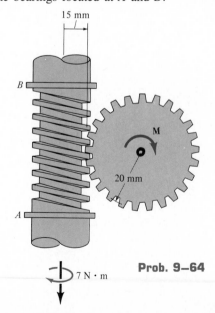

Prob. 9–64

9–65. The braking mechanism consists of two pinned arms and a square-threaded screw with left and right-hand threads. Thus, when turned, the screw draws the two arms together. If the pitch of the screw is 4 mm, the mean diameter 12 mm, and the coefficient of friction is $\mu = 0.35$, determine the tension in the screw when a torque of 5 N · m is applied to the screw. If the coefficient of friction between the brake pads A and B and the circular shaft is $\mu' = 0.5$, determine the maximum torque M the brake can resist.

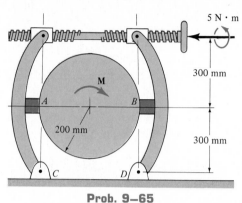

Prob. 9–65

9–66. The fixture clamp consists of a square-threaded screw having a coefficient of friction of $\mu = 0.3$, mean diameter of 3 mm, and a pitch of 1 mm. The five points indicated are pin connections. Determine the clamping force at the smooth blocks D and E when a torque of $M = 0.08$ N · m is applied to the handle of the screw.

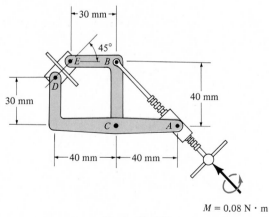

$M = 0.08$ N · m

Prob. 9–66

9–67. A turnbuckle, similar to that shown in Fig. 9–20, is used to tension member AB of the truss. The coefficient of friction between the square-threaded screws and the turnbuckle is $\mu = 0.5$ and the screws have a mean radius of 6 mm and a pitch of 3 mm. If a moment of $M = 10$ N · m is applied to the turnbuckle, determine the force in each member of the truss. No external forces act on the truss.

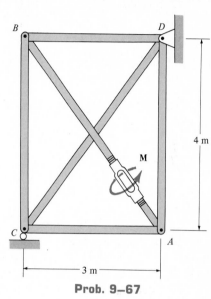

Prob. 9–67

*9.5 Frictional Forces on Flat Belts

In the design of belt drives or band brakes it is necessary to determine the frictional forces developed between a belt and its contacting surface. Consider the flat belt passing over a fixed drum of radius r, as shown in Fig. 9–21a. The *total* angle of belt contact in radians is β and the coefficient of friction between the two surfaces is μ. If it is known that the tension acting in the belt on the right of the drum is $\mathbf{T}_1$, let it be required to find the tension $\mathbf{T}_2$ needed to pull the belt counterclockwise over the surface of the drum. *Obviously, T_2 must be greater than T_1 since the belt must overcome the resistance of friction at the surface of contact.*

Frictional Analysis. A free-body diagram of the belt is shown in Fig. 9–21b. Here the normal force $\mathbf{N}$ and the frictional force $\mathbf{F}$, acting at different points along the contacting surface of the belt, will vary both in magnitude and direction. Due to this unknown force distribution, the analysis of the problem will proceed on the basis of initially studying the forces acting on a differential element of the belt.

A free-body diagram of an element having a length $ds = r\,d\theta$ is shown in Fig. 9–21c. Assuming either impending motion or motion of the belt, the

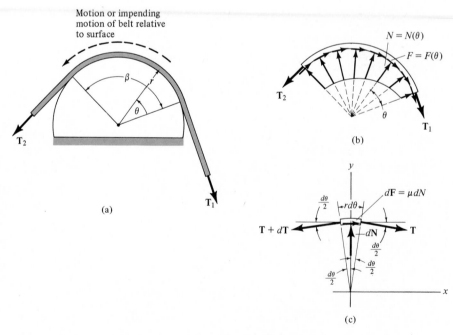

Fig. 9–21

magnitude of frictional force $dF = \mu \, dN$. This force opposes the sliding motion of the belt and thereby increases the magnitude of tensile force acting in the belt by dT. Applying the two force equations of equilibrium, we have

$$\xrightarrow{+} \Sigma F_x = 0; \quad T \cos (d\theta/2) + \mu \, dN - (T + dT) \cos (d\theta/2) = 0$$

$$+\uparrow \Sigma F_y = 0; \quad dN - (T + dT) \sin (d\theta/2) - T \sin (d\theta/2) = 0$$

Since $d\theta$ is of *infinitesimal size,* $\sin (d\theta/2)$ and $\cos (d\theta/2)$ can be replaced by $d\theta/2$ and 1, respectively. Also, the *product* of two infinitesimals dT and $d\theta/2$ may be neglected when compared to infinitesimals of the first order. The above two equations therefore reduce to

$$\mu \, dN = dT$$

and

$$dN = T \, d\theta$$

Eliminating dN yields

$$\frac{dT}{T} = \mu \, d\theta$$

Integrating this equation between all the points of contact that the belt makes with the drum, and noting that $T = T_1$ at $\theta = 0$ and $T = T_2$ at $\theta = \beta$, yields

$$\int_{T_1}^{T_2} \frac{dT}{T} = \mu \int_0^\beta d\theta$$

or

$$\ln \frac{T_2}{T_1} = \mu\beta$$

Solving for T_2, we obtain

$$T_2 = T_1 e^{\mu\beta} \qquad (9\text{--}5)$$

where T_2, T_1 = belt tensions; $\mathbf{T}_1$ opposes the direction of motion (or impending motion) of the belt, while $\mathbf{T}_2$ acts in the direction of belt motion (or impending motion); because of friction, $T_2 > T_1$

μ = coefficient of static or kinetic friction between the belt and the surface of contact

β = angle of belt to surface contact, measured in radians

e = 2.718. . . , base of the natural logarithm

Note that Eq. 9–5 is *independent* of the *radius* of the drum and instead depends upon the angle of belt to surface contact, β. Furthermore, as indicated by the integration, this equation is valid for flat belts placed on *any shape* of contacting surface.

Example 9–9

The maximum tension that can be developed in the belt shown in Fig. 9–22a is 500 N. If the pulley at *A* is free to rotate and the coefficient of static friction at the fixed drums *B* and *C* is $\mu_s = 0.25$, determine the largest mass of the cylinder that can be lifted by the belt. Assume that the force **T** applied at the end of the belt is directed vertically downward, as shown.

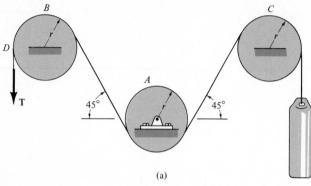

Fig. 9–22

(a)

Solution

Lifting the cylinder, which has a weight $W = mg$, causes the belt to move counterclockwise over the drums at *B* and *C*; hence, the maximum tension T_2 in the belt occurs at *D*. Thus, $T_2 = 500$ N. A section of the belt passing over the drum at *B* is shown in Fig. 9–22b. Since $180° = \pi$ rad, the angle of contact between the drum and the belt is $\beta = (135°/180°)\pi = 3\pi/4$ rad. Using Eq. 9–5, we have

$$T_2 = T_1 e^{\mu_s \beta}; \qquad 500 \text{ N} = T_1 e^{0.25[(3/4)\pi]}$$

Hence,

$$T_1 = \frac{500 \text{ N}}{e^{0.25[(3/4)\pi]}} = \frac{500}{1.80} = 277.4 \text{ N}$$

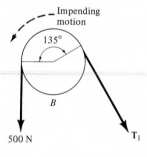

(b)

Since the pulley at *A* is free to rotate, equilibrium requires that the tension in the belt remains the *same* on both sides of the pulley.

The section of the belt passing over the drum at *C* is shown in Fig. 9–22c. The load $W < 277.4$ N. Why? Applying Eq. 9–5, we obtain

$$T_2 = T_1 e^{\mu_s \beta}; \qquad 277.4 = W e^{0.25[(3/4)\pi]}$$

$$W = 153.9 \text{ N}$$

so that

$$m = \frac{W}{g} = \frac{153.9}{9.81}$$

$$= 15.7 \text{ kg} \qquad \qquad \textit{Ans.}$$

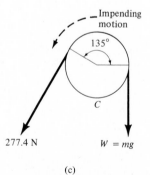

(c)

Problems

*9–68. A mass of 250 kg is to be supported by the cord, which wraps over the pipe. Determine the smallest vertical force **F** needed to support the load if the cord passes (a) over the pipe, $\beta = \pi$, and (b) two times over the pipe, $\beta = 3\pi$.

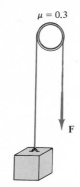

$\mu = 0.3$

F

Prob. 9–68

9–69. Determine the *minimum* tension in the rope at points A and B that is necessary to maintain equilibrium. Take $\mu = 0.3$ between the rope and the fixed post D. The rope is wrapped only once around the post.

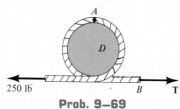

A

D

250 lb

B   **T**

Prob. 9–69

9–70. The truck, which has a mass of 3.4 Mg, is to be lowered down the slope by a rope that is wrapped around a tree. If the wheels are free to roll and the man at A can resist a pull of 300 N, determine the minimum number of turns the rope should be wrapped around the tree to lower the truck at a constant speed. The coefficient of friction between the tree and rope is $\mu = 0.3$.

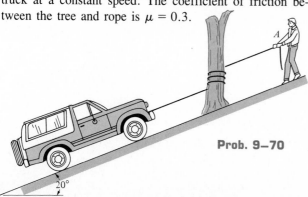

A

$20°$

Prob. 9–70

9–71. A cord having a weight of 0.5 lb/ft and a total length of 10 ft is suspended over a peg P as shown. If the coefficient of friction between the peg and cord is $\mu = 0.5$, determine the shortest length h which one side of the suspended cord can have without causing motion. Neglect the size of the peg and the length of cord draped over it.

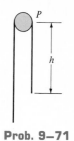

P

h

Prob. 9–71

*9–72. The cord supporting the 6-kg block passes around three pegs, A, B, C, where $\mu = 0.2$. Determine the range of values of **P** for which the block will not move up or down.

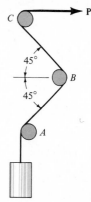

C   **P**

$45°$

B

$45°$

A

Prob. 9–72

9-73. A boy weighing 100 lb attempts to pull himself up a tree using a sling and rope which is draped over the limb at *B*. If $\mu = 0.4$, determine the smallest force at which he must pull on the rope to lift himself.

Prob. 9-73

9-74. The wheel is subjected to a torque of $M = 50 \text{ N} \cdot \text{m}$. If the coefficient of friction between the band brake and the rim of the wheel is $\mu = 0.3$, determine the smallest horizontal force *P* that must be applied to the lever to stop the wheel.

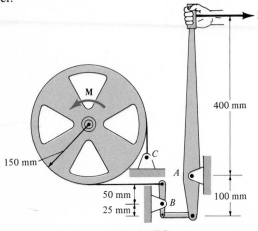

Prob. 9-74

9-75. The cylinder weighs 10 lb and is held in equilibrium by the belt and wall. If slipping does not occur at the wall, determine the minimum vertical force *P* which must be applied to the belt for equilibrium. The coefficient of friction between the belt and cylinder is $\mu = 0.25$.

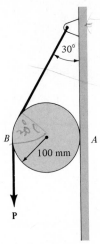

Prob. 9-75

*9-76. Show that the frictional relationship between the belt tensions, the coefficient of friction μ, and the angular contacts α and β for the V-belt is $T_2 = T_1 e^{\mu\beta/\sin(\alpha/2)}$.

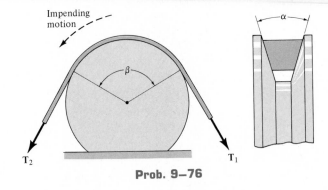

Prob. 9-76

9–77. A belt is used to drive the pulley. Two situations are to be considered, one with a V-belt such that $\alpha = 45°$ (see Prob. 9–76) and the other with a flat belt. Both have a coefficient of friction of $\mu = 0.6$ with the pulley. If the belt wraps 180° around the pulley in each case, determine which situation is most effective in resisting friction along the periphery of the pulley.

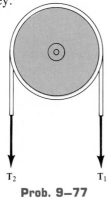

T_2 T_1

Prob. 9–77

9–78. The 60° V-fan belt of an automobile engine passes around the hub H of a generator G and over the housing F of a fan. If the generator locks, causing the axle of the hub to become fixed, determine the maximum possible torque **M** resisted by the axle as the belt slips over the hub. The belt can sustain a maximum tension of $T_{max} = 225$ lb. Assume that slipping of the belt occurs only at H and that the coefficient of friction for the hub is $\mu = 0.25$. Use the result of Prob. 9–76.

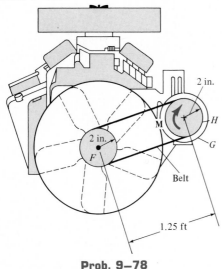

2 in.

M   H

2 in.   G

F

Belt

1.25 ft

Prob. 9–78

■9–79. The smooth pipe is being hoisted using a rope which is wrapped around the pipe and passes through a ring as shown. If the end of the rope is subjected to a tension **T** and the coefficient of friction between the rope and ring is $\mu = 0.3$, determine the angle θ for equilibrium.

T

θ

Prob. 9–79

*9–80. The uniform beam has a weight of 80 kg and is supported by a cord which is wrapped over two pegs for which $\mu = 0.5$. Determine the minimum distance x for the placement of a 1-kN force without causing the beam to move.

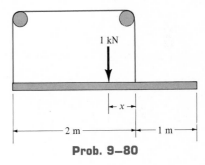

1 kN

x

2 m   1 m

Prob. 9–80

429

9–81. A cable is attached to an 80-lb plate B, passes over a fixed disk at C, and is attached to the block at A. Using the coefficients of friction shown in the figure, determine the smallest weight of block A that will prevent sliding motion of B down the plane.

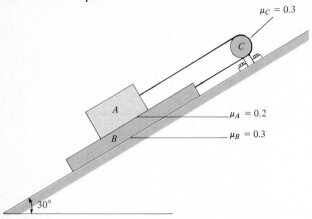

$\mu_C = 0.3$

$\mu_A = 0.2$

$\mu_B = 0.3$

30°

Prob. 9–81

9–82. Block A has a weight of 100 lb and rests on a surface for which $\mu = 0.25$. If the coefficient of friction between the cord and the fixed peg at C is $\mu' = 0.3$, determine the greatest weight of the suspended cylinder B without causing motion.

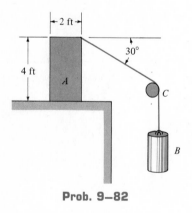

2 ft

4 ft

30°

A

C

B

Prob. 9–82

9–83. Blocks A and B have a mass of 7 kg and 10 kg, respectively. Using the coefficients of friction indicated, determine the largest vertical force P which can be applied to the cord without causing motion.

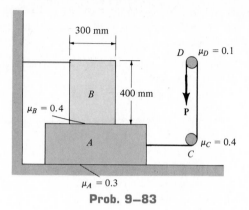

300 mm

D $\mu_D = 0.1$

B 400 mm

$\mu_B = 0.4$

P

A

$\mu_C = 0.4$

C

$\mu_A = 0.3$

Prob. 9–83

*9–84. Blocks A and B weigh 50 lb and 30 lb, respectively. Using the coefficients of friction indicated, determine the greatest weight of block E without causing motion.

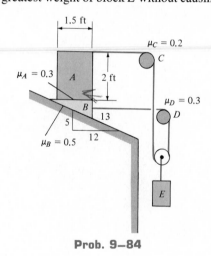

1.5 ft

$\mu_C = 0.2$

C

$\mu_A = 0.3$ A 2 ft

$\mu_D = 0.3$

B 13 D

5

$\mu_B = 0.5$ 12

E

Prob. 9–84

Frictional Forces on Collar Bearings, Pivot Bearings, and Disks

9.6 *

Pivot and *collar bearings* are commonly used to support the *axial load* on a rotating shaft. These two types of support are shown in Fig. 9–23. Provided the bearings are not lubricated, or only partially lubricated, the laws of dry friction may be applied to determine the moment **M** needed to turn the shaft when the shaft supports an axial force **P.**

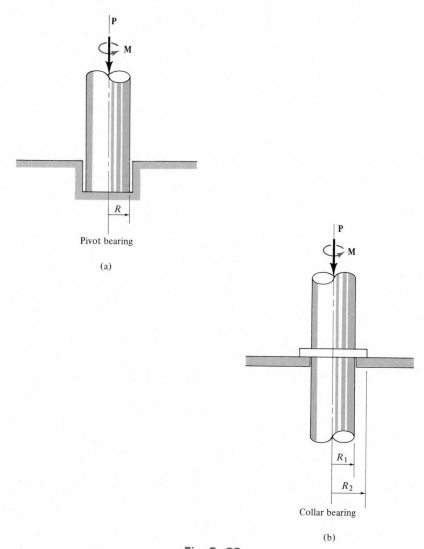

Pivot bearing

(a)

Collar bearing

(b)

Fig. 9–23

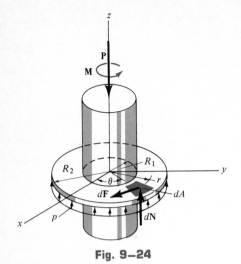

Fig. 9–24

Frictional Analysis. Consider, for example, the end of a collar-bearing shaft shown in Fig. 9–24, which is subjected to an axial force **P** and has a total bearing or contact area $\pi(R_2^2 - R_1^2)$. In the following analysis, the normal pressure p is considered to be *uniformly distributed* over this area—a reasonable assumption provided the bearing is new and evenly supported. Since $\Sigma F_z = 0$, p, measured as a force per unit area, is $p = P/\pi(R_2^2 - R_1^2)$.

The moment needed to cause impending rotation of the shaft can be determined from moment equilibrium of the frictional forces dF developed at the bearing surface by applying $\Sigma M_z = 0$. A small area element $dA = (r\, d\theta)(dr)$, shown in Fig. 9–24, is subjected to both a normal force $dN = p\, dA$ and an associated frictional force

$$dF = \mu_s\, dN = \mu_s\, p\, dA = \frac{\mu_s P}{\pi(R_2^2 - R_1^2)}\, dA$$

The normal force does not create a moment about the z axis of the shaft; however, the frictional force does, namely $dM = r\, dF$. Integration is needed to compute the total moment created by all the frictional forces acting on differential areas dA. Therefore, for impending rotational motion,

$$\Sigma M_z = 0; \qquad\qquad M - \int_A r\, dF = 0$$

Using $dA = (r\, d\theta)(dr)$ and integrating over the entire bearing area yields

$$M = \int_{R_1}^{R_2} \int_0^{2\pi} r \left[\frac{\mu_s P}{\pi(R_2^2 - R_1^2)} \right] (r\, d\theta\, dr) = \frac{\mu_s P}{\pi(R_2^2 - R_1^2)} \int_{R_1}^{R_2} r^2\, dr \int_0^{2\pi} d\theta$$

or

$$M = \tfrac{2}{3}\mu_s P \left(\frac{R_2^3 - R_1^3}{R_2^2 - R_1^2} \right) \qquad\qquad (9\text{–}6)$$

This equation gives the magnitude of moment required for impending rotation of the shaft. The frictional moment developed at the end of the shaft, when it is *rotating* at constant speed, can be found by substituting μ_k for μ_s in Eq. 9–6.

When $R_2 = R$ and $R_1 = 0$, as in the case of a pivot bearing, Fig. 9–23a, Eq. 9–6 reduces to

$$M = \tfrac{2}{3}\mu_s PR \qquad\qquad (9\text{–}7)$$

Recall from the initial assumption that both Eqs. 9–6 and 9–7 apply only for bearing surfaces subjected to *constant pressure*. If the pressure is not uniform, a variation of the pressure as a function of the bearing area must be determined before integrating to obtain the moment. The following example illustrates this concept.

Example 9–10

The uniform bar shown in Fig. 9–25a has a total mass m. If it is assumed that the normal pressure acting at the contacting surface varies linearly along the length of the bar as shown, determine the couple $\mathbf{M}$ required to rotate the bar. Assume that the bar's width a is negligible in comparison to its length l. The coefficient of friction is equal to μ_s.

Solution

A free-body diagram of the bar is shown in Fig. 9–25b. Since the bar has a total weight of $W = mg$, the intensity w_o of the distributed load at the center $(x = 0)$ is determined from vertical force equilibrium, Fig. 9–25a.

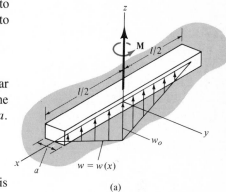

(a)

$$+\uparrow \Sigma F_z = 0; \quad -mg + 2\left[\frac{1}{2}\left(\frac{l}{2}\right)w_o\right] = 0 \qquad w_o = \frac{2mg}{l}$$

Since $w = 0$ at $x = l/2$, the distributed load expressed as a function of x is

$$w = w_o\left(1 - \frac{2x}{l}\right) = \frac{2mg}{l}\left(1 - \frac{2x}{l}\right)$$

The magnitude of the normal force acting on a segment of area having a length dx is therefore

$$dN = w\,dx = \frac{2mg}{l}\left(1 - \frac{2x}{l}\right)dx$$

The magnitude of the frictional force acting on the same element of area is

$$dF = \mu\,dN = \frac{2\mu mg}{l}\left(1 - \frac{2x}{l}\right)dx$$

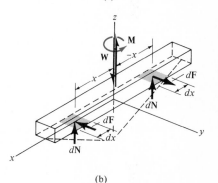

(b)

Fig. 9–25

Hence, the moment created by this force about the z axis is

$$dM = x\,dF = \frac{2\mu mg}{l}x\left(1 - \frac{2x}{l}\right)dx$$

The summation of moments about the z axis of the bar is determined by integration, which yields

$$\Sigma M_z = 0; \qquad M - 2\int_0^{l/2}\frac{2\mu mg}{l}x\left(1 - \frac{2x}{l}\right)dx = 0$$

$$M = \frac{4\mu mg}{l}\left(\frac{x^2}{2} - \frac{2x^3}{3l}\right)\Big|_0^{l/2}$$

$$M = \frac{\mu mgl}{6} \qquad\qquad Ans.$$

★9.7 Frictional Forces on Journal Bearings

When a shaft or axle is subjected to lateral loads, a *journal bearing* is commonly used for support. Well-lubricated journal bearings are subjected to the laws of fluid mechanics, in which the viscosity of the lubricant, the speed of rotation, and the amount of clearance between the shaft and bearing are needed to determine the frictional resistance of the bearing. When the bearing is not lubricated or only partially lubricated, however, a reasonable analysis of the frictional resistance can be based on the laws of dry friction.

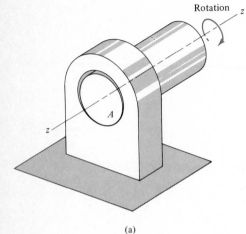

(a)

Frictional Analysis. A typical journal-bearing support is shown in Fig. 9–26a. As the shaft rotates in the direction shown in the figure, it rolls up against the wall of the bearing to some point A where slipping occurs. If the lateral load acting at the end of the shaft is **W**, it is necessary that the bearing reactive force **R** acting at A be equal and opposite to **W**, Fig. 9–26b. The moment needed to maintain constant rotation of the shaft can be found by summing moments about the z axis of the shaft; i.e.,

$$\Sigma M_z = 0; \qquad -M + (R \sin \phi_k)r = 0$$

or

$$M = Rr \sin \phi_k$$

where ϕ_k is the angle of kinetic friction defined by $\tan \phi_k = F/N = \mu_k N/N = \mu_k$. In Fig. 9–26c, it is seen that $r \sin \phi_k = r_f$. The dashed circle with radius r_f is called the *friction circle,* and as the shaft rotates, the reaction **R** will always be tangent to it. If the bearing is partially lubricated, μ_k is small, and therefore $\mu_k = \tan \phi_k \approx \sin \phi_k \approx \phi_k$. Under these conditions, a reasonable approximation to the moment needed to overcome the frictional resistance becomes

$$M \approx Rr\mu_k \qquad (9-8)$$

The following example illustrates a common application of this equation.

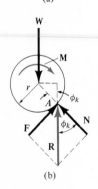

(b)

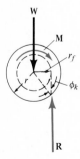

(c)

Fig. 9–26

Example 9–11

The 100-mm-diameter pulley shown in Fig. 9–27a fits loosely on a 10-mm-diameter shaft for which the coefficients of static and kinetic friction are $\mu = 0.4$. Determine the minimum tension T in the belt to (a) raise the 100-kg block at constant velocity, and (b) lower the block at constant velocity. Assume no slipping occurs between the belt and pulley and neglect the weight of the pulley.

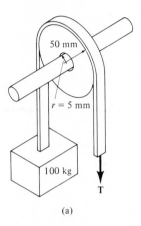

(a)

Fig. 9–27

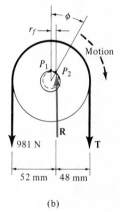

(b)

Solution

Part (a). A free-body diagram of the pulley is shown in Fig. 9–27b. When the pulley is subjected to cable tensions of 981 N each, the pulley makes contact with the shaft at point P_1. As the tension T is *increased,* the pulley will roll around the shaft to point P_2 before motion impends. From the figure, the friction circle has a radius $r_f = r \sin \phi$. Using the simplification $\sin \phi \approx \phi$, $r_f \approx r\mu = (5 \text{ mm})(0.4) = 2 \text{ mm}$, so that summing moments about P_2 gives

$$\zeta + \Sigma M_{P_2} = 0; \qquad 981 \text{ N}(52 \text{ mm}) - T(48 \text{ mm}) = 0$$
$$T = 1062.8 \text{ N} \qquad\qquad Ans.$$

Part (b). When the block is lowered, the resultant force **R** acting on the shaft passes through point P_3, as shown in Fig. 9–27c. Summing moments about this point yields

$$\zeta + \Sigma M_{P_3} = 0; \qquad 981 \text{ N}(48 \text{ mm}) - T(52 \text{ mm}) = 0$$
$$T = 905.5 \text{ N} \qquad\qquad Ans.$$

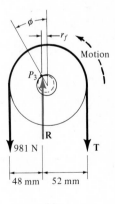

(c)

Problems

9–85. The collar bearing uniformly supports an axial force of $P = 500$ lb. If the coefficient of friction is $\mu = 0.3$, determine the torque **M** acting at the bearing surface, required to overcome friction.

9–87. The annular ring bearing is subjected to a thrust of 800 lb. If $\mu = 0.35$, determine the torque **M** that must be applied to the shaft to turn it.

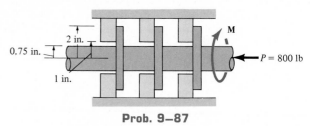

Prob. 9–87

*9–88.** The plate clutch consists of a plate A that slides over the rotating shaft S. The shaft is fixed to the driving plate gear B. If the gear C, which is in mesh with B, is subjected to a torque of $M = 0.6$ N · m, determine the force **P**, which must be applied via the control arm, to stop the rotation. The coefficient of friction between the plates A and D is $\mu = 0.4$. Assume the bearing pressure between A and D to be uniform.

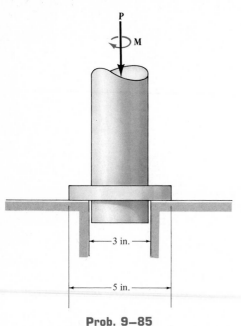

Prob. 9–85

9–86. Assuming the pressure of the sanding disk D on the surface to be uniform, determine the vertical couple forces developed at the handles which are necessary to hold it in equilibrium. The disk is pushed against the surface with a total horizontal force of 18 lb. Take $\mu = 0.59$.

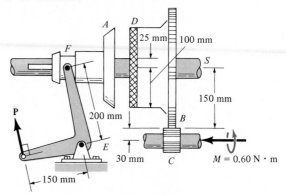

Prob. 9–88

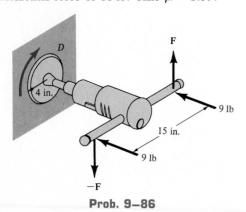

Prob. 9–86

9–89. Assuming that the variation of pressure at the bottom of the pivot bearing is defined as $p = p_o(R_2/r)$, determine the torque **M** needed to rotate the shaft if the applied axial force is **P**. The coefficient of friction is μ. For the solution, it is necessary to determine p_o in terms of P and the bearing dimensions R_1 and R_2.

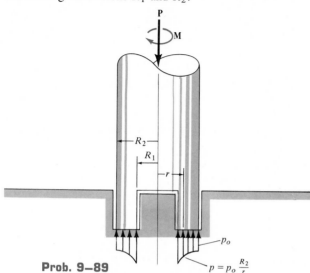

Prob. 9–89

9–90. Because of wearing at the edges, the pivot bearing is subjected to a conical pressure distribution at its surface of contact. Determine the torque **M** required to turn the shaft which supports an axial force **P**. The coefficient of friction is μ. For the solution, it is necessary to determine the peak pressure p_o in terms of P and the bearing radius R.

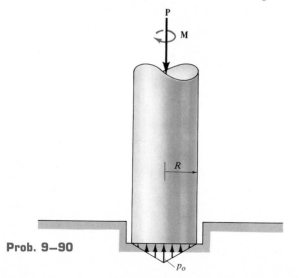

Prob. 9–90

9–91. The *Morse taper shank* of angle θ is used on a lathe. If the contact pressure between the socket B and the shank is uniform, and the coefficient of friction is μ, determine the torque **M** needed to rotate the shank if the axial force **P** is applied to the shaft.

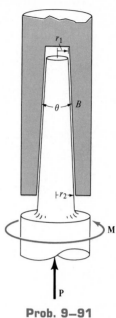

Prob. 9–91

*9–92. The pivot bearing is subjected to a pressure distribution at its surface of contact which varies as shown. If the coefficient of friction is μ, determine the torque **M** required to turn the shaft if it supports an axial force **P**.

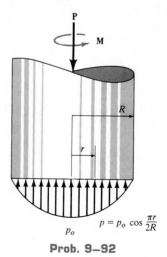

Prob. 9–92

437

9–93. A 200-mm-diameter post is driven 3 m into sand for which $\mu = 0.3$. If the normal pressure acting *completely around the post* varies linearly with depth as shown, determine the frictional torque **M** that must be overcome to rotate the post.

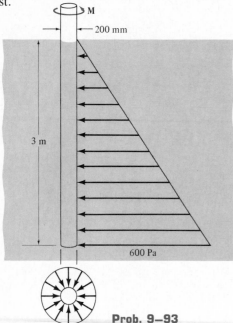

Prob. 9–93

9–94. A wheel having a diameter of 400 mm and a mass of 10 kg is supported loosely at its center by a fixed shaft having a diameter of 75 mm. If forces of 200 N and 220 N act vertically downward on the ends of a cord draped over its periphery, determine the coefficient of friction between the shaft and wheel. The wheel is on the verge of rotating.

9–95. A disk having an outer diameter of 8 in. fits loosely over a fixed shaft having a diameter of 3 in. If the coefficient of friction between the disk and the shaft is $\mu = 0.15$, determine the smallest vertical force **F**, acting on the rim, which must be applied to the disk to cause it to slip over the shaft. The disk weighs 10 lb.

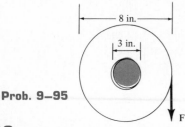

Prob. 9–95

*9–96. The pulley has a radius of 12 in. and fits loosely on the 3-in.-diameter shaft. If the loadings acting on the belt cause the pulley to rotate uniformly, determine the frictional force and the coefficient of friction. The pulley weighs 18 lb.

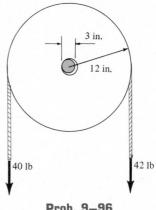

Prob. 9–96

9–97. The collar bushing B fits loosely over the fixed shaft S, which has a radius of 20 mm. Determine the smallest force $\mathbf{T}_A$ needed to pull the belt downward at A. Also determine the resultant normal and frictional components of force developed on the collar bushing if $\mu = 0.3$ between the collar and the shaft. Assume that the belt does not slip on the collar; rather, the collar slips on the shaft.

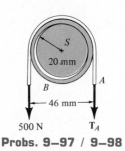

Probs. 9–97 / 9–98

9–98. If the smallest tension force $T_A = 600$ N is required to pull the belt downward at A over the shaft S, determine the coefficient of friction between the loosely fitting collar bushing B and the shaft. Assume that the belt does not slip on the collar; rather, the collar slips on the shaft.

9–99. The collar fits *loosely* around a fixed shaft that has a radius of 2 in. If the coefficient of friction between the shaft and the collar is $\mu = 0.3$, determine the tension **T** in the horizontal segment of the belt so that the belt can be lowered in the direction of the 20-lb force with a constant speed. Assume that the belt does not slip on the collar; rather, the collar slips on the shaft. Neglect the weight and thickness of the belt and collar. The radius, measured from the center of the collar to the mean thickness of the belt, is 2.25 in.

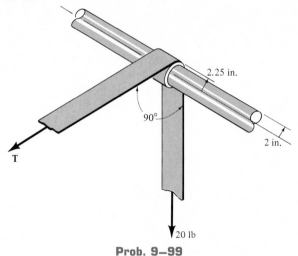

Prob. 9–99

9–101. The axle of the pulley fits loosely in a 50-mm-diameter pin hole. If $\mu = 0.30$, determine the minimum tension **T** required to raise the 50-kg block. Neglect the weight of the pulley and assume that the cord does not slip on the pulley.

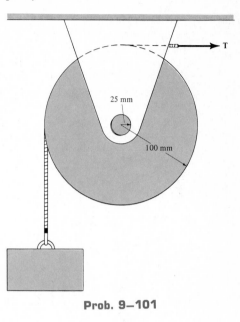

Prob. 9–101

*9–100. A pulley having a diameter of 80 mm and mass of 1.25 kg is supported loosely on a shaft having a diameter of 20 mm. Determine the moment **M** that must be applied to the pulley to cause it to rotate with constant motion. The coefficient of friction between the shaft and pulley is $\mu_k = 0.4$. Also calculate the angle θ which the normal force at the point of contact makes with the horizontal. The shaft itself cannot rotate.

Prob. 9–100

439

*9.8 Rolling Resistance

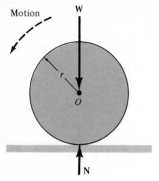

Motion W

N

Rigid surface of contact

(a)

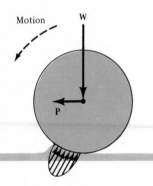

Motion W

P

Soft surface of contact

(b)

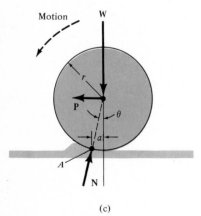

Motion W

P

θ

a

A

N

(c)

Fig. 9–28

If a *rigid* cylinder of weight **W** rolls at constant velocity along a *rigid* surface, the normal force exerted by the surface on the cylinder acts at the tangent point of contact, as shown in Fig. 9–28*a*. Under these conditions, provided the cylinder does not encounter frictional resistance from the air, motion would continue indefinitely. Actually, however, no materials are perfectly rigid; and therefore the reaction of the surface on the cylinder consists of a *distribution* of normal pressure. If we assume that the cylinder is much harder than the surface, then a small "hill" is formed which is caused by the indentation of the cylinder into the surface, Fig. 9–28*b*. Since the cylinder can never really "climb" over this "hill," the normal pressure is always present, and consequently, to maintain motion, a driving force **P** must be applied to the cylinder. For analysis, the normal pressure distribution can be replaced by its *resultant* force **N**, which acts at an angle θ from the vertical, Fig. 9–28*c*. To keep the cylinder in equilibrium, i.e., moving at constant velocity, it is necessary that **N** be *concurrent* with the driving force **P** and the weight **W**. Summing moments about point A gives $Wa = P(r \cos \theta)$. Since the deformations are generally small in relation to the radius, $\cos \theta \approx 1$; hence,

$$Wa \approx Pr$$

or

$$P \approx \frac{Wa}{r} \qquad (9\text{--}9)$$

The distance a is termed the *coefficient of rolling resistance,* which has the dimension of length. For instance, $a \approx 0.5$ mm for a mild steel wheel rolling on a steel rail. For hardened steel ball bearings on steel $a \approx 0.1$ mm. Experimentally, though, this factor is difficult to measure, since it depends upon such parameters as the rate of rotation of the cylinder, the elastic properties of the contacting surfaces, and the surface finish. For this reason, little reliance is placed on the data for determining a. The analysis presented here does, however, indicate why a heavy load offers greater resistance to motion than a light load under the same conditions. Furthermore, since the force needed to *roll* the cylinder over the surface is much less than that needed to *slide* the cylinder across the surface, the analysis indicates why roller or ball bearings are often used to minimize the frictional resistance between moving parts.

Example 9–12

A 10-kg steel wheel shown in Fig. 9–29a has a radius of 100 mm and rests on an inclined plane made of wood. If θ is increased so that the wheel begins to roll down the incline with constant velocity when $\theta = 1.2°$, determine the coefficient of rolling resistance.

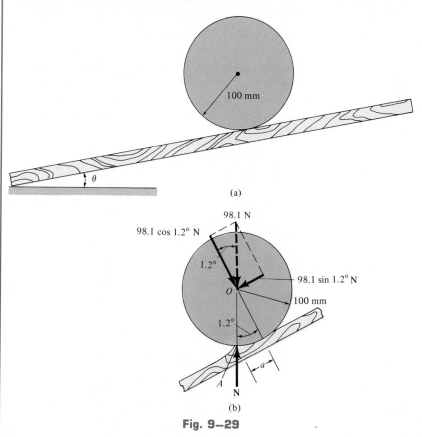

(a)

(b)

Fig. 9–29

Solution

As shown on the free-body diagram, Fig. 9–29b, when the wheel has impending motion, the normal reaction **N** acts at point A defined by the dimension a. Resolving the weight into rectangular components parallel and perpendicular to the incline, and summing moments about point A, yields (approximately)

$\zeta + \Sigma M_A = 0;$ $98.1 \cos 1.2°(a) - 98.1 \sin 1.2°(100) = 0$

Solving, we obtain

$$a = 2.1 \text{ mm} \qquad \qquad Ans.$$

Problems

9–102. A tank car has a weight of 40 kip and is supported by eight wheels, each of which has a diameter of 30 in. If the coefficient of rolling resistance is 0.015 in. between the tracks and each wheel, determine the magnitude of horizontal force required to overcome the rolling resistance of the wheels.

9–103. The hand cart has wheels with a diameter of 80 mm. If a crate having a mass of 500 kg is placed on the cart so that each wheel carries an equal load, determine the horizontal force **P** that must be applied to the handle needed to overcome the rolling resistance. The coefficient of rolling resistance is 2 mm. Neglect the mass of the cart.

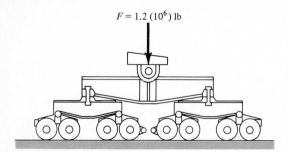

Prob. 9–103

***9–104.** The carriage is used to support a crane load of $F = 1.2(10^6)$ lb. The seven axles shown in the arrangement serve to equalize the load to each of the eight wheels. (By comparison, note that a single supporting beam would *deform* and the wheels would share an unequal amount of load.) Determine the horizontal force that must be applied to the carriage in order to overcome the resistance to rolling. The coefficient of rolling resistance is 0.007 in. and each wheel has a diameter of 36 in.

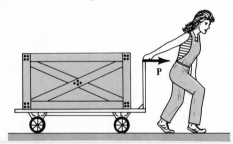

$F = 1.2 (10^6)$ lb

Prob. 9–104

9–105. Experimentally, it is found that a cylinder having a diameter of 120 mm rolls with a constant speed down an inclined plane having a slope of 12 mm/m. Determine the coefficient of rolling resistance for the cylinder.

9–106. The car has a weight of 2600 lb and center of gravity at G. Determine the horizontal force P that must be applied to overcome the rolling resistance of the wheels. The coefficient of rolling resistance is 0.5 in. The tires have a diameter of 2.75 ft.

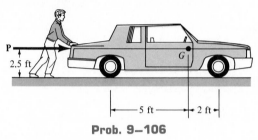

P

2.5 ft

G

5 ft — 2 ft

Prob. 9–106

9–107. The pipe is subjected to a load that has a weight **W**. If the coefficients of rolling resistance for the pipe's top and bottom surfaces are a_A and a_B, respectively, show that a force having a magnitude of $P = [W(a_A + a_B)]/2r$ is required to move the load and thereby roll the pipe forward. Neglect the weight of the pipe.

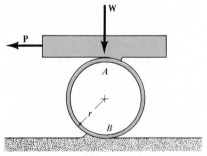

W

P

A

r

B

Prob. 9–107

*9–108. A large stone having a mass of 200 kg is moved along the floor using a series of 150-mm-diameter rollers for which the coefficient of rolling resistance is 3 mm at the ground and 7 mm at the bottom surface of the stone. Determine the horizontal force P needed to push the stone forward at a constant speed. *Hint:* Use the result of Prob. 9–107.

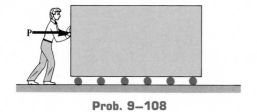

Prob. 9–108

9–109. The 1.4-Mg machine is to be moved over a level surface using a series of rollers for which the coefficient of rolling resistance is 0.5 mm at the ground and 0.2 mm at the bottom surface of the machine. Determine the appropriate diameter of the rollers so that the machine can be pushed forward with a horizontal force of $P = 250$ N. *Hint:* Use the result of Prob. 9–107.

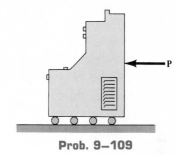

Prob. 9–109

Review Problems

9–110. A cable is wrapped around the inner core of a spool. Determine the magnitude of the vertical force **P** that must be applied to the end of the cable to rotate the spool. The coefficient of friction between the spool and the contacting surfaces at A and B is $\mu = 0.6$. The spool has a mass of 100 kg.

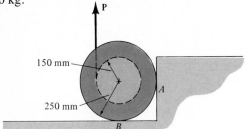

Prob. 9–110

9–111. The two pin-connected rods each have a weight of 15 lb. If the coefficient of static friction at C is $\mu_s = 0.5$, determine the maximum angle θ of spread for equilibrium.

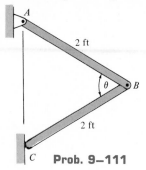

Prob. 9–111

*9–112. If the smallest tension force $T_A = 480$ N is required to pull the belt downward at A over the shaft S, determine the coefficient of friction between the loosely fitting collar bushing B and the shaft. Assume that the belt does not slip on the collar; rather, the collar slips on the shaft.

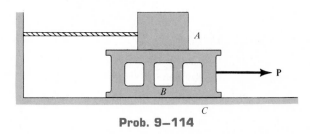

Prob. 9–112

9–113. The four wheels on an automobile each have a diameter of 2.5 ft. If the coefficient of rolling resistance is 0.013 ft with the road, and the car weighs 4000 lb, determine the horizontal force **P** that is required to overcome the rolling resistance of the wheels.

9–114. Blocks A and B weigh 10 lb and 25 lb, respectively. If the coefficient of friction between A and B is $\mu_{AB} = 0.4$ and between B and the floor C, $\mu_{BC} = 0.2$, determine the maximum horizontal force **P** that can be applied without causing motion.

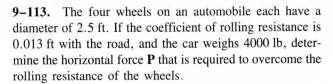

Prob. 9–114

9–115. The *double-collar bearing* is subjected to an axial force of $P = 4$ kN. Assuming that collar A supports $0.6P$ and collar B supports $0.4P$, both with a uniform distribution of pressure, determine the maximum frictional torque **M** that may be resisted by the bearing. $\mu_s = 0.2$ for both collars.

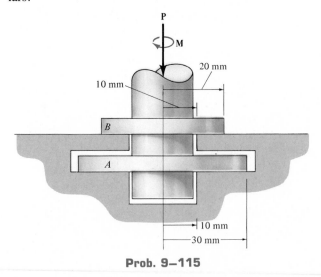

Prob. 9–115

*9–116. A man attempts to lift the uniform 40-lb ladder to an upright position by applying a force **F** perpendicular to the ladder at rung R. Determine the coefficient of friction between the ladder and the ground at A if the ladder begins to slip on the ground when his hands reach a height of 6 ft.

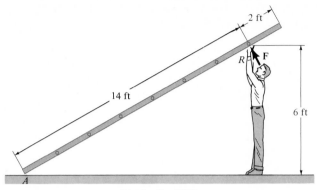

Prob. 9–116

9–117. If the coefficient of friction between the 500-lb drum and the inclined planes is $\mu = 0.3$, determine the couple moment **M** needed to rotate the drum.

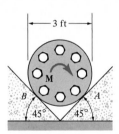

Prob. 9–117

9–118. A boy having a mass of 70 kg attempts to walk on a board which is supported by a pin at A and a post BC. If $\mu_B = 0.3$ and $\mu_C = 0.5$, determine the maximum angle θ of alignment for the post so that he can safely reach the other side. Neglect the mass of the board and post and the thickness of the post in the calculation. *Hint:* The post is a two-force member.

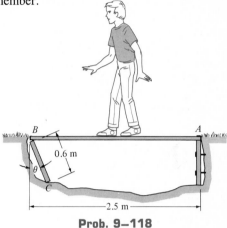

Prob. 9–118

9–119. A 12-in.-diameter disk fits loosely over a 3-in.-diameter shaft for which the coefficient of friction is $\mu = 0.15$. If the disk weighs 100 lb, determine the vertical tangential force which must be applied to the disk to cause it to slip over the shaft.

*9–120. If the coefficient of friction between blocks A and B is $\mu_{AB} = 0.8$ and between B and the floor $\mu_{BC} = 0.1$, determine the minimum force F that will cause block A to move.

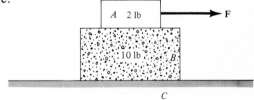

Probs. 9–120 / 9–121

9–121. If the coefficient of friction between blocks A and B is $\mu_{AB} = 0.8$, determine the coefficient of friction between block B and the floor so that when the force **F** is increased the blocks slip at all their contacting surfaces at the same time. What is the magnitude of force **F** needed to move the blocks?

9-122. The boat has a weight of 500 lb and is held in position off the side of a ship by the spars at A and B. A man having a weight of 130 lb gets in the boat, wraps a rope around an overhead boom at C, and ties it to the ends of the boat as shown. If the boat is disconnected from the spars, determine the *minimum number* of *half turns* the rope must make around the boom so that the boat can be safely lowered into the water. The coefficient of friction between the rope and the boom is $\mu = 0.15$. *Hint:* The problem requires that the normal force between the man's feet and the boat be as small as possible.

Prob. 9-122

Moments of Inertia

10

In Chapter 7 we determined the *centroid* for an area by considering the *first moment* of the area about an axis, i.e., for the computation it is necessary to evaluate an integral of the form $\int x \, dA$. There are many important topics in engineering practice, however, which require evaluation of an integral of the *second moment* of an area, i.e., $\int x^2 \, dA$. This integral is referred to as the moment of inertia for an area. The terminology ''moment of inertia'' as used here is actually a misnomer; however, it has been adopted because of the similarity with integrals of the same form related to mass.

Methods used to determine both the area and mass moments of inertia will be discussed in this chapter. In particular, a body's mass moment of inertia is frequently used to solve problems in dynamics.

Definition of Moments of Inertia for Areas **10.1**

The moment of inertia of an area originates whenever one computes the moment of a distributed load that varies linearly from the moment axis. A common example of this type of loading occurs in the study of fluid mechanics. It was pointed out in Sec. 7.6 that the pressure, or force per unit area, exerted at a point located a distance z below the surface of a liquid is $p = \gamma z$, Eq. 7–17, where γ is the specific weight of the liquid. Thus, the magnitude of force exerted by a liquid on the area dA of the submerged plate shown in Fig. 10–1 is $dF = p \, dA = \gamma z \, dA$. The moment of this force about the x axis of the plate is $dM = dFz = \gamma z^2 \, dA$ and, therefore, the moment created by the entire distributed (pressure) loading is $M = \gamma \int z^2 \, dA$. Here the integral represents the moment of inertia of the area of the plate about the x axis. Since integrals of

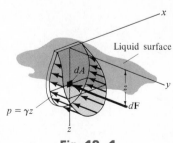

Fig. 10–1

this form often arise in formulas used in fluid mechanics, mechanics of materials, structural mechanics, and machine design, the engineer should become familiar with the methods used for their computation.

Moment of Inertia. Consider the area A, shown in Fig. 10–2, which lies in the x-y plane. By definition, the moments of inertia of the differential planar area dA about the x and y axes are $dI_x = y^2\, dA$ and $dI_y = x^2\, dA$, respectively. For the entire area the *moment of inertia* is determined by integration, i.e.,

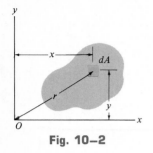

Fig. 10–2

$$I_x = \int y^2\, dA$$

$$I_y = \int x^2\, dA$$

(10–1)

We can also formulate the second moment of the differential area dA about the pole O or z axis, Fig. 10–2. This is referred to as the polar moment of inertia, $dJ_O = r^2\, dA$. Here r is the perpendicular distance from the pole (z axis) to the element dA. For the entire area the *polar moment of inertia* is

$$J_O = \int r^2\, dA = I_x + I_y$$

(10–2)

The relationship between J_O and I_x, I_y is possible since $r^2 = x^2 + y^2$, Fig. 10–2.

From the above formulations it is seen that I_x, I_y, and J_O will *always* be *positive,* since they involve the product of distance squared and area. Furthermore, the units for moment of inertia involve length raised to the fourth power, e.g., m^4, mm^4, or ft^4, in^4.

10.2 Parallel-Axis Theorem for an Area

If the moment of inertia for an area is known about an axis passing through its centroid, we can determine the moment of inertia of the area about a corresponding parallel axis using the *parallel-axis theorem.* To derive this theorem, consider finding the moment of inertia of the shaded area shown in Fig. 10–3 about the x axis. In this case, a differential element dA is located at an arbitrary distance y from the centroidal $\bar{x}$ axis, whereas the *fixed distance* between the parallel x and $\bar{x}$ axes is defined as d_y. Since the moment of inertia of dA about the x axis is $dI_x = (y + d_y)^2\, dA$, then for the entire area,

$$I_x = \int (y + d_y)^2\, dA$$

$$= \int y^2\, dA + 2d_y \int y\, dA + d_y^2 \int dA$$

The first term on the right represents the moment of inertia of the area about

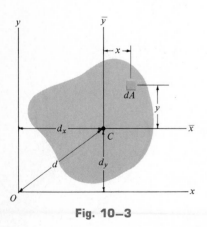

Fig. 10–3

the $\bar{x}$ axis, $\bar{I}_{\bar{x}}$. The second term is zero since the $\bar{x}$ axis passes through the area's centroid C, i.e., $\int y \, dA = \bar{y}A = 0$. The final result is therefore

$$I_x = \bar{I}_{\bar{x}} + A d_y^2 \qquad (10\text{--}3)$$

A similar expression can be written for I_y, i.e.,

$$I_y = \bar{I}_{\bar{y}} + A d_x^2 \qquad (10\text{--}4)$$

And finally, for the polar moment of inertia about an axis perpendicular to the x-y plane and passing through the pole O (z axis), Fig. 10–3, we have

$$J_O = \bar{J}_C + A d^2 \qquad (10\text{--}5)$$

The form of each of these equations states that *the moment of inertia of an area about an axis is equal to the area's moment of inertia about a parallel axis passing through the centroid plus the product of the area and the square of the perpendicular distance between the axes.*

Radius of Gyration of an Area   10.3

The *radius of gyration* of a planar area is often used for column design in structural mechanics. Provided the area and moments of inertia are *known,* the radii of gyration are determined from the formulas

$$k_x = \sqrt{\frac{I_x}{A}} \qquad k_y = \sqrt{\frac{I_y}{A}} \qquad k_O = \sqrt{\frac{J_O}{A}} \qquad (10\text{--}6)$$

Note that the form of these equations is easily remembered since it is similar to that for finding the moment of inertia of a differential area about an axis. For example, $I_x = k_x^2 A$; whereas for a differential area, $dI_x = y^2 \, dA$.

Moments of Inertia for an Area   10.4
by Integration

When the boundaries for a planar area can be expressed by mathematical functions, Eqs. 10–1 may be integrated to determine the moments of inertia for the area. If the element of area chosen for integration has a differential size in two directions as shown in Fig. 10–2, a double integration must be performed to evaluate the moment of inertia. Most often, however, it is easier to choose an element having a differential size or thickness in only one direction, because then the evaluation requires only a single integration.

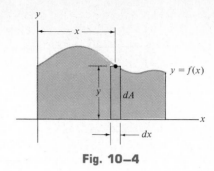

Fig. 10–4

PROCEDURE FOR ANALYSIS

If a single integration is performed to determine the moment of inertia of an area about an axis, it will first be necessary to determine the most effective way to *orient* the element with respect to the axis, so that the process of integration is made as simple as possible. In this regard, two possibilities exist:

Case 1. The length of the element can be *parallel* to the axis. This situation occurs when using the rectangular element shown in Fig. 10–4 for computing I_y of the area. *Direct* application of Eq. 10–1, $I_y = \int x^2\, dA$, can be made in this case since *all parts* of the element lie at the *same* moment-arm distance x from the y axis.*

Case 2. The length of the element can be *perpendicular* to the axis. Here Eq. 10–1 *does not apply* since all parts of the element *do not* lie at the same moment-arm distance from the axis. Instead, the *parallel-axis theorem* must be used. Hence, if the rectangular element in Fig. 10–4 is used for computing I_x for the area, it will first be necessary to compute the moment of inertia of the *element* about a horizontal axis passing through its centroid and then determine the moment of inertia of the *element* about the x axis by using the parallel-axis theorem. Integration of this result will yield I_x.

The following examples illustrate the above procedure for each case.

---

*In the case of the element $dA = dx\, dy$, Fig. 10–2, the moment arms y and x are appropriate for the formulation of I_x and I_y (Eq. 10–1) since the *entire* element, because of its "smallness," lies at the specified y and x perpendicular distances from the x and y axes.

Example 10–1

Determine the moment of inertia for the rectangular area shown in Fig. 10–5 with respect to (a) the centroidal $\bar{x}$ axis, (b) the axis x_b passing through the base of the rectangle, and (c) the pole or $\bar{z}$ axis perpendicular to the $\bar{x}$-$\bar{y}$ plane and passing through the centroid C.

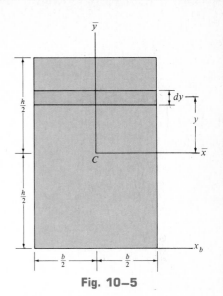

Fig. 10–5

Solution (Case 1)

Part (a). The differential element shown in Fig. 10–5 is chosen for integration. Because of its location and orientation, the *entire element* is at a distance y from the $\bar{x}$ axis. Here it is necessary to integrate from $y = -h/2$ to $y = h/2$. Since $dA = b\,dy$, then

$$\bar{I}_x = \int y^2\,dA = \int_{-h/2}^{h/2} y^2(b\,dy) = b\int_{-h/2}^{h/2} y^2\,dy$$

$$= \frac{1}{12}bh^3 \qquad\qquad Ans.$$

Part (b). The moment of inertia about an axis passing through the base of the rectangle can be obtained by using the result of part (a) and applying the parallel-axis theorem, Eq. 10–3.

$$I_{x_b} = \bar{I}_x + Ad_y^2$$

$$= \frac{1}{12}bh^3 + bh\left(\frac{h}{2}\right)^2 = \frac{1}{3}bh^3 \qquad\qquad Ans.$$

Part (c). The moment of inertia $\bar{I}_y$ may be found by interchanging the dimensions b and h in the result of part (a), in which case

$$\bar{I}_y = \frac{1}{12}hb^3$$

Using Eq. 10–2, the polar moment of inertia about C is

$$\bar{J}_C = \bar{I}_x + \bar{I}_y = \frac{1}{12}bh(h^2 + b^2) \qquad\qquad Ans.$$

Example 10–2

Compute the moment of inertia of the shaded area shown in Fig. 10–6a about the x axis.

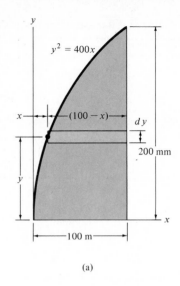

(a)

Solution I *(Case 1)*

The differential element of area that is *parallel* to the x axis, as shown in Fig. 10–6a, is chosen for integration. Since the element has a thickness dy and intersects the curve at the arbitrary point (x, y), the area is $dA = (100 - x)\,dy$. Furthermore, all parts of the element lie at the same distance y from the x axis. Hence, integrating with respect to y, from $y = 0$ to $y = 200$ mm, yields

$$I_x = \int y^2\,dA = \int y^2 (100 - x)\,dy$$

$$= \int_0^{200} y^2 \left(100 - \frac{y^2}{400}\right) dy = 100 \int_0^{200} y^2\,dy - \frac{1}{400}\int_0^{200} y^4\,dy$$

$$= 106.7(10^6)\ \text{mm}^4 \qquad\qquad\qquad Ans.$$

Solution II *(Case 2)*

A differential element *perpendicular* to the y axis, as shown in Fig. 10–6b, is chosen for integration. In this case, however, all parts of the element do *not* lie at the same distance from the x axis, and therefore the parallel-axis theorem must be used to determine the *moment of inertia of the element* with respect to the axis. For a rectangle having a base b and height h, the moment of inertia about its centroidal axis has been computed in part (a) of Example 10–1. There it was found that $\bar{I}_{\bar{x}} = \frac{1}{12}bh^3$. For the differential element shown in Fig. 10–6b, $b = dx$ and $h = y$, and thus $d\bar{I}_{\bar{x}} = \frac{1}{12}\,dx\,y^3$. Since the centroid of the element is at $\tilde{y} = y/2$ from the x axis, the moment of inertia of the element about the x axis is

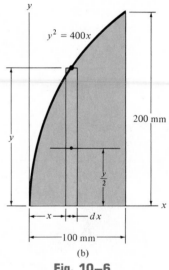

(b)

Fig. 10–6

$$dI_x = d\bar{I}_{\bar{x}} + dA\,\tilde{y}^2 = \frac{1}{12}\,dx\,y^3 + y\,dx \left(\frac{y}{2}\right)^2 = \frac{1}{3}\,y^3\,dx$$

[This result can also be concluded from part (b) of Example 10–1]. Integrating with respect to x, from $x = 0$ to $x = 100$ mm, yields

$$I_x = \int dI_x = \int \frac{1}{3}\,y^3\,dx = \int_0^{100} \frac{1}{3}(400x)^{3/2}\,dx$$

$$= 106.7(10^6)\ \text{mm}^4 \qquad\qquad\qquad Ans.$$

Example 10–3

Determine the moment of inertia with respect to the x axis of the circular area shown in Fig. 10–7a.

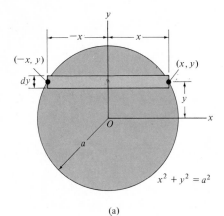

(a)

Fig. 10–7

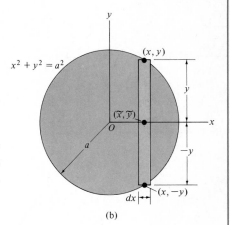

(b)

Solution I *(Case 1)*

Using the differential element shown in Fig. 10–7a, since $dA = 2x\,dy$, we have

$$I_x = \int y^2\,dA = \int y^2(2x)\,dy$$

$$= \int_{-a}^{a} y^2(2\sqrt{a^2 - y^2})\,dy = \frac{\pi a^4}{4} \qquad Ans.$$

Solution II *(Case 2)*

When the differential element is chosen as shown in Fig. 10–7b, the parallel-axis theorem must be used. Why? The centroid for the element lies on the x axis. Hence, $\widetilde{y} = 0$. Applying Eq. 10–3, noting that $dA = 2y\,dx$, we have

$$dI_x = d\bar{I}_{\tilde{x}} + dA\,\widetilde{y}^2$$

$$= \frac{1}{12}\,dx\,(2y)^3 + 2y\,dx\,(0)$$

$$= \frac{2}{3}y^3\,dx$$

Therefore, integrating with respect to x yields

$$I_x = \int_{-a}^{a} \frac{2}{3}(a^2 - x^2)^{3/2}\,dx = \frac{\pi a^4}{4} \qquad Ans.$$

Example 10–4

Determine the moment of inertia of the shaded area shown in Fig. 10–8 about the x axis.

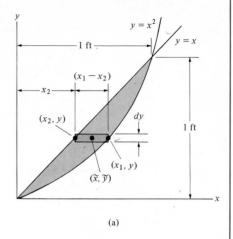

Solution I (Case 1)

The differential element of area parallel to the x axis is chosen for integration, Fig. 10–8a. The element intersects the curve at the arbitrary points (x_2, y) and (x_1, y). Consequently, its area is $dA = (x_1 - x_2) \, dy$. Since all parts of the element lie at the same distance y from the x axis, we have

$$I_x = \int y^2 \, dA = \int_0^1 y^2 (x_1 - x_2) \, dy = \int_0^1 y^2 (\sqrt{y} - y) \, dy$$

$$I_x = \frac{2}{7} y^{7/2} - \frac{1}{4} y^4 \Big|_0^1 = 0.0357 \text{ ft}^4 \qquad \text{Ans.}$$

Solution II (Case 2)

The differential element of area parallel to the y axis is shown in Fig. 10–8b. It intersects the curves at the arbitrary points (x, y_2) and (x, y_1). Since all parts of its entirety do *not* lie at the same distance from the x axis, we must first use the parallel-axis theorem to find the element's moment of inertia about the x axis, then integrate this result to determine I_x. Thus,

$$dI_x = d\bar{I}_{\bar{x}} + dA \, \widetilde{y}^2 = \frac{1}{12} dx \, (y_2 - y_1)^3 + (y_2 - y_1) \, dx \left(y_1 + \frac{y_2 - y_1}{2} \right)^2$$

$$= \frac{1}{3} (y_2^3 - y_1^3) \, dx = \frac{1}{3} (x^3 - x^6) \, dx$$

$$I_x = \frac{1}{3} \int_0^1 (x^3 - x^6) \, dx = \frac{1}{12} x^4 - \frac{1}{21} x^7 \Big|_0^1 = 0.0357 \text{ ft}^4 \qquad \text{Ans.}$$

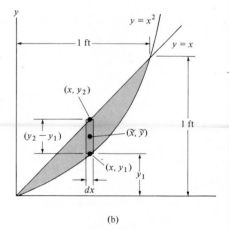

(b)

Fig. 10–8

By comparison, Solution I requires much less computation. Realize, however, that if an integral using a particular element appears difficult to evaluate, one should try solving the problem using an element oriented in the other direction.

Problems

10–1. The polar moment of inertia of the area is $\bar{J}_C = 28\ \text{in}^4$, computed about the z axis passing through the centroid C. The moment of inertia about the x axis is $17\ \text{in}^4$ and the moment of inertia about the y' axis is $56\ \text{in}^4$. Determine the area A.

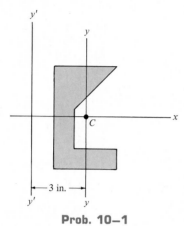

Prob. 10–1

10–2. The area has a moment of inertia about the yy axis of $25(10^6)\ \text{mm}^4$. If the area is $15(10^3)\ \text{mm}^2$, determine the moment of inertia of the area about the $y'y'$ axis. The $\overline{yy}$ axis passes through the centroid C of the area.

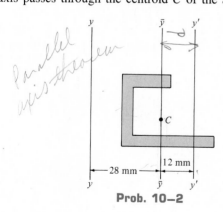

Prob. 10–2

10–3. Determine the moment of inertia of the shaded area about the x axis.

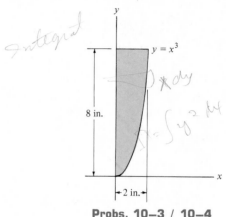

Probs. 10–3 / 10–4

*10–4. Determine the moment of inertia of the shaded area about the y axis.

10–5. Determine the moment of inertia of the shaded area about the x axis.

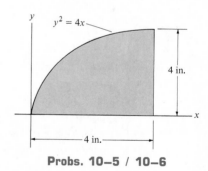

Probs. 10–5 / 10–6

10–6. Determine the moment of inertia of the shaded area about the y axis.

455

10–7. Determine the moment of inertia of the shaded area about the y axis.

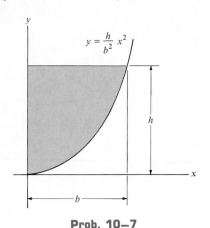

$$y = \frac{h}{b^2} x^2$$

Prob. 10–7

*10–8. Determine the moment of inertia of the area about the x axis. Solve the problem in two ways, using rectangular differential elements: (a) having a thickness dx, and (b) having a thickness dy.

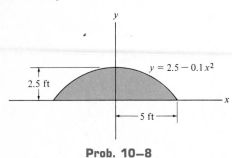

$$y = 2.5 - 0.1 x^2$$

Prob. 10–8

10–9. Compute the moment of inertia of the triangular area about (a) the x axis, and (b) the centroidal $\bar{x}$ axis.

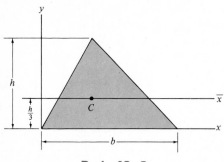

Prob. 10–9

10–10. Determine the moment of inertia of the shaded area about the x axis.

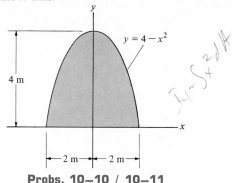

$$y = 4 - x^2$$

Probs. 10–10 / 10–11

10–11. Determine the moment of inertia of the shaded area about the y axis.

*10–12. Determine the moments of inertia I_x and I_y of the area having a boundary defined by the cosine curve.

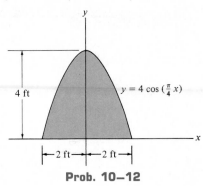

$$y = 4 \cos \left(\frac{\pi}{4} x \right)$$

Prob. 10–12

10–13. Determine the moment of inertia of the shaded area about the x axis. Use differential elements having a thickness of dx.

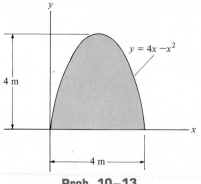

$$y = 4x - x^2$$

Prob. 10–13

■**10–14.** Determine the moment of inertia of the area about the y axis. Use Simpson's rule to evaluate the integral.

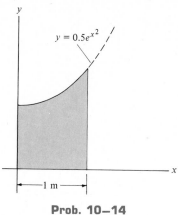

$y = 0.5e^{x^2}$

1 m

Prob. 10–14

10–15. Determine the moment of inertia of the semicircular area about the x axis. Then, using the parallel-axis theorem, compute the moment of inertia about the $\bar{x}$ axis that passes through the centroid C.

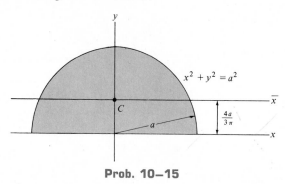

$x^2 + y^2 = a^2$

C

$\dfrac{4a}{3\pi}$

a

Prob. 10–15

****10–16.** Determine the moments of inertia I_x and I_y of the shaded elliptical area. What is the polar moment of inertia about the origin O?

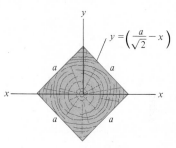

a

$\dfrac{x^2}{a^2} + \dfrac{y^2}{b^2} = 1$

O

b

Prob. 10–16

10–17. Determine the moment of inertia of the beam's cross-sectional area about the xx axis.

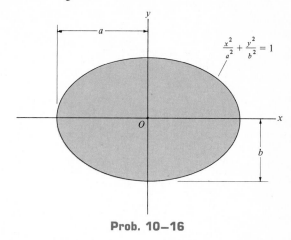

$y = \left(\dfrac{a}{\sqrt{2}} - x\right)$

a a

a a

Prob. 10–17

10.5 Moments of Inertia for Composite Areas

A composite area consists of a series of connected "simpler" area shapes, such as semicircles, rectangles, and triangles. Provided the moment of inertia of each of these shapes is known or can be computed about a common axis, then the moment of inertia of the composite area equals the *algebraic sum* of the moments of inertia of all its composite parts.

PROCEDURE FOR ANALYSIS

The following procedure provides a method for determining the moment of inertia of a composite area about a reference axis.

Composite Parts. Using a sketch, divide the area into its composite parts and indicate the perpendicular distance from the *centroid* of each part to the reference axis.

Parallel-Axis Theorem. The moment of inertia of each part should be computed about its centroidal axis, which is parallel to the reference axis. For the calculation use the table given on the inside back cover of this book. If the centroidal axis does not coincide with the reference axis, the parallel-axis theorem, $I = \bar{I} + Ad^2$, should be used to determine the moment of inertia of the part about the reference axis.

Summation. The moment of inertia of the entire area about the reference axis is determined by summing the results of its composite parts. In particular, if a composite part has a "hole," the moment of inertia for the composite is found by "subtracting" the moment of inertia for the hole from the moment of inertia of the entire area including the hole.

Example 10–5

Compute the moment of inertia of the composite area shown in Fig. 10–9a about the x axis.

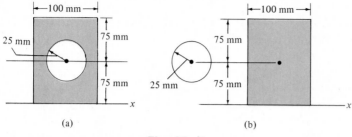

Fig. 10–9

Solution

Composite Parts. The composite area is obtained by *subtracting* the circle from the rectangle as shown in Fig. 10–9b. The centroid of each area is located in the figure.

Parallel-Axis Theorem. The moments of inertia about the x axis are computed using the parallel-axis theorem and the data contained in the table on the inside back cover.

Circle

$$I_x = \bar{I}_{\bar{x}} + A d_y^2$$
$$= \frac{1}{4}\pi(25)^4 + \pi(25)^2(75)^2 = 11.4(10^6) \text{ mm}^4$$

Rectangle

$$I_x = \bar{I}_{\bar{x}} + A d_y^2$$
$$= \frac{1}{12}(100)(150)^3 + (100)(150)(75)^2 = 112.5(10^6) \text{ mm}^4$$

Summation. The moment of inertia for the composite area is thus

$$I_x = -11.4(10^6) + 112.5(10^6)$$
$$= 101(10^6) \text{ mm}^4 \qquad\qquad \textit{Ans.}$$

Example 10-6

Compute the moments of inertia of the beam's cross-sectional area shown in Fig. 10–10a about the x and y centroidal axes.

Solution

Composite Parts. The cross section can be considered as three composite rectangular areas A, B, and D shown in Fig. 10–10b. For the calculation, the centroid of each of these rectangles is located in the figure.

Parallel-Axis Theorem. From the table on the inside back cover, or Example 10–1, the moment of inertia of a rectangle about its centroidal axis is $\bar{I} = \frac{1}{12}bh^3$. Hence, using the parallel-axis theorem for rectangles A and D, the computations are as follows:

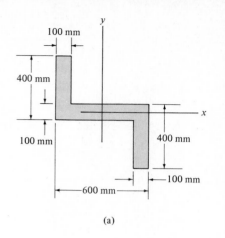

(a)

Rectangle A

$$I_x = \bar{I}_{\bar{x}} + A d_y^2 = \frac{1}{12}(100)(300)^3 + (100)(300)(200)^2$$

$$= 1.425(10^9) \text{ mm}^4$$

$$I_y = \bar{I}_{\bar{y}} + A d_x^2 = \frac{1}{12}(300)(100)^3 + (100)(300)(250)^2$$

$$= 1.90(10^9) \text{ mm}^4$$

Rectangle B

$$I_x = \frac{1}{12}(600)(100)^3 = 0.05(10^9) \text{ mm}^4$$

$$I_y = \frac{1}{12}(100)(600)^3 = 1.80(10^9) \text{ mm}^4$$

Rectangle D

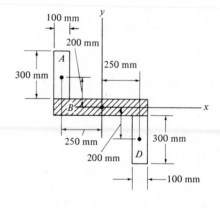

(b)

Fig. 10–10

$$I_x = \bar{I}_{\bar{x}} + A d_y^2 = \frac{1}{12}(100)(300)^3 + (100)(300)(200)^2$$

$$= 1.425(10^9) \text{ mm}^4$$

$$I_y = \bar{I}_{\bar{y}} + A d_x^2 = \frac{1}{12}(300)(100)^3 + (100)(300)(250)^2$$

$$= 1.90(10^9) \text{ mm}^4$$

Summation. The moments of inertia for the entire cross section are thus

$$I_x = 1.425(10^9) + 0.05(10^9) + 1.425(10^9)$$

$$= 2.90(10^9) \text{ mm}^4 \qquad \textit{Ans.}$$

$$I_y = 1.90(10^9) + 1.80(10^9) + 1.90(10^9) = 5.60(10^9) \text{ mm}^4 \quad \textit{Ans.}$$

Problems

10–18. Determine the moments of inertia I_x and I_y of the beam's cross-sectional area.

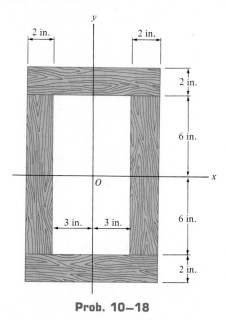

Prob. 10–18

10–19. Determine the location of the centroid $\bar{y}$ of the beam constructed from the two channels and the cover plate. If each channel has a cross-sectional area of $A_c = 11.8 \text{ in}^2$ and a moment of inertia about a horizontal axis passing through its own centroid, C_c, of $(\bar{I}_x)_{C_c} = 349 \text{ in}^4$, determine the moment of inertia of the beam about the $\bar{x}\bar{x}$ axis.

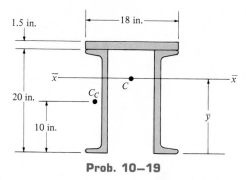

Prob. 10–19

*10–20. Determine the radius of gyration $\bar{k}_{\bar{x}}$ for the column's cross-sectional area.

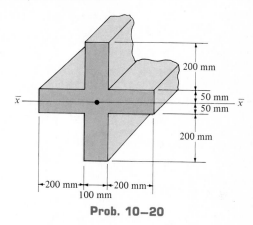

Prob. 10–20

10–21. Compute the polar moments of inertia J_O for the cross-sectional area of the solid shaft and tube. What percentage of J_O is contributed by the tube to that of the solid shaft?

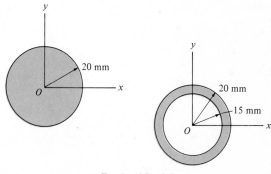

Prob. 10–21

10–22. Determine the moment of inertia of the beam's cross-sectional area with respect to the $\bar{x}\bar{x}$ axis passing through the centroid C of the cross section. Neglect the size of the corner welds at A and B for the calculation. $\bar{y} = 104.3$ mm.

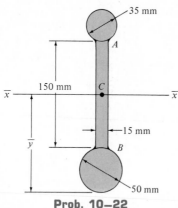

Prob. 10–22

10–23. The composite cross section for the column consists of two cover plates riveted to two channels. Determine the radius of gyration $k_{\bar{x}}$ with respect to the centroidal $\bar{x}\bar{x}$ axis. Each channel has a cross-sectional area of $A_c = 11.8$ in^2 and a moment of inertia $(\bar{I}_x)_c = 349$ in^4.

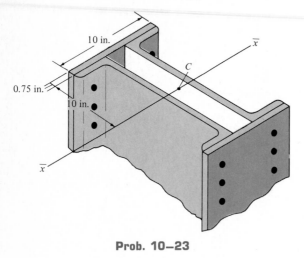

Prob. 10–23

*10–24. The composite beam consists of a wide-flange beam and a cover plate welded together as shown. Determine the moment of inertia of the cross-sectional area with respect to the $\bar{x}\bar{x}$ centroidal axis. $\bar{y} = 77.8$ mm.

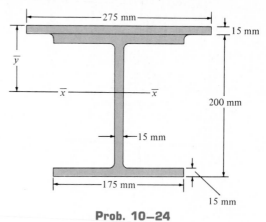

Prob. 10–24

10–25. Determine the moment of inertia I_x of the shaded area about the x axis.

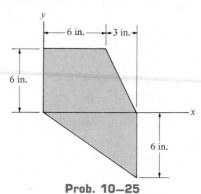

Prob. 10–25

10–26. Locate the centroid $\bar{y}$ of the channel's cross-sectional area and then determine the moment of inertia with respect to the $\bar{x}\bar{x}$ axis passing through the centroid.

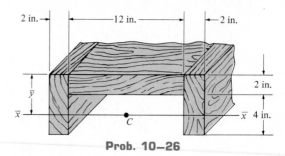

Prob. 10–26

10–27. Determine the moments of inertia I_x and I_y of the Z-section. The origin of coordinates is at the centroid C.

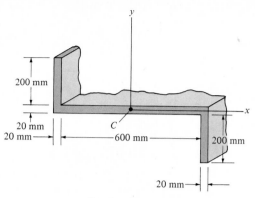

Prob. 10–27

*10–28. Compute the polar radius of gyration, k_O, for the ring and show that for small thicknesses, $t = r_2 - r_1$, k_O is equal to the mean radius r_m of the ring.

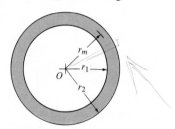

Prob. 10–28

10–29. Determine the centroid $\bar{y}$ for the beam's cross-sectional area, then find $\bar{I}_{\bar{x}}$.

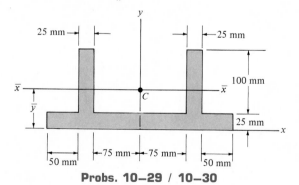

Probs. 10–29 / 10–30

10–30. Determine I_y for the beam having the cross-sectional area shown.

10–31. Determine $\bar{y}$, which locates the centroidal axis $\bar{x}\bar{x}$, and then find the moments of inertia $\bar{I}_{\bar{x}}$ and $\bar{I}_{\bar{y}}$ for the T-beam.

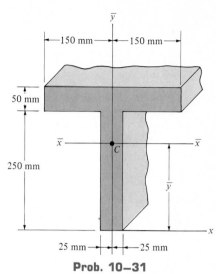

Prob. 10–31

*10–32. Determine the moments of inertia $\bar{I}_{\bar{x}}$ and $\bar{I}_{\bar{y}}$ for the channel section. $\bar{x} = 33.9$ m, $\bar{y} = 150$ mm.

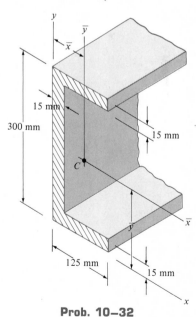

Prob. 10–32

10–33. Locate the centroid C for the area and compute the moments of inertia $\bar{I}_{\bar{x}}$ and $\bar{I}_{\bar{y}}$ for the shaded area about the centroidal axes.

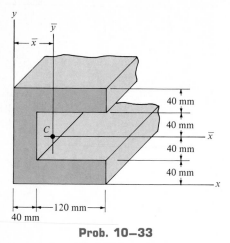

Prob. 10–33

10–34. Determine the location $(\bar{x}, \bar{y})$ of the centroid C of the cross-sectional area for the angle. Then find the moments of inertia $\bar{I}_{\bar{x}}$ and $\bar{I}_{\bar{y}}$ about the $\bar{x}$ and $\bar{y}$ centroidal axes.

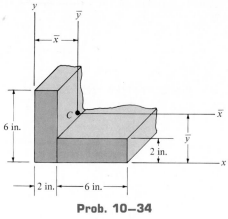

Prob. 10–34

★10.6 Product of Inertia for an Area

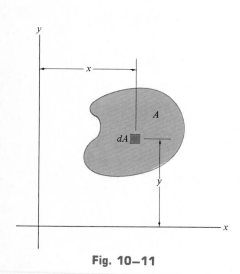

Fig. 10–11

In general, the moment of inertia for an area is different for every axis about which it is computed. In some applications of structural design it is necessary to know the orientation of those axes which give, respectively, the maximum and minimum moments of inertia for the area. The method for determining this is discussed in Sec. 10.7. To use this method, however, one must first compute the product of inertia for the area as well as its moments of inertia for given x, y axes.

The product of inertia for an element of area located at point (x, y) as shown in Fig. 10–11 is defined as $dI_{xy} = xy\, dA$. Thus, for the entire area A, the *product of inertia* is

$$I_{xy} = \int xy\, dA \qquad (10\text{–}7)$$

If the element of area chosen has a differential size in two directions, as shown in Fig. 10–11, a double integration must be performed to evaluate I_{xy}. Most often, however, it is easier to choose an element having a differential size or thickness in only one direction, in which case the evaluation requires only a single integration (see Example 10–7).

Like the moment of inertia, the product of inertia has units of length raised to the fourth power, e.g., m^4, mm^4 or ft^4, in^4. However, since x or y may be a negative quantity, while the element of area is always positive, the product of inertia may be positive, negative, or zero, depending upon the location and orientation of the coordinate axes. For example, the product of inertia I_{xy} for an area will be *zero* if either the x or y axis is an axis of *symmetry* for the area.

To show this, consider the shaded area in Fig. 10–12, where for every element dA located at point (x, y) there is a corresponding element dA located at $(x, -y)$. Since the products of inertia for these elements are respectively $xy\,dA$ and $-xy\,dA$, in their algebraic sum or integration all the elements of area that are chosen in this way will cancel each other. Consequently, the product of inertia for the total area becomes zero. It also follows from the definition of I_{xy} that the "sign" of this quantity depends upon the quadrant where the area is located, Fig. 10–13.

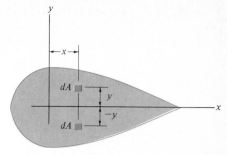

Fig. 10–12

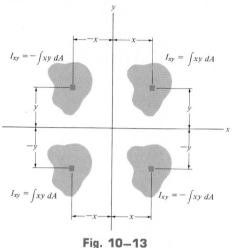

Fig. 10–13

Parallel-Axis Theorem. Consider the shaded area shown in Fig. 10–14, where $\bar{x}$ and $\bar{y}$ represent a set of axes passing through the *centroid* of the area, and x and y represent a corresponding set of parallel axes. Since the product of inertia of dA with respect to the x and y axes is $dI_{xy} = (x + d_x)(y + d_y)\,dA$, then for the entire area,

$$I_{xy} = \int (x + d_x)(y + d_y)\,dA$$

$$= \int xy\,dA + d_x \int y\,dA + d_y \int x\,dA + d_x d_y \int dA$$

The first term on the right represents the product of inertia of the area with respect to the centroidal axis, $\bar{I}_{\bar{x}\bar{y}}$. The integrals in the second and third terms are zero since the moments of the area are taken about the centroidal axis. Realizing that the fourth integral represents the total area A, we therefore have as the final result,

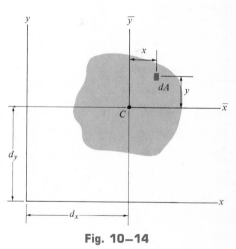

Fig. 10–14

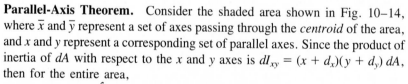

$$I_{xy} = \bar{I}_{\bar{x}\bar{y}} + A d_x d_y \qquad (10\text{–}8)$$

The similarity between this equation and the parallel-axis theorem for moments of inertia should be noted. In particular, it is important that the *algebraic signs* for d_x and d_y be maintained when applying Eq. 10–8. As illustrated in Example 10–8, the parallel-axis theorem finds important application in determining the product of inertia of a *composite area* with respect to a set of x, y axes.

465

Example 10–7

Determine the product of inertia I_{xy} of the triangle shown in Fig. 10–15a.

Solution I

Consider the differential element that has a thickness dx and area $dA = y\,dx$, Fig. 10–15b. The product of inertia of the element about the x, y axes is determined using the parallel-axis theorem.

$$dI_{xy} = d\bar{I}_{\tilde{x}\tilde{y}} + dA\,\tilde{x}\,\tilde{y}$$

where $(\tilde{x}, \tilde{y})$ locates the *centroid* of the element. Since $dI_{\tilde{x}\tilde{y}} = 0$, due to symmetry, and $\tilde{x} = x$, $\tilde{y} = y/2$, then

$$dI_{xy} = 0 + (y\,dx)x\left(\frac{y}{2}\right) = \left(\frac{h}{b}x\,dx\right)x\left(\frac{h}{2b}x\right)$$

$$= \frac{h^2}{2b^2}x^3\,dx$$

Integrating with respect to x from $x = 0$ to $x = b$ yields

$$I_{xy} = \frac{h^2}{2b^2}\int_0^b x^3\,dx = \frac{b^2 h^2}{8} \qquad \textit{Ans.}$$

Solution II

Consider a differential element that has a thickness dy and area $dA = (b - x)\,dy$, as shown in Fig. 10–15c. The *centroid* is located at point $\tilde{x} = x + (b - x)/2 = (b + x)/2$, $\tilde{y} = y$, so that the product of inertia of the element becomes

$$dI_{xy} = d\bar{I}_{\tilde{x}\tilde{y}} + dA\,\tilde{x}\,\tilde{y}$$

$$= 0 + (b - x)\,dy\left(\frac{b + x}{2}\right)y$$

$$= \left(b - \frac{b}{h}y\right)dy\left(\frac{b + \frac{b}{h}y}{2}\right)y = \frac{1}{2}y\left(b^2 - \frac{b^2}{h^2}y^2\right)dy$$

Integrating with respect to y from $y = 0$ to $y = h$ yields

$$I_{xy} = \frac{1}{2}\int_0^h y\left(b^2 - \frac{b^2}{h^2}y^2\right)dy = \frac{b^2 h^2}{8} \qquad \textit{Ans.}$$

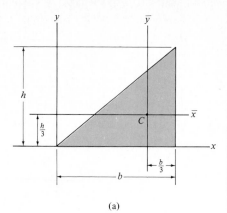

(a)

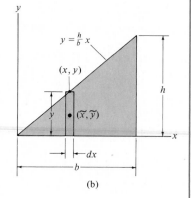

(b)

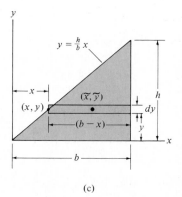

(c)

Fig. 10–15

Example 10–8

Compute the product of inertia of the beam's cross-sectional area, shown in Fig. 10–16a, about the x and y centroidal axes.

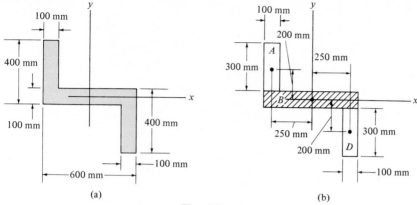

(a)

(b)

Fig. 10–16

Solution

As in Example 10–6, the cross section can be considered as three composite rectangular areas A, B, and D, Fig. 10–16b. The coordinates for the centroid of each of these rectangles are shown in the figure. Due to symmetry, the product of inertia of *each rectangle* is *zero* about a set of $\bar{x}$, $\bar{y}$ axes that pass through the rectangle's centroid. Hence, application of the parallel-axis theorem to each of the rectangles yields

Rectangle A

$$I_{xy} = \bar{I}_{\bar{x}\bar{y}} + A d_x d_y$$
$$= 0 + (300)(100)(-250)(200)$$
$$= -1.50(10^9) \text{ mm}^4$$

Rectangle B

$$I_{xy} = \bar{I}_{\bar{x}\bar{y}} + A d_x d_y$$
$$= 0 + 0$$
$$= 0$$

Rectangle D

$$I_{xy} = \bar{I}_{\bar{x}\bar{y}} + A d_x d_y$$
$$= 0 + (300)(100)(250)(-200)$$
$$= -1.50(10^9) \text{ mm}^4$$

The product of inertia for the entire cross section is thus

$$I_{xy} = [-1.50(10^9)] + 0 + [-1.50(10^9)] = -3.00(10^9) \text{ mm}^4 \quad \textit{Ans.}$$

*10.7 Moments of Inertia for an Area About Inclined Axes

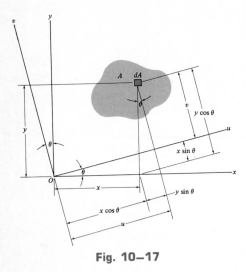

Fig. 10–17

In structural mechanics, it is sometimes necessary to calculate the moments and product of inertia I_u, I_v, and I_{uv} for an area with respect to a set of inclined u and v axes when the values for θ, I_x, I_y, and I_{xy} are *known*. From Fig. 10–17 the perpendicular distance from the area element dA may be related to the axes of the two coordinate systems by using the *transformation equations*

$$u = x \cos \theta + y \sin \theta$$
$$v = y \cos \theta - x \sin \theta$$

Using these equations, the moments and product of inertia of dA about the u and v axes become

$$dI_u = v^2 \, dA = (y \cos \theta - x \sin \theta)^2 \, dA$$
$$dI_v = u^2 \, dA = (x \cos \theta + y \sin \theta)^2 \, dA$$
$$dI_{uv} = uv \, dA = (x \cos \theta + y \sin \theta)(y \cos \theta - x \sin \theta) \, dA$$

Expanding each expression and integrating, realizing that $I_x = \int y^2 \, dA$, $I_y = \int x^2 \, dA$, and $I_{xy} = \int xy \, dA$, we obtain

$$I_u = I_x \cos^2 \theta + I_y \sin^2 \theta - 2I_{xy} \sin \theta \cos \theta$$
$$I_v = I_x \sin^2 \theta + I_y \cos^2 \theta + 2I_{xy} \sin \theta \cos \theta$$
$$I_{uv} = I_x \sin \theta \cos \theta - I_y \sin \theta \cos \theta + I_{xy}(\cos^2 \theta - \sin^2 \theta)$$

These equations may be simplified by using the trigonometric identities $\sin 2\theta = 2 \sin \theta \cos \theta$ and $\cos 2\theta = \cos^2 \theta - \sin^2 \theta$, in which case

$$I_u = \frac{I_x + I_y}{2} + \frac{I_x - I_y}{2} \cos 2\theta - I_{xy} \sin 2\theta$$

$$I_v = \frac{I_x + I_y}{2} - \frac{I_x - I_y}{2} \cos 2\theta + I_{xy} \sin 2\theta \qquad (10-9)$$

$$I_{uv} = \frac{I_x - I_y}{2} \sin 2\theta + I_{xy} \cos 2\theta$$

Note that if the first and second equations are added together, it is seen that the polar moment of inertia about the z axis passing through point O is *independent* of the orientation of the u and v axes, i.e.,

$$J_O = I_u + I_v = I_x + I_y$$

Principal Moments of Inertia. From Eqs. 10–9, it may be seen that I_u, I_v, and I_{uv} depend upon the angle of inclination, θ, of the u, v axes. We will now determine the orientation of the u, v axes about which the moments of inertia for the area, I_u and I_v, are maximum and minimum. This particular set of axes is called the *principal axes* of the area, and the corresponding moments of inertia with respect to these axes are called the *principal moments of inertia*. In general, there is a set of principal axes for every chosen origin O, although in structural mechanics the area's centroid is an important location for O.

The angle $\theta = \theta_p$, which defines the orientation of the principal axes for the area, may be found by differentiating the first of Eqs. 10–9 with respect to θ and setting the result equal to zero. Thus,

$$\frac{dI_u}{d\theta} = -2\left(\frac{I_x - I_y}{2}\right)\sin 2\theta - 2I_{xy}\cos 2\theta = 0$$

Therefore, at $\theta = \theta_p$,

$$\tan 2\theta_p = \frac{-I_{xy}}{\dfrac{I_x - I_y}{2}} \tag{10–10}$$

This equation has two roots, θ_{p_1} and θ_{p_2}, which specify the inclination of the principal axes. Because of the nature of the tangent, the values of $2\theta_{p_1}$ and $2\theta_{p_2}$ are 180° apart, so that θ_{p_1} and θ_{p_2} are 90° apart. Assuming that I_{xy} and $(I_x - I_y)$ are both positive quantities, the sine and cosine of $2\theta_{p_1}$ and $2\theta_{p_2}$ can be obtained from the triangles shown in Fig. 10–18, which are based upon Eq. 10–10.

For θ_{p_1}

$$\sin 2\theta_{p_1} = -I_{xy}\bigg/\sqrt{\left(\frac{I_x - I_y}{2}\right)^2 + I_{xy}^2}$$

$$\cos 2\theta_{p_1} = \left(\frac{I_x - I_y}{2}\right)\bigg/\sqrt{\left(\frac{I_x - I_y}{2}\right)^2 + I_{xy}^2}$$

For θ_{p_2}

$$\sin 2\theta_{p_2} = I_{xy}\bigg/\sqrt{\left(\frac{I_x - I_y}{2}\right)^2 + I_{xy}^2}$$

$$\cos 2\theta_{p_2} = -\left(\frac{I_x - I_y}{2}\right)\bigg/\sqrt{\left(\frac{I_x - I_y}{2}\right)^2 + I_{xy}^2}$$

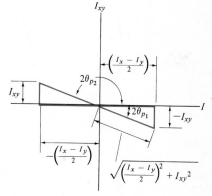

Fig. 10–18

If these two sets of trigonometric relations are substituted into the first or second of Eqs. 10–9 and simplified, the result is

$$I_{\substack{\max \\ \min}} = \frac{I_x + I_y}{2} \pm \sqrt{\left(\frac{I_x - I_y}{2}\right)^2 + I_{xy}^2} \tag{10–11}$$

Depending upon the sign chosen, this result gives the maximum or minimum moment of inertia for the area. Furthermore, if the above trigonometric relations for θ_{p_1} and θ_{p_2} are substituted into the third of Eqs. 10–9, it may be seen that $I_{uv} = 0$; that is, the *product of inertia with respect to the principal axes is zero*. Since it was indicated in Sec. 10.6 that the product of inertia is zero with respect to any symmetrical axis, it therefore follows that *any symmetrical axis represents a principal axis of inertia for the area.*

Example 10–9

Determine the principal moments of inertia for the beam's cross-sectional area shown in Fig. 10–19a with respect to an axis passing through the centroid.

Solution

The moments and product of inertia of the cross section with respect to the x, y axes have been computed in Examples 10–6 and 10–8. The results are

$$I_x = 2.90(10^9) \text{ mm}^4 \qquad I_y = 5.60(10^9) \text{ mm}^4 \qquad I_{xy} = -3.00(10^9) \text{ mm}^4$$

Using Eq. 10–10, the angles of inclination of the principal axes u and v are

$$\tan 2\theta_p = \frac{-I_{xy}}{\dfrac{I_x - I_y}{2}} = \frac{3.00(10^9)}{\dfrac{2.90(10^9) - 5.60(10^9)}{2}} = -2.22$$

$$2\theta_{p_1} = -65.8° \qquad \text{and} \qquad 2\theta_{p_2} = 114.2°$$

Thus, as shown in Fig. 10–19b,

$$\theta_{p_1} = -32.9° \qquad \text{and} \qquad \theta_{p_2} = 57.1°$$

The principal moments of inertia with respect to the u and v axes are determined by using Eq. 10–11. Hence,

$$I_{\substack{max \\ min}} = \frac{I_x + I_y}{2} \pm \sqrt{\left(\frac{I_x - I_y}{2}\right)^2 + I_{xy}^2}$$

$$= \frac{2.90(10^9) + 5.60(10^9)}{2} \pm \sqrt{\left[\frac{2.90(10^9) - 5.60(10^9)}{2}\right]^2 + [-3.00(10^9)]^2}$$

$$= 4.25(10^9) \pm 3.29(10^9)$$

or

$$(I_u)_{max} = 7.54(10^9) \text{ mm}^4 \qquad (I_v)_{min} = 0.960(10^9) \text{ mm}^4 \qquad \textit{Ans.}$$

Specifically, the maximum moment of inertia, $(I_u)_{max} = 7.54(10^9) \text{ mm}^4$, occurs with respect to the u axis, since *by inspection* most of the cross-sectional area is farthest away from this axis. Stated in a general manner, $(I_u)_{max}$ occurs about an axis located within $\pm45°$ of the axis (x or y) which has the largest I (I_x or I_y). (To conclude this mathematically, substitute the data with $\theta = 57.1°$ into the first of Eqs. 10–9.)

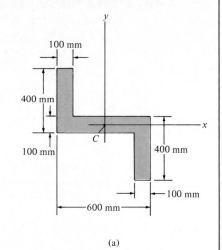

(a)

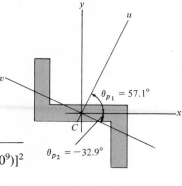

(b)

Fig. 10–19

Mohr's Circle for Moments of Inertia $\mathbf{10.8}^{\star}$

Equations 10–9 to 10–11 have a graphical solution that is convenient to use and generally easy to remember. Squaring the first and third of Eqs. 10–9 and adding, it is found that

$$\left(I_u - \frac{I_x + I_y}{2}\right)^2 + I_{uv}^2 = \left(\frac{I_x - I_y}{2}\right)^2 + I_{xy}^2 \qquad (10\text{–}12)$$

In a given problem, I_u and I_{uv} are *variables*, and I_x, I_y, and I_{xy} are *known constants*. Thus, Eq. 10–12 may be written in compact form as

$$(I_u - a)^2 + I_{uv}^2 = R^2$$

When this equation is plotted, the resulting graph represents a *circle* of radius

$$R = \sqrt{\left(\frac{I_x - I_y}{2}\right)^2 + I_{xy}^2}$$

having its center located at point $(a, 0)$, where $a = (I_x + I_y)/2$. The circle so constructed is called *Mohr's circle*, named after the German engineer Otto Mohr (1835–1918).

PROCEDURE FOR ANALYSIS

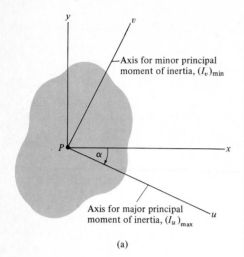

(a)

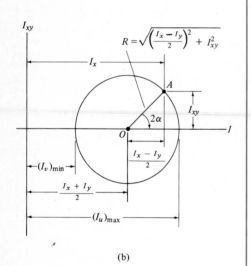

(b)

Fig. 10–20

There are several methods for plotting Mohr's circle as defined by Eq. 10–12. The main purpose in using the circle here is to have a convenient means for transforming I_x, I_y, and I_{xy} into the principal moments of inertia. The following procedure provides a method for doing this.

Compute I_x, I_y, I_{xy}. Establish the x, y axes for the area, with the origin located at the point P of interest, and determine I_x, I_y, and I_{xy}, Fig. 10–20a.

Construct the Circle. Construct a rectangular coordinate system such that the abscissa represents the moment of inertia I, and the ordinate represents the product of inertia I_{xy}, Fig. 10–20b. Determine the center of the circle, O, which is located at a distance $(I_x + I_y)/2$ from the origin, and plot the "controlling point" A having coordinates (I_x, I_{xy}). By definition, I_x is always positive, whereas I_{xy} will be either positive or negative. Connect the controlling point A with the center of the circle, and determine the distance OA by trigonometry. This distance represents the radius of the circle, Fig. 10–20b. Finally, draw the circle.

Principal Moments of Inertia. The points where the circle intersects the abscissa give the values of the principal moments of inertia $(I_v)_{min}$ and $(I_u)_{max}$. Notice that the *product of inertia will be zero at these points*, Fig. 10–20b.

Principal Axes. To find the direction of the major principal axis, determine by trigonometry the angle 2α, *measured from the radius OA to the direction line of the positive abscissa*, Fig. 10–20b. This angle represents twice the angle from the x axis of the area in question to the axis of maximum moment of inertia $(I_u)_{max}$, Fig. 10–20a. Both the angle on the circle, 2α, and the angle on the area, α, *must be measured in the same sense*, as shown in Fig. 10–20. The axis for minimum moment of inertia $(I_v)_{min}$ is perpendicular to the axis for $(I_u)_{max}$.

Using trigonometry, the above procedure may be verified to be in accordance with the equations developed in Sec. 10.7.

Example 10–10

Using Mohr's circle, determine the principal moments of inertia for the beam's cross-sectional area, shown in Fig. 10–21a, with respect to an axis passing through the centroid.

Fig. 10–21

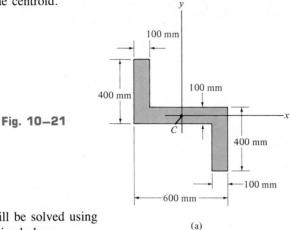

(a)

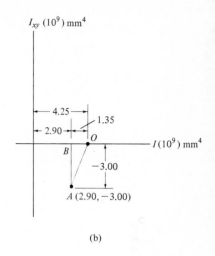

(b)

Solution

The problem will be solved using the procedure outlined above.

Compute I_x, I_y, I_{xy}. The moments of inertia and the product of inertia have been determined in Examples 10–6 and 10–8 with respect to the x, y axes shown in Fig. 10–21a. The results are $I_x = 2.90(10^9)$ mm^4, $I_y = 5.60(10^9)$ mm^4, and $I_{xy} = -3.00(10^9)$ mm^4.

Construct the Circle. The I and I_{xy} axes are shown in Fig. 10–21b. The center of the circle, O, lies at a distance $(I_x + I_y)/2 = (2.90 + 5.60)/2 = 4.25$ from the origin. When the controlling point $A(2.90, -3.00)$ is connected to point O, the radius OA is determined from the triangle OBA using the Pythagorean theorem.

$$OA = \sqrt{(1.35)^2 + (-3.00)^2} = 3.29$$

The circle is constructed in Fig. 10–21c.

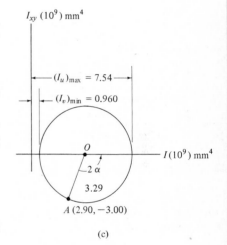

(c)

Principal Moments of Inertia. The circle intersects the I axis at points $(7.54, 0)$ and $(0.960, 0)$. Hence,

$$(I_u)_{max} = 7.54(10^9) \text{ mm}^4 \qquad Ans.$$
$$(I_v)_{min} = 0.960(10^9) \text{ mm}^4 \qquad Ans.$$

Principal Axes. As shown in Fig. 10–21c, the angle 2α is determined from the circle by measuring counterclockwise from OA to the direction of the *positive* I axis. Hence,

$$2\alpha = 180° - \sin^{-1}\left(\frac{|BA|}{|OA|}\right) = 180° - \sin^{-1}\left(\frac{3.00}{3.29}\right) = 114.2°$$

The principal axis for $(I_u)_{max} = 7.54(10^9)$ mm^4 is therefore oriented at an angle $\alpha = 57.1°$, measured *counterclockwise*, from the *positive* x axis to the *positive* u axis. The v axis is perpendicular to this axis. The results are shown in Fig. 10–21d.

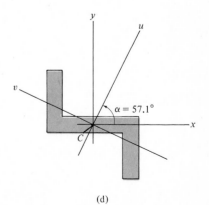

(d)

Problems

10–35. Determine the product of inertia of the shaded area with respect to the x and y axes.

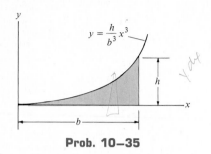

Prob. 10–35

*10–36. Determine the product of inertia of the shaded area with respect to the x and y axes.

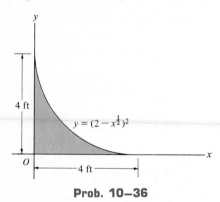

Prob. 10–36

10–37. Determine the product of inertia of the shaded area of the ellipse with respect to the x and y axes.

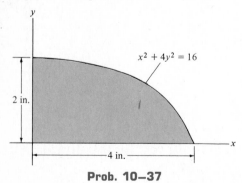

Prob. 10–37

10–38. Determine the product of inertia of the quarter circular area with respect to the x and y axes.

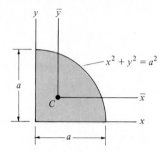

Probs. 10–38 / 10–39

10–39. Determine the product of inertia of the quarter circular area with respect to the $\bar{x}$ and $\bar{y}$ axes which pass through the centroid C.

*10–40. Determine the product of inertia of the parabolic area with respect to the x and y axes.

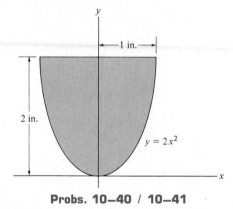

Probs. 10–40 / 10–41

10–41. Determine the product of inertia I_{xy} of the right half of the parabolic area in Prob. 10–40, bounded by the lines $y = 2$ in. and $x = 0$.

474

10–42. The area of the cross section of an airplane wing has the following properties about the $\bar{x}$ and $\bar{y}$ axes passing through the centroid C: $\bar{I}_{\bar{x}} = 450$ in⁴, $\bar{I}_{\bar{y}} = 1730$ in⁴, $\bar{I}_{\bar{x}\bar{y}} = 138$ in⁴. Determine the orientation of the principal axes and the principal moments of inertia.

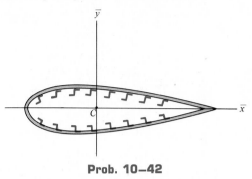

Prob. 10–42

10–43. Determine the product of inertia $\bar{I}_{\bar{x}\bar{y}}$ of the cross-sectional area of the channel with respect to the $\bar{x}$ and $\bar{y}$ axes. $\bar{x} = 33.9$ mm, $\bar{y} = 150$ mm.

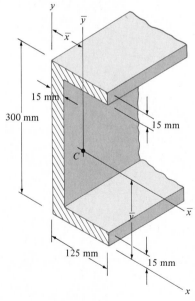

Prob. 10–43

*10–44. Determine the product of inertia of the cross-sectional area with respect to the x and y axes that have their origin located at the centroid C.

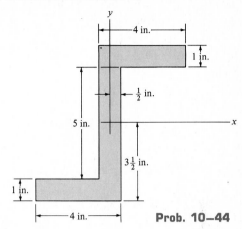

Prob. 10–44

10–45. Determine the product of inertia of the area with respect to the x and y axes having their origin at O.

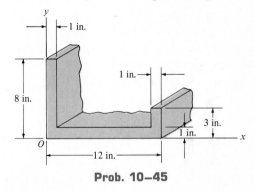

Prob. 10–45

10–46. Locate the position $\bar{x}, \bar{y}$ for the centroid C of the cross-sectional area and then compute the product of inertia with respect to the $\bar{x}$ and $\bar{y}$ axes.

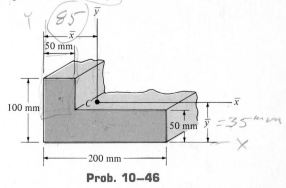

Prob. 10–46

475

10–47. Locate the position $(\bar{x}, \bar{y})$ for the centroid C of the angle's cross-sectional area and then compute the product of inertia with respect to the $\bar{x}$ and $\bar{y}$ axes. Assume that all corners are square.

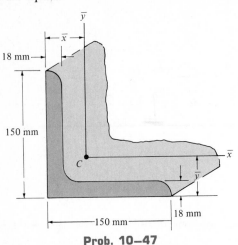

Prob. 10–47

*10–48. Locate the centroid, $\bar{y}$, and compute the moments of inertia I_u and I_v of the channel section. The u and v axes have their origin at the centroid C. For the calculation, assume all corners to be square.

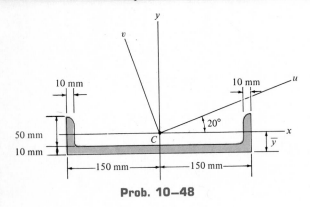

Prob. 10–48

10–49. Determine the moments of inertia I_u and I_v and the product of inertia I_{uv} for the semicircular area.

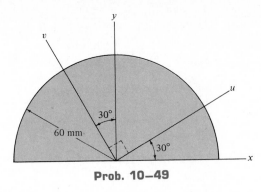

Prob. 10–49

10–50. Determine the principal moments of inertia of the cross-sectional area about the principal axes that have their origin located at the centroid C. Use the equations developed in Sec. 10.7. For the calculation, assume all corners to be square.

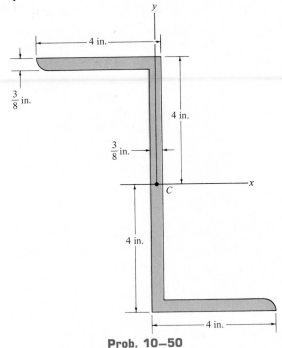

Prob. 10–50

10–51. Compute the moments of inertia I_u and I_v of the shaded area.

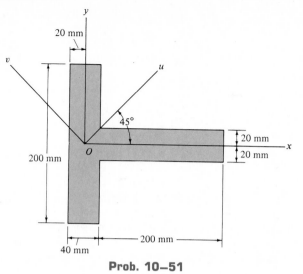

Prob. 10–51

***10–52.** Determine the principal moments of inertia for the angle's cross-sectional area with respect to a set of principal axes that have their origin located at the centroid C. Use the equations developed in Sec. 10.7. For the calculation, assume all corners to be square.

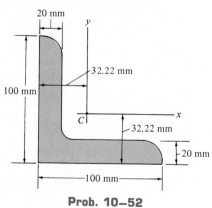

Prob. 10–52

10–53. Locate the centroid, $\bar{y}$, and determine the orientation of the principal centroidal axes for the composite area. What are the moments of inertia with respect to these axes?

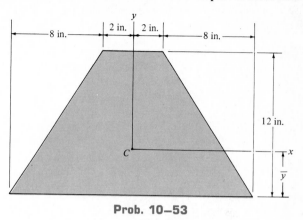

Prob. 10–53

10–54. Determine the moments of inertia I_u and I_v and the product of inertia I_{uv} for the rectangular area. The u and v axes pass through the centroid C.

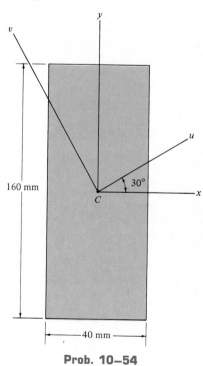

Prob. 10–54

10–55. Determine the principal moments of inertia of the composite area with respect to a set of principal axes that have their origin located at the centroid C. Use the equations developed in Sec. 10.7. $I_{xy} = -12.96(10^6)$ mm^4.

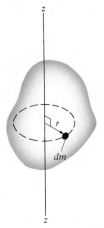

Prob. 10–55

*10–56. Using Mohr's circle, determine the principal moments of inertia of the triangular area and the orientation of the principal axes of inertia having an origin at point O. $I_{xy} = 60(10^3)$ mm^4. Use the results of Prob. 10–9.

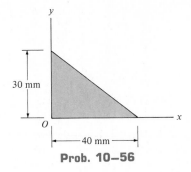

Prob. 10–56

10–57. Solve Prob. 10–50 using Mohr's circle.

10–58. Solve Prob. 10–52 using Mohr's circle.

10–59. Construct Mohr's circle for the shaded area in Prob. 10–53.

10.9 Mass Moment of Inertia

The mass moment of inertia of a body is a property that measures the resistance of the body to angular acceleration. Since it is used throughout the study of rotational dynamics, methods for its calculation will now be discussed.

We define the *mass moment of inertia* as the integral of the "second moment" about an axis of all the differential-size elements of mass dm which compose the body.* For example, consider the rigid body shown in Fig. 10–22. The body's moment of inertia about the z axis is

$$I = \int_m r^2 \, dm \tag{10–13}$$

Here the "moment arm" r is the perpendicular distance from the axis to the arbitrary element dm. Since the formulation involves r, the value of I is *unique* for each axis z about which it is computed. The axis which is generally

Fig. 10–22

---

*Another property of the body which measures the symmetry of the body's mass with respect to a coordinate system is the mass product of inertia. This property most often applies to the three-dimensional motion of a body and is described in *Engineering Mechanics: Dynamics* (Chapter 21).

chosen for analysis, however, passes through the body's mass center G. The moment of inertia computed about this axis will be defined as I_G. Note that because r is squared in Eq. 10–13, the mass moment of inertia is always a *positive quantity*. Common units used for its measurement are $kg \cdot m^2$ or $slug \cdot ft^2$.

PROCEDURE FOR ANALYSIS

For integration, we will only consider bodies having surfaces which are generated by revolving a curve about an axis. An example of such a body which is generated about the z axis is shown in Fig. 10–23.

If the body consists of material having a variable mass density, $\rho = \rho(x, y, z)$, the elemental mass dm of the body may be expressed in terms of its density and volume as $dm = \rho\, dV$. Substituting in Eq. 10–13, the body's moment of inertia is then computed using *volume elements* for integration, i.e.,

$$I = \int_V r^2 \rho\, dV \qquad (10\text{–}14)$$

In the special case of ρ being a *constant*, this term may be factored out of the integral and the integration is then purely a function of geometry,

$$I = \rho \int_V r^2\, dV \qquad (10\text{–}15)$$

When the elemental volume chosen for integration has differential sizes in all three directions, e.g., $dV = dx\, dy\, dz$, Fig. 10–23a, the moment of inertia of the body must be computed using "triple integration." The integration process can, however, be simplified to a *single integration* provided the chosen elemental volume has a differential size or thickness in only *one direction*. Shell or disk elements are often used for this purpose.

Shell Element. If a *shell element* having a height z, radius $r = y$, and thickness dy is chosen for integration, Fig. 10–23b, then the volume is $dV = (2\pi y)(z)\, dy$. This element may be used in Eq. 10–14 or 10–15 for computing the moment of inertia I_z of the body about the z axis, since the *entire element*, due to its "thinness," lies at the *same* perpendicular distance $r = y$ from the z axis (see Example 10–11).

Disk Element. If a disk element having a radius $r = y$ and a thickness dz is chosen for integration, Fig. 10–23c, then the volume is $dV = (\pi y^2)\, dz$. In this case, however, the element is *finite* in the radial direction, and consequently parts of it *do not* all lie at the *same radial distance* r from the z axis. As a result, Eq. 10–14 or 10–15 *cannot* be used to determine I_z. Instead, to perform the integration using this element, it is first necessary to determine the moment of inertia *of the element* about the z axis and then integrate this result (see Example 10–12).

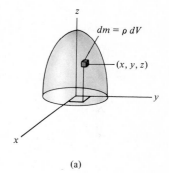

(a)

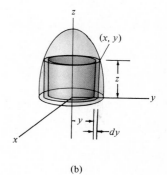

(b)

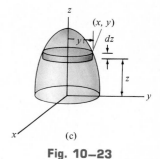

(c)

Fig. 10–23

Example 10–11

Determine the moment of inertia of the right circular cylinder shown in Fig. 10–24a about the z axis. The mass density ρ of the material is constant.

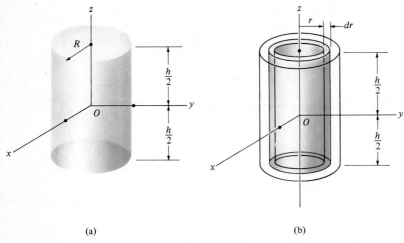

(a)

(b)

Fig. 10–24

Solution

Shell Element. This problem may be solved using the *shell element* in Fig. 10–24b and single integration. The volume of the element is $dV = (2\pi r)(h)\ dr$, so that the mass is $dm = \rho\ dV = \rho(2\pi hr\ dr)$. Since the *entire element* lies at the same distance r from the z axis, the moment of inertia *of the element* is

$$dI_z = r^2\ dm = \rho 2\pi hr^3\ dr$$

Integrating over the entire region of the cylinder yields

$$I_z = \int_m r^2\ dm = \rho 2\pi h \int_0^R r^3\ dr = \frac{\rho\pi}{2}R^4 h$$

The mass of the cylinder is

$$m = \int_m dm = \rho 2\pi h \int_0^R r\ dr = \rho\pi hR^2$$

so that

$$I_z = \frac{1}{2}mR^2 \qquad\qquad Ans.$$

Example 10–12

A solid is formed by revolving the shaded area shown in Fig. 10–25a about the y axis. If the mass density of the material is 5 slug/ft^3, determine the moment of inertia about the y axis.

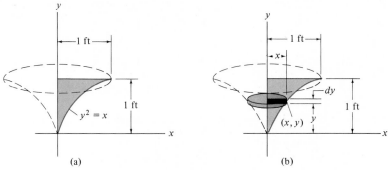

(a)   (b)

Fig. 10–25

Solution

Disk Element. The moment of inertia will be computed using a *disk element,* as shown in Fig. 10–25b. Here the element intersects the curve at the arbitrary point (x, y) and has a mass

$$dm = \rho\, dV = \rho\, (\pi x^2)\, dy$$

Although all portions of the element are *not* located at the same distance from the y axis, it is still possible to determine the moment of inertia dI_y *of the element* about the y axis. In Example 10–11 it was shown that the moment of inertia of a cylinder about its longitudinal axis is $I = \frac{1}{2}mR^2$, where m and R are the mass and radius of the cylinder. Since the height of the cylinder is not involved in this formula, the moment of inertia of the disk element in Fig. 10–25b is

$$dI_y = \tfrac{1}{2}(dm)x^2 = \tfrac{1}{2}[\rho(\pi x^2)\, dy]x^2$$

Substituting $x = y^2$, $\rho = 5$ slug/ft^3, and integrating with respect to y, from $y = 0$ to $y = 1$ ft, yields the moment of inertia for the entire solid.

$$I_y = \frac{\pi 5}{2} \int_0^1 x^4\, dy = \frac{\pi 5}{2} \int_0^1 y^8\, dy = 0.873 \text{ slug} \cdot \text{ft}^2 \qquad Ans.$$

Parallel-Axis Theorem. If the moment of inertia of the body about an axis passing through the body's mass center is known, then the body's moment of inertia may be determined about any other *parallel axis* by using the *parallel-axis theorem*. This theorem can be derived by considering the body shown in Fig. 10–26. The $\bar{z}$ axis passes through the body's mass center G, whereas the corresponding *parallel axis z* lies at a constant distance d away. Selecting the differential element of mass dm which is located at point (x, y) and using the Pythagorean theorem, $r^2 = (d + x)^2 + y^2$, we can express the moment of inertia of the body computed about the z axis as

$$I = \int_m r^2 \, dm = \int_m [(d + x)^2 + y^2] \, dm$$

$$= \int_m (x^2 + y^2) \, dm + 2d \int_m x \, dm + d^2 \int_m dm$$

Since $r_O^2 = x^2 + y^2$, the first integral represents I_G. The second integral equals *zero*, since the $\bar{z}$ axis passes through the body's mass center, i.e., $\int x \, dm = \bar{x} \int dm$ since $\bar{x} = 0$. Finally, the third integral represents the total mass m of the body. Hence, the moment of inertia about the z axis can be written as

$$I = I_G + md^2 \qquad\qquad (10\text{–}16)$$

where I_G = moment of inertia about the $\bar{z}$ axis passing through the mass center G

m = mass of the body

d = perpendicular distance between the parallel axes

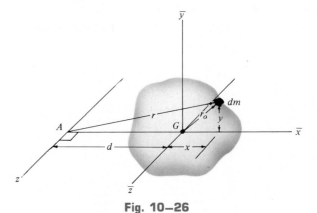

Fig. 10–26

Radius of Gyration. Occasionally, the moment of inertia of a body about a specified axis is reported in handbooks using the *radius of gyration, k*. This value depends only on the geometry of the body. And, when it and the body's mass *m* are known, the moment of inertia is determined from the equation

$$I = mk^2 \quad \text{or} \quad k = \sqrt{\frac{I}{m}} \tag{10-17}$$

Note the *similarity* between the definition of k in this formula and r in the equation $dI = r^2\,dm$, which defines the moment of inertia of an elemental mass dm of the body about an axis.

Composite Bodies. If a body is constructed from a number of simple shapes such as disks, spheres, and rods, the moment of inertia of the body about any axis z can be determined by adding algebraically the moments of inertia of all the composite shapes computed about the z axis. Algebraic addition is necessary since a composite part must be considered as a negative quantity if it has already been included within another part—for example, a "hole" subtracted from a solid plate. The parallel-axis theorem is needed for the calculations if the center of gravity of each composite part does not lie on the z axis. For the calculation, then, $I = \Sigma(I_G + md^2)$, where I_G for each of the composite parts is computed by integration or can be determined from a table, such as the one given on the inside back cover of this book.

Example 10–13

If the plate shown in Fig. 10–27a has a density of 8000 kg/m^3 and a thickness of 10 mm, compute its moment of inertia about an axis directed perpendicular to the page and passing through point O.

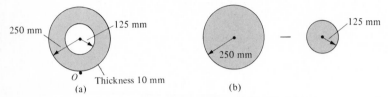

250 mm   125 mm

O Thickness 10 mm

(a)

250 mm   125 mm

(b)

Fig. 10–27

Solution

The plate consists of two composite parts, the 250-mm-radius disk *minus* a 125-mm-radius disk, Fig. 10–27b. The moment of inertia about O can be determined by computing the moment of inertia of each of these parts about O and then *algebraically* adding the results. The computations are performed by using the parallel-axis theorem in conjunction with the data listed in the table on the inside back cover.

Disk. The moment of inertia of a thin disk about an axis perpendicular to the plane of the disk is $I_G = \frac{1}{2}mr^2$. The mass center of the disk is located at a distance of 0.25 m from point O. Thus,

$$m_d = \rho_d V_d = 8000 \text{ kg/m}^3[\pi(0.25 \text{ m})^2(0.01 \text{ m})] = 15.71 \text{ kg}$$
$$(I_O)_d = \frac{1}{2}m_d r_d^2 + m_d d^2$$
$$= \frac{1}{2}(15.71 \text{ kg})(0.25 \text{ m})^2 + (15.71 \text{ kg})(0.25 \text{ m})^2$$
$$= 1.473 \text{ kg} \cdot \text{m}^2$$

Hole. For the 125-mm-radius disk (hole), we have

$$m_h = \rho_h V_h = 8000 \text{ kg/m}^3[\pi(0.125 \text{ m})^2(0.01 \text{ m})] = 3.93 \text{ kg}$$
$$(I_O)_h = \frac{1}{2}m_h r_h^2 + m_h d^2$$
$$= \frac{1}{2}(3.93 \text{ kg})(0.125 \text{ m})^2 + (3.93 \text{ kg})(0.25 \text{ m})^2$$
$$= 0.276 \text{ kg} \cdot \text{m}^2$$

The moment of inertia of the plate about point O is therefore

$$I_O = (I_O)_d - (I_O)_h$$
$$= 1.473 - 0.276$$
$$= 1.197 \text{ kg} \cdot \text{m}^2 \qquad\qquad \textit{Ans.}$$

484

Example 10–14

The pendulum consists of two thin rods each having a weight of 10 lb and suspended from point O as shown in Fig. 10–28. Compute the pendulum's moment of inertia about an axis passing through (a) the pin at O, and (b) the mass center G of the pendulum.

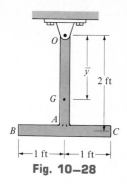

Fig. 10–28

Solution

Part (a). Using the table on the inside back cover, the moment of inertia of rod OA about an axis perpendicular to the page and passing through the end point O of the rod is $I_O = \frac{1}{3}ml^2$. Hence,

$$(I_O)_{OA} = \frac{1}{3}ml^2 = \frac{1}{3}\left(\frac{10}{32.2}\right)(2)^2 = 0.414 \text{ slug} \cdot \text{ft}^2$$

The same value may be computed using $I_G = \frac{1}{12}ml^2$ and the parallel-axis theorem; i.e.,

$$(I_O)_{OA} = \frac{1}{12}ml^2 + md^2 = \frac{1}{12}\left(\frac{10}{32.2}\right)(2)^2 + \frac{10}{32.2}(1)^2$$

$$= 0.414 \text{ slug} \cdot \text{ft}^2$$

For rod BC we have

$$(I_O)_{BC} = \frac{1}{12}ml^2 + md^2 = \frac{1}{12}\left(\frac{10}{32.2}\right)(2)^2 + \frac{10}{32.2}(2)^2$$

$$= 1.346 \text{ slug} \cdot \text{ft}^2$$

The moment of inertia of the pendulum is therefore

$$I_O = 0.414 + 1.346 = 1.76 \text{ slug} \cdot \text{ft}^2 \qquad \textit{Ans.}$$

Part (b). The mass center G will be located relative to the pin at O. Assuming this distance to be $\bar{y}$, Fig. 10–28, and using the formula for determining the mass center, we have

$$\bar{y} = \frac{\Sigma \tilde{y}m}{\Sigma m} = \frac{1\left(\dfrac{10}{32.2}\right) + 2\left(\dfrac{10}{32.2}\right)}{\left(\dfrac{10}{32.2}\right) + \left(\dfrac{10}{32.2}\right)} = 1.50 \text{ ft}$$

The moment of inertia I_G may be computed in the same manner as I_O, which requires successive applications of the parallel-axis theorem in order to transfer the moments of inertia of rods OA and BC to G. A more direct solution, however, involves applying the parallel-axis theorem using the result for I_O, i.e.,

$$I_O = I_G + md^2; \qquad 1.76 = I_G + \left(\frac{20}{32.2}\right)(1.50)^2$$

$$I_G = 0.362 \text{ slug} \cdot \text{ft}^2 \qquad \textit{Ans.}$$

Problems

*10–60. Determine the moment of inertia I_y for the slender rod. The rod's density ρ and cross-sectional area A are constant. Express the result in terms of the rod's total mass m.

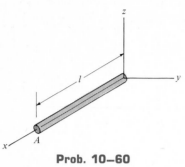

Prob. 10–60

10–61. The right circular cone is formed by revolving the shaded area around the x axis. Determine the moment of inertia I_x and express the result in terms of the total mass m of the cone. The cone has a constant density ρ.

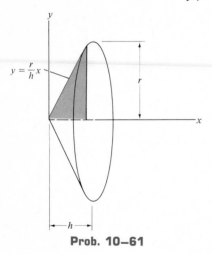

Prob. 10–61

10–62. The solid is formed by revolving the shaded area around the x axis. Determine the radius of gyration k_x. The mass density of the material is $\rho = 5 \text{ Mg/m}^3$.

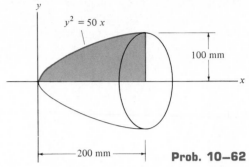

Prob. 10–62

10–63. An ellipsoid is formed by rotating the shaded area about the x axis. Determine the moment of inertia of this body with respect to the x axis and express the result in terms of the mass m of the solid. The density of the material, ρ, is constant.

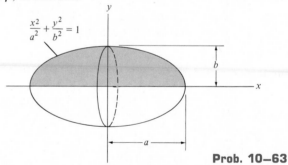

Prob. 10–63

*10–64. The sphere is formed by revolving the shaded area around the x axis. Determine the moment of inertia I_x and express the result in terms of the total mass m of the sphere. The sphere has a constant density ρ.

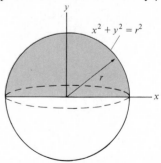

Prob. 10–64

10–65. A semiellipsoid is formed by rotating the shaded area about the x axis. Determine the moment of inertia of this body with respect to the x axis and express the result in terms of the mass m of the solid. The density of the material, ρ, is constant.

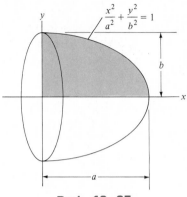

$$\frac{x^2}{a^2} + \frac{y^2}{b^2} = 1$$

Prob. 10–65

10–66. Determine the moment of inertia of the homogeneous triangular prism with respect to the y axis. Express the result in terms of the mass m of the prism. *Hint:* For integration, use thin plate elements parallel to the x-y plane and having a thickness of dz.

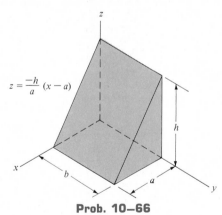

$$z = \frac{-h}{a}(x - a)$$

Prob. 10–66

10–67. The concrete solid is formed by rotating the shaded area about the y axis. Determine the moment of inertia I_y. The specific weight of concrete is $\gamma = 150 \text{ lb/ft}^3$.

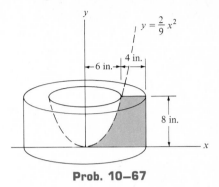

$$y = \frac{2}{9}x^2$$

6 in. · 4 in. · 8 in.

Prob. 10–67

*10–68. The wheel consists of a thin ring having a mass of 10 kg and four spokes made from slender rods and each having a mass of 2 kg. Determine the wheel's moment of inertia about an axis perpendicular to the page and passing through point A.

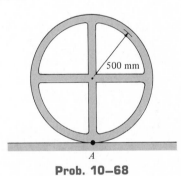

500 mm

A

Prob. 10–68

10–69. The pendulum consists of a plate having a weight of 12 lb and a slender rod having a weight of 4 lb. Determine the radius of gyration of the pendulum about an axis perpendicular to the page and passing through point O.

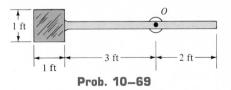

1 ft · 1 ft · 3 ft · 2 ft · O

Prob. 10–69

487

10–70. The pendulum consists of the 3-kg slender rod and the 5-kg thin plate. Determine the location $\bar{y}$ of the center of mass G of the pendulum, then calculate the moment of inertia of the pendulum about an axis perpendicular to the page and passing through G.

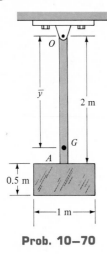

Prob. 10–70

10–71. Determine the moment of inertia of the thin plate about an axis perpendicular to the page and passing through the pin at O. The plate has a hole in its center. Its thickness is 50 mm and the material has a density of $\rho = 50$ kg/m³.

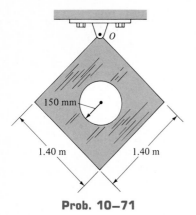

Prob. 10–71

*10–72. Determine the location $\bar{y}$ of the center of mass G of the assembly and then calculate the moment of inertia about an axis perpendicular to the page and passing through G. The block has a mass of 3 kg and the mass of the semicylinder is 5 kg.

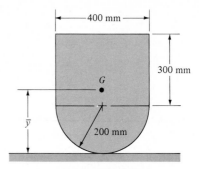

Prob. 10–72

10–73. Determine the moment of inertia of the wheel about the z axis, which passes through the center point O. The material has a specific gravity of $\gamma = 90$ lb/ft³.

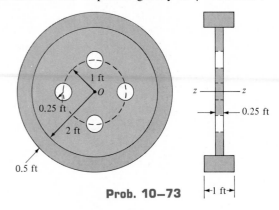

Prob. 10–73

10–74. Determine the moment of inertia I_z of the frustum of the cone which has a conical depression. The material has a density of 200 kg/m³.

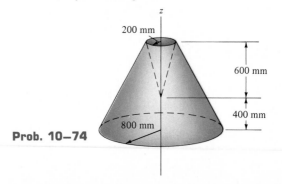

Prob. 10–74

Review Problems

10–75. Determine the moments of inertia I_x and I_y of the shaded area.

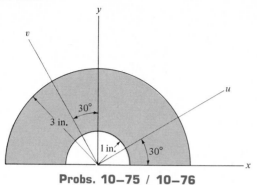

Probs. 10–75 / 10–76

*10–76. Determine the moments of inertia I_u and I_v and the product of inertia I_{uv} for the shaded area.

10–77. Determine the moment of inertia of the beam's cross-sectional area with respect to the $\bar{x}\bar{x}$ axis passing through the centroid C. $\bar{y} = 0.875$ in.

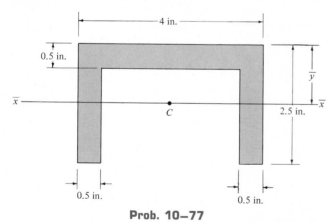

Prob. 10–77

10–78. Locate the center of mass G of the baseball bat by determining $\bar{y}$. Then, compute the moment of inertia of the bat about an axis which passes through G and is perpendicular to the plane of the page. For the calculation, consider the bat to be composed of a truncated cone and cylinder. Neglect the size of the lip at A. The density of wood is $\rho_w = 750$ kg/m^3.

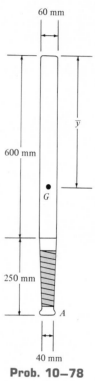

Prob. 10–78

10–79. Determine the directions of the principal axes with origin located at point O, and the principal moments of inertia for the rectangular area about these axes.

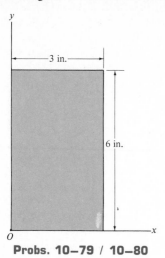

Probs. 10–79 / 10–80

*10–80. Determine the principal moments of inertia using Mohr's circle.

10–81. Determine the moment of inertia of the homogeneous pyramid of mass m with respect to the z axis. The density of the material is ρ. *Suggestion:* Use a rectangular plate element having a volume of $dV = (2x)(2y)\ dz$.

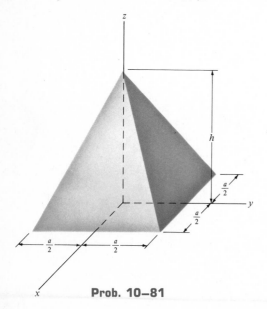

Prob. 10–81

10–82. Determine the moments of inertia I_x and I_y of the shaded area.

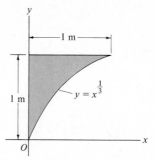

Probs. 10–82 / 10–83 / 10–84 / 10–85

10–83. Determine the product of inertia I_{xy} of the shaded area with respect to the x and y axes.

*10–84. Determine the directions of the principal axes with origin located at point O, and the principal moments of inertia for the area about these axes. $I_x = 0.1667$ in^4, $I_y = 0.0333$ in^4, $I_{xy} = 0.0625$ in^4.

10–85. Determine the principal moments of inertia for the area using Mohr's circle. $I_x = 0.1667$ in^4, $I_y = 0.0333$ in^4, $I_{xy} = 0.0625$ in^4.

Virtual Work

In the previous chapters, statics problems have been solved using the equations of equilibrium. It is also possible to solve these problems using the principle of virtual work or the method of potential energy. These methods have been found to be very useful for determining the equilibrium configuration or position of a series of connected rigid bodies. Although application requires more mathematical sophistication, using calculus, than the conventional vector approach, using the equations of equilibrium, it will be shown that, once the equation of virtual work or the potential-energy function is established, the solution may be obtained *directly*, without having to dismember the system to obtain relationships between forces occurring at the connections. Furthermore, by using the potential-energy method, we will be able to investigate the "type" of equilibrium or the stability of the configuration.

Definition of Work and Virtual Work   11.1

Work of a Force. In mechanics a force **F** does work only when it undergoes a displacement in the direction of the force. For example, consider the force **F** in Fig. 11–1, which has a location on the path s specified by the position vector **r.** If the force moves along the path to a new position $\mathbf{r}' = \mathbf{r} + d\mathbf{r}$, the displacement is then $d\mathbf{r}$ and therefore the work dU is a *scalar quantity,* defined by the dot product

$$dU = \mathbf{F} \cdot d\mathbf{r}$$

The magnitude of $d\mathbf{r}$ is represented by ds, the differential arc segment along the path. If the angle between the tails of $d\mathbf{r}$ and **F** is θ, Fig. 11–1, then by

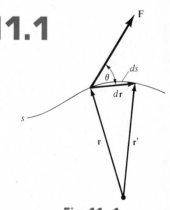

Fig. 11–1

definition of the dot product, the above equation may also be written as

$$dU = F \, ds \cos \theta$$

Work as expressed by this equation may be interpreted in one of two ways: either as the product of **F** and the component of displacement in the direction of the force, i.e., $ds \cos \theta$, or as the product of ds and the component of force in the direction of displacement, i.e., $F \cos \theta$. Note that if $0° \leqslant \theta < 90°$, then the vectors with magnitudes of $F \cos \theta$ and ds have the *same sense,* so that the work is *positive;* whereas if $90° < \theta \leqslant 180°$, these vectors have an *opposite sense,* and therefore the work is *negative.* Also, $dU = 0$ if the force is *perpendicular* to displacement, since $\cos 90° = 0$, or if the force is applied at a *fixed point,* in which case the displacement $ds = 0$.

The basic unit for work combines the units of force and displacement. In the SI system a *joule* (J) is equivalent to the work done by a force of 1 newton which moves 1 meter in the direction of the force ($1 \, J = 1 \, N \cdot m$). In the FPS system work is defined in units of ft $\cdot$ lb. The moment of a force has the same combination of units; however, the concepts of moment and work are in no way related. A moment is a vector quantity, whereas work is a scalar.

Work of a Couple. The two forces of a couple do work when the couple *rotates* about an axis perpendicular to the plane of the couple. To show this, consider the body in Fig. 11–2*a*, which is subjected to a couple having a magnitude of $M = Fr$. Any general differential displacement of the body can be considered as a combination of a translation and rotation. When the body *translates* such that the *component of displacement* along the line of action of each force is ds_t, clearly the "positive" work of one force ($F ds_t$) *cancels* the "negative" work of the other ($-F ds_t$), Fig. 11–2*b*. Consider now a differential *rotation* $d\theta$ of the body about an axis perpendicular to the plane of the couple, which intersects the plane at the midpoint O, Fig. 11–2*c*. (For the derivation, any other point in the plane may also be considered.) As shown, each force undergoes a displacement $ds_\theta = (r/2) \, d\theta$ in the direction of the force; hence, the total work done is

$$dU = F\left(\frac{r}{2} \, d\theta\right) + F\left(\frac{r}{2} \, d\theta\right) = (Fr) \, d\theta$$

or

$$dU = M \, d\theta$$

The resultant work is *positive* when the sense of **M** is the *same* as that of $d\boldsymbol\theta$, and negative when they have an opposite sense. As in the case of the moment vector, the *direction and sense* of $d\boldsymbol\theta$ are defined by the right-hand rule, where the fingers of the right hand follow the rotation or "curl" and the thumb indicates the direction of $d\boldsymbol\theta$. Hence, the line of action of $d\boldsymbol\theta$ must be *parallel* to the line of action of **M.** This is always the case if movement of the body occurs in the *same plane.* If the body rotates in space, however, the *component* of $d\boldsymbol\theta$ in the direction of **M** is required; i.e., the work done is defined by the dot product, $dU = \mathbf{M} \cdot d\boldsymbol\theta.$

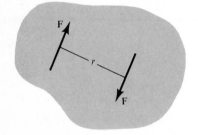

(a)

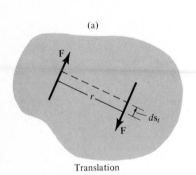

Translation

(b)

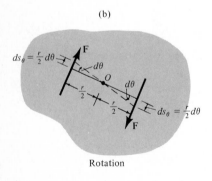

Rotation

(c)

Fig. 11–2

Virtual Work. The definitions of the work of a force and a couple have been presented in terms of *actual movements* expressed by differential displacements having magnitudes of ds and $d\theta$. Consider now an *imaginary* or *virtual movement,* which indicates a displacement or rotation that is *assumed* and *does not actually exist.* These movements are first-order differential quantities and will be denoted by the symbols δs and $\delta\theta$ (delta s and delta θ), respectively. The *virtual work* done by a force undergoing a virtual displacement is

$$\delta U = F \cos\theta\, \delta s \tag{11-1}$$

Similarly, when a couple undergoes a virtual rotation in the plane of the couple forces, the *virtual work* is

$$\delta U = M\, \delta\theta \tag{11-2}$$

Principle of Virtual Work for a Particle and a Rigid Body   11.2

If a particle is in equilibrium, the resultant of a force system acting on it must be equal to zero. Hence, if the particle undergoes an imaginary or virtual displacement in the x, y, or z direction, the virtual work (δU) done by the force system must be equal to zero since the components $\Sigma F_x = 0$, $\Sigma F_y = 0$, $\Sigma F_z = 0$. Alternatively, this may be expressed as

$$\delta U = 0$$

For example, if the particle in Fig. 11–3 is given a virtual displacement $\delta\mathbf{x}$, only the x components of the forces acting on the particle do work. (No work is done by the y and z components since they are perpendicular to the displacement.) The virtual work equation is therefore

$$\delta U = 0; \qquad F_{1x}\, \delta x + F_{2x}\, \delta x + F_{3x}\, \delta x = 0$$

Factoring out δx, which is common to every term, yields

$$(F_{1x} + F_{2x} + F_{3x})\, \delta x = 0$$

Since $\delta x \neq 0$, this equation is satisfied only if the sum of the force components in the x direction is equal to zero, i.e., $\Sigma F_x = 0$. Two other virtual work equations can be written by assuming virtual displacements $\delta\mathbf{y}$ and $\delta\mathbf{z}$ in the y and z directions, respectively. Doing this, however, amounts to satisfying the equilibrium equations $\Sigma F_y = 0$ and $\Sigma F_z = 0$ for the particle.

In a similar manner, a rigid body that is subjected to a coplanar force system will be in equilibrium provided $\Sigma F_x = 0$, $\Sigma F_y = 0$, and $\Sigma M_O = 0$. We can also write a set of three virtual work equations for the body, each of which requires $\delta U = 0$. If these equations involve separate virtual translations in the x and y directions and a virtual rotation about an axis perpendicular to the x-y plane and passing through point O, then it can be shown that they will corre-

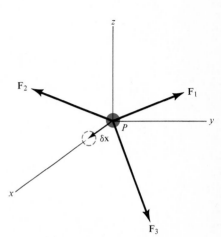

Fig. 11–3

spond to the above mentioned three equilibrium equations. When writing these equations, it is *not necessary* to include the work done by the *internal forces* acting on the body since a rigid body *does not deform* when subjected to an external loading, and furthermore, when the body moves through a virtual displacement, the internal forces occur in equal but opposite collinear pairs, so that the corresponding work done by each pair of forces *cancels*.

As in the case of a particle, however, no added advantage would be gained by solving rigid-body equilibrium problems using the principle of virtual work. This is because for each application of the virtual-work equation the virtual displacement, common to every term, factors out, leaving an equation that could have been obtained in a more *direct manner* by applying the equations of equilibrium.

11.3 Principle of Virtual Work for a System of Connected Rigid Bodies

The method of virtual work is most suitable for solving equilibrium problems that involve a system of several *connected* rigid bodies such as shown in Fig. 11–4. For each of these systems, the number of *independent virtual-work equations* that can be written depends upon the number of *independent virtual displacements* that can be made by the system.

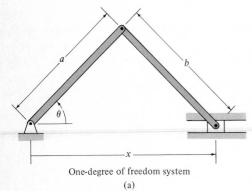

One-degree of freedom system
(a)

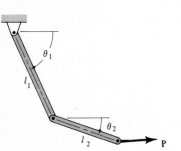

Two-degree of freedom system
(b)

Fig. 11–4

Degrees of Freedom. In general, *for a system of connected bodies, the number of independent virtual displacements equals the minimum number of independent coordinates q needed to specify completely the location of all members of the system.* Each independent coordinate gives the system a *degree of freedom* that must be consistent with the constraining action of the supports. Thus, an *n*-degree-of-freedom system requires *n* independent coordinates q_n to specify the location of all its members with respect to a fixed reference point. For example, the link and sliding-block arrangement shown in Fig. 11–4a is an example of a one-degree-of-freedom system. The independent coordinate $q = \theta$ may be used to specify the location of the two connecting links and the block. The coordinate x could also be used as the independent coordinate. However, since the block is constrained to move within the slot, x is not independent of θ; rather it can be related to θ using the cosine law, $b^2 = a^2 + x^2 - 2ax \cos \theta$. The double-link arrangement, shown in Fig. 11–4b, is an example of a two-degree-of-freedom system. To specify the location of each link, the coordinate angles θ_1 and θ_2 must be known, since a rotation of one link is independent of a rotation of the other.

Principle of Virtual Work. The principle of virtual work for a system of rigid bodies whose connections are *frictionless* may be stated as follows: *A system of connected rigid bodies is in equilibrium provided the virtual work done by all the external forces and couples acting on the system is zero for each independent virtual displacement of the system.* Mathematically, this may be expressed as

$$\delta U = 0 \qquad\qquad (11\text{-}3)$$

where δU represents the virtual work of all the external forces (and couples) acting on the system during any independent virtual displacement.

It has already been pointed out that if a system has n degrees of freedom it takes n independent coordinates q_n to completely specify the location of the system. Hence, for the system it is possible to write n independent virtual-work equations; one for each independent coordinate, while the remaining $n - 1$ coordinates are held *fixed*.

PROCEDURE FOR ANALYSIS

The following procedure provides a method for applying the equation of virtual work to solve problems involving a system of frictionless connected rigid bodies having a single degree of freedom.

Free-Body Diagram. Draw the free-body diagram of the entire system of connected bodies and define the *independent coordinate q*. Sketch the "deflected position" of the system on the free-body diagram when the system undergoes a *positive* virtual displacement δq. From this, specify the "active" forces and couples, that is, those that do work.

Virtual Displacements. Indicate *position coordinates s_i*, measured from a *fixed point* on the free-body diagram to each of the i number of "active" forces and couples. Each coordinate axis should be in the *same direction* as the line of action of the "active" force to which it is directed.

Relate each of the position coordinates s_i to the independent coordinate q; then *differentiate* these expressions in order to express the virtual displacements δs_i in terms of δq.

Virtual-Work Equation. Write the *virtual-work equation* for the system assuming that, whether possible or not, all the position coordinates s_i undergo *positive* virtual displacements δs_i. Using the relations for δs_i, express the work of *each* "active" force and couple in the equation in terms of the single independent virtual displacement δq. By factoring out this common displacement, one is left with an equation that generally can be solved for an unknown force, couple, or equilibrium position.

If the system contains n degrees of freedom, n independent coordinates q_n must be specified. In this case, follow the above procedure and let *only one* of the independent coordinates undergo a virtual displacement δq_n, while the remaining $n - 1$ coordinates are held fixed. In this way, n virtual-work equations can be written, one for each independent coordinate.

The following examples should help to clarify application of this procedure.

Example 11–1

Determine the angle θ for equilibrium of the two-member linkage shown in Fig. 11–5a. Each member has a mass of 10 kg.

(a)

Solution

Free-Body Diagram. The system has only one degree of freedom, since the location of both links may be specified by the single independent coordinate $(q =)\ \theta$. As shown on the free-body diagram in Fig. 11–5b, when θ undergoes a *positive* (clockwise) virtual rotation $\delta\theta$, only the active forces, **F** and the two 98.1-N weights, do work. (The reactive forces $\mathbf{D}_x$ and $\mathbf{D}_y$ are fixed, and $\mathbf{B}_y$ does not move along its line of action.)

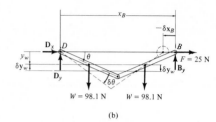

(b)

Fig. 11–5

Virtual Displacements. If the origin of coordinates is established at the *fixed* pin support D, the location of **F** and **W** may be specified by the position coordinates x_B and y_w, as shown in the figure. In order to compute the work, note that these coordinates are in the *same direction* as the lines of action of their associated forces.

Expressing the position coordinates in terms of the independent coordinate θ and taking the derivatives yields

$$x_B = 2(1\ \cos\ \theta)\ \text{m} \qquad \delta x_B = -2\ \sin\ \theta\ \delta\theta\ \text{m} \qquad (1)$$

$$y_w = \tfrac{1}{2}(1\ \sin\ \theta)\ \text{m} \qquad \delta y_w = 0.5\ \cos\ \theta\ \delta\theta\ \text{m} \qquad (2)$$

It is seen by the *signs* of these equations, and indicated in Fig. 11–5b, that an *increase* in θ (i.e., $\delta\theta$) causes a *decrease* in x_B and an *increase* in y_w.

Virtual-Work Equation. If the virtual displacements δx_B and δy_w were *both positive*, then the forces **W** and **F** would do positive work since the forces and their corresponding displacements would be in the same sense. Hence, the virtual-work equation for the displacement $\delta\theta$ is

$$\delta U = 0; \qquad W\ \delta y_w + W\ \delta y_w + F\ \delta x_B = 0 \qquad (3)$$

Substituting Eqs. (1) and (2) into Eq. (3) in order to relate the virtual displacements to the common virtual displacement $\delta\theta$ yields

$$98.1(0.5\ \cos\ \theta\ \delta\theta) + 98.1(0.5\ \cos\ \theta\ \delta\theta) + 25(-2\ \sin\ \theta\ \delta\theta) = 0$$

Notice that the "negative work" done by **F** (force in the opposite sense to displacement) has been *accounted for* in the above equation by the "negative sign" of Eq. (1). Factoring out the *common displacement* $\delta\theta$ and solving for θ, noting that $\delta\theta \neq 0$, yields

$$(98.1\ \cos\ \theta - 50\ \sin\ \theta)\ \delta\theta = 0$$

$$\theta = \tan^{-1}\frac{98.1}{50} = 63.0° \qquad\qquad Ans.$$

If this problem had been solved using the equations of equilibrium, it would have been necessary to dismember the links and apply three scalar equations to *each* link. The principle of virtual work, by means of calculus, has eliminated this task so that the answer is obtained in a very direct manner.

Example 11–2

Using the principle of virtual work, determine the angle θ required to maintain equilibrium of the mechanism shown in Fig. 11–6a. Neglect the weight of the links. The spring is unstretched when $\theta = 0°$ and it maintains a horizontal position due to the roller.

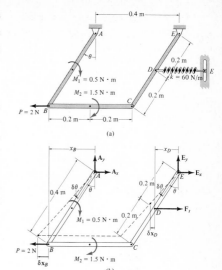

Fig. 11–6

Solution

Free-Body Diagram. The mechanism has one degree of freedom, and therefore the location of each member may be specified using the independent coordinate θ. When θ undergoes a *positive* virtual displacement $\delta\theta$, as shown on the free-body diagram in Fig. 11–6b, links AB and EC rotate by the same amount since they have the same length, and link BC only translates. Since a couple does work *only* when it rotates, the work done by M_2 is zero. The reactive forces at A and E do no work since the supports do not translate.

Virtual Displacements. The position coordinates x_B and x_D are *parallel* to the lines of action of P and F and locate these forces with respect to the *fixed points* A and E. From Fig. 11–6b,

$$x_B = 0.4 \sin\theta \text{ m}$$
$$x_D = 0.2 \sin\theta \text{ m}$$

Thus,

$$\delta x_B = 0.4 \cos\theta \, \delta\theta \text{ m}$$
$$\delta x_D = 0.2 \cos\theta \, \delta\theta \text{ m}$$

Virtual-Work Equation. Applying the equation of virtual work, noting that F_s is opposite to *positive* δx_D displacement and hence does negative work, we obtain

$$\delta U = 0; \qquad M_1 \, \delta\theta + P \, \delta x_B - F_s \, \delta x_D = 0$$

Relating each of the virtual displacements to the *common* virtual displacement $\delta\theta$ yields

$$0.5 \, \delta\theta + 2(0.4 \cos\theta \, \delta\theta) - F_s(0.2 \cos\theta \, \delta\theta) = 0$$
$$(0.5 + 0.8 \cos\theta - 0.2 F_s \cos\theta) \, \delta\theta = 0 \qquad (1)$$

For the arbitrary angle θ, the spring is stretched a distance of $x_D = (0.2 \sin\theta)$ m; and therefore, $F_s = 60$ N/m$(0.2 \sin\theta)$ m $= (12 \sin\theta)$ N. Substituting into Eq. (1) and noting that $\delta\theta \neq 0$, we have

$$0.5 + 0.8 \cos\theta - 0.2(12 \sin\theta) \cos\theta = 0$$

Since $\sin 2\theta = 2 \sin\theta \cos\theta$, then

$$1 = 2.4 \sin 2\theta - 1.6 \cos\theta$$

Solving for θ by trial and error yields

$$\theta = 36.3° \qquad\qquad Ans.$$

Example 11-3

Using the principle of virtual work, determine the force that the spring must exert in order to hold the mechanism shown in Fig. 11–7a in equilibrium when $\theta = 45°$. Neglect the weight of the members.

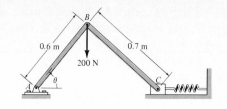

(a)

Solution

Free-Body Diagram. As shown on the free-body diagram, Fig. 11–7b, the system has one degree of freedom, defined by the independent coordinate θ. When θ undergoes a *positive* virtual displacement $\delta\theta$, only $\mathbf{F}_s$ and the 200-N force do work.

Virtual Displacements. Forces $\mathbf{F}_s$ and 200 N are located from the fixed origin A using position coordinates y_B and x_C. From Fig. 11–7b, x_C can be related to θ by the "law of cosines." Hence,

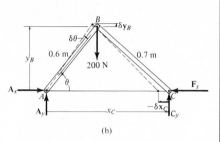

(b)

Fig. 11–7

$$(0.7)^2 = (0.6)^2 + x_C^2 - 2(0.6)x_C \cos \theta \qquad (1)$$

$$0 = 0 + 2x_C \, \delta x_C - 1.2 \, \delta x_C \cos \theta + 1.2 x_C \sin \theta \, \delta\theta$$

$$\delta x_C = \frac{1.2 x_C \sin \theta}{1.2 \cos \theta - 2x_C} \delta\theta \qquad (2)$$

Also,

$$y_B = 0.6 \sin \theta$$

$$\delta y_B = 0.6 \cos \theta \, \delta\theta \qquad (3)$$

Virtual-Work Equation. When y_B and x_C undergo *positive* virtual displacements δy_B and δx_C, $\mathbf{F}_s$ and 200 N do *negative work,* since they both act in the opposite sense to δy_B and δx_C. Hence,

$$\delta U = 0; \qquad -200 \, \delta y_B - F_s \, \delta x_C = 0$$

Substituting Eqs. (2) and (3) into this equation, factoring out $\delta\theta$, and solving for F_s yields

$$-200(0.6 \cos \theta \, \delta\theta) - F_s \frac{1.2 x_C \sin \theta}{1.2 \cos \theta - 2x_C} \delta\theta = 0$$

$$F_s = \frac{-120 \cos \theta (1.2 \cos \theta - 2x_C)}{1.2 x_C \sin \theta}$$

At the required equilibrium position $\theta = 45°$, the corresponding value of x_C can be found by using Eq. (1), in which case

$$x_C^2 - 1.2 \cos 45° \, x_C - 0.13 = 0$$

Solving for the positive root yields

$$x_C = 0.981 \text{ m}$$

Thus,

$$F_s = \frac{-120 \cos 45° \, [1.2 \cos 45° - 2(0.981)]}{1.2(0.981) \sin 45°} = 113.5 \text{ N} \qquad Ans.$$

Example 11–4

Using the principle of virtual work, determine the equilibrium position of the two-bar linkage shown in Fig. 11–8a. Neglect the weight of the links.

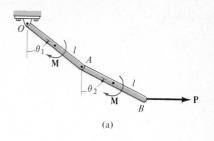

Solution

The system has two degrees of freedom, since the *independent coordinates* θ_1 and θ_2 must be known to locate the position of both links. The position coordinate x_B, measured from the fixed point O, is used to specify the location of **P**, Fig. 11–8b and c.

If θ_1 is held *fixed* and θ_2 varies by an amount $\delta\theta_2$, as shown in Fig. 11–8b, the virtual-work equation becomes

$$[\delta U = 0]_{\theta_2}; \qquad P(\delta x_B)_{\theta_2} - M\,\delta\theta_2 = 0 \qquad (1)$$

where P and M represent the magnitudes of the applied force and couple acting on link AB.

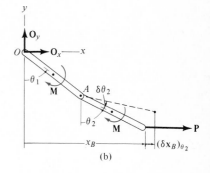

When θ_2 is held *fixed* and θ_1 varies by an amount $\delta\theta_1$, as shown in Fig. 11–8c, the virtual-work equation becomes

$$[\delta U = 0]_{\theta_1}; \qquad P(\delta x_B)_{\theta_1} - M\,\delta\theta_1 - M\,\delta\theta_1 = 0 \qquad (2)$$

The *position coordinate* x_B may be related to the independent coordinates θ_1 and θ_2 by the equation

$$x_B = l \sin\theta_1 + l \sin\theta_2 \qquad (3)$$

To obtain the variation of δx_B in terms of $\delta\theta_2$, it is necessary to take the *partial derivative* of x_B with respect to θ_2 since x_B is a function of both θ_1 and θ_2. Hence,

$$\frac{\partial x_B}{\partial\theta_2} = l\cos\theta_2 \qquad (\delta x_B)_{\theta_2} = l\cos\theta_2\,\delta\theta_2$$

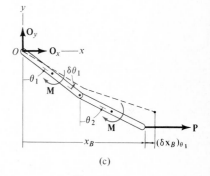

Substituting into Eq. (1), we have

$$(Pl\cos\theta_2 - M)\,\delta\theta_2 = 0$$

Since $\delta\theta_2 \neq 0$, then

$$\theta_2 = \cos^{-1}\left(\frac{M}{Pl}\right) \qquad \textit{Ans.}$$

Fig. 11–8

Using Eq. (3) to obtain the variation of x_B with θ_1 yields

$$\frac{\partial x_B}{\partial\theta_1} = l\cos\theta_1 \qquad (\delta x_B)_{\theta_1} = l\cos\theta_1\,\delta\theta_1$$

Substituting into Eq. (2), we have

$$(Pl\cos\theta_1 - 2M)\,\delta\theta_1 = 0$$

Since $\delta\theta_1 \neq 0$, then

$$\theta_1 = \cos^{-1}\left(\frac{2M}{Pl}\right) \qquad \textit{Ans.}$$

Problems

11–1. The thin rod of weight W rests against the smooth wall and floor. Determine the magnitude of force **P** needed to hold it in equilibrium for a given angle θ.

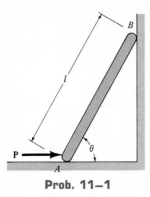

Prob. 11–1

■11–2. The spring has an unstretched length of 0.3 m. Determine the angle θ for equilibrium if the uniform links each have a mass of 5 kg.

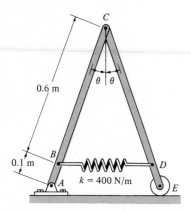

Prob. 11–2

11–3. If each of the three links of the mechanism has a weight of 20 lb, determine the angle θ for equilibrium. The spring, which always remains horizontal, is unstretched when $\theta = 0°$.

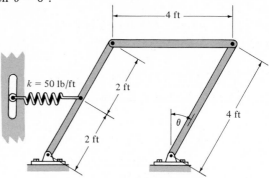

Prob. 11–3

*11–4. Determine the force **F** needed to lift the block having a weight of 100 lb. *Hint:* First show that the coordinates s_A and s_B are related to the *constant* vertical length l of the cord by the equation $l = s_A + 2s_B$.

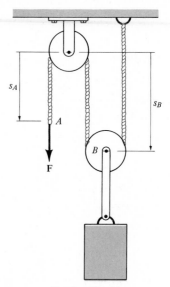

Prob. 11–4

11–5. The crankshaft is subjected to a torque of $M = 50$ N $\cdot$ m. Determine the horizontal compressive force **F** applied to the piston for equilibrium when $\theta = 60°$.

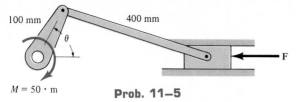

100 mm   400 mm

θ

$M = 50 \cdot$ m   **F**

Prob. 11–5

11–6. The disk has a weight of 10 lb and is subjected to a tangential force $F = 5$ lb and a couple moment $M = 8$ lb $\cdot$ ft. Determine the disk's rotation if the end of the spring wraps around the periphery of the disk as the disk turns. The spring is originally unstretched.

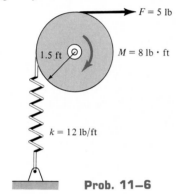

$F = 5$ lb

1.5 ft $M = 8$ lb $\cdot$ ft

$k = 12$ lb/ft

Prob. 11–6

11–7. If a force $P = 40$ N is applied perpendicular to the handle of the toggle press, determine the vertical compressive force developed at C; $\theta = 30°$.

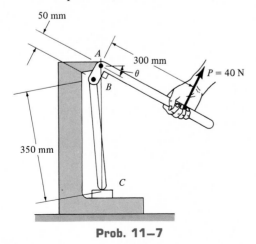

50 mm

A

300 mm

θ

B $P = 40$ N

350 mm

C

Prob. 11–7

*11–8.** The pin-connected mechanism is constrained at A by a pin and at B by a smooth sliding block. If $P = 2$ lb, determine the angle θ for equilibrium.

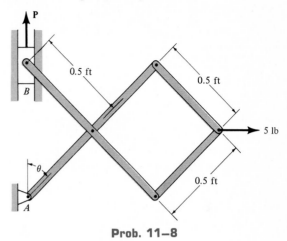

P

0.5 ft   0.5 ft

B

5 lb

θ

0.5 ft

A

Prob. 11–8

■**11–9.** Each member of the pin-connected mechanism has a mass of 8 kg. If the spring is unstretched when $\theta = 0°$, determine the angle θ for equilibrium.

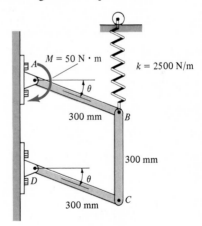

$M = 50$ N $\cdot$ m

$k = 2500$ N/m

A

θ

300 mm B

300 mm

D θ

300 mm C

Prob. 11–9

11–10. The spring has a unstretched length of 0.5 ft. Determine the angle θ for equilibrium. Neglect the weight of the members.

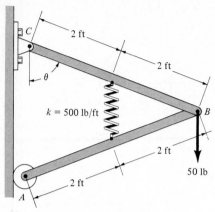

Probs. 11–10 / 11–11

11–11. The spring has an unstretched length of 0.5 ft. Determine the angle θ for equilibrium. Each member weighs 15 lb.

*11–12. The uniform bar AB weighs 10 lb. If the attached spring is unstretched when $\theta = 90°$, determine the angle θ for equilibrium. Note that the spring always remains in the vertical position since the top end is attached to a roller which moves freely along the horizontal guide.

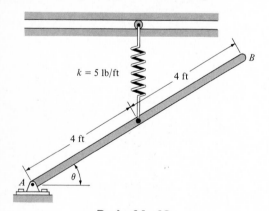

Prob. 11–12

■11–13. Determine the angle θ for equilibrium of the uniform 7-kg bar AC. Due to the roller guide at B, the spring remains horizontal and is unstretched when $\theta = 0°$.

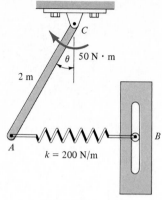

Prob. 11–13

■11–14. The 4-ft members of the mechanism are pin-connected at their centers. If vertical forces act at C and E as shown, determine the angle θ for equilibrium. The spring is unstretched when $\theta = 45°$. Neglect the weight of the members.

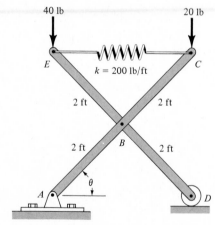

Prob. 11–14

11–15. The chain puller is used to draw two ends of a chain together in order to attach the "master link." The device is operated by turning the screw S, which pushes the bar AB downward, thereby drawing the tips C and D towards one another. If the sliding contacts at A and B are smooth, determine the force $\mathbf{F}$ maintained by the screw at E which is required to develop a drawing tension of 120 lb in the chains.

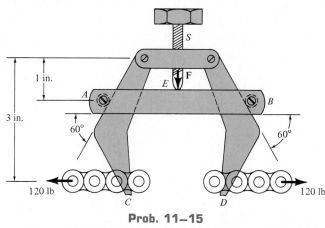

Prob. 11–15

*11–16. A horizontal force acts on the end of the link as shown. Determine the angles θ_1 and θ_2 for equilibrium of the two links. Each link is uniform and has a mass m.

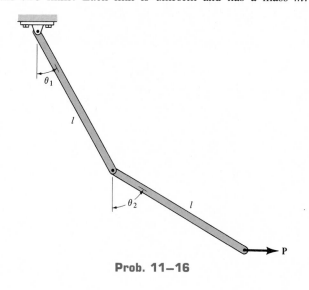

Prob. 11–16

11–17. Rods AB and BC have a center of gravity located at their midpoints. If all contacting surfaces are smooth and BC has a mass of 100 kg, determine the appropriate mass of AB required for equilibrium.

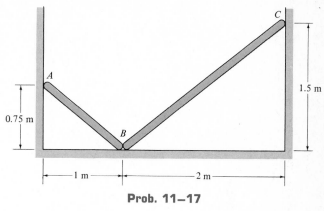

Prob. 11–17

■11–18. A 3-lb disk is attached to the end of rod ABC. If the rod is supported by a smooth slider block at C and rod BD, determine the angle θ for equilibrium. Neglect the weight of the rods and the slider.

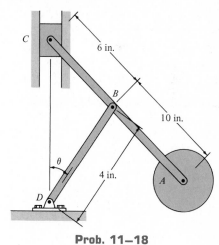

Prob. 11–18

503

11–19. Determine the mass of A and B required to hold the 400-g desk lamp in balance for any angle θ or ϕ. Neglect the weight of the mechanism and the size of the lamp.

Prob. 11–19

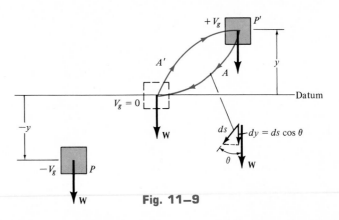

$^\star\mathbf{11.4}$ Conservative Forces

The work done by a force when it undergoes a *differential displacement* has been defined as $dU = F \cos \theta \, ds$. If the force is displaced over a path that has a *finite length s*, the work is determined by integrating over the path, i.e.,

$$U = \int_s F \cos \theta \, ds$$

To evaluate the integral, it is necessary to obtain a relationship between F and the component of displacement $ds \cos \theta$. In some instances, however, the work done by a force will be *independent* of its path and, instead, will depend only upon the initial and final locations of the force along the path. A force that has this property is called a *conservative force*.

Weight. The weight of a body is a conservative force, since the work done by the weight depends *only* on the body's *vertical displacement*. To show this, consider the body in Fig. 11–9, which is initially at P'. If the body is moved

Fig. 11–9

down along the *arbitrary path A* to the dashed position, then, for a given displacement *ds* along the path, the displacement component in the direction of **W** has a magnitude of $dy = ds \cos \theta$, as shown. Since both the force and displacement are in the same direction, the work is positive; hence,

$$U = \int_s W \cos \theta \, ds = \int_0^y W \, dy$$

or

$$U = Wy$$

In a similar manner, the work done by the weight when the body moves up a distance *y* back to *P'*, along the arbitrary path *A'*, is

$$U = -Wy$$

Why is the work negative?

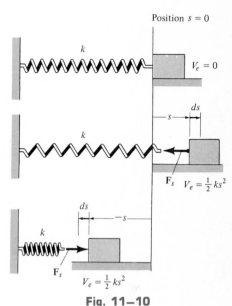

Position $s = 0$

Fig. 11–10

Elastic Spring. The force developed by an elastic spring ($F_s = ks$) is also a conservative force. If the spring is attached to a body and the body is displaced along *any path,* such that it causes the spring to elongate or compress from a position s_1 to a further position s_2, the work is

$$U = \int_{s_1}^{s_2} F_s \, ds = \int_{s_1}^{s_2} (-ks) \, ds$$
$$= -(\tfrac{1}{2}ks_2^2 - \tfrac{1}{2}ks_1^2)$$

For either extension or compression, the work is negative, since the spring exerts a force **F**$_s$ *on the body* that is opposite to the body's displacement δs, Fig. 11–10.

Friction. In contrast to a conservative force, consider the force of *friction* exerted on a moving body by a fixed surface. The work done by the frictional force depends upon the path; the longer the path, the greater the work. Consequently, frictional forces are *nonconservative,* and the work done is dissipated from the body in the form of heat.

Potential Energy 11.5*

When a conservative force acts on a body, it gives the body the capacity to do work. This capacity, measured as *potential energy,* depends upon the location of the body.

Gravitational Potential Energy. If a body is located a distance *y above* a fixed horizontal reference or datum, Fig. 11–9, the weight of the body has *positive* gravitational potential energy V_g since **W** has the capacity of doing positive work when the body is moved back down to the datum. Likewise, if the body is located a distance y *below* the datum, V_g is *negative* since the weight does negative work when the body is moved back up to the datum. At the datum $V_g = 0$.

Measuring y as *positive upward*, the gravitational potential energy of the body's weight **W** is thus

$$V_g = Wy \qquad (11\text{–}4)$$

Elastic Potential Energy. The elastic potential energy V_e that a spring produces on an attached body when the spring is elongated or compressed from an unstretched position ($s = 0$) to a final position s is

$$V_e = \tfrac{1}{2}ks^2 \qquad (11\text{–}5)$$

Here V_e is *always positive*, since in the deformed position the spring has the capacity of doing *positive work* in *returning* the body back to the spring's undeformed position, Fig. 11–10.

Potential Function. In the general case, if a body is subjected to *both* gravitational and elastic forces, the *potential energy or potential function V* of the body can be expressed as the algebraic sum

$$V = V_g + V_e \qquad (11\text{–}6)$$

where measurement of V depends upon the location of the body with respect to a selected datum in accordance with Eqs. 11–4 and 11–5.

In general, if a system of frictionless connected rigid bodies has a *single degree of freedom* such that its position is defined by the independent coordinate q, then the potential energy for the system can be expressed as $V = V(q)$, which is defined by Eq. 11–6. The work done by all the conservative forces acting on the system in moving the system from q_1 to q_2 is measured by the *difference* in V, i.e.,

$$U_{1\text{–}2} = V(q_1) - V(q_2) \qquad (11\text{–}7)$$

For example, the potential function for a system consisting of a block of weight **W** supported by a spring, Fig. 11–11a, can be expressed in terms of its independent coordinate ($q =$) y, measured from a fixed datum located at the unstretched length of the spring, i.e.,

$$\begin{aligned} V &= V_g + V_e \\ &= -Wy + \tfrac{1}{2}ky^2 \end{aligned} \qquad (11\text{–}8)$$

If the block moves from y_1 to a further downward position y_2, then the work of **W** and $\mathbf{F}_s$ is

$$U_{1\text{–}2} = V(y_1) - V(y_2) = -W[y_1 - y_2] + \tfrac{1}{2}ky_1^2 - \tfrac{1}{2}ky_2^2$$

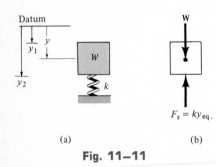

(a)

(b)

Fig. 11–11

Potential-Energy Criterion for Equilibrium 11.6*

System Having One Degree of Freedom. When the displacement of a frictionless connected system is *infinitesimal*, i.e., from q to $q + dq$, Eq. 11–7 becomes

$$dU = V(q) - V(q + dq)$$

or

$$dU = -dV$$

Furthermore, if the independent coordinate undergoes a *virtual displacement* δq, rather than an actual displacement dq, then $\delta U = -\delta V$. For equilibrium, the principle of virtual work requires $\delta U = 0$ and therefore, provided the potential function for the system is known, this also requires that

$$\delta V = \frac{dV}{dq} \delta q = 0 \qquad (11\text{–}9)$$

Hence, *when a system of connected rigid bodies is in equilibrium, the first variation or change in V is zero*. This change is computed by taking the *first derivative* of the potential function and setting it equal to zero. For example, using Eq. 11–8 to determine the equilibrium position for the spring and block in Fig. 11–11a, we have

$$\delta V = \frac{dV}{dy} \delta y = (W - ky)\, \delta y = 0$$

Since $\delta y \neq 0$, the term in parentheses must be zero. Hence, the equilibrium position $y = y_{eq}$ is

$$y_{eq} = \frac{W}{k}$$

Of course, the *same result* is obtained by applying $\Sigma F_y = 0$ to the forces acting on the free-body diagram of the block, Fig. 11–11b.

System Having n Degrees of Freedom. When the system of connected bodies has n degrees of freedom, the total potential energy stored in the system will be a function of n independent coordinates q_n; i.e., $V = V(q_1, q_2, \cdots, q_n)$. In order to apply the equilibrium criterion $\delta V = 0$, it is necessary to compute the change in potential energy δV by using the "chain rule" of differential calculus; i.e., the total variation or total differential of V must be equal to zero. Thus,

$$\delta V = \frac{\partial V}{\partial q_1} \delta q_1 + \frac{\partial V}{\partial q_2} \delta q_2 + \cdots + \frac{\partial V}{\partial q_n} \delta q_n = 0$$

Since the virtual displacements δ_1, δq_2, $\cdots$, δq_n are independent of one another, it is necessary that

$$\frac{\partial V}{\partial q_1} = 0, \quad \frac{\partial V}{\partial q_2} = 0, \quad \cdots, \quad \frac{\partial V}{\partial q_n} = 0$$

As for the case of virtual work, *it is possible to write n independent potential functions for a system having n degrees of freedom.*

*11.7 Stability of Equilibrium

Once the equilibrium configuration for a body or system of connected bodies is defined, it is sometimes important to investigate the "type" of equilibrium or the stability of the configuration. For example, consider the position of a ball resting at a point on each of the three paths shown in Fig. 11–12. Each situation represents an equilibrium state for the ball. When the ball is at A, it is said to be in *stable equilibrium* because if it is given a small displacement up the hill, it will always *return* to its original, lowest, position. At A, its total potential energy is a *minimum*. When the ball is at B, it is in *neutral equilibrium*. A small displacement either to the left or right of B will not alter this condition. The ball *remains* in equilibrium in the displaced position, and therefore its potential energy is *constant*. When the ball is at C, it is in *unstable equilibrium*. Here a small displacement will cause the ball's potential energy to be *decreased*, and so it will roll farther *away* from its original, highest position. At C, the potential energy of the ball is a *maximum*.

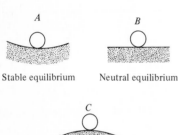

Stable equilibrium   Neutral equilibrium

Unstable equilibrium

Fig. 11–12

Type of Equilibrium. The above example illustrates that one of three types of equilibrium positions can be specified for a body or system of connected bodies.

1. *Stable equilibrium* occurs when a small displacement of the system causes the system to return to its original position. In this case the original potential energy of the system is a minimum.
2. *Neutral equilibrium* occurs when a small displacement of the system causes the system to remain in its displaced state. In this case the potential energy of the system remains constant.
3. *Unstable equilibrium* occurs when a small displacement of the system causes the system to move farther away from its original position. In this case, the original potential energy of the system is a maximum.

System Having One Degree of Freedom. For *equilibrium* of a system having a single degree of freedom, defined by the independent coordinate q, it has been shown that the first derivative of the potential function for the system must be equal to zero; i.e., $dV/dq = 0$. If the potential function $V = V(q)$ is plotted, Fig. 11–13, the first derivative (equilibrium position) is represented as the slope dV/dq, which is zero when the function is maximum, minimum, or an inflection point.

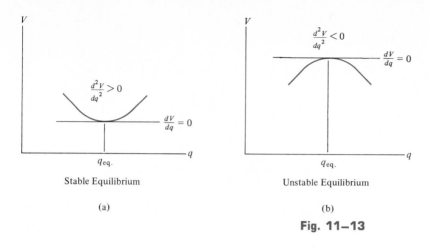

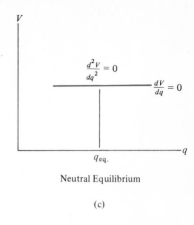

Stable Equilibrium

Unstable Equilibrium

Neutral Equilibrium

(a)

(b)

(c)

Fig. 11–13

If the *stability* of a body at the equilibrium position is to be investigated, it is necessary to compute the *second derivative* of V and evaluate it at the equilibrium position $q = q_{eq}$. As shown in Fig. 11–13a, if $V = V(q)$ is a *minimum*, then

$$\frac{dV}{dq} = 0, \frac{d^2V}{dq^2} > 0 \qquad \text{stable equilibrium} \qquad (11\text{--}10)$$

If $V = V(q)$ is a *maximum*, Fig. 11–13b, then

$$\frac{dV}{dq} = 0, \frac{d^2V}{dq^2} < 0 \qquad \text{unstable equilibrium} \qquad (11\text{--}11)$$

If the second derivative is zero, it will be necessary to investigate *higher-order* derivatives to determine the stability. In particular, stable equilibrium will occur if the order of the lowest remaining nonzero derivative is *even* and the sign of this nonzero derivative is positive when it is evaluated at $q = q_{eq}$; otherwise, it is unstable.

If the system is in neutral equilibrium, Fig. 11–13c, it is required that

$$\frac{dV}{dq} = \frac{d^2V}{dq^2} = \frac{d^3V}{dq^3} = \cdots = 0 \qquad \text{neutral equilibrium} \qquad (11\text{--}12)$$

since then V must be constant at and around the "neighborhood" of q_{eq}.

System Having Two Degrees of Freedom. A criterion for investigating stability becomes increasingly complex as the number of degrees of freedom for the system increases. For a system having two degrees of freedom, with independent coordinates (q_1, q_2), it may be verified (using the calculus of functions of two variables) that equilibrium and stability occur at a point (q_{1eq}, q_{2eq}) when

$$\frac{\partial V}{\partial q_1} = \frac{\partial V}{\partial q_2} = 0$$

$$\left[\left(\frac{\partial^2 V}{\partial q_1 \, \partial q_2} \right)^2 - \left(\frac{\partial^2 V}{\partial q_1^2} \right) \left(\frac{\partial^2 V}{\partial q_2^2} \right) \right] < 0$$

$$\left(\frac{\partial^2 V}{\partial q_1^2} + \frac{\partial^2 V}{\partial q_2^2} \right) > 0$$

Both equilibrium and instability occur when

$$\frac{\partial V}{\partial q_1} = \frac{\partial V}{\partial q_2} = 0$$

$$\left[\left(\frac{\partial^2 V}{\partial q_1 \, \partial q_2} \right)^2 - \left(\frac{\partial^2 V}{\partial q_1^2} \right) \left(\frac{\partial^2 V}{\partial q_2^2} \right) \right] < 0$$

$$\left(\frac{\partial^2 V}{\partial q_1^2} + \frac{\partial^2 V}{\partial q_2^2} \right) < 0$$

PROCEDURE FOR ANALYSIS

Using potential-energy methods, the equilibrium positions and the stability of a body or a system of connected bodies having a single degree of freedom can be obtained by applying the following procedure.

Potential Function. Formulate the potential function $V = V_g + V_e$ for the system. To do this, sketch the system so that it is located at some *arbitrary position* specified by the independent coordinate q. A horizontal *datum* is established through a *fixed point,** and the *gravitational potential energy* V_g is expressed in terms of the weight W of each member and its vertical distance y from the datum, $V_g = Wy$, Eq. 11–4. The elastic potential energy V_e of the system is expressed in terms of the stretch or compression, s, of any connecting spring and the spring's stiffness k, $V_e = \frac{1}{2}ks^2$, Eq. 11–5. Once V has been established, express the *position coordinates* y and s in terms of the independent coordinate q.

Equilibrium Position. The equilibrium position is determined by taking the first derivative of V and setting it equal to zero, $\delta V = 0$, Eq. 11–9.

Stability. Stability at the equilibrium position is determined by evaluating the second or higher-order derivatives of V as indicated by Eqs. 11–10 to 11–12.

The following examples numerically illustrate this procedure.

*The location of the datum is *arbitrary* since only the *changes* or differentials of V are required for investigation of the equilibrium position and its stability.

Example 11–5

The uniform link shown in Fig. 11–14a has a mass of 10 kg. The spring is unstretched when $\theta = 0°$. Determine the angle θ for equilibrium and investigate the stability at the equilibrium position.

Solution

Potential Function. The datum is established at the top of the link when the *spring is unstretched*, Fig. 11–14b. When the link is located at the arbitrary position θ, the spring increases its potential energy by stretching and the weight decreases its potential energy. Hence,

$$V = V_e + V_g = \frac{1}{2}ks^2 - W\left(s + \frac{l}{2}\cos\theta - \frac{l}{2}\right)$$

Since $l = s + l\cos\theta$ or $s = l(1 - \cos\theta)$, then

$$V = \frac{1}{2}kl^2(1 - \cos\theta)^2 - \frac{Wl}{2}(1 - \cos\theta)$$

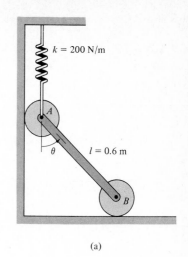

(a)

Equilibrium Position. The derivative of V gives

$$\delta V = 0; \quad \frac{dV}{d\theta}\delta\theta = \left[kl^2(1 - \cos\theta)\sin\theta - \frac{Wl}{2}\sin\theta\right]\delta\theta = 0$$

Since $\delta\theta \neq 0$, then

$$l\left[kl(1 - \cos\theta) - \frac{W}{2}\right]\sin\theta = 0$$

This equation is satisfied provided

$$\sin\theta = 0 \qquad \theta = 0° \qquad\qquad \textit{Ans.}$$

$$\theta = \cos^{-1}\left(1 - \frac{W}{2kl}\right) = \cos^{-1}\left[1 - \frac{10(9.81)}{2(200)(0.6)}\right] = 53.8° \quad \textit{Ans.}$$

Stability. Computing the second derivative of V gives

$$\frac{d^2V}{d\theta^2} = kl^2(1 - \cos\theta)\cos\theta + kl^2\sin\theta\sin\theta - \frac{Wl}{2}\cos\theta$$

$$= kl^2(\cos\theta - \cos 2\theta) - \frac{Wl}{2}\cos\theta$$

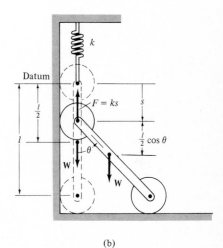

(b)

Fig. 11–14

Substituting values for the constants, with $\theta = 0°$ and $\theta = 53.8°$, yields

$$\left.\frac{d^2V}{d\theta^2}\right|_{\theta=0°} = 200(0.6)^2(\cos 0° - \cos 0°) - \frac{10(9.81)(0.6)}{2}\cos 0°$$

$$= -29.4 < 0 \qquad \text{(unstable equilibrium at } \theta = 0°) \quad \textit{Ans.}$$

$$\left.\frac{d^2V}{d\theta^2}\right|_{\theta=53.8°} = 200(0.6)^2(\cos 53.8° - \cos 107.6°) - \frac{10(9.81)(0.6)}{2}\cos 53.8°$$

$$= 46.9 > 0 \qquad \text{(stable equilibrium at } \theta = 53.8°) \qquad \textit{Ans.}$$

Example 11–6

Determine the mass m of the block required for equilibrium of the uniform 10-kg rod shown in Fig. 11–15a when $\theta = 20°$. Investigate the stability at the equilibrium position.

Solution

Potential Function. The datum is established through point A, Fig. 11–15b. When $\theta = 0°$, the block is assumed to be suspended $(y_W)_1$ below the datum. Hence,

$$V = V_e + V_g = 98.1\left(\frac{1.5 \sin \theta}{2}\right) - m(9.81)(\Delta y) \quad (1)$$

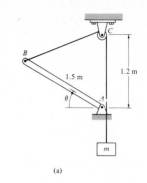

(a)

The distance Δy may be related to the independent coordinate θ by measuring the difference in cord lengths $B'C$ and BC. Since

$$B'C = \sqrt{(1.5)^2 + (1.2)^2} = 1.92$$
$$BC = \sqrt{(1.5 \cos \theta)^2 + (1.2 - 1.5 \sin \theta)^2} = \sqrt{3.69 - 3.60 \sin \theta}$$

then

$$\Delta y = B'C - BC = 1.92 - \sqrt{3.69 - 3.60 \sin \theta}$$

Substituting the above result into Eq. (1) yields

$$V = 98.1\left(\frac{1.5 \sin \theta}{2}\right) - m(98.1)(1.92 - \sqrt{3.69 - 3.60 \sin \theta}) \quad (2)$$

Equilibrium Position

$$\frac{dV}{d\theta}\,\delta\theta = \left\{73.6 \cos \theta - \left[\frac{m(98.1)}{2}\right]\left(\frac{3.60 \cos \theta}{\sqrt{3.69 - 3.60 \sin \theta}}\right)\right\}\delta\theta = 0$$

$$\left.\frac{dV}{d\theta}\right|_{\theta=20°}\delta\theta = (69.16 - 10.58m)\,\delta\theta = 0$$

$$m = \frac{69.16}{10.58} = 6.54 \text{ kg} \qquad Ans.$$

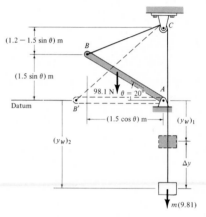

(b)

Fig. 11–15

Stability. Taking the second derivative of Eq. (2), we obtain

$$\frac{d^2V}{d\theta^2} = -73.6 \sin \theta - \left[\frac{m(9.81)}{2}\right]\left(\frac{-1}{2}\right)\frac{(3.60 \cos \theta)^2}{(3.69 - 3.60 \sin \theta)^{3/2}}$$

$$- \frac{m(9.81)}{2}\left(\frac{-3.60 \sin \theta}{\sqrt{3.69 - 3.60 \sin \theta}}\right)$$

For the equilibrium position $\theta = 20°$, $m = 6.54$ kg, hence

$$\frac{d^2V}{d\theta^2} = 47.6 > 0 \qquad \text{(stable equilibrium at } \theta = 20°\text{)} \qquad Ans.$$

Example 11–7

A homogeneous block having a mass m rests on the top surface of a cylinder, Fig. 11–16a. Show that this is a condition of unstable equilibrium if $h > 2R$.

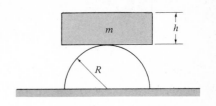

(a)

Solution

Potential Function. The datum is established at the base of the cylinder, Fig. 11–16b. If the block is displaced by an amount θ from the equilibrium position, the potential function may be written in the form

$$V = V_e + V_g$$
$$= 0 + mgy$$

From Fig. 11–16b,

$$y = \left(R + \frac{h}{2}\right) \cos \theta + R\theta \sin \theta$$

Thus,

$$V = mg\left[\left(R + \frac{h}{2}\right) \cos \theta + R\theta \sin \theta\right]$$

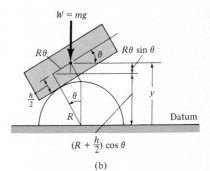

(b)

Fig. 11–16

Equilibrium Position

$$\delta V = \frac{dV}{d\theta} \delta\theta = mg\left[-\left(R + \frac{h}{2}\right) \sin \theta + R \sin \theta + R\theta \cos \theta\right] \delta\theta = 0$$

$$= mg\left(-\frac{h}{2} \sin \theta + R\theta \cos \theta\right) \delta\theta = 0$$

Obviously $\theta = 0°$ is the equilibrium position that satisfies this equation.

Stability. Taking the second derivative of V yields

$$\frac{d^2V}{d\theta^2} = mg\left(-\frac{h}{2} \cos \theta + R \cos \theta - R\theta \sin \theta\right)$$

At $\theta = 0°$,

$$\left.\frac{d^2V}{d\theta^2}\right|_{\theta=0°} = -mg\left(\frac{h}{2} - R\right)$$

Since all the constants are positive, the block is in unstable equilibrium if $h > 2R$, for then $d^2V/d\theta^2 < 0$.

Problems

*11–20. If the potential function for a conservative one-degree-of-freedom system is $V = (6x^3 - x^2 - 2x + 10)$ J, where x is given in meters, determine the positions for equilibrium and investigate the stability at each of these positions.

11–21. If the potential function for a conservative one-degree-of-freedom system is $V = (10 \cos 2\theta + 25 \sin \theta)$ J, where $0° < \theta < 180°$, determine the positions for equilibrium and investigate the stability at each of these positions.

■11–22. Solve Prob. 11–2 using the principle of potential energy, and investigate the stability at the equilibrium position.

11–23. Solve Prob. 11–3 using the principle of potential energy, and investigate the stability at the equilibrium position.

*11–24. Solve Prob. 11–12 using the principle of potential energy, and investigate the stability at the equilibrium position.

11–25. The uniform beam has a weight W. If the contacting surfaces are smooth, determine the angle θ for equilibrium. The spring is unstretched when $\theta = 0°$.

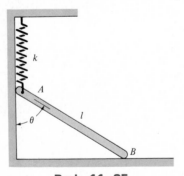

Prob. 11–25

11–26. The piston block weighs W and is supported by the two links AB and BC. Determine the necessary spring stiffness k required to hold the system in neutral equilibrium. The springs are subjected to an initial compression Δ when the links are vertical as shown.

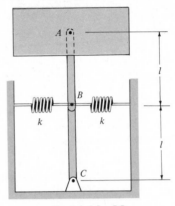

Prob. 11–26

■11–27. The two bars each have a weight of 8 lb. Determine the angle θ for equilibrium and investigate the stability at the equilibrium position. The spring has an unstretched length of 1 ft.

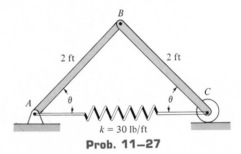

Prob. 11–27

*11–28. Each of the two springs has an unstretched length of 500 mm. Determine the mass m needed to hold the springs in the equilibrium position shown, i.e., $y = 1$ m.

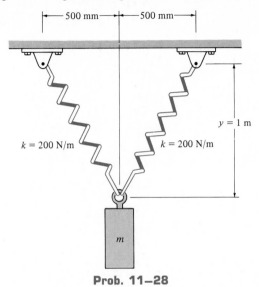

500 mm — 500 mm

$k = 200$ N/m $y = 1$ m $k = 200$ N/m

m

Prob. 11–28

11–29. Two smooth bars of length 500 mm are pin-connected. Determine the angle θ for equilibrium and investigate the stability at this position. The bars each have a mass of 3 kg and the suspended block D has a mass of 7 kg. Cord DC has a total length of 1 m.

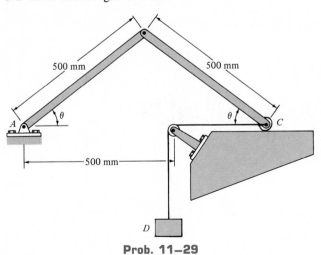

500 mm   500 mm

θ θ C

A

500 mm

D

Prob. 11–29

11–30. The spring of the scale has an unstretched length of a. Determine the angle θ for equilibrium when a weight W is supported on the platform. Neglect the weights of the members. What value of W would be required to keep the scale in neutral equilibrium when $\theta = 0°$?

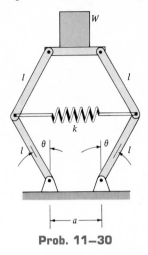

W

l l

k

l θ θ l

a

Prob. 11–30

11–31. The *Roberval balance* is in equilibrium when no weights are placed on the pans A and B. If two masses m_A and m_B are placed at *any* location a and b on the pans, show that neutral equilibrium is maintained if $m_A d_A = m_B d_B$.

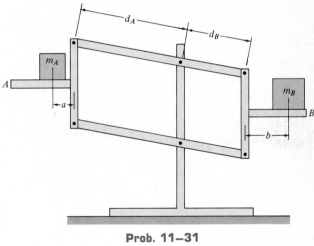

d_A d_B

m_A

A

a b B

m_B

Prob. 11–31

515

*11–32. The truck is weighed on the highway inspection scale. If a known mass m is placed a distance s from the fulcrum B of the scale, determine the mass of the truck m_t for equilibrium if its center of gravity is located at a distance d from point C. When the scale is empty, the weight of the lever ABC balances the scale CDE.

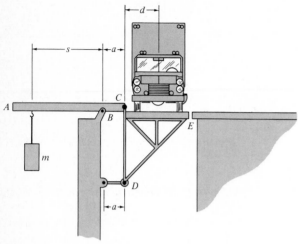

Prob. 11–32

■11–33. The tightrope walker will always maintain a stable equilibrium position on the rope provided the combined center of gravity of the pole and himself is located *below* line aa. In neutral equilibrium, the combined center of gravity must be *coincident* with line aa. If the tightrope walker has a mass of 80 kg and the distance to his center of gravity G, measured from aa, is 0.75 m, determine the required length of the pole for neutral equilibrium if it has a mass of 9 kg/m. Assume that the pole, when held at its center, deflects into a circular arc of radius 8 m as shown. *Hint:* Use the table on the inside back cover to determine the location of the centroid for a circular arc.

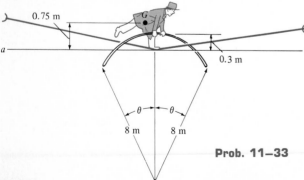

Prob. 11–33

11–34. The rod supports a disk B having a weight of 800 N. It is held in equilibrium using four springs for which $k = 400$ N/m. Determine the minimum distance d between the springs so that the rod remains in stable equilibrium when it is vertical. Neglect the weight of the rod. It rests on a roller at A.

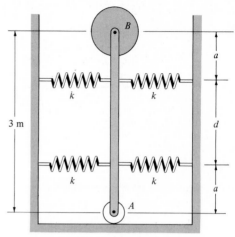

Prob. 11–34

11–35. A uniform square book having a mass m is suspended by a cord as shown. Assuming that it remains closed and the wall is smooth, determine the angle θ at which the cord hangs from the wall for equilibrium. Is the book in stable equilibrium?

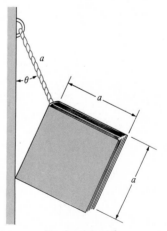

Prob. 11–35

*11–36. The cup has a hemispherical bottom and a mass *m*. Determine the position *h* of the center of mass *G* so that the cup is in neutral equilibrium.

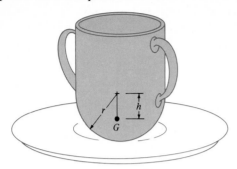

Prob. 11–36

11–37. The assembly shown consists of a semicircular cylinder and a triangular prism. If the prism weighs 8 lb and the cylinder weighs 2 lb, investigate the stability when the assembly is resting in the equilibrium position.

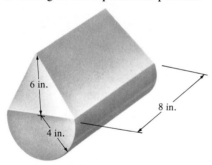

6 in.

8 in.

4 in.

Prob. 11–37

11–38. The homogeneous cone has a conical cavity cut into it as shown. Determine the depth *d* of the cavity so that the cone balances on the pivot and remains in neutral equilibrium.

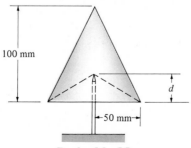

100 mm

d

50 mm

Prob. 11–38

11–39. A homogeneous cone rests on top of the cylindrical surface. Derive a relationship between the radius *r* of the cylinder and the height *h* of the cone for neutral equilibrium. *Hint:* Establish the potential function for a *small* angle θ of tilt of the cone, i.e., approximate $\sin \theta \approx 0$ and $\cos \theta \approx 1 - \theta^2/2$.

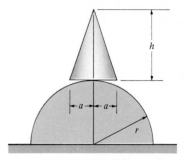

h

a *a*

r

Prob. 11–39

*11–40. The homogeneous block has a mass of 10 kg and rests on the smooth corners of two ledges. Determine the angle θ for placement that will cause the block to be stable.

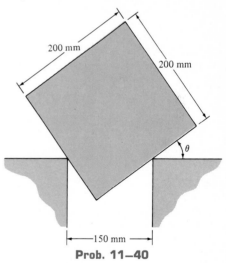

200 mm

200 mm

θ

150 mm

Prob. 11–40

517

11–41. Two uniform bars, each having a weight W, are pin-connected at their ends. If they are placed over a smooth cylindrical surface, show that the angle θ for equilibrium must satisfy the equation $\cos \theta / \sin^3 \theta = a/2r$.

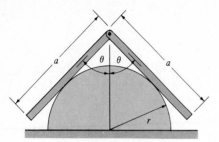

Prob. 11–41

11–42. The triangular block of weight W rests on the smooth corners which are a distance a apart. If the block has three equal sides of length d, determine the angle θ for equilibrium.

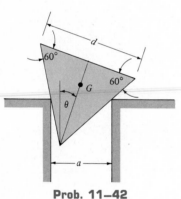

Prob. 11–42

11–43. If the potential function for a conservative two-degree-of-freedom system is $V = (6y^2 + 12x^2)$ J, where x and y are given in meters, determine the equilibrium position and investigate the stability at this position.

*11–44. The two blocks each have a mass of 15 kg. Determine the stretch of the springs, x_1 and x_2, for equilibrium.

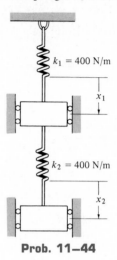

$k_1 = 400$ N/m

x_1

$k_2 = 400$ N/m

x_2

Prob. 11–44

Review Problems

11–45. The uniform rod AB has a weight of 10 lb. If the spring DC is unstretched when $\theta = 90°$, determine the angle θ for equilibrium using the principle of virtual work. The spring always acts in the horizontal position because of the roller guide at D.

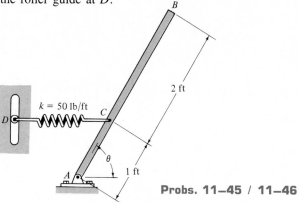

Probs. 11–45 / 11–46

11–46. Solve Prob. 11–45 using the principle of potential energy. Investigate the stability of the rod when it is in the equilibrium position.

11–47. Determine the angles θ for equilibrium of the 4-lb disk using the principle of virtual work. Neglect the weight of the rod. The spring is unstretched when $\theta = 0°$ and always remains in the vertical position due to the roller guide.

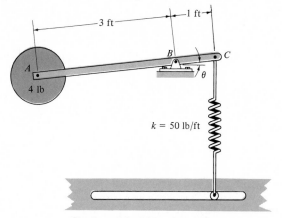

Probs. 11–47 / 11–48

*11–48.** Solve Prob. 11–47 using the principle of potential energy. Investigate the stability of the assembly when it is in each equilibrium position.

11–49. The 60-lb hemisphere supports a cylinder having a specific weight of 311 lb/ft³. If the radii of the cylinder and hemisphere are both 5 in., determine the maximum height h of the cylinder which will produce neutral equilibrium in the position shown.

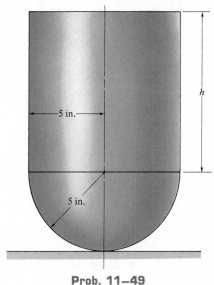

Prob. 11–49

11–50. Determine the force $\mathbf{P}$ that must be applied to the cord wrapped around the shaft at C, which is necessary to lift the mass m. As the mass is lifted, the pulley rolls on a cord that winds up on shaft B and unwinds from shaft A.

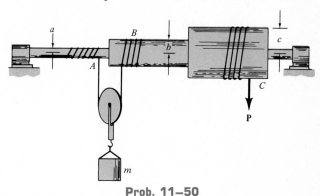

Prob. 11–50

519

11-51. The machine shown is used for forming metal plates. It consists of two toggles *ABC* and *DEF*, which are operated by the hydraulic cylinder *H*. The toggles push the movable bar *G* forward, pressing the plate *p* into the cavity. If the force which the plate exerts on the head is $P = 8$ kN, determine the force **F** in the hydraulic cylinder when $\theta = 30°$.

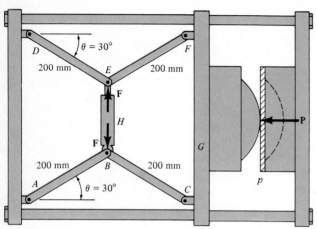

Prob. 11-51

11-53. Determine the force **F** acting on the cord which is required to maintain equilibrium of the horizontal 20-kg bar *CB*. *Hint:* First show that the coordinates s_A and s_B are related to the *constant* vertical length *l* of the cord by the equation $5s_B - s_A = l$.

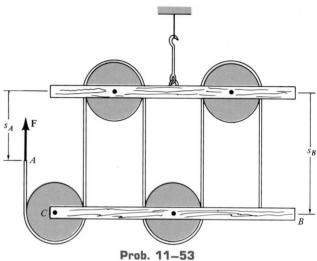

Prob. 11-53

*11-52. Determine the horizontal force **F** required to maintain equilibrium of the slider mechanism when $\theta = 60°$.

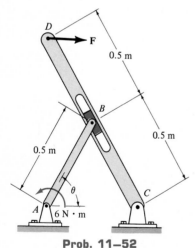

Prob. 11-52

Mathematical Expressions

Quadratic Formula

If $ax^2 + bx + c = 0$, then $x = \dfrac{-b \pm \sqrt{b^2 - 4ac}}{2a}$

Hyperbolic Functions

$$\sinh x = \frac{e^x - e^{-x}}{2}, \quad \cosh x = \frac{e^x + e^{-x}}{2}, \quad \tanh x = \frac{\sinh x}{\cosh x}$$

Trigonometric Identities

$\sin^2 \theta + \cos^2 \theta = 1$

$\sin (\theta \pm \phi) = \sin \theta \cos \phi \pm \cos \theta \sin \phi$

$\sin 2\theta = 2 \sin \theta \cos \theta$

$\cos (\theta \pm \phi) = \cos \theta \cos \phi \mp \sin \theta \sin \phi$

$\cos 2\theta = \cos^2 \theta - \sin^2 \theta$

$\cos \theta = \pm\sqrt{\dfrac{1 + \cos 2\theta}{2}}, \quad \sin \theta = \pm\sqrt{\dfrac{1 - \cos 2\theta}{2}}$

$\tan \theta = \dfrac{\sin \theta}{\cos \theta}$

$1 + \tan^2 \theta = \sec^2 \theta$

$1 + \operatorname{ctn}^2 \theta = \csc^2 \theta$

Power-Series Expansions

$$\sin x = x - \frac{x^3}{3!} + \frac{x^5}{5!} - \frac{x^7}{7!} + \cdots$$

$$\cos x = 1 - \frac{x^2}{2!} + \frac{x^4}{4!} - \frac{x^6}{6!} + \cdots$$

$$\sinh x = x + \frac{x^3}{3!} + \frac{x^5}{5!} + \cdots$$

$$\cosh x = 1 + \frac{x^2}{2!} + \frac{x^4}{4!} + \cdots$$

Derivatives

$$\frac{d}{dx}(u^n) = nu^{n-1}\frac{du}{dx}$$

$$\frac{d}{dx}(uv) = u\frac{dv}{dx} + v\frac{du}{dx}$$

$$\frac{d}{dx}\left(\frac{u}{v}\right) = \frac{v\frac{du}{dx} - u\frac{dv}{dx}}{v^2}$$

$$\frac{d}{dx}(\sin u) = \cos u\frac{du}{dx}$$

$$\frac{d}{dx}(\cos u) = -\sin u\frac{du}{dx}$$

$$\frac{d}{dx}(\tan u) = \sec^2 u\frac{du}{dx}$$

$$\frac{d}{dx}(\cot u) = -\csc^2 u\frac{du}{dx}$$

$$\frac{d}{dx}(\sec u) = \tan u \sec u\frac{du}{dx}$$

$$\frac{d}{dx}(\csc u) = -\csc u \cot u\frac{du}{dx}$$

$$\frac{d}{dx}(\sinh u) = \cosh u\frac{du}{dx}$$

$$\frac{d}{dx}(\cosh u) = \sinh u\frac{du}{dx}$$

Integrals

$$\int x^n\, dx = \frac{x^{n+1}}{n+1} + C,\ n \neq -1$$

$$\int \frac{dx}{a + bx} = \frac{1}{b}\ln(a + bx) + C$$

$$\int \sqrt{a + bx}\, dx = \frac{2}{3b}\sqrt{(a + bx)^3} + C$$

$$\int x\sqrt{a + bx}\, dx = \frac{-2(2a - 3bx)\sqrt{(a + bx)^3}}{15b^2} + C$$

$$\int x^2\sqrt{a + bx}\, dx = \frac{2(8a^2 - 12abx + 15b^2x^2)\sqrt{(a + bx)^3}}{105b^3} + C$$

$$\int \sqrt{a^2 - x^2}\, dx = \frac{1}{2}\left[x\sqrt{a^2 - x^2} + a^2 \sin^{-1}\frac{x}{a}\right] + C,\ a > 0$$

$$\int x\sqrt{a^2 - x^2}\, dx = -\frac{1}{3}\sqrt{(a^2 - x^2)^3} + C$$

$$\int x^2\sqrt{a^2 - x^2}\, dx = -\frac{x}{4}\sqrt{(a^2 - x^2)^3} + \frac{a^2}{8}\left(x\sqrt{a^2 - x^2} + a^2\sin^{-1}\frac{x}{a}\right) + C,\ a > 0$$

$$\int \sqrt{x^2 \pm a^2}\, dx = \frac{1}{2}\left[x\sqrt{x^2 + a^2} \pm a^2\ln(x + \sqrt{x^2 \pm a^2})\right] + C$$

$$\int x\sqrt{x^2 \pm a^2}\, dx = \frac{1}{3}\sqrt{(x^2 \pm a^2)^3} + C$$

$$\int \frac{x\, dx}{\sqrt{x^2 \pm a^2}} = \sqrt{x^2 \pm a^2} + C$$

$$\int \frac{dx}{\sqrt{a + bx + cx^2}} = \frac{1}{\sqrt{c}}\ln\left[\sqrt{a + bx + cx^2} + x\sqrt{c} + \frac{b}{2\sqrt{c}}\right] + C,\ c > 0$$

$$= \frac{1}{\sqrt{-c}}\sin^{-1}\left(\frac{-2cx - b}{\sqrt{b^2 - 4ac}}\right) + C,\ c < 0$$

$$\int \sin x \, dx = -\cos x + C$$

$$\int \cos x \, dx = \sin x + C$$

$$\int x \cos(ax) \, dx = \frac{1}{a^2} \cos(ax) + \frac{x}{a} \sin(ax) + C$$

$$\int x^2 \cos(ax) \, dx = \frac{2x}{a^2} \cos(ax) + \frac{a^2 x^2 - 2}{a^3} \sin(ax) + C$$

$$\int e^{ax} \, dx = \frac{1}{a} e^{ax} + C$$

$$\int \sinh x \, dx = \cosh x + C$$

$$\int \cosh x \, dx = \sinh x + C$$

B

Numerical and Computer Analysis

Occasionally the application of the laws of mechanics will lead to a system of equations for which a closed-form solution is difficult or impossible to obtain. When confronted with this situation, engineers will often use a numerical method which in most cases can be programmed on a microcomputer or "programmable" pocket calculator. Here we will briefly present a computer program for solving a set of linear algebraic equations and three numerical methods which can be used to solve an algebraic or transcendental equation, evaluate a definite integral, and solve an ordinary differential equation. Application of each method will be explained by example, and an associated computer program written in Microsoft BASIC, which is designed to run on most personal computers, is provided.* A text on numerical analysis should be consulted for further discussion regarding a check of the accuracy of each method and the inherent errors that can develop.

Linear Algebraic Equations. Application of the equations of static equilibrium or the equations of motion sometimes requires solving a set of linear algebraic equations. The computer program listed in Fig. B–1 can be used for this purpose. It is based on the method of a Gauss elimination and can solve at most ten equations with ten unknowns. To do so, the equations should first be written in the following general format:

$$A_{11}x_1 + A_{12}x_2 + \cdots + A_{1n}x_n = B_1$$
$$A_{21}x_1 + A_{22}x_2 + \cdots + A_{2n}x_n = B_2$$

.

.

.

$$A_{n1}x_1 + A_{n2}x_2 + \cdots + A_{nn}x_n = B_n$$

The "A" and "B" coefficients are "called" for when running the program. The output presents the unknowns $x_1, \ldots, x_n$.

---

*Similar types of programs can be written or purchased for programmable pocket calculators.

```
1 PRINT"Linear system of equations":PRINT
2 DIM A(10,11)
3 INPUT"Input number of equations : ",N
4 PRINT
5 PRINT"A  coefficients"
6 FOR I = 1 TO N
7 FOR J = 1 TO N
8 PRINT "A(";I;",";J;
9 INPUT")=",A(I,J)
10 NEXT J
11 NEXT I
12 PRINT
13 PRINT"B  coefficients"
14 FOR I = 1 TO N
15 PRINT "B(";I;
16 INPUT")=",A(I,N+1)
17 NEXT I
18 GOSUB 25
19 PRINT
20 PRINT"Unknowns"
21 FOR I = 1 TO N
22 PRINT "X(";I;")=";A(I,N+1)
23 NEXT I
24 END
25 REM Subroutine Guassian
26 FOR M=1 TO N
27 NP=M
28 BG=ABS(A(M,M))
29 FOR I = M TO N
30 IF ABS(A(I,M))<=BG THEN 33
31 BG=ABS(A(I,M))
32 NP=I
33 NEXT I
34 IF NP=M THEN 40
35 FOR I = M TO N+1
36 TE=A(M,I)
37 A(M,I)=A(NP,I)
38 A(NP,I)=TE
39 NEXT I
40 FOR I = M+1 TO N
41 FC=A(I,M)/A(M,M)
42 FOR J = M+1 TO N+1
43 A(I,J)=A(I,J)-FC*A(M,J)
44 NEXT J
45 NEXT I
46 NEXT M
47 A(N,N+1)=A(N,N+1)/A(N,N)
48 FOR I = N-1 TO 1 STEP -1
49 SM=0
50 FOR J=I+1 TO N
51 SM=SM+A(I,J)*A(J,N+1)
52 NEXT J
53 A(I,N+1)=(A(I,N+1)-SM)/A(I,I)
54 NEXT I
55 RETURN
```

Fig. B—1

Example B–1

Solve the two equations

$$3x_1 + x_2 = 4$$
$$2x_1 - x_2 = 10$$

Solution

When the program begins to run, it first calls for the number of equations (2), then the A coefficients in the following sequence: $A_{11} = 3$, $A_{12} = 1$, $A_{21} = 2$, $A_{22} = -1$ and finally the B coefficients $B_1 = 4$, $B_2 = 10$. The output appears as:

Unknowns

$X(1) = 2.8$ *Ans.*

$X(2) = -4.4$ *Ans.*

Simpson's Rule. This numerical method can be used to determine the area under a curve given as a graph or as an explicit function $y = f(x)$. Likewise it can be used to compute the value of a definite integral which involves the function $y = f(x)$. To do so, the area must be subdivided into an *even number* of strips or intervals having a width h. The curve between three consecutive ordinates is approximated by a parabola, and the entire area or definite integral is then determined from the formula

$$\int_{x_0}^{x_n} f(x)\, dx \simeq \frac{h}{3}[y_0 + 4(y_1 + y_3 + \cdots + y_{n-1})$$
$$+ 2(y_2 + y_4 + \cdots + y_{n-2}) + y_n] \quad \text{(B–1)}$$

The computer program for this equation is given in Fig. B–2. For its use, one must first enter the function (line 6), then the upper and lower limits of the integral (lines 7 and 8), and finally the number of intervals (line 9). The value of the integral is then given as the output.

```
 1 PRINT"Simpson's rule":PRINT
 2 PRINT" To execute this program :":PRINT
 3 PRINT"    1- Modify right-hand side of the equation given below,
 4 PRINT"       then press RETURN key"
 5 PRINT"    2- Type   RUN 6":PRINT:EDIT 6
 6 DEF FNF(X)=LOG(X)
 7 PRINT:INPUT" Enter Lower Limit = ",A
 8 INPUT" Enter Upper Limit = ",B
 9 INPUT" Enter Number (even) of Intervals = ",N%
10 H=(B-A)/N%:AR=FNF(A):X=A+H
11 FOR J%=2 TO N%
12 K=2*(2-J%+2*INT(J%/2))
13 AR=AR+K*FNF(X)
14 X=X+H:NEXT J%
15 AR=H*(AR+FNF(B))/3
16 PRINT" Integral = ",AR
17 END
```

Fig. B–2

Example B–2

Evaluate the definite integral $\int_{2}^{5} \ln x \, dx$

Solution

The interval $x_0 = 2$ to $x_6 = 5$ will be divided into six equal parts ($n = 6$), each having a width $h = (5 - 2)/6 = 0.5$. We then compute $y = f(x) = \ln x$ at each point of subdivision.

n	x_n	y_n
0	2	0.693
1	2.5	0.916
2	3	1.099
3	3.5	1.253
4	4	1.386
5	4.5	1.504
6	5	1.609

Thus, Eq. B–1 becomes

$$\int_{2}^{5} \ln x \, dx \simeq \frac{0.5}{3}[0.693 + 4(0.916 + 1.253 + 1.504)$$
$$+ 2(1.099 + 1.386) + 1.609]$$
$$\simeq 3.66 \qquad\qquad Ans.$$

This answer is equivalent to the exact answer to three significant figures. Obviously, accuracy to a larger number of significant figures can be improved by selecting a smaller interval h (or larger n).

Using the computer program, we enter the function $\ln x$, line 6 in Fig. B–2, the upper and lower limits 2 and 5, and the number of intervals $n = 6$. The output appears as

$$\text{Integral} \doteq 3.66082 \qquad\qquad Ans.$$

The Secant Method. This numerical method is used to find the real roots of an algebraic or transcendental equation $f(x) = 0$. The method derives its name from the fact that the formula used is established from the slope of the secant line to the graph $y = f(x)$. This slope is $(f(x_n) - f(x_{n-1}))/(x_n - x_{n-1})$ and the secant formula is

$$x_{n+1} = x_n - f(x_n) \left(\frac{x_n - x_{n-1}}{f(x_n) - f(x_{n-1})} \right) \qquad (B-2)$$

For application it is necessary to provide two initial guesses, x_0 and x_1, and thereby evaluate x_2 from Eq. B–2 ($n = 1$). One then proceeds to reapply Eq. B–2 with x_1 and the calculated value of x_2 and obtain x_3 ($n = 2$), etc., until the value $x_{n+1} \simeq x_n$. One can see this will occur if x_n is approaching the root of the function $f(x) = 0$ since the correction term on the right of Eq. B–2 will tend toward zero. In particular, the larger the slope the smaller the correction to x_n and the faster the root will be found. Likewise, if the slope is very small in the neighborhood of the root, the method leads to large corrections for x_n and convergence to the root is slow and may even lead to a failure to find it. In such cases other numerical techniques must be used for solution.

A computer program based on Eq. B–2 is listed in Fig. B–3. The user must supply the function (line 7) along with two initial guesses x_0 and x_1 used to approximate the solution. The output specifies the value of the root. If it cannot be determined it is so stated.

```
1 PRINT"Secant method":PRINT
2 PRINT" To execute this program :":PRINT
3 PRINT"    1) Modify right hand side of the equation given below,"
4 PRINT"       then press RETURN key."
5 PRINT"    2) Type   RUN 7"
6 PRINT:EDIT 7
7 DEF FNF(X)=.5*SIN(X)-2*COS(X)+1.3
8 INPUT"Enter point #1 =",X
9 INPUT"Enter point #2 =",X1
10 IF X=X1 THEN 14
11 EP=.00001:TL=2E-20
12 FP=(FNF(X1)-FNF(X))/(X1-X)
13 IF ABS(FP)>TL THEN 15
14 PRINT"Root can not be found.":END
15 DX=FNF(X1)/FP
16 IF ABS(DX)>EP THEN 19
17 PRINT "Root = ";X1;"        Function evaluated at this root = ";FNF(X1)
18 END
19 X=X1:X1=X1-DX
20 GOTO 12
```

Fig. B–3

Example B–3

Determine the root of the equation

$$f(x) = 0.5 \sin x - 2 \cos x + 1.30 = 0$$

Solution

Guesses of the initial roots will be $x_0 = 45°$ and $x_1 = 30°$. Applying Eq. B–2,

$$x_2 = 30° - (-0.1821)\frac{(30° - 45°)}{(-0.1821 - 0.2393)} = 36.48°$$

Using this value in Eq. B–2, along with $x_1 = 30°$, we have

$$x_3 = 36.48° - (-0.0108)\frac{36.48° - 30°}{(-0.0108 + 0.1821)} = 36.89°$$

Repeating the process with this value and $x_2 = 36.48°$ yields

$$x_4 = 36.89° - (0.0005)\left(\frac{36.89° - 36.48°}{(0.0005 + 0.0108)}\right) = 36.87°$$

Thus $x = 36.9°$ is appropriate to three significant figures.

If the problem is solved using the computer program, we enter the function, line 7 in Fig. B–3. The first and second guesses must be in radians. Choosing these to be 0.8 rad, and 0.5 rad, the result appears as Root = 0.6435022.

Function evaluated at this root = 1.66893E–06.

This result converted from radians to degrees is therefore $x = 36.9°$.

Runge-Kutta Method. This numerical method is used to solve an ordinary differential equation. It consists of applying a set of formulas which are used to find specific values of y for corresponding incremental values h in x. The formulas given in general form are as follows:

First-Order Equation: To integrate $\dot{x} = f(t, x)$ step-by-step use

$$x_{i+1} = x_i + \frac{1}{6}(k_1 + 2k_2 + 2k_3 + k_4) \tag{B-3}$$

where

$$k_1 = hf(t_i, x_i)$$
$$k_2 = hf\left(t_i + \frac{h}{2}, x_i + \frac{k_1}{2}\right)$$
$$k_3 = hf\left(t_i + \frac{h}{2}, x_i + \frac{k_2}{2}\right) \tag{B-4}$$
$$k_4 = hf(t_i + h, x_i + k_3)$$

Second-Order Equation. To integrate $\ddot{x} = f(t, x, \dot{x})$

$$x_{i+1} = x_i + h\left[\dot{x}_i + \frac{1}{6}(k_1 + k_2 + k_3)\right]$$
$$\dot{x}_{i+1} = \dot{x}_i + \frac{1}{6}(k_1 + 2k_2 + 2k_3 + k_4) \tag{B-5}$$

where

$$k_1 = hf(t_i, x_i, \dot{x}_i)$$
$$k_2 = hf\left(t_i + \frac{h}{2}, x_i + \frac{h}{2}\dot{x}_i, \dot{x}_i + \frac{k_1}{2}\right)$$
$$k_3 = hf\left(t_i + \frac{h}{2}, x_i + \frac{h}{2}\dot{x}_i + \frac{h}{4}k_1, \dot{x}_i + \frac{k_2}{2}\right) \tag{B-6}$$
$$k_4 = hf\left(t_i + h, x_i + h\dot{x}_i + \frac{h}{2}k_2, \dot{x}_i + k_3\right)$$

To use these equations, one starts with initial values $t_i = t_0$, $x_i = x_0$ and $\dot{x}_i = \dot{x}_0$ (for the second-order equation). Choosing an increment h for t_0 the four constants k are computed and these results are substituted into Eq. B–3 or B–5 in order to compute $x_{i+1} = x_1$, $\dot{x}_{i+1} = x_1$, corresponding to $t_{i+1} = t_1 = t_0 + h$. Repeating this process using t_1, x_1, $\dot{x}_1$ and h, values for x_2, $\dot{x}_2$ and $t_2 = t_1 + h$ are then computed, etc.

Computer programs which solve first and second-order differential equations by this method are listed in Figs. B–4 and B–5, respectively. In order to use these programs, the operator specifies the function $\dot{x} = f(t, x)$ or $\ddot{x} = f(t, x, \dot{x})$ (line 7), the initial values t_0, x_0, $\dot{x}_0$ (for second-order equation), the final time t_n, and the step size h. The output gives the values of t, x, and $\dot{x}$ for each time increment until t_n is reached.

```
1 PRINT"Runge-Kutta Method for 1-st order Differential Equation":PRINT
2 PRINT" To execute this program :":PRINT
3 PRINT"   1) Modify right hand side of the equation given below,"
4 PRINT"      then Press RETURN key"
5 PRINT"   2) Type  RUN 7"
6 PRINT:EDIT 7
7 DEF FNF(T,X)=5*T+X
8 CLS:PRINT" Initial Conditions":PRINT
9 INPUT"Input   t   = ",T
10 INPUT"        x   = ",X
11 INPUT"Final   t   = ",T1
12 INPUT"step size  = ",H:PRINT
13 PRINT"          t              x"
14 IF T>=T1+H THEN 23
15 PRINT USING"######.#####";T;X
16 K1=H*FNF(T,X)
17 K2=H*FNF(T+.5*H,X+.5*K1)
18 K3=H*FNF(T+.5*H,X+.5*K2)
19 K4=H*FNF(T+H,X+K3)
20 T=T+H
21 X=X+(K1+K2+K2+K3+K3+K4)/6
22 GOTO 14
23 END
```

Fig. B—4

```
1 PRINT"Runge-Kutta Method for 2-nd order Differential Equation":PRINT
2 PRINT" To execute this program :":PRINT
3 PRINT"   1) Modify right hand side of the equation given below,"
4 PRINT"      then Press RETURN key"
5 PRINT"   2) Type  RUN 7"
6 PRINT:EDIT 7
7 DEF FNF(T,X,XD)=
8 INPUT"Input   t   = ",T
9 INPUT"        x   = ",X
10 INPUT"      dx/dt = ",XD
11 INPUT"Final   t   = ",T1
12 INPUT"step size  = ",H:PRINT
13 PRINT"          t              x              dx/dt"
14 IF T>=T1+H THEN 24
15 PRINT USING"######.#####";T;X;XD
16 K1=H*FNF(T,X,XD)
17 K2=H*FNF(T+.5*H,X+.5*H*XD,XD+.5*K1)
18 K3=H*FNF(T+.5*H,X+(.5*H)*(XD+.5*K1),XD+.5*K2)
19 K4=H*FNF(T+H,X+H*XD+.5*H*K2,XD+K3)
20 T=T+H
21 X=X+H*XD+H*(K1+K2+K3)/6
22 XD=XD+(K1+K2+K2+K3+K3+K4)/6
23 GOTO 14
24 END
```

Fig. B—5

531

Example B–4

Solve the differential equation $\dot{x} = 5t + x$. Obtain the results for two steps using time increments of $h = 0.02$ s. At $t_0 = 0$, $x_0 = 0$.

Solution

This is a first-order equation, so Eqs. B–3 and B–4 apply. Thus, for $t_0 = 0$, $x_0 = 0$, $h = 0.02$, we have

$$k_1 = 0.02[0 + 0] = 0$$
$$k_2 = 0.02[5(0.1) + 0] = 0.001$$
$$k_3 = 0.02[5(0.01) + 0.0005] = 0.00101$$
$$k_4 = 0.02[5(0.02) + 0.00101] = 0.00202$$

$$x_1 = 0 + \frac{1}{6}[0 + 2(0.001) + 2(0.00101) + 0.00202] = 0.00101$$

Using the values $t_1 = 0 + 0.02 = 0.02$ and $x_1 = 0.00101$ with $h = 0.02$, the value for x_2 is now computed from Eqs. B–3 and B–4.

$$k_1 = 0.02[5(0.02) + 0.00101] = 0.00202$$
$$k_2 = 0.02[5(0.03) + 0.00202] = 0.00304$$
$$k_3 = 0.02[5(0.03) + 0.00253] = 0.00305$$
$$k_4 = 0.02[5(0.04) + 0.00406] = 0.00408$$

$$x_2 = 0.001 + \frac{1}{6}[0.00202 + 2(0.00304) + 2(0.00305) + 0.00408]$$

$$x_2 = 0.00405 \hspace{3cm} Ans.$$

To solve this problem using the computer program in Fig. B–4, the function is first entered on line 7, then the data $t_0 = 0$, $x_0 = 0$, $t_n = 0.04$, and $h = 0.02$ is specified. The results appear as

t	x
0.00000	0.00000
0.02000	0.00101
0.04000	0.00405

Ans.

Answers*

Chapter 1

1–1. (a) 78.5 N, (b) 0.392 mN, (c) 7.46 MN.

1–2. 2.42 Mg/m³.

1–5. (a) 0.185 Mg², (b) 4 μg², (c) 0.0122 km³.

1–6. (a) kN · m, (b) Gg/m, (c) μN/s², (d) GN/s.

1–7. (a) GN/s, (b) Gg/N, (c) GN/(kg · s).

1–9. (a) 27.1 N · m, (b) 70.7 kN/m³, (c) 1.27 mm/s.

1–10. $16.4(10^3)$ mm³.

1–11. 238 lb.

1–13. 4.96 μN.

1–14. $F = 0.1$ nN, $W_1 = 78.5$ N, $W_2 = 118$ N.

1–15. (a) 4.81 slug, (b) 70.2 kg, (c) 689 N, (d) 25.5 lb, (e) 70.2 kg.

Chapter 2

2–1. $F_R = 11.6$ kN, $\theta = 17.6°$ ∡θ.

2–2. $F_R = 45.8$ lb, $\phi = 162°$ ↖φ.

2–3. $F_x = -3.46$ kN, $F_y = 2.00$ kN.

2–5. $F_u = 186$ N, $F_v = 98.7$ N.

2–6. $F_u = 80$ N, $F_v = 54.7$ N.

2–7. $F_u = 320$ N, $F_v = 332$ N.

2–9. $F_\parallel = 46.5$ lb, $F_\perp = 99.7$ lb.

2–10. $F_R = 10.8$ kN, $\phi = 3.16°$ ↘φ.

2–11. $\theta = 54.9°$, $F_R = 10.4$ kN.

2–13. 744 lb, 23.8°.

2–14. $T = 54.7$ lb, $P = 42.6$ lb.

2–15. $\theta = 60°$, $P = 40$ lb, $T = 69.3$ lb.

2–17. 38.3°.

2–18. 60°.

2–19. $F_R = 2 F \cos (\theta/2)$, $\phi = \theta/2$.

2–21. $F_R = 19.2$ N, $\theta = 2.37°$ ↘θ.

2–22. $F_A = 3.66$ kN, $F_B = 7.07$ kN.

2–23. $\theta = 60°$, $F_A = 8.66$ kN, $F_B = 5$ kN.

2–25. $F_x = -15$ kN, $F_y = 26.0$ kN.

2–26. $F_x = 144$ lb, $F_y = -42$ lb.

2–27. $F_R = 4.46$ kN, $\theta = 26.2°$ ↘θ.

2–29. $F_R = 97.8$ N, $\theta = 46.5°$ ↘θ.

2–30. $F_R = 60.3$ kN, $\theta = 15.0°$ ↘θ.

2–31. $F_x = -13.2$ kN, $F_y = 18.8$ kN.

2–33. 24.9 N ↓ .

2–34. $F_{1x} = -516$ lb, $F_{1y} = 0$, $F_{2x} = 257$ lb, $F_{2y} = -306$ lb, $F_{3x} = 530$ lb, $F_{3y} = 306$ lb, $F_{4x} = -271$ lb, $F_{4y} = 0$.

2–35. $\theta = 29.1°$, $F_1 = 275$ N.

2–37. $\theta = 37.0°$, $F_1 = 889$ N.

2–38. $F_R = 717$ N, $\theta = 37.1°$.

2–39. $\theta = 117°$, $F_3 = 1.12F_1$.

2–41. $F_R = 9.33$ kN, $\alpha = 57.6°$, $\beta = 41.4°$, $\gamma = 113°$.

2–42. $F_R = 39.4$ lb, $\alpha = 52.8°$, $\beta = 141°$, $\gamma = 99.5°$.

2–43. $\mathbf{F}_1 = \{175\mathbf{i} + 175\mathbf{j} - 247\mathbf{k}\}$ N, $\mathbf{F}_2 = \{173\mathbf{i} - 100\mathbf{j} + 150\mathbf{k}\}$ N.

2–45. $\mathbf{F}_1 = \{176\mathbf{j} - 605\mathbf{k}\}$ lb, $\mathbf{F}_2 = \{125\mathbf{i} - 177\mathbf{j} + 125\mathbf{k}\}$ lb, $F_R = 496$ lb, $\alpha = 75.4°$, $\beta = 90.0°$, $\gamma = 165°$.

2–46. $\mathbf{F}_1 = \{225\mathbf{j} + 268\mathbf{k}\}$ N, $\mathbf{F}_2 = \{70.7\mathbf{i} + 50\mathbf{j} - 50\mathbf{k}\}$ N, $\mathbf{F}_3 = \{125\mathbf{i} - 177\mathbf{j} + 125\mathbf{k}\}$ N, $F_R = 407$ N, $\alpha = 61.3°$, $\beta = 76.0°$, $\gamma = 32.5°$.

2–47. $\alpha_1 = 45.6°$, $\beta_1 = 53.1°$, $\gamma_1 = 66.4°$.

*Note: Answers to every fourth problem are omitted.

2–49. 102 N.

2–50. $F_x = 40$ N, $F_y = 40$ N, $F_z = 56.6$ N.

2–51. $F = 32.7$ N, $F_y = 16.3$ N.

2–53. $F_2 = 180$ N, $\alpha_2 = 147°$, $\beta_2 = 119°$, $\gamma_2 = 75°$.

2–54. $F = 113$ lb, $\alpha = 45°$, $\beta = 69.3°$, $\gamma = 128°$.

2–55. $F_R = 32.4$ lb, $\alpha_2 = 122°$, $\beta_2 = 74.5°$, $\gamma_2 = 144°$.

2–57. $\mathbf{r} = \{-3\mathbf{i} - 6\mathbf{j} + 2\mathbf{k}\}$ m, $r = 7$ m, $\alpha = 115°$, $\beta = 149°$, $\gamma = 73.4°$.

2–58. $r = 6.83$ ft, $\alpha = 79.3°$, $\beta = 116°$, $\gamma = 28.5°$.

2–59. 732 mm.

2–61. 6.63 m.

2–62. $x = 4.42$ m, $y = 4.42$ m.

2–63. 12.4 km.

2–65. $\mathbf{F} = \{-37.7\mathbf{i} - 15.6\mathbf{j} + 44.0\mathbf{k}\}$ lb, $\alpha = 129°$, $\beta = 105°$, $\gamma = 42.9°$.

2–66. $\mathbf{F} = \{452\mathbf{i} + 371\mathbf{j} - 136\mathbf{k}\}$ lb, $\alpha = 41.1°$, $\beta = 51.9°$, $\gamma = 103°$.

2–67. $F_R = 822$ N, $\alpha = 72.8°$, $\beta = 83.3°$, $\gamma = 162°$.

2–69. $\mathbf{F}_1 = \{-26.2\mathbf{i} - 41.9\mathbf{j} + 62.9\mathbf{k}\}$ lb, $\mathbf{F}_2 = \{13.4\mathbf{i} - 26.7\mathbf{j} - 40.1\mathbf{k}\}$ lb, $F_R = 73.5$ lb, $\alpha = 100°$, $\beta = 159°$, $\gamma = 71.9°$.

2–70. $\mathbf{F} = \{98.1\mathbf{i} + 269\mathbf{j} - 201\mathbf{k}\}$ lb.

2–71. $\mathbf{F}_A = \{254\mathbf{i} + 50.7\mathbf{j} - 152\mathbf{k}\}$ lb, $\mathbf{F}_B = \{260\mathbf{i} + 260\mathbf{j} - 156\mathbf{k}\}$ lb, $F_R = 675$ lb, $\alpha = 40.4°$, $\beta = 62.6°$, $\gamma = 117°$.

2–73. $r_{AB} = 492$ mm, $\mathbf{F} = \{-13.2\mathbf{i} - 18.3\mathbf{j} + 19.8\mathbf{k}\}$ N.

2–74. $x = -6.41$ m, $y = 13.4$ m.

2–75. $F_R = 1.50$ kN, $\alpha = 77.6°$, $\beta = 90.6°$, $\gamma = 167°$.

2–77. $x = 7.65$ ft, $y = 4.24$ ft, $z = 3.76$ ft.

2–78. $x = 8.67$ ft, $y = 1.89$ ft.

2–79. $\mathbf{F} = \{-34.3\mathbf{i} + 22.9\mathbf{j} - 68.6\mathbf{k}\}$ lb.

2–81. 121°.

2–82. 109°.

2–83. 2.67 m.

2–85. $F_A = 293$ N, $F_B = 566$ N, $F_{P_{OA}} = 693$ N, $F_{P_{OB}} = 773$ N.

2–86. 40.9 lb.

2–87. 54.0 N.

2–89. 70.5°.

2–90. 65.8°.

2–91. 31.1 N.

2–93. $\theta = 74.4°$, $\phi = 55.4°$.

2–94. $(F_{AC})_P = 22.7$ N, $(F_{AB})_P = 14.8$ N.

2–95. $F_1 = 333$ lb, $F_2 = 373$ lb.

2–97. $\mathbf{F}_1 = \{90\mathbf{i} - 120\mathbf{j}\}$ lb, $\mathbf{F}_2 = \{-275\mathbf{j}\}$ lb, $\mathbf{F}_3 = \{-37.5\mathbf{i} - 65.0\mathbf{j}\}$ lb, $F_R = 463$ lb.

2–98. $F_{AB} = 735$ N, $F_{CA} = 820$ N.

2–99. (a) $F_x = 35.4$ lb, $F_y = 35.4$ lb, (b) $F_x = 14.9$ lb, $F_{y'} = 40.8$ lb.

2–101. $F_R = 3.44$ kN, same direction, $F_{R\max} = 4.50$ kN.

2–102. $\mathbf{F}_1 = \{F_1 \cos \theta \mathbf{i} + F_1 \sin \theta \mathbf{j}\}$ N, $\mathbf{F}_2 = \{350\mathbf{i}\}$ N, $\mathbf{F}_3 = \{-100\mathbf{j}\}$ N, $\theta = 67.0°$, $F_1 = 434$ N.

2–103. $\mathbf{F}_{EA} = \{12\mathbf{i} - 8\mathbf{j} - 24\mathbf{k}\}$ kN, $\mathbf{F}_{EB} = \{12\mathbf{i} + 8\mathbf{j} - 24\mathbf{k}\}$ kN, $\mathbf{F}_{EC} = \{-12\mathbf{i} + 8\mathbf{j} - 24\mathbf{k}\}$ kN, $\mathbf{F}_{ED} = \{-12\mathbf{i} - 8\mathbf{j} - 24\mathbf{k}\}$ kN, $\Sigma \mathbf{F} = \mathbf{F}_R = \{-96\mathbf{k}\}$ kN.

2–105. $F_{AC} = 43.9$ lb, $F_{AB} = 53.8$ lb.

2–106. $\theta = 79.5°$, $F_{AC} = 90.8$ lb.

2–107. $F_R = 407$ N, $\theta = 26.5°$ $\theta\nearrow$.

Chapter 3

3–1. $F_1 = 439$ N, $F_2 = 233$ N.

3–2. $\theta = 16.4°$, $F_1 = 137$ lb.

3–3. $\theta = 31.8°$, $F = 4.94$ kN.

3–5. $\theta = 78.7°$, $F_{CD} = 157$ lb.

3–6. $T_{BC} = 2.99$ kN, $T_{AB} = 3.78$ kN.

3–7. $F = 3.59$ kN, $T = 0.536$ kN.

3–9. $T = 10.2$ kN, $F = 7.93$ kN.

3–10. $\theta = 43.1°$, $m_A = 20.5$ kg.

3–11. 11.5°.

3–13. 7.2 ft.

3–14. $x_{AD} = 0.491$ m, $x_{AC} = 0.793$ m, $x_{AB} = 0.467$ m.

3–15. 158 N.

3–17. $\theta = 20°$, $T = 305$ lb.

3–18. 1.13 mN.

3–19. 2.65 ft.

3–21. 88.8 lb.

3–22. 35.0°.

3–23. 43.0°.

3–25. 7.18°.

3–26. 40.8 lb.

3–27. 6.59 m.

3–29. $F_1 = 266$ lb, $F_3 = 80.1$ lb, $F_2 = 82.2$ lb.

3–30. $F_1 = 583$ N, $F_2 = 136$ N, $F_3 = 553$ N.

3–31. $F_{AD} = 1200$ N, $F_{AC} = 400$ N, $F_{AB} = 800$ N.

3–33. $F_{AB} = 0.980$ kN, $F_{AC} = 0.463$ kN, $F_{AD} = 1.55$ kN.

3–34. $F_{AB} = 441$ N, $F_{AC} = 515$ N, $F_{AD} = 221$ N.

3–35. $F_{AB} = 35.9$ lb, $F_{AC} = F_{AD} = 25.4$ lb.

3–37. $F_{AC} = 574$ lb, $F_{AB} = 500$ lb, $F_{CD} = 1.22$ kip, $F_{CE} = F_{CF} = 416$ lb.

3–38. $F_{AD} = F_{AC} = 104$ N, $F_{AB} = 220$ N.

3–39. 1.64 ft.

3–41. (CD) 106 lb.

3–42. 5.33 ft.

3–43. 6 lb.

3–45. $F_{AB} = 3.55$ mN, $F_{AE} = 5.03$ mN, $F_{BD} = 2.67$ mN, $F_{BC} = 1.60$ mN.

3–46. $F_{AB} = 175$ lb, $\ell = 2.34$ ft.

3–47. $T_B = 109$ lb, $T_C = 47.4$ lb, $T_D = 87.8$ lb.

3–49. $F_{AB} = 2.52$ kN, $F_{CB} = 2.52$ kN, $F_{BD} = 3.65$ kN.

3–50. $F_C = 736$ N, $F_B = 441$ N.

Chapter 4

4–5. 2.30 kip · ft $\downarrow$.

4–6. $\mathbf{M}_P = \{-57\mathbf{i} + 27\mathbf{j} + 23\mathbf{k}\}$ kN · m.

4–7. $\mathbf{M}_P = \{343\mathbf{i} + 274\mathbf{j} + 309\mathbf{k}\}$ lb · ft.

4–9. (a) 73.9 N · m $\downarrow$, (b) 82.2 N $\leftarrow$.

4–10. $(M_A)_1 = 3$ kip · ft $\downarrow$, $(M_A)_2 = 5.60$ kip · ft $\downarrow$, $(M_A)_3 = 2.59$ kip · ft $\downarrow$.

4–11. $(M_B)_1 = 4.12$ kip · ft $\uparrow$, $(M_B)_2 = 2$ kip · ft $\uparrow$, $(M_B)_3 = 40$ lb · ft $\uparrow$.

4–13. 768 lb · ft $\downarrow$, 636 lb · ft $\uparrow$, clockwise.

4–14. 39.8 lb.

4–15. (a) $(M_A)_{max} = 330$ lb · ft $\downarrow$, $\theta = 76.0°$, (b) $(M_A)_{min} = 0$, $\theta = 166°$.

4–17. $M_P = 17.1$ lb · ft $\uparrow$, $M_F = 23.8$ lb · ft $\downarrow$, No.

4–18. $(M_A)_1 = 13.6$ N · m $\uparrow$, $(M_A)_2 = 11.7$ N · m $\downarrow$.

4–19. $F = 1.33$ kip, $F = 1.63$ kip.

4–21. 33.6 °.

4–22. $y = 2$ m, $z = 1$ m.

4–23. $y = 1$ m, $z = 3$ m, $d = 1.15$ m.

4–25. $(\mathbf{M}_A)_1 = \{-2.4\mathbf{i} + 3.6\mathbf{j}\}$ kN · m, $(\mathbf{M}_A)_2 = \{1.2\mathbf{i} + 1.2\mathbf{j}\}$ kN · m, $(\mathbf{M}_A)_3 = \{0.5\mathbf{i}\}$ kN · m.

4–26. $\mathbf{M}_C = \{-35.4\mathbf{i} - 128\mathbf{j} - 222\mathbf{k}\}$ lb · ft.

4–27. $\mathbf{M}_A = \{-5.39\mathbf{i} + 13.1\mathbf{j} + 11.4\mathbf{k}\}$ N · m.

4–29. $\mathbf{M}_A = \{321\mathbf{i} - 464\mathbf{j} - 32.1\mathbf{k}\}$ N · m.

4–30. $\mathbf{M}_{Oa} = \{218\mathbf{j} + 163\mathbf{k}\}$ N · m.

4–31. $\mathbf{M}_P = \{-45.3\mathbf{i} - 49.1\mathbf{j} + 15.1\mathbf{k}\}$ kN · m.

4–33. 321 N · m.

4–34. 391 N · m.

4–35. 6.25 lb · ft.

4–37. 14.8 N · m.

4–38. 20.2 N.

4–39. 18.3 kN · m $\uparrow$.

4–41. $\mathbf{M}_C = \{56\mathbf{i} - 15\mathbf{j} + 57\mathbf{k}\}$ kN · m.

4–42. $\mathbf{M}_C = \{557\mathbf{i} - 7.14\mathbf{j} + 857\mathbf{k}\}$ N · m.

4–43. No, no effect.

4–45. 1.54 m.

4–46. $\theta = 32.9°$, $M_3 = 651$ N · m.

4–47. $F = 300$ N, $P = 500$ N, $d = 3.85$ m.

4–49. $M_R = 59.9$ N · m, $\alpha = 99.0°$, $\beta = 106°$, $\gamma = 18.3°$.

4–50. $\mathbf{M}_{C_R} = \{63.6\mathbf{i} - 170\mathbf{j} + 264\mathbf{k}\}$ N · m.

4–51. $\mathbf{M}_C = \{-50\mathbf{i} + 60\mathbf{j}\}$ lb · ft, $M_C = 78.1$ lb · ft.

4–53. $F_P = 375$ N $30°$, $M_P = 737$ N · m $\uparrow$.

4–54. $F_R = 80$ lb, $\theta = 50°$, $M_{R_P} = 696$ lb · ft $\downarrow$.

4–55. $F_R = 6$ kN, $\theta = 67.4°\angle\theta$, $M_{R_P} = 27.2$ kN · m $\downarrow$.

4–57. $F_R = 375$ lb $\uparrow$, $x = 15.5$ ft.

4–58. $F_R = 115$ lb, $\theta = 73.2°\,\theta$, $x = -4.35$ ft.

4–59. $F_R = 5.87$ kN, $\theta = 77.4°\,\theta$, $x = -2.33$ m.

4–61. $\mathbf{F}_R = \{-300\mathbf{i} + 200\mathbf{j} - 500\mathbf{k}\}$ N, $\mathbf{M}_{R_P} = \{-3800\mathbf{i} - 7200\mathbf{j} - 600\mathbf{k}\}$ N · m.

4–62. $\mathbf{F}_R = \{-80\mathbf{i} - 80\mathbf{j} + 40\mathbf{k}\}$ lb, $\mathbf{M}_{R_P} = \{-240\mathbf{i} + 720\mathbf{j} + 960\mathbf{k}\}$ lb · ft.

4–63. $\mathbf{F}_R = \{8\mathbf{i} + 6\mathbf{j} + 8\mathbf{k}\}$ kN, $\mathbf{M}_{R_P} = \{-46\mathbf{i} + 66\mathbf{j} - 56\mathbf{k}\}$ kN · m.

4–65. $F = 40$ N $60°$, $M_O = 11.4$ N · m $\downarrow$.

4–66. $F_R = 1.35$ kN, $\theta = 33.3°\,\theta$, $M_O = 3.02$ kN · m $\downarrow$.

4–67. $F_R = 69.3$ lb, $\theta = 19.0°\,\theta$, $d = 3.22$ ft.

4–69. $F_R = 12.1$ kip, $d = 10.1$ ft.

4–70. $F_R = 1.03$ kip, $\theta = 54.4°\,\theta$, $d = 5.94$ ft.

4–71. $F_R = 4.43$ kN, $\theta = 71.6°\,\theta$, $d = 3.52$ m.

4–73. $F_R = 197.0$ lb, $\theta = 42.6°\angle\theta$, $d_{AB} = 5.24$ ft.

4–74. $F_R = 197.0$ lb, $\theta = 42.6°\angle\theta$, $d_{BC} = 0.824$ ft.

4–75. $F_2 = 25.9$ lb, $F_1 = 68.1$ lb, $\theta = 18.1°\,\theta$.

4–77. $F_R = 65.9$ lb, $\theta = 49.8°\,\theta$, $d_{BC} = 4.62$ ft.

4–78. $F_R = 65.9$ lb, $\theta = 49.8°\,\theta$, $(M_R)_A = 106$ lb · ft $\downarrow$.

4–79. $\mathbf{F} = \{14.3\mathbf{i} + 42.9\mathbf{j} - 21.4\mathbf{k}\}$ N, $\mathbf{M}_A = \{-129\mathbf{i} + 50.0\mathbf{j} + 14.3\mathbf{k}\}$ N · m.

4–81. $\mathbf{F}_R = \{1.15\mathbf{i} + 0.620\mathbf{j} - 1.21\mathbf{k}\}$ kN, $\mathbf{M}_{R_O} = \{-7.44\mathbf{i} + 13.8\mathbf{j}\}$ kN · m.

4–82. $\mathbf{F}_R = \{0.232\mathbf{i} + 5.06\mathbf{j} + 12.4\mathbf{k}\}$ kN, $\mathbf{M}_O = \{36\mathbf{i} - 26.1\mathbf{j} + 12.3\mathbf{k}\}$ kN · m.

4–83. $F_R = 140$ kN, $y = 7.14$ m, $x = 5.71$ m.

4–85. $F_R = 35$ kN $\downarrow$, $y = -0.476$ m, $x = -0.518$ m.

4–86. $F_R = 40$ lb $\uparrow$, $y = 0.5$ ft.

4–87. $\mathbf{F}_R = \{33.8\mathbf{i} + 30.0\mathbf{j} + 50.9\mathbf{k}\}$ lb,
$\mathbf{M}_{R_O} = \{96.9\mathbf{i} - 88.7\mathbf{k}\}$ lb $\cdot$ ft.

4–89. $z = 0.28d$, $\mathbf{F} = \{0.6F\mathbf{i} + 0.8F\mathbf{j}\}$,
$\mathbf{M} = \{-0.576Fd\mathbf{i} - 0.768Fd\mathbf{j}\}$.

4–90. 18.0 N $\cdot$ m, $\theta = 120°$.

4–91. 464 lb $\cdot$ ft.

4–93. $F = 450$ N, $\theta = 27.4°$ $\searrow_\theta$, $M = 1.71$ kN $\cdot$ m.

4–94. $F_R = 4.50$ kN, $d = 2.22$ m.

4–95. $\mathbf{M}_O = \{-1.77\mathbf{i} - 1.10\mathbf{j}\}$ kN $\cdot$ m.

4–97. $M_{x'} = 26.0$ kip $\cdot$ ft, $M_{y'} = 15.5$ kip $\cdot$ ft.

4–98. (a) $\mathbf{F} = \{-1.50\mathbf{k}\}$ kip, $\mathbf{M}_O = \{-1.50\mathbf{i}\}$ kip $\cdot$ ft,
(b) $\mathbf{F} = \{-1.50\mathbf{k}\}$ kip, $\mathbf{M}_A = \{-2.38\mathbf{i} - 0.750\mathbf{j}\}$ lb $\cdot$ ft.

4–99. $F_C = 400$ lb, $F_D = 500$ lb.

4–101. 360 N $\cdot$ m.

4–102. $\mathbf{M}_O = \{1.06\mathbf{i} + 1.06\mathbf{j} - 4.03\mathbf{k}\}$ N $\cdot$ m, $\alpha = 75.7°$,
$\beta = 75.7°$, $\gamma = 159°$.

4–103. 4.03 N $\cdot$ m.

Chapter 5

5–1. R_A, R_B, R_C.

5–2. N_A, N_B.

5–3. A_x, A_y, B_y.

5–5. A_x, A_y, T_B.

5–6. A_x, A_y.

5–7. A_x, A_y, B_y.

5–9. A_x, A_y, F_B.

5–10. A_x, A_y, F_{BC}.

5–11. C_x, C_y, F_{AB}.

5–13. (a) A_x, A_y, B_y; (b) C_x, C_y, M_C; (c) T_{AB}, T_{EB}, T_{CD};
(d) (housing) A_x, A_y, B_x, B_y, (wheel) A_x, A_y, N_A.

5–14. (a) T_{AC}, T_{BD}, (b) A_x, A_y, T_{BC}, (c) B_x, B_y, T_{CD},
(d) N_A, N_B.

5–15. $F = 19.4$ kN, $R_B = 9.66$ kN, $R_A = 8.54$ kN.

5–17. $F_A = 48.7$ N, $\theta = 82.5°$ $\angle_\theta$.

5–18. $T_{BD} = 2.55$ kN, $A_y = 0.490$ kN, $A_x = 2.55$ kN.

5–19. $F_B = 3.76$ kip, $A_x = 3.26$ kip, $A_y = 3.12$ kip.

5–21. $F_{AB} = 340$ lb, $C_x = 204$ lb, $C_y = 672$ lb.

5–22. $B_x = 18$ lb, $A_x = 18$ lb, $A_y = 30$ lb.

5–23. $R_C = 493$ N, $R_A = 247$ N, $R_B = 554$ N.

5–25. $A_y = 390$ N, $B_x = 0$, $B_y = 60$ kN.

5–26. $F_{BC} = 80$ kN, $A_x = 54$ kN, $A_y = 16$ kN.

5–27. $N_a = 105$ lb, $B_x = 97.4$ lb, $B_y = 269$ lb.

5–29. $R_A = 1.06$ kN, $F_{CD} = 0.501$ kN, $B_y = 1.42$ kN.

5–30. $B_y = 642$ N, $A_y = 180$ N, $A_x = 192$ N.

5–31. $N_B = 2.14$ kip, $A_x = 1.29$ kip, $A_y = 1.49$ kip.

5–33. $B_x = 989$ N, $A_x = 989$ N, $B_y = 186$ N.

5–34. $F_{CD} = 486$ kip, $A_x = 372$ kip, $A_y = 188$ kip.

5–35. $A_y = 400$ lb, $B_x = 533$ lb, $A_x = 533$ lb.

5–37. $F = 981$ N, $F_A = F_B = 425$ N.

5–38. $26.4°$.

5–39. $T = 1.01$ kN, $D_x = 0.840$ kN, $D_y = 0.508$ kN.

5–41. 105 N $\uparrow$.

5–42. $C_x = 333$ lb, $C_y = 722$ lb.

5–43. ($x = 10$ ft), $B_x = 4.17$ kip, $A_x = 4.17$ kip,
$A_y = 5.00$ kip, ($x = 4$ ft) $B_x = 1.67$ kip,
$A_x = 1.67$ kip, $A_y = 5.00$ kip.

5–45. $P = 30.2$ N, $P = 20.0$ N.

5–46. 1.64 N.

5–47. $n = 5$, $x = 8$ in.

5–49. $8.55°$.

5–50. $W_L = L(S_1 - S_2)/d$.

5–51. $d = \frac{3}{4}a$.

5–53. 568 mm.

5–54. $47.5°$.

5–57. $T_2 = 3.01$ kN, $T_1 = 2.52$ kN, $\bar{x} = 2.30$ m.

5–59. $\phi = \tan^{-1}(0.5 \cot \theta)$.

5–61. (a) I_x, I_y, I_z, M_x, M_y, M_z; (b) A_y, A_z, C_x, C_y, C_z;
(c) T_{BE}, T_{BD}, A_x, A_z, C_x, C_z; (d) y axis along
auger, x axis along pin at C, C_x, C_y, C_z, M_y, M_z, T_{AB}.

5–62. $P = 62.5$ lb, $B_x = 22.3$ lb, $A_x = 84.8$ lb, $B_z = 25$ lb,
$A_z = 25$ lb, $B_y = 0$.

5–63. $B_z = 373$ N, $A_x = 0$, $A_y = 0$, $A_z = 333$ N,
$T_{CD} = 43.5$ N.

5–65. $T_A = 235$ lb, $T_B = 293$ lb, $T_C = 372$ lb.

5–66. $R_F = 13.7$ kip, $R_D = 21.1$ kip, $R_E = 24.1$ kip.

5–67. $A_x = 360$ N, $A_y = 240$ N, $A_z = 720$ N,
$(M_A)_x = 7.20$ kN $\cdot$ m, $(M_A)_y = 10.8$ kN $\cdot$ m,
$(M_A)_z = 0$.

5–69. $T_{BD} = T_{BC} = 21.9$ kN, $A_x = 0$, $A_y = 18.8$ kN,
$A_z = 5.50$ kN.

5–70. $T_{ED} = 21.3$ kip, $T_{AC} = T_{BC} = 8.89$ kip.

5–71. $F_{CD} = 1.02$ kN, $A_z = 208$ N, $B_z = 139$ N,
$A_y = 573$ N, $B_y = 382$ N.

5–73. $F_{BC} = 0$, $A_y = 0$, $A_z = 800$ lb, $M_{A_x} = 4,800$ lb $\cdot$ ft,
$M_{A_y} = 0$, $M_{A_z} = 0$.

5–74. $A_x = 0$, $A_y = 1500$ lb, $A_z = 750$ lb, $T = 919$ lb.

5–75. $C_z = 80.3$ N, $D_z = 39.7$ N, $T = 58.0$ N,
$C_y = 27.0$ N, $D_y = 74.3$ N, $D_x = 0$.

5–77. $A_z = 320$ N, $A_y = 240$ N, $T_{BC} = 535$ N, $D_x = 0$,
$D_y = 240$ N, $D_z = 285$ N.

5–78. $A_z = 5$ kN, $A_x = 0$, $A_y = 16.7$ kN.

5–79. $T_{BC} = 205$ N, $T_{ED} = 629$ N, $A_x = 32.4$ N,
$A_y = 107$ N, $A_z = 1.28$ kN.

5–81. $A_x = 24.2$ lb, $A_y = 240$ lb, $A_z = 32.3$ lb,
$T_{BCD} = 77.5$ lb, $T_{CE} = 137$ lb.

5–82. $A_x = 8.0$ kN, $A_y = 0$, $A_z = 24.4$ kN,
$M_y = 20$ kN · m, $M_x = 572$ kN · m, $M_z = 64$ kN · m.

5–83. $A_y = 8$ kN, $B_y = 5$ kN, $B_x = 5.20$ kN.

5–85. $F_s = 1.5$ N, $P = 1.5$ N, $F = 0.3$ N.

5–86. $P = 75$ lb, $A_y = 0$, $B_z = 75$ lb,
$A_z = 75$ lb, $B_x = 112$ lb, $A_x = 37.5$ lb.

5–87. $F_A = 30$ lb, $F_B = 36.2$ lb, $F_C = 9.38$ lb.

5–89. $F_D = 110$ kip, $F_L = 170$ kip, $s = 11.4$ ft.

5–90. $T_{BD} = T_{CD} = 70.2$ lb, $A_x = 37.5$ lb, $A_y = 0$,
$A_z = 37.5$ lb.

5–91. $A_x = 107$ lb, $A_y = 0$, $A_z = 107$ lb, $F = 428$ lb.

Chapter 6

6–1. $F_{AB} = 286$ lb (T), $F_{BC} = 808$ lb (T),
$F_{AC} = 571$ lb (C).

6–2. $F_{AD} = 849$ lb (C), $F_{AB} = 600$ lb (T),
$F_{BC} = 600$ lb (T), $F_{BD} = 400$ lb (C),
$F_{DC} = 1.41$ kip (T), $F_{DE} = 1.60$ kip (C).

6–3. $F_{AD} = 9.90$ kN (C), $F_{AB} = 7$ kN (T),
$F_{DC} = 14.8$ kN (C), $F_{DB} = 4.95$ kN (T),
$F_{BC} = 10.5$ kN (T).

6–5. $F_{CD} = F_{AD} = 0.687P$ (T), $F_{CB} = F_{AB} = 0.943P$ (C),
$F_{DB} = 1.33P$ (T).

6–6. $F_{CD} = F_{BC} = 2.67$ kN (C), $F_{CG} = F_{CF} = 13.3$ kN (C),
$F_{AG} = F_{EF} = 13.3$ kN (C), $F_{BA} = F_{DE} = 8.43$ kN (C),
$F_{BG} = F_{DF} = 0$.

6–7. $F_{CB} = 3$ kN (T), $F_{CD} = 2.60$ kN (C),
$F_{DE} = F_{CD} = 2.60$ kN (C), $F_{DB} = 2$ kN (T),
$F_{BE} = 2$ kN (C), $F_{BA} = 5$ kN (T).

6–9. $F_{AE} = 372$ N (C), $F_{AB} = 332$ N (T),
$F_{BC} = 332$ N (T), $F_{BE} = 196$ N (C),
$F_{EC} = 557$ N (T), $F_{ED} = 929$ N (T),
$F_{DC} = 416$ N (T).

6–10. $F_{AB} = 8.21$ kN (C), $F_{AG} = 4.56$ kN (T),
$F_{BG} = 8.21$ kN (T), $F_{BC} = 9.11$ kN (C),
$F_{GC} = 1.40$ kN (C), $F_{GF} = 8.33$ kN (T),
$F_{CF} = 1.40$ kN (C), $F_{CD} = 7.55$ kN (C),
$F_{FD} = 6.81$ kN (T), $F_{FE} = 3.77$ kN (T),
$F_{DE} = 6.81$ kN (C).

6–11. $F_{BC} = 40.4$ lb (C), $F_{CD} = F_{AB} = 57.7$ lb (C),
$F_{DE} = F_{AE} = 28.9$ lb (T), $F_{CE} = F_{BE} = 23.1$ lb (T).

6–13. $F_{JI} = 0$, $F_{JH} = 300$ lb (C), $F_{GI} = F_{IH} = 0$,
$F_{HF} = 0$, $F_{HG} = 300$ lb (C), $F_{GF} = 780$ lb (T),
$F_{GE} = 720$ lb (C), $F_{EF} = 0$.

6–14. BL, BK, CK, FH, FI, EI, EJ, $F_{BC} = 14.9$ kN (C),
$F_{JI} = 10.7$ kN (T), $F_{CJ} = 3.77$ kN (C).

6–15. $F_{BA} = 0$, $F_{BC} = 0$, $F_{AC} = P \cos 2\theta / \sin \theta$ (T),
$F_{AD} = P \cos 2\theta$ ctn θ (C), $F_{DC} = P$ ctn θ (C).

6–17. $F_{BH} = F_{DF} = 0$, $F_{AB} = F_{DE} = 594$ lb (C),
$F_{EF} = F_{FG} = F_{HG} = F_{AH} = 356$ lb (T),
$F_{DC} = F_{BC} = 396$ lb (C), $F_{DG} = F_{BG} = 2.84$ lb (C),
$F_{CG} = 155$ lb (T).

6–18. $F_{AB} = F_{BC} = 3.75$ kip (C), $F_{AH} = 3$ kip (T),
$F_{BH} = 1$ kip (C), $F_{EF} = 3$ kip (T),
$F_{ED} = F_{DC} = 3.75$ kip (C), $F_{DF} = 1$ kip (C),
$F_{HC} = 1$ kip (T), $F_{HG} = 2$ kip (T),
$F_{CF} = 1$ kip (T), $F_{CG} = 2$ kip (C), $F_{FG} = 2$ kip (T).

6–19. NB, NC, OD, OC, LE, JH, JG, $F_{CD} = 5.62$ kN (T),
$F_{CM} = 2$ kN (T).

6–21. $F_{AB} = F_{BC} = F_{CD} = 1.51$ kN (C), $F_{AI} = F_{IH} = $
3.31 kN (T), $F_{FG} = F_{FE} = F_{ED} = 6.49$ kN (C),
$F_{GH} = 5.62$ kN (T), $F_{DH} = 4.62$ kN (T).

6–22. $F_{HG} = 29$ kN (C), $F_{BC} = 20.5$ kN (T),
$F_{HC} = 12.0$ kN (T).

6–23. $F_{GF} = 29$ kN (C), $F_{CD} = 23.5$ kN (T),
$F_{CF} = 7.78$ kN (T).

6–25. $F_{HI} = 35$ kN (C), $F_{BC} = 50$ kN (T),
$F_{HB} = 21.2$ kN (C).

6–26. $F_{KJ} = 11.2$ kip (T), $F_{CD} = 9.38$ kip (C),
$F_{CJ} = 3.12$ kip (C), $F_{DJ} = 0$.

6–27. $F_{JI} = 7.50$ kip (T), $F_{EI} = 2.50$ kip (C).

6–29. $F_{CD} = 2.50$ kip (T), $F_{FE} = 3.50$ kip (T),
$F_{CE} = 2.50$ kip (C).

6–30. $F_{BC} = 0$, $F_{FE} = 3.50$ kip (T), $F_{FC} = 3.54$ kip (T).

6–31. $F_{EF} = P$ (C), $F_{CB} = 1.12P$ (T), $F_{BE} = 0.5P$ (T).

6–33. $F_{BC} = 700$ lb (T), $F_{FC} = 76.9$ lb (T).

6–34. $F_{BC} = 8$ kN (T), $F_{HG} = 10.1$ kN (C),
$F_{BG} = 1.80$ kN (T).

6–35. $F_{CD} = 8$ kN (T), $F_{GD} = 0$, $F_{GC} = 7$ kN (T).

6–37. $F_{KJ} = 3.07$ kip (C), $F_{CD} = 3.07$ kip (T),
$F_{NJ} = 167$ lb (C), $F_{ND} = 167$ lb (T).

6–38. $F_{JI} = 2.13$ kip (C), $F_{DE} = 2.13$ kip (T).

6–39. $F_{CD} = 7$ kN (T), $F_{KJ} = 5.59$ kN (C),
$F_{KD} = 2.83$ kN (C).

6–41. $F_{ML} = 38.4$ kN (T), $F_{DE} = 37.1$ kN (C),
$F_{DL} = 3.84$ kN (C).

6–42. $F_{EF} = 37.1$ kN (C), $F_{EL} = 6$ kN (T).

6–43. $F_{FG} = 15.3$ kN (C), $F_{BC} = 3.67$ kN (T),
$F_{BF} = 11.8$ kN (T).

6–45. $F_{EB} = E_y = 4.80$ kN (T), $F_{AB} = 6.46$ kN (T),
$F_{AC} = F_{AD} = 1.50$ kN (C),
$F_{BC} = F_{BD} = 3.70$ kN (C).

6–46. $F_{DC} = F_{DA} = 2.59$ kN (C), $F_{DB} = 3.85$ kN (C),
$F_{BC} = F_{BA} = 0.890$ kN (T), $F_{AC} = 0.617$ kN (T).

6–47. $F_{BC} = 0$, $F_{CD} = 0$, $F_{CF} = 8$ kN (C), $F_{BD} = 0$,
$F_{BA} = 6$ kN (C), $F_{AD} = 0$, $F_{DF} = 0$,
$F_{DE} = 9$ kN (C), $F_{EF} = 0$, $F_{AE} = 0$, $F_{AF} = 0$.

6–49. $F_{CE} = 707$ lb (T), $F_{BC} = 0$, $F_{CD} = 500$ lb (C),
$F_{EB} = 0$, $F_{BF} = 1.13$ kip (T), $F_{BA} = 800$ lb (C),
$F_{ED} = 500$ lb (C), $F_{EF} = 0$, $F_{FD} = 625$ lb (T),
$F_{AD} = 375$ lb (C), $F_{AF} = 1.30$ kip (C).

6–50. $F_{CD} = 919$ N (C), $F_{DB} = 0$.

6–51. (a) DA, 3 unks., (b) ABC, 5 unks., (c) BD,
3 unks., (d) IE, FH, 9 unks., (e) 5 unks., (f) AB,
CD, 6 unks., (g) AB, DB, FB, EK, LI, 8 unks.,
(h) IB, HC, HD, HF, 8 unks.

6–53. $P = 10$ lb, $R_A = 20$ lb, $R_B = 10$ lb.

6–54. $P = 18.8$ lb, $R_A = 27.5$ lb, $R_B = 18.8$ lb.

6–55. $P = 21.8$ N, $R_A = 43.6$ N, $R_B = 43.6$ N, $R_C = 131$ N.

6–57. 100 lb, $\theta = 14.6°$.

6–58. $R_B = 26.7$ lb, $A_x = 0$, $A_y = 34.7$ lb.

6–59. $A_x = 2.33$ kN, $A_y = 600$ N.

6–61. $B_y = 449$ lb, $A_x = 92.3$ lb, $A_y = 737$ lb,
$M_A = 4.05$ kip · ft.

6–62. $E_y = 80$ lb, $B_y = 10.8$ kip, $A_y = 96.7$ lb.

6–63. $A_y = 360$ lb, $B_y = 940$ lb, $C_x = 250$ lb, $C_y = 900$ lb,
$M_C = 7800$ lb · ft.

6–65. $B_x = A_x = 6.67$ kN, $B_y = A_y = 5.00$ kN,
$C_x = 6.67$ kN, $C_y = 0$, $D_x = 4$ kN, $D_y = 5$ kN,
$M_D = 24.0$ kN.

6–66. $A_x = B_x = 66.7$ lb, $A_y = B_y = 40$ lb.

6–67. 20.7 Mg, No.

6–69. $A_x = 120$ lb, $A_y = 0$, $N_C = 15.0$ lb.

6–70. $B_y = 112$ lb, $A_y = 62.5$ lb, $A_x = 133$ lb, $B_x = 183$ lb,
$D_x = 50$ lb, $D_y = 50$ lb.

6–71. $B_x = 341$ lb, $B_y = 500$ lb, $E_y = 520$ lb,
$E_x = 1.00$ kip, $C_x = 662$ lb, $C_y = 20$ lb.

6–73. $C_x = 75$ lb, $C_y = 100$ lb.

6–74. $B_x = 75$ lb, $B_y = 300$ lb, $E_x = 225$ lb, $E_y = 600$ lb,
$D_x = 300$ lb, $D_y = 300$ lb.

6–75. $A_x = 0$, $A_y = 34$ N, $C_y = 6.54$ N, $C_x = 0$,
$x = 291$ mm, $B_x = 0$, $B_y = 1.06$ N.

6–77. (a) $F = 175$ lb, $N_C = 350$ lb, (b) $F = 87.5$ lb,
$N_C = 87.5$ lb.

6–78. (a) $F = 205$ lb, $N_C = 380$ lb, (b) $F = 102$ lb,
$N_C = 72.5$ lb.

6–79. $N_C = 20$ lb, $B_x = 34$ lb, $B_y = 62$ lb,
$M_A = 336$ lb · ft, $A_y = 12$ lb, $A_x = 34$ lb.

6–81. $F_{AC} = 2.51$ kip, $F_{AB} = 3.08$ kip, $F_{AD} = 3.43$ kip.

6–82. $F_{AB} = 14.0$ kN, $C_y = 7.01$ kN, $C_x = 10.7$ kN,
$M_D = 9.79$ kN · m, $D_x = 0$, $D_y = 1.96$ kN.

6–83. 51.2 lb.

6–85. $B_x = 150$ lb, $B_y = 150$ lb, $C'_y = 50$lb, $A_y = 100$ lb,
$C_y = 200$ lb, $C_x = 217$ lb, $D_y = 50$ lb, $D_x = 217$ lb,
$C'_x = 66.7$ lb.

6–86. 4.36 in.

6–87. $B_x = 200$ lb, $C_x = 0$, $C_y = 200$ lb.

6–89. 13.9 N.

6–90. 222 lb.

6–91. $F_x = G_x = 333$ lb, $F_y = G_y = 250$ lb.

6–93. 19.6 kN.

6–94. $B_x = 84.2$ lb, $B_y = 45.1$ lb, $R_A = 30.9$ lb.

6–95. $N_C = 192$ lb, $C_x = 154$ lb, $C_y = 115$ lb, $B_x = 0$,
$B_y = 17.1$ lb, $A_x = 154$ lb, $A_y = 18.1$ lb.

6–97. $N_H = 218$ lb, $C_x = 245$ lb, $C_y = 625$ lb.

6–98. 286 N.

6–99. $\Delta = 160$ mm, $\Delta = 240$ mm.

6–101. $F_{IJ} = 9.06$ kN (T), $F_{BC} = 15.4$ kN (C).

6–102. 33.6°.

6–103. $P = 513$ N, $F_{CE} = 5.01$ kN.

6–105. $A_x = 1500$ N, $A_z = 0$, $T_{ED} = 5.30$ kN,
$C_z = 2.25$ kN, $C_x = 2.25$ kN, $C_y = 3.00$ kN.

6–106. 22.2 lb.

6–107. $A_x = 172$ N, $A_y = 115$ N, $A_z = 0$, $C_x = 47.3$ N,
$C_y = 61.9$ N, $C_z = 125$ N.

6–109. 133 lb.

6–110. $F_{AG} = 471$ lb (C), $F_{AB} = 333$ lb (T), $F_{BC} = 333$ lb (T),
$F_{DE} = 943$ lb (C), $F_{DC} = 667$ lb (T), $F_{EC} = 667$ lb (T),
$F_{EG} = 667$ lb (C), $F_{GC} = 471$ lb (T).

6–111. $A_x = C_x = 100$ lb, $A_y = C_y = 100$ lb.

6–113. $F_{CA} = F_{CB} = 122$ lb (C), $F_{CD} = 173$ lb (T),
$F_{BD} = 86.6$ lb (T), $F_{BA} = 0$, $F_{AD} = 86.6$ lb (T).

6–114. 2.60 kip (T).

6–115. 2 kip (C).

6–117. $B_x = 4.17$ kN, $B_y = 4.90$ kN, $C_x = 5.15$ kN,
$C_y = 4.90$ kN, $A_x = 5.15$ kN, $A_y = 3.92$ kN.

6–118. $F_{BF} = 155$ lb (C), $F_{FD} = 0$.

6–119. $F_{AB} = 1.75$ kN (T), $F_{AC} = 3.40$ kN (C),
$F_{BC} = 425$ N (C).

Chapter 7

7–1. $\bar{x} = (r \sin \alpha)/\alpha$.

7–2. $\bar{x} = 0.546$ m, $O_x = 0$, $O_y = 7.06$ N,
$M_O = 3.86$ N · m.

7–3. $\bar{x} = 0.620$ ft, $A_y = 0.442$ lb, $B_x = 0$, $B_y = 0.720$ lb.

7–5. $\bar{x} = 0$, $\bar{y} = 1.6$ ft.

7–6. $\bar{x} = \frac{3}{4}b$, $\bar{y} = \frac{3}{10}h$.

7–7. $\bar{x} = \frac{3}{8}b$, $\bar{y} = \frac{2}{5}h$.

7–9. $\bar{x} = (2r \sin \alpha)/3\alpha$.

7–10. $\bar{x} = 0$, $\bar{y} = 4r/3\pi$.

7–11. $\bar{x} = 1.26$ m, $\bar{y} = 0.143$ m, $N_B = 48.0$ kN,
$A_x = 34.0$ kN, $A_y = 73.8$ kN.

7–13. 1.23 ft.

7–14. 0.980 ft.

7–15. 0.587 m.

7–17. $\bar{z} = \frac{2}{3}h$.

7–18. $\bar{x} = a/2$, $\bar{y} = \bar{z} = 0$.

7–19. $\bar{x} = \bar{z} = 0$, $\bar{y} = 0.833a$.

7–21. $\bar{z} = 0.675a$.

7–22. $\bar{y} = \bar{z} = 0$, $\bar{x} = 0.343$ m, $A_x = 2.56$ kN,
$B_x = C_x = 1.28$ kN, $A_y = 7.46$ kN, $A_z = 0$.

7–23. $\bar{y} = 0$, $\bar{z} = 0$, $\bar{x} = 0.4a$.

7–25. 67.9 mm.

7–26. 179 mm.

7–27. $\bar{x} = 6.5$ in., $\bar{y} = 4$ in., $\theta = 10.6°$.

7–29. $\bar{x} = 1.60$ ft, $\bar{y} = 7.04$ ft, $A_x = 0$, $A_y = 149$ lb,
$M_A = 502$ lb · ft.

7–30. 3 m.

7–31. $\bar{y} = 2.91$ ft, $T_{CE} = T_{DF} = 34.8$ lb.

7–33. $\bar{x} = 1.63$ ft, $\bar{y} = 3.56$ ft.

7–34. $\bar{x} = 4.74$ in., $\bar{y} = 2.99$ in.

7–35. $\bar{x} = 3.52$ ft, $\bar{y} = 4.09$ ft, $W = 42.8$ kips.

7–37. 2 in.

7–38. $\bar{y} = \dfrac{h}{3}\left(\dfrac{2b_1 + b_2}{b_1 + b_2}\right)$.

7–39. $\bar{y} = 150$ mm, $\bar{x} = 33.9$ mm.

7–41. $\bar{x} = 1.25$ ft, $\bar{y} = 2.20$ ft, $B_x = 265$ lb, $C_x = 265$ lb,
$C_y = 704$ lb.

7–42. 101 mm.

7–43. $\bar{x} = \left(\frac{2}{3}r \sin^3 \alpha\right)\Big/\left(\alpha - \dfrac{\sin 2\alpha}{2}\right)$.

7–45. 0.7 ft.

7–46. 58.1 mm.

7–47. $\bar{z} = 2.48$ ft, $\theta = 38.9°$.

7–49. $h = 80$ mm, or $h = 48$ mm.

7–50. $\theta = 70.4°$, $\ell = 265$ mm.

7–51. 0.667 ft^2, 0.375 ft, 1.57 ft^3.

7–53. 1.54 m^2, 0.420 m, 4.07 m^3.

7–54. 16.8 lb.

7–55. $3.12(10^6)$ lb.

7–57. $1.27(10^6)$ mm^3.

7–58. $4.25(10^6)$ mm^3.

7–59. 3.49 m^3.

7–61. 2.68 kg.

7–62. 29.3 kip.

7–63. 2.26 gallons.

7–65. 36.3 kN.

7–66. 3.28 ft.

7–67. 106 mm.

7–69. $F_R = 51$ kN $\downarrow$, $d = 17.9$ m.

7–70. $F_R = 30$ kN $\downarrow$, $d = 4.10$ m.

7–71. $b = 5.62$ m, $a = 1.54$ m.

7–73. $A_x = 0$, $B_y = 7$ kN, $A_y = 11$kN.

7–74. 2.18 kN/m.

7–75. $W_B = 0.438$ kip/ft, $W_A = 2.14$ kip/ft.

7–77. (a) $W_1 = 2P/L$, $W_2 = 4P/L$, (b) $W_1 = 83.3$ lb/ft,
$W_2 = 167$ lb/ft.

7–78. $C_x = 216$ lb, $B_x = 206$ lb, $B_y = 275$ lb, $A_x = 9.37$ lb,
$A_y = 50$ lb.

7–79. $E_x = B_x = 394$ lb, $E_y = B_y = 296$ lb, $C_y = 102$ lb,
$C_x = 419$ lb, $D_x = 91.6$ lb, $D_y = 0$, $A_x = 66.6$ lb,
$A_y = 193$ lb.

7–81. $F_R = 410$ lb, $\bar{y} = 64$ ft.

7–82. $F_R = 1.87$ kip $\downarrow$, $\bar{x} = 3.66$ ft.

7–83. $A_x = 0$, $A_y = 3.64$ kN, $B_y = 11.2$ kN.

7–85. $F_R = 744$ lb, $\bar{z} = 16.9$ ft.

7–86. $F = 30$ kip, $\bar{z} = 8$ ft.

7–87. $F = 2.56$ kN, $\bar{z} = 3$ m, $N_C = N_B = 960$ N,
$N_A = 640$ N.

7–89. $F_W = 14.7$ kN, $h = 2.11$ m.

7–90. $F_D = 101$ kN, $C_x = 46.6$ kN, $C_y = 0$.

7–91. 486 kips $\angle_{34°}$.

7–93. 3.85 m.

7–94. 369 kg/m^3.

7–95. $F_R = 45.9$ kN, $d = 0.898$ m.

7–97. 15.7 m.

7–98. $F_x = 628$ kN, $F_y = 538$ kN.

7–99. 391 kN/m.

7–101. $\bar{x} = 1.47$ in., $\bar{y} = 2.68$ in., $\bar{z} = 2.84$ in.

7–102. (0, 0.0667 m, 0).

7–103. 0.0133 m^2, 0.150 m, 0.0126 m^3.

7–105. $A_x = 0$, $A_y = 2.75$ kip, $B_y = 1$ kip.

7–106. (0.0883 ft, 0, 0).

7–107. 3.33 ft^2, 1.20 ft, 25.1 ft^3.

7–109. $F_B = 1.53$ kip, $F_C = 0.350$ kip.

7–110. 331 kN/m.

7–111. $\bar{x} = 2.40$ in.

Chapter 8

8–1. **(a)** moment increases.

8–2. $T = 1000$ lb, $V_b = 1000$ lb.

8–3. $V_{AB} = 4$ kN, $V_{BC} = 2$ kN.

8–5. 20.4 kN.

8–6. 2.45 kN/m.

8–7. $A_A = 12$ kN, $V_A = 0$, $M_A = 0$, $A_B = 20$ kN, $V_B = 0$, $M_B = 1.20$ kN · m.

8–9. $A_A = 550$ lb, $A_B = 250$ lb, $A_C = 950$ lb.

8–10. $T_C = 40$ lb · ft, $T_D = 55$ lb · ft, $T_E = 10$ lb · ft.

8–11. $T_C = 0$, $T_D = 400$ N · m, $T_E = 600$ N · m.

8–13. $V_A = 0$, $A_A = 5$ kN, $M_A = 0.8$ kN · m, $V_B = 0$, $A_B = 7.5$ kN, $M_B = 0.2$ kN · m.

8–14. $A_C = 0$, $V_C = 108$ lb, $M_C = 433$ lb · ft.

8–15. 25 kip · ft.

8–17. $A_A = 85$ lb, $V_A = 0$, $M_A = 21.2$ lb · ft.

8–18. $A_D = 1.33$ kN, $V_D = 0$, $M_D = 2$ kN · m.

8–19. $M_C = 800$ lb · ft, $V_C = 0$, $A_C = 0$, $M_D = 1,600$ lb · ft, $V_D = 800$ lb, $A_D = 0$.

8–21. $M_C = 62.5$ kN · m, $V_C = 9$ kN, $A_C = 0$, $M_B = 184$ kN · m, $V_B = 27.5$ kN, $A_B = 0$.

8–22. $M_C = 24$ kip · ft.

8–23. $A_C = 2.20$ kip, $V_C = 0.336$ kip, $M_C = 1.76$ kip · ft.

8–25. $M_C = 4.23$ kip · ft, $V_C = 649$ lb, $A_C = 265$ lb, $M_D = 3.18$ kip · ft, $V_D = 637$ lb, $A_D = 265$ lb.

8–26. $V_C = 5.79$ kip, $M_C = 4.95$ kip · ft.

8–27. $A_D = 326$ lb, $V_D = 0$, $M_D = 0$, $M_E = 977$ lb · ft, $V_E = 0$, $A_E = 91.1$ lb.

8–29. $A_C = 4.32$ kip, $V_C = 1.35$ kip, $M_C = 4.72$ kip · ft.

8–30. $A_C = 400$ N, $V_C = 96$ N, $M_C = 144$ N · m.

8–31. $A_E = 0$, $V_E = 50$ N, $M_E = 100$ N · m, $A_D = 0$, $V_D = 750$ N, $M_D = 1300$ N · m.

8–33. $A_A = 0$, $V_A = 0$, $M_A = 0$, $A_B = 14.1$ lb, $V_B = 14.1$ lb, $M_B = 28.3$ lb · in.

8–34. $\mathbf{F} = \{-20\mathbf{i} - 10\mathbf{j} + 140\mathbf{k}\}$ lb, $\mathbf{M} = \{120\mathbf{i} - 500\mathbf{j} - 90\mathbf{k}\}$ lb · in.

8–35. $V_x = -200$ lb, $V_y = 400$ lb, $A_z = 450$ lb, $M_x = -800$ lb · ft, $M_y = -400$ lb · ft, $M_z = 350$ lb · ft.

8–37. $(V_C)_x = 10$ lb, $(A_C)_y = 0$, $(V_C)_z = 104$ lb, $(M_C)_x = 238$ lb · ft, $(M_C)_y = -72$ lb · ft, $(M_C)_z = -20$ lb · ft.

8–38. $A_x = 0$, $V_y = -164$ N, $V_z = 0$, $M_x = 0$, $M_y = 0$, $M_z = 81.8$ N · m.

8–39. **(b)** $x = 0$, $V = 267$, $M = 0$; $x = 12$, $V = -533$, $M = 0$.

8–41. $x = 0$, $V = 177$, $M = 0$; $x = 2$, $M = 354$; $x = 4$, $V = -177$, $M = 0$.

8–42. **(b)** $x = 0$, $V = -600$, $M = 0$; $x = 4$, $M = -2400$; $x = 8$, $M = -1600$; $x = 12$, $V = 400$, $M = 0$.

8–43. $x = 0$, $V = 1.5$; $M = 0$; $x = 3^-$, $V = 1.5$, $M = 4.5$, $x = 3^+$, $V = -4.5$, $M = 13.5$; $x = 6$, $V = -4.5$, $M = 0$.

8–45. **(b)** $x = 0$, $V = 800$, $M = 0$; $x = 5^-$, $V = 800$, $M = 4000$.

8–46. $x = 0$, $V = 18$, $M = -65$; $x = 5$, $V = 8$, $M = 0$.

8–47. **(b)** $x = 0$, $V = 4.8$, $M = 0$; $x = 6$, $V = 0$, $M = 14.4$.

8–49. $x = 0$, $V = 2.5$, $M = 0$; $x = 5^+$, $V = -7.5$, $M = 37.5$.

8–50. **(b)** $x = 0$, $V = 37.5$, $M = -125$; $x = 7.5$, $V = 0$, $M = 15.6$.

8–51. $x = 0$, $V = 5.29$, $M = 0$, $x = 2^-$, $V = 4.90$, $M = 10.2$.

8–53. **(b)** $x = 12$, $V = -1500$, $M = -6000$.

8–54. $x = 6.93$, $V = 0$, $M = 2.31$.

8–55. $x = 6.58$, $V = 0$, $M = 1.36$.

8–57. $x = 3^-$, $V = -1100$, $M = -2.40$.

8–58. $A = -25x$, $V = -43.3x$, $M = 25x - 21.7x^2$.

8–59. $x = 3.9$ m, $M_E = 11.7$ kN · m.

8–61. $V = -200 \cos \theta$, $A = 200 \sin \theta$, $M = 200 \sin \theta - 150$, $\theta = 48.6°$, $A = 0$, $V = 200$, $M = -200x - 150$.

8–62. $A = P(2 \cos \theta + \sin \theta)$, $V = P(2 \sin \theta - \cos \theta)$, $M = Pr(2 \cos \theta + \sin \theta - 2)$, $A = P(\sin \theta + 3 \cos \theta)$, $V = P(3 \sin \theta - \cos \theta)$, $M = Pr(\sin \theta + 3 \cos \theta - 2)$.

8–63. $V = (th\gamma x^2)/2d$, $M = (-th\gamma x^3)/6d$.

8–65. $a = L/3$.

8–66. $a = \frac{1}{3}L$.

8–67. $V_x = 0$, $A_y = 0$, $V_z = \pm 4(6 - y)$ lb,
$M_x = \pm 2(y^2 - 12y + 32)$ lb $\cdot$ ft, $M_y = \pm 8$ lb $\cdot$ ft,
$M_z = 0$.

8–74. $x = 10$, $V = 0$, $M = 43$.

8–75. $x = 1.25^+$, $V = -9.17$, $M = 16.5$; $x = 3^+$, $V = 15$,
$M = -7.5$.

8–77. $x = 3$, $V = 0$, $M = 3$.

8–78. $x = 9^+$, $V = -1.38$, $M = 25.9$.

8–79. $x = 10$, $V = 0$, $M = 600$.

8–81. $x = 3$, $V = -12$, $M = 12$.

8–82. $x = 0$, $V = 22$, $M = -59$.

8–83. $x = 3^-$, $V = -1.5$, $M = 45.5$, $x = 3^+$, $V = -7.5$,
$M = 47.5$.

8–85. $x = 7^-$, $V = -10.6$, $M = -21.2$; $x = 13$, $V = 0$,
$M = 6$.

8–86. $x = 3^+$, $V = -8.43$, $M = 19.7$; $x = 15$, $V = 0$, $M = 9$.

8–87. $T_{DB} = 78.2$ lb, $T_{CD} = 43.7$ lb, $T_{AC} = 74.7$ lb,
$\ell = 15.7$ ft.

8–89. 5.65 ft.

8–90. $T_{AB} = 787$ lb, $y_B = 8.79$ ft, $T_{ED} = 830$ lb,
$y_D = 6.10$ ft, $T_{CD} = 718$ lb, $T_{CB} = 657$ lb.

8–91. $T_{\max} = 8.08$ kN, $P = 0.714$ kN, $y_D = 2.80$ m.

8–93. 11.3 m.

8–94. 564 N/m.

8–95. $T_C = 4.04$ kip, $T_B = 6.83$ kip, $T_A = 6.04$ kip.

8–97. $\ell = 55.6$ ft, $h = 10.6$ ft.

8–98. $F_H = 384$ lb, $d = 10.4$ ft.

8–99. 4.00 kip, $T_{\max} = 2.01$ kip.

8–101. $\mathcal{L} = 30.2$ m, $T_{\max} = 376$ N.

8–103. $\dfrac{h}{L} = 0.141$.

8–105. $T_C = 16.4$ lb, $T_B = 52.4$ lb, $T_A = 37.3$ lb.

8–106. 18.2 ft.

8–107. 44.8 lb.

8–109. $y = 0.741(10^{-3})x^3$, $T_{\max} = 5.03$ kip.

8–110. $y = 20(1 - \cos 1.80x)$, $T_{\max} = 83.4$ lb.

8–111. $y = 3.90(10^{-3})x^3 + x$, $T = 616$ lb.

8–113. $A_E = 0$, $V_E = -50$ N, $M_E = -100$ N $\cdot$ m, $A_D = 0$,
$V_D = 550$ N, $M_D = -900$ N $\cdot$ m.

8–114. $V_F = -447$ N, $A_F = -224$ N, $M_F = 224$ N $\cdot$ m.

8–115. $x = 3^-$, $V = -4.5$, $M = -9$; $x = 10^+$, $V = 19.7$,
$M = -54.5$.

8–117. $B_x = -100$ lb, $B_y = 0$, $B_z = 50$ lb, $(M_B)_x = 100$ lb $\cdot$ ft,
$(M_B)_y = 50$ lb $\cdot$ ft, $(M_B)_z = 200$ lb $\cdot$ ft.

8–118. $x = 22^-$, $V = -56$, $M = -696$.

8–119. $F_H = 247$ N, $h = 50.3$ ft.

Chapter 9

9–1. (a) 4 unks, (b) 8 unks, (c) 4 unks, (d) 6 unks.

9–2. 537 mm.

9–3. $P = 72$ lb, $x = 0.5$ ft.

9–5. $P = 246$ N, $F_C = 346$ lb.

9–6. 2.60 kN.

9–7. $F = 22.5$ lb, $\mu_m = 0.15$.

9–9. 26.7 ft.

9–10. $F_A = 100$ lb, $F_V = 173$ lb.

9–11. (a) $W = 318$ lb, (b) $W = 360$ lb.

9–13. 1.53 kip.

9–14. 1.14 kN.

9–15. $F_A = 444$ N, $N_A = 1.47$ kN, $N_B = 1.24$ N.

9–17. $31.0°$.

9–18. Truck does not slip or tip.

9–19. 377 lb.

9–21. $21.8°$.

9–22. $\mu = 0.268$.

9–23. No, No.

9–25. drum slips, $O_x = 100$ N, $O_y = 100$ N.

9–26. drum does not slip, $O_x = 245$ N, $O_y = 267$ N.

9–27. 6.33 N $\cdot$ m.

9–29. $A_y = 474$ lb, $B_y = 232$ lb, $B_x = 36.0$ lb.

9–30. $F_D = 36.9$ lb, $A_y = 468$ lb, $B_y = 228$ lb, $B_x = 34.6$ lb.

9–31. $11.0°$.

9–33. $32.5°$.

9–34. $P = 60$ lb for two cartons, $P = 90$ lb for three
cartons.

9–35. 7.84 lb.

9–37. 39.8 N.

9–38. $80.9°$.

9–39. 234 N.

9–41. 28.1 N.

9–42. 13.3 lb.

9–43. $46.4°$.

9–45. 107 N.

9–46. 0.344.

9–47. (a) $\theta = \tan^{-1}\left(\dfrac{1 - \mu_A\mu_B}{2\mu_A}\right)$, (b) $\theta = 66.5°$.

9–49. $\theta = 26.6°$, $P = 855$ N.
9–50. 65.2°.
9–51. 60.9 lb.
9–53. 76.7 lb.
9–54. 25.3°.
9–55. 38.6°.
9–57. 1.90 kN.
9–58. 625 lb.
9–59. 66.6 lb.
9–61. 51.2 N.
9–62. 1.98 kN.
9–63. 961 lb, $P = 18.3$ lb.
9–65. 880 N, 352 N · m.
9–66. 72.7 N.
9–67. $F_{BA} = 138$ N (T), $F_{BD} = 82.8$ N (C), $F_{BC} =$ 110 N (C), $F_{AD} = 110$ N (C), $F_{AC} = 82.8$ N (C), $F_{CD} = 138$ N (T).
9–69. $T_A = 97.4$ lb, $T_B = 38.0$ lb.
9–70. Approx. 2 turns (695°).
9–71. 1.72 ft.
9–73. 77.8 lb.
9–74. 53.6 N.
9–75. 78.7 lb.
9–77. V-Belt is almost 30 times more effective.
9–78. 29.7 lb · ft.
9–79. 99.2°.
9–81. 8.89 lb.
9–82. 39.5 lb.
9–83. 223 N.
9–85. 25.5 lb · ft.
9–86. 1.89 lb.
9–87. 36.3 lb · ft.
9–89. $M = \frac{1}{2}\mu P(R_2 + R_1)$.
9–90. $M = \frac{1}{2}\mu PR$.
9–91. $M = \frac{2}{3}\left(\frac{\mu P}{\sin\frac{\theta}{2}}\right)\frac{(r_2^3 - r_1^3)}{(r_2^2 - r_1^2)}$.
9–93. 17.0 N · m.
9–94. 0.206.
9–95. 0.596 lb.
9–97. $T_A = 853$ N, $N_s = 1.30$ kN, $F = 389$ N.
9–98. 0.104.
9–99. 13.6 lb.
9–101. 545 N.
9–102. 40 lb.
9–103. 245 N.
9–105. 0.72 mm.
9–106. 78.8 lb.
9–109. 38.5 mm.
9–110. 531 N.
9–111. 28.1°.
9–113. 41.6 lb.
9–114. 11 lb.
9–115. 15.4 N · m.
9–117. 292 lb · ft.
9–118. 16.7°.
9–119. 3.90 lb.
9–121. $F = 1.6$ lb, $\mu_{BC} = 0.133$.
9–122. 3 half turns.

Chapter 10

10–1. 5.00 in^2.
10–2. 15.4(10^6) mm^4.
10–3. 307 in^4.
10–5. 34.1 in^4.
10–6. 73.1 in^4.
10–7. $I_y = 2hb^3/15$.
10–9. **(a)** $I_x = bh^3/12$, **(b)** $\bar{I}_x = bh^3/36$.
10–10. 39.0 m^4.
10–11. 8.53 m^4.
10–13. 39.0 m^4.
10–14. 0.314 m^4.
10–15. $I_x = \pi a^4/8$, $\bar{I}_x = 0.110a^4$.
10–17. $I_x = a^4/12$.
10–18. $I_x = 2.55(10^3)$ in^4, $I_y = 1.12(10^3)$ in^4.
10–19. $\bar{y} = 15.7$ in., $\bar{I}_x = 2.16(10^3)$ in^4.
10–21. shaft $J_O = 251(10^3)$ mm^4, tube $J_O = 172(10^3)$ mm^4, 68.4%.
10–22. 30.2(10^6) mm^4.
10–23. 7.74 in.
10–25. 648 in^4.
10–26. $\bar{y} = 2$ in., $\bar{I}_x = 128$ in^4.
10–27. $I_x = 124(10^6)$ mm^4, $I_y = 1.21(10^9)$ mm^4.
10–29. $\bar{y} = 37.5$ mm, $\bar{I}_x = 12.4(10^6)$ mm^4.
10–30. $I_y = 94.8(10^6)$ mm^4.
10–31. $\bar{y} = 207$ mm, $\bar{I}_x = 222(10^6)$ mm^4, $\bar{I}_y = 115(10^6)$ mm^4.

10–33. $\bar{x} = 68.0$ mm, $\bar{y} = 80$ mm, $\bar{I}_x = 49.5(10^6)$ mm^4, $\bar{I}_y = 36.9(10^6)$ mm^4.

10–34. $\bar{x} = 3.0$ in., $\bar{y} = 2.0$ in., $\bar{I}_x = 64$ in^4, $\bar{I}_y = 136$ in^4.

10–35. $I_{xy} = h^2 b^2 / 16$.

10–37. 8 in^4.

10–38. $I_{xy} = a^4 / 8$.

10–39. $-0.0165 a^4$.

10–41. 0.667 in^4.

10–42. $\theta = 6.08°$, $I_{max} = 1.74(10^3)$ in^4, $I_{min} = 435$ in^4.

10–43. 0.

10–45. 97.8 in^4.

10–46. $\bar{x} = 85$ mm, $\bar{y} = 35$ mm, $7.50(10^6)$ mm^4.

10–47. $\bar{y} = 44.1$ mm, $\bar{x} = 44.1$ mm, $6.26(10^6)$ mm^4.

10–49. $I_u = 5.09(10^6)$ mm^4, $I_v = 5.09(10^6)$ mm^4, $I_{uv} = 0$.

10–50. $I_{max} = 64.1$ in^4, $I_{min} = 5.33$ in^4.

10–51. $I_u = 85.3(10^6)$ mm^4, $I_v = 85.3(10^6)$ mm^4.

10–53. x and y are principal axes, $\bar{y} = 4.67$ in., $I_x = 1.47(10^3)$ in^4, $I_y = 2.50(10^3)$ in^4.

10–54. $I_u = 10.5(10^6)$ mm^4, $I_v = 4.05(10^6)$ mm^4, $I_{uv} = 5.54(10^6)$ mm^4.

10–55. $\theta = 8.70°$, $I_{max} = 116(10^6)$ mm^4, $I_{min} = 29.7(10^6)$ mm^4.

10–57. $I_{max} = 64.1$ in^4, $I_{min} = 5.33$ in^4.

10–58. $I_{max} = 4.92(10^6)$ mm^4, $I_{min} = 1.36(10^6)$ mm^4.

10–59. $I_{max} = 2.50(10^3)$ in^4, $I_{min} = 1.47(10^3)$ in^4.

10–61. $I_x = \frac{3}{10} m r^2$.

10–62. 57.7 mm.

10–63. $I_x = \frac{2}{5} m b^2$.

10–65. $I_x = \frac{2}{5} m b^2$.

10–66. $I_y = \frac{1}{6} m [a^2 + h^2]$.

10–67. 2.25 slug · ft^2.

10–69. 3.15 ft.

10–70. $\bar{y} = 1.78$ m, $I_G = 4.45$ kg · m^2.

10–71. 6.23 kg · m^2.

10–73. 118 slug · ft^2.

10–74. 34.2 kg · m^2.

10–75. $I_x = 31.4$ in^4, $I_y = 31.4$ in^4.

10–77. 2.27 in^4.

10–78. $\bar{y} = 0.393$ m, $I_G = 0.0885$ kg · m^2.

10–79. $\theta_{P_1} = -22.5°$, $\theta_{P_2} = 67.5°$, $I_{max} = 250$ in^4, $I_{min} = 20.4$ in^4.

10–81. $I_z = \frac{1}{10} m a^2$.

10–82. $I_x = 0.167$ in^4, $I_y = 0.0333$ in^4.

10–83. 0.0625 in^4.

10–85. $I_{max} = 0.191$ in^4, $I_{min} = 0.00859$ in^4.

Chapter 11

11–1. $P = W/(2 \tan \theta)$.

11–2. 15.5°.

11–3. $\theta = 0°$, $\theta = 36.9°$.

11–5. 512 N.

11–6. 32.9°.

11–7. 427 N.

11–9. 27.4°.

11–10. 81.4°.

11–11. 80.5°.

11–13. 3.30°.

11–14. 16.6°.

11–15. 312 lb.

11–17. 100 kg.

11–18. $\theta = 0°$, $\theta = 33.1°$.

11–19. $m_B = 2$ kg, $m_A = 1.6$ kg.

11–21. $\theta = 38.7°$, unstable; $\theta = 90°$, stable.

11–22. $\theta = 15.5°$, stable.

11–23. $\theta = 0°$, stable; $\theta = 36.9°$, unstable.

11–25. $\theta = 0°$, $\theta = \cos^{-1} \left(\dfrac{W}{2k\ell} + 1 \right)$.

11–26. $k = W/\ell$.

11–27. $\theta = 2.55°$, unstable.

11–29. $\theta = 12.1°$, unstable.

11–30. $\theta = 0°$, $\theta = \cos^{-1} \dfrac{W}{4k\ell}$, $W = 2k\ell$.

11–33. 12.8 ft.

11–34. 2.45 m.

11–35. $\theta = 18.4°$, unstable.

11–37. $\theta = 0°$, unstable.

11–38. $d = h/3$.

11–39. $\theta = 0°$, $r = h/4$.

11–42. $\theta = 0°$, $\theta = \cos^{-1} \left(\dfrac{d}{4a} \right)$.

11–43. $x = 0$, $y = 0$, stable.

11–45. $\theta = 90°$, $\theta = 17.5°$.

11–46. $\theta = 90°$, stable; $\theta = 17.5°$, unstable.

11–47. $\theta = 90°$, $\theta = 13.9°$.

11–49. 3.99 in.

11–50. $P = \left(\dfrac{b - a}{2c} \right) mg$.

11–51. 4.62 kN.

11–53. 39.2 N.

Index

Accuracy, 8
Active force, 71, 495
Addition
  of couples, 120
  of forces, 16, 24
  of vectors, 14
Angle
  of kinetic friction, 395
  of pitch, 417
  of repose, 395
  of static friction, 395
Axial force, 339
Axis
  of symmetry, 286, 464
  of a wrench, 135

Ball-and-socket joint, 190
Band brake, 424
Beam, 356
Bearings
  collar, 431
  journal, 434
  pivot, 431
Belt friction, 424
Bending moment, 339
Bending-moment diagrams, 356
Bending-moment function, 356

Cables
  sag, 373
  span, 378
  subject to concentrated loads, 373
  subject to distributed load, 376
  subject to weight, 380
Cantilevered beam, 356
Cartesian vectors, 33
  addition, 37

  cross product, 95
  direction, 35
  dot product, 58
  magnitude, 35
  subtraction, 37
  unit, 34
Catenary, 383
Center of gravity
  of a body, 156, 280
  of a system of particles, 279
Center of mass, 280, 281
Center of pressure, 325
Centroid, 281
  area, 282
  composite body, 296
  line, 282
  volume, 281
Coefficient
  of kinetic friction, 393
  of rolling resistance, 440
  of static friction, 393
Collar bearings, 431
Commutative law, 58, 96
Components
  of force, 16, 24
  of vector, 15, 33
Composite
  area, 458
  body, 296, 483
Compression, 215
Concurrent forces, 134
Conservative force, 504
Constraints
  improper, 197
  partial, 198
  redundant, 195
Coordinate direction angles, 35

Coplanar force system, 74
Coplanar vectors, 134
Cosine law, 16
Coulomb friction, 391
Couple, 119
  addition, 120
  equivalent, 120
  resultant, 120
Cross product, 95

Datum, 506
Degree of freedom, 494
Density, 281
Determinant, 98
Diagram
  bending-moment, 356
  free-body, 70, 153, 189, 243
  shear, 356
Differential elements
  disk, 479
  shell, 479
Dimensional homogeneity, 7
Direction angles, 35
Direction cosines, 36
Disk element, 479
Displacement, 492
  virtual, 493
Distributed load, 313
Distributive law, 58, 96
Dot product, 58
Dynamics, 1

Elastic spring, 70
Energy, potential, 505
Engineering notation, 7
Equations of equilibrium
  for particle, 69

for rigid body, 152, 167, 194
Equilibrium
  necessary conditions, 69, 153
  neutral, 508
  particle, 69
  rigid-body, 152
  stable, 508
  sufficient conditions, 69, 153
  unstable, 508
Equipollent, 133
Equivalent force systems, 113
External forces, 152, 156

Fixed support, 155
Fluid, 324
Fluid friction, 391
Fluid pressure, 324
Foot, 5
Force, 2
  active, 71
  addition, 16, 24
  axial, 339
  component, 16, 24
  compressive, 215
  concentrated, 3
  conservative, 504
  definition, 2
  distributed, 313
  external, 152, 156
  friction, 391
  gravitational, 3
  internal, 151, 156, 337, 494
  normal, 391
  reactive, 71
  shear, 339
  tensile, 215
Frames, 243
Free-body diagram
  particle, 70
  rigid body, 153, 189, 191, 339
  frame, 243
  machine, 243
Friction, 391
  angles, 395
  bearing, 434
  belt, 424
  circle, 434
  Coulomb, 391
  disks, 431
  dry, 391
  fluid, 391
  kinetic, 394
  problems, 395
  rolling, 440
  screws, 417
  static, 394
  wedge, 415

Gravitational force, 3
Guldin, Paul, 306

Gussett plate, 213
Gyration, radius of, 449, 483

Hinge, 189
Homogeneous body, 281

Impending motion, 392
Improper constraints, 197
Inclined axis, moments of inertia, 468
Internal forces, 151, 156, 337, 494

Jack, 417
Joints, method of, 216, 239
Joule, 492
Journal bearings, 434

Kilogram, 5
Kinetic friction force, 394

Law
  of cosines, 16
  of gravitation, 3
  Newton's, 3
  of sines, 16
Length, 2
Linear algebraic equations, 524
Limiting static frictional force, 394
Line of action of a vector, 14
Link, 154
Loading function, 313

Machines, 243
Magnitude of a vector, 13
Mass, 2
Mass moment of inertia, 478
Mathematical expressions, 521
Mechanics, definition, 1
Meter, 5
Method
  of joints, 216, 239
  of sections, 229, 239, 337
Mixed triple product, 114
Mohr, Otto, 471
Mohr's circle, 471
Moment arm, 99
Moment diagram, 356
Moment of a couple, 119
Moment of a force, 95, 98, 115
  about a point, 98
  about a specified axis, 113
Moment of inertia, 447
  composite areas, 458
  determined by integration, 449, 479
  mass, 478
  Mohr's circle, 471
  polar, 448
  principal, 468
  product, 464
Multiforce member, 243, 339

Negative vector, 14
Neutral equilibrium, 508
Newton, 5
Newton, Isaac, 2
Newton's laws
  of gravitational attraction, 3
  of motion, 3
Nonconservative force, 505
Numerical accuracy, 8

Pappus, 306
Parabolic cable, 378
Parallel-axis theorem
  area moment of inertia, 448
  mass moment of inertia, 482
  polar moment of inertia, 449
  product of inertia, 465
Parallelogram law, 14
Partial constraints, 198
Particle, 2
  equilibrium, 69
  free-body diagram, 70
Pascal's law, 324
Pin support, 154, 189
Pitch, 417
Pitch angle, 417
Pivot bearing, 431
Polar moment of inertia, 448
Position vector, 45
Potential energy, 505
  criteria for equilibrium, 507
  elastic, 506
  function, 506
  gravitational, 505
Pound, 5
Prefixes, 5
Pressure, 313, 324
Principal axis, 468
Principal moments of inertia, 468
Principle of moments, 105
Principle of transmissibility, 105
Principle of virtual work
  particle, 493
  rigid body, 493
  system of bodies, 494
Product of inertia, 464
Pulley, 70
Purlin, 213

Radius of gyration, 449, 483
Reactive force, 71
Redundant constraints, 195
Resultant
  couple, 120
  force, 16
Right-hand rule, 96
Right-handed coordinate system, 45
Rigid body, 3
Rocker support, 154
Roller support, 154
Rolling resistance, 440

Rotation
    differential, 492
    virtual, 493
Rounding-off, 8
Runge-Kutta method, 530

Sag, 378
Scalar, 13
Scientific notation, 7
Screws, 417
    self-locking, 418
Secant method, 528
Second, 5
Sections, method of, 229, 239
Sense, 14
Shear diagram, 356
Shear force, 339
Shell element, 479
Significant figures, 7
Simple truss, 215
Simply-supported beam, 356
Simpson's Rule, 526
Sine law, 16
Slug, 5
Specific weight, 281
Space truss, 239
Spring, 70
Square-threaded screw, 417
Stability, 508
Static friction force, 394
Statically determinant, 195
Statically indeterminant, 195
Statics, definition, 1
Stiffness, spring, 70
Stress, 337
Subtraction of vectors, 15
Support reactions, 153

Surface of revolution, 306
Symmetry, axis of, 286, 464
System
    of connected bodies, 494
    of forces and couples, 129
    of units, 5

Tail, 14
Tension, 215
Tetrahedron, 239
Three-force members, 177
Time, 2
Tip, 14
Tipping, 392
Torque, 98
Transmissibility, principle of, 105
Triangle addition, 14
Trusses
    plane, 213
    simple, 215
    space, 239
Two-force member, 176

Unit vector, 34
Units, 4
    base, 4
    conversion, 8
    derived, 4
    international system, 5
    U.S. customary, 5
Unstable equilibrium, 508

V belt, 429
Varignon's theorem, 105
Vector, 13
    addition, 14
    Cartesian, 33

    collinear, 25
    components, 15, 33
    concurrent, 37
    coplanar, 24
    cross product, 95
    direction, 13
    division by a scalar, 14
    dot product, 58
    free, 128
    magnitude, 13
    multiplication by a scalar, 14
    negative, 14
    position, 45
    projection, 59
    resolution, 15
    sense, 14
    sliding, 128
    subtraction, 14
    unit, 34
Virtual work
    for connected bodies, 494
    for particle, 493
    for rigid body, 493

Wedge, 415
Weight, 4, 156
Work
    of a couple, 492
    of a force, 491
    of a force exerted by a spring, 505
    of a gravitational force, 505
    virtual, 493
Wrench, 135

Zero-force member, 222

Geometric Properties of Line and Area Elements

Centroid Location	Centroid Location	Area Moment of Inertia

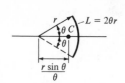

Circular arc segment

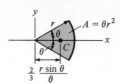

Circular sector area

$$I_x = \tfrac{1}{4}r^4(\theta - \tfrac{1}{2}\sin 2\theta)$$
$$I_y = \tfrac{1}{4}r^4(\theta + \tfrac{1}{2}\sin 2\theta)$$

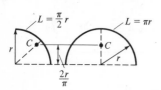

Quarter and semicircular arcs

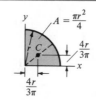

Quarter circular area

$$I_x = \tfrac{1}{16}\pi r^4$$
$$I_y = \tfrac{1}{16}\pi r^4$$

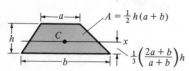

Trapezoidal area

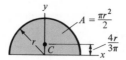

Semicircular area

$$I_x = \tfrac{1}{8}\pi r^4$$
$$I_y = \tfrac{1}{8}\pi r^4$$

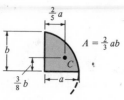

Semiparabolic area

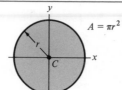

Circular area

$$I_x = \tfrac{1}{4}\pi r^4$$
$$I_y = \tfrac{1}{4}\pi r^4$$

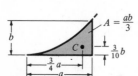

Exparabolic area

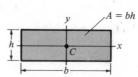

Rectangular area

$$I_x = \tfrac{1}{12}bh^3$$
$$I_y = \tfrac{1}{12}hb^3$$

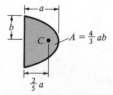

Parabolic area

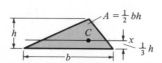

Triangular area

$$I_x = \tfrac{1}{36}bh^3$$